CYBER FORENSICS

Dejey

*Assistant Professor
Department of Computer Science and Engineering
Anna University Regional Campus
Tirunelveli*

S. Murugan IPS

*Inspector General of Police
Tamil Nadu*

Oxford University Press is a department of the University of Oxford.
It furthers the University's objective of excellence in research, scholarship,
and education by publishing worldwide. Oxford is a registered trade mark of
Oxford University Press in the UK and in certain other countries.

Published in India by
Oxford University Press
Ground Floor, 2/11, Ansari Road, Daryaganj, New Delhi 110002, India

© Oxford University Press 2018

The moral rights of the author/s have been asserted.

First published in 2018

All rights reserved. No part of this publication may be reproduced, stored in
a retrieval system, or transmitted, in any form or by any means, without the
prior permission in writing of Oxford University Press, or as expressly permitted
by law, by licence, or under terms agreed with the appropriate reprographics
rights organization. Enquiries concerning reproduction outside the scope of the
above should be sent to the Rights Department, Oxford University Press, at the
address above.

You must not circulate this work in any other form
and you must impose this same condition on any acquirer.

ISBN-13: 978-0-19-948944-2
ISBN-10: 0-19-948944-0

Typeset in ACaslonPro-Regular
by Archetype, New Delhi 110063
Printed in India by Magic International (P) Ltd., Greater Noida

Cover image: Elena Abrazhevich / Shutterstock

Third-party website addresses mentioned in this book are provided
by Oxford University Press in good faith and for information only.
Oxford University Press disclaims any responsibility for the material contained therein.

Dedicated to

*Our Parents, Teachers,
and the Almighty*

Preface

Incidents of cybercrime are on the upward trend in the recent years and create challenges not only to law enforcement agencies, but also for business firms and common netizens. Cyber criminals are preying the computer systems of government and private organizations for exploiting their services and their ability to penetrate networks all over the world has increased at an exponential rate. Though the field of cyber security attempts to implement and maintain robust approaches to defend cyberattacks, even the strongest defense mechanisms such as firewall, intrusion detection, and encryption are at times insufficient to achieve computer security. The ultimate aim of cyber forensics is to identify whether any cybercrime has been executed and to trace its source from the digital evidence obtained and to recover data if in case it is compromised. Cyber forensics tools are useful in the investigation of cybercrimes. The success of a cybercrime investigation relies on accurate acquisition, in-depth analysis, and structured presentation of digital evidence before the appropriate forum or court of law.

Cyber law plays an important role in safeguarding the privacy of data. It plays a pivotal role to penalize the offenders of cyber space. A cyber user must be aware of the legal implications of cybercrime and its associated forensic activities. The field of cyber forensics opens up plenty of job opportunities, which benefit government organizations, private organizations, and educational institutions as it focuses on the protection of digital assets and intelligence. The computer and information security sector is bound to witness tremendous growth as everything in this world is going digital; the same is the case with risks and cybercrimes. Hence, security measures and risk management policies should be in place to respond to threats to cyber security.

The intended audience of this book includes undergraduate and postgraduate students in computer science, computer applications, computer science and engineering, information technology, as well as professional instructors and researchers. The book may also be useful for cyber forensic professionals and cybercrime investigators to understand cybercrime investigations, forensic investigation tools, and prosecution of cybercrimes. It will be useful for computer professionals for implementing security measures to protect their digital assets.

ABOUT THE BOOK

This book is intended to make the reader aware of the growing cyber threats, cybercrimes, and how they are committed. It also provides the ways to combat cybercrime and provides a detailed description on the investigation from digital evidence. It highlights cyber forensics and also gives an insight to the present and future trends of cybercrime and forensics. Further, it attempts to elucidate the national and international laws governing cybercrime with various cases, case laws, and case studies.

The language of the text is simple and lucid, while at the same time conveying all the concepts clearly. The theory is well balanced with a number of illustrations and case studies. Most of the illustrations have been drawn in a manner that helps a student to easily understand the theory. Facts relevant to the concept discussed have been provided in the form of boxed items in each chapter to help promote clarity among readers. Demonstrations with appropriate tools are presented to expose the readers to how real-world cybercrime cases are cracked.

SALIENT FEATURES

- Contains a set of review questions and application exercises at the end of every chapter to be attempted by the student independently. Around 200 multiple-choice questions with answers have also been provided.
- Serves as a single-point resource work addressing cybercrime, cyber forensics, and associated laws and also covers the basics of networks and network security.
- Explains state-of-the-art technologies such as cryptocurrencies and block chain with examples, since cybercrime issues are more prevalent in the deep web than in the surface.

- Discusses ransomware, the crime that shook the world in 2017, and also the challenges faced by law enforcing agencies (LEA) in addressing deep web issues such as Silk Road and issues of anonymity.
- Illustrates concepts through cases, case studies, and case laws for the students to better relate to the relevant discussions.

ONLINE RESOURCES

The following resources are available to support the faculty and students using this text:

For faculty
- Lecture PPTs
- Instructor's manual (hints/answers to chapter-end application exercises)

For students
- Colour illustrations from the book
- Video resources on email tracing, tracking, and recovery of deleted files

CONTENTS AND COVERAGE

Each chapter commences with a list of learning objectives, which inform the reader what he/she is going to learn in the chapter. All the chapters are organized in such a way that the reader gets a thorough knowledge of the fundamentals, terminologies, and the advanced concepts clearly.

Chapter 1 explains the networking architecture and technologies, the OSI model and its characteristics, functions, and associated protocols, LAN technologies, network topologies, networking devices, and TCP and IP suites. Besides this, the chapter presents the security vulnerabilities in the TCP/IP suite and the security mechanism developed for networking layers as well as the security options and protocols at the network, transport, and application layers. Finally, firewall and intrusion detection system and their use in network security are explained.

Chapter 2 deals with cybercrime, its types and associated concepts. Cybercrimes are presented in three categories namely cybercrimes against person, property, and nation. The role of electronic communication device (ECD) and mobile ECD in the commission of cybercrime and the tools that facilitate them are presented. Challenges with cybercrime are explored along with cybercrime prevention strategies. Cybercrime incidents on the national and global arena are highlighted.

Chapter 3 presents a deeper insight into various categories of cybercrime, the ways to handle them as well as the preventive measures to be adopted. Some of the incidents of cybercrime reported in the country and around the globe are included as cases.

Chapter 4 discusses the current scenario and future of cybercrime—cyber war, an overview of cryptocurrencies that facilitate a ransomware attack, block chain, its underlying technology, ransomware, and the dark side of web. The challenges with dark web and deep web are also discussed.

Chapter 5 presents the components of security, and the definitions of cyber forensics. It highlights forensic investigation and forensic examination processes besides describing their benefits. The chapter explains the various types of forensics; incident and incident handling approaches, and the role of CSIRT are also discussed.

Chapter 6 explains the digital evidence collection procedure and the obstacles to this process. The chapter highlights the sources of evidences, namely various operating systems and their artifacts, the Windows registry, and various file systems. It also explains the sources of digital evidence in mobile devices and the Internet. Challenges associated with digital evidence are also discussed.

Chapter 7 lists out the tools used for forensic investigation, namely free and open-source, as well as proprietary forensic suites, imaging and validation tools, integrity verification tools, tools for data recovery and RAM analysis, tools for analysis of registry, encryption and decryption, password recovery, and network analysis. Other miscellaneous tools used for forensic investigation in the UNIX system, as well as forensic analysis tools for mobile devices and email are summarized. The current requirement for forensic investigators is explained with the career prospects in this field and the available certifications and training are elucidated.

Chapter 8 discusses the preliminaries of electronic evidence, an overview on how to acquire evidence, a detailed insight into the seizure process, and a deeper insight into acquiring evidence from computers, email, the Internet, and mobile devices. Besides these, it explains the process involved in acquiring evidence from other devices and media, as well as from third-party organizations. Finally, it explains the handling of digital evidence.

Chapter 9 provides an insight into forensic copying, computation of hash, analysis of files stored in storage media, and identification and retrieval of deleted files. It presents the steps associated with live forensics. It introduces the working of various forensic tools such as FTK Imager, Autopsy, Volatility, and WinHex. Further, it explains email tracking and tracing with necessary tools along with the role of a forensic analyst for report preparation.

Chapter 10 explains how to document the collected evidence and present it in the court of law. It provides the basics of electronic records supported by law, admissibility of electronic records in accordance with the rules, and categorization of evidence. The chapter also discusses the steps involved in presenting the evidence, namely reporting and testimony, with relevant guidelines and challenges. The court presentation system and a summary of the investigation process are also discussed.

Chapter 11 provides some of the cybercrimes as case studies. A gist of the case is presented first. This is followed by an explanation on investigation and analysis along with evidence gathering. Finally, applicable sections of the law under the Indian Penal Code (IPC) and Information Technology (Amendment) Act (ITAA) 2008 are listed for every case.

Chapter 12 introduces cyber laws and their need. The laws and their legal issues have been discussed. The chapter also offers an introduction to cyber security and the strategies to be adopted. How the laws are applicable to minimize risk is discussed, along with the initiatives taken by the government.

Chapter 13 explains the laws governing different domains such as intellectual property rights, cyber space, and Internet with respect to privacy. The chapter provides a deeper insight into the laws that are part of the ITA 2000 and IPC to handle cybercrimes and categorizes them as crime against individual, property, and nation. Besides this, the cyber laws associated with ensuring cyber security are also presented. The other laws related to cyber security and cybercrime investigations are discussed. The amendments carried out to the Indian Evidence Act (IEA) 1872 and the Banker's Book Evidence Act 1891 are also furnished in this chapter.

Chapter 14 exposes the reader with the cyber laws in force at the international level. It provides an insight into the cyber laws of representative countries such as the United States of America, the United Kingdom, the Netherlands, Malaysia, and Australia. A comparison of the cybercrime legislations in representative countries for certain specific cybercrimes is presented. Case studies, as applicable, are highlighted wherever necessary.

ACKNOWLEDGEMENTS

The authors thank Advocate Jemila Samerin and Dr A.S. Vijila Samerin, Assistant Professor in English, for their help in the process of proofreading and editing. They sincerely thank the talented team of editors at Oxford University Press, India and all the reviewers who provided prompt guidance and very valuable feedback, which helped immensely in improving the contents of this book.

Dr Dejey thanks the co-author Dr S. Murugan for his unconditional support and the authorities of Anna University, Chennai for granting permission and providing the necessary infrastructure for this venture. She conveys her special thanks to her parents, family, friends, and relatives who supported and encouraged her, while she was working long hours on the book. She also thanks her student and scholar Ms M. Kaviya Elakkiya for her consistent support in the different phases of creation of this book.

Dr S. Murugan thanks the co-author Dr Dejey who travelled with him in the journey of writing this book. Further, he thanks his family members for their support and encouragement and his mother for her blessings. He also acknowledges the Tamil Nadu Police Department for support and encouragement.

Dejey
Murugan IPS

FEATURES OF

Classification of Cybercrime — 3

Cyber Forensics — The Present and the Future — 7

Cyber Laws in India and Case Studies — 13

Addresses important segments
A single-point resource work addressing cybercrime, forensics, and associated laws

Covers basics of networks and security
Provides an introductory chapter on networking and cyber security

Networks and Network Security — 1

Interesting examples
Presents brief cases and boxed items containing additional information throughout all chapters

---

Box 8.4 Forensic Boot Disk

It is used to boot the suspect system safely. It contains a file system and statically linked utilities such as ls, fdisk, ps, nc, dd, and ifconfig. It places the suspect media in a locked or read-only state. It does not swap any data to the suspect media. Some of the open source bootable images include FIRE (http://biatchux.dmzs.com/?section=main), Linuxcare Bootable Business Cards (http://lbt.linuxcare.com/index.epl), and Trinux (http://trinux/sourceforge.net/).

Case 2: Cyber Stalking

In the first successful prosecution under the California (USA) cyber stalking law, prosecutors obtained a guilty plea from a 50-year-old former security guard who used the Internet to solicit rape of a woman who rejected his romantic advances.

He terrorized the 28-year-old victim by impersonating her in various Internet chat rooms and online bulletin boards, where he posted, along with her telephone number and address, messages that she fantasized about being raped. On at least six occasions, sometimes in the middle of the night, men knocked at the woman's door saying that they wanted to rape her.

THE BOOK

POINTS TO REMEMBER
- Evidence may be oral or documentary, and documentary evidence may be primary or secondary.
- Digital evidence is unique and can be judged as neither primary nor secondary evidence.

MULTIPLE-CHOICE QUESTIONS

1. The rules and regulations of cyber laws are covered under _____.
   (a) Information Technology (Certifying Authorities) Rules 2000
   (b) Information Technology (Security Procedure)

Rich pedagogy

Chapter-end learning features include points to remember, key terms, multiple-choice questions with answers, review questions, and application exercises that cover all the concepts learnt in each chapter and help provide a quick recap

KEY TERMS

Digital evidence This refers to information or data stored on, transmitted, or received by an electronic device in binary form that is of value in a criminal

REVIEW QUESTIONS

1. List the types of cybercrime and their probable location of digital evidence.
2. ...are the steps to be followed by an IO during ...e scene investigation?

APPLICATION EXERCISES

1. Imagine that you are the forensic examiner for a crime reported as 'theft of intellectual property' by an organization, say, X. The IO had reported after a thorough investigation that an employee, say, Y had become disgruntled in recent months

CYBER LAWS IN INDIA AND CASE STUDIES — 13

INTERNATIONAL CYBER LAWS AND CASE STUDIES — 14

Balanced discussion of both Indian and International cyber laws

Provides a balanced discussion of both international and Indian laws according to the context

Brief Contents

Preface v
Features of the Book viii
Detailed Contents xi

1. Networks and Network Security	1
2. Introduction to Cybercrime	44
3. Classification of Cybercrime	78
4. Cybercrime—The Present and the Future	121
5. Introduction to Cyber Forensics	149
6. Digital Evidence	184
7. Cyber Forensics—The Present and the Future	232
8. Acquisition and Handling of Digital Evidence	269
9. Analysis of Digital Evidence	301
10. Admissibility of Digital Evidence	350
11. Cybercrime Case Studies	371
12. Introduction to Cyber Laws	391
13. Cyber Laws in India and Case Studies	399
14. International Cyber Laws and Case Studies	439

Appendix 474

Index 477

Detailed Contents

Preface v
Features of the Book viii
Brief Contents x

1. Networks and Network Security 1

1.1 Network 1
1.2 Networking Architecture 1
1.3 Networking Technologies 2
1.4 Network Models 4
    1.4.1 *OSI Model 4*
    1.4.2 *Internet Model 6*
1.5 Networking Devices 6
1.6 LAN Technologies 8
1.7 Networking Topologies 11
1.8 Network Protocols—TCP/IP Protocol Suite 12
1.9 Physical Layer 13
1.10 Data Link Layer 14
    1.10.1 *Functions 14*
    1.10.2 *Error Control 15*
    1.10.3 *Flow Control 17*
1.11 Network Layer 17
    1.11.1 *Network Addressing 17*
    1.11.2 *Routing in Network Layer 17*
1.12 Network Layer Protocols 19
    1.12.1 *Address Resolution Protocol and Reverse Address Resolution Protocol 20*
    1.12.2 *Internet Control Message Protocol 20*
    1.12.3 *Internet Protocol Version 4 20*
    1.12.4 *Internet Protocol Version 6 21*
1.13 Transport Layer 21
    1.13.1 *Transmission Control Protocol 22*
    1.13.2 *User Datagram Protocol 25*
1.14 Application Layer 26
    1.14.1 *Application Layer Protocols 26*
1.15 Security Vulnerabilities in TCP/IP Suite 28
1.16 Security Mechanisms in Networking Layers 28
1.17 Network Security at Network Layer with Internet Protocol Security 28
    1.17.1 *IPSec Communication 29*
    1.17.2 *Internet Key Exchange 30*
1.18 Network Security at Transport Layer 30
    1.18.1 *Secure Socket Layer 31*
    1.18.2 *Transport Layer Security Protocol 32*
    1.18.3 *HTTPS 32*
    1.18.4 *Secure Shell Protocol 33*
1.19 Network Security at Application Layer 33
    1.19.1 *Pretty Good Privacy 34*
    1.19.2 *Secure MIME 34*
    1.19.3 *DNSSec 35*
1.20 Network Security with Firewall 36
1.21 Network Security with Intrusion Detection System and Intrusion Detection and Prevention System 36

2. Introduction to Cybercrime 44

2.1 Introduction 44
    2.1.1 *Definition 45*
2.2 Role of Electronic Communication Devices and Information and Communication Technologies in Cybercrime 45
2.3 Mens rea and Actusreus in Cybercrime 46
2.4 Types of Cybercrime 47
2.5 Cybercrime against Individuals 47
2.6 Cybercrime against Property 48
2.7 Cybercrime against Nation 52
    2.7.1 *Content-related Offences 53*
2.8 Crimes Associated with Mobile Electronic Communication Devices 54
2.9 Classification of Cybercriminals 55
2.10 Execution of Cybercrime 56
2.11 Tools used in Cybercrime 57
2.12 Factors Influencing Cybercrime 59
2.13 Challenges to Cybercrime 60
2.14 Strategies to Prevent Cybercrimes 61
    2.14.1 *Indian Perspective 62*
    2.14.2 *Global Best Practices 64*

2.15 Extent of Cybercrime 65
    2.15.1 *Cybercrime Statistics and World* 65
    2.15.2 *Cybercrime Statistics in India* 66
    2.15.3 *Recent Sensitive Cybercrimes* 67
    2.15.4 *Latest incidents of Cybercrime in India* 68
2.16 Terms and Terminologies Associated with Cybercrime 69

3. Classification of Cybercrime 78

3.1 Introduction 78
3.2 Cybercrime against Individuals 78
    3.2.1 *Internet Grooming* 79
    3.2.2 *Cyber Stalking* 79
    3.2.3 *Cyber Harassment* 80
    3.2.4 *Cyber Extortion* 81
    3.2.5 *Online Pedophilia* 82
3.3 Cybercrime against Property 83
    3.3.1 *Illegal Access—Hacking and Cracking* 83
    3.3.2 *Illegal Data Acquisition—Data Espionage* 84
    3.3.3 *Illegal Interception* 87
    3.3.4 *Data Interference* 90
    3.3.5 *System Interference—Computer Threats* 97
    3.3.6 *Copyright- and Trademark-related Offences* 103
    3.3.7 *Computer-related Offences* 105
3.4 Cybercrime against Nation 109
    3.4.1 *Cyber Terrorism* 109
    3.4.2 *Cyber Warfare* 110
    3.4.3 *Cyber Laundering* 110
    3.4.4 *Content-related Offences* 111

4. Cybercrime—The Present and the Future 121

4.1 Introduction to Cyber War—The Present and the Future of Cybercrime 121
4.2 Cryptocurrency 122
    4.2.1 *Characteristics* 123
    4.2.2 *Types* 124
4.3 Bitcoin 125
    4.3.1 *Bitcoin Cash* 127
4.4 Ethereum 127
4.5 Comparison between Bitcoin and Ethereum 128
4.6 Blockchain 129
    4.6.1 *Association between Bitcoin and Blockchain* 131

4.7 Ransomware 131
    4.7.1 *Evolution of Ransomware* 131
    4.7.2 *Types of Ransomware* 132
    4.7.3 *Entities Affected by Ransomware Attack* 133
    4.7.4 *Mode of Infection of Users with Ransomware* 133
    4.7.5 *Events in Ransomware Attack* 134
    4.7.6 *Post-delivery of Ransomware* 135
    4.7.7 *Preventing Ransomware from Full Execution* 135
    4.7.8 *Steps to Carry Out in Event of Infection with Ransomware* 135
    4.7.9 *Best Practices to Adopt* 135
    4.7.10 *Role of Antivirus* 136
    4.7.11 *Prevention and Response Team* 136
4.8 Deep Web and Dark Web 138
4.9 Deep Web and its Challenges 138
    4.9.1 *The Internet* 138
    4.9.2 *Accessing Dark Web* 140
    4.9.3 *Size and Scale of Deep Web* 140
    4.9.4 *Onion Router—TOR* 141
    4.9.5 *Search Engines vs Deep Web* 141
    4.9.6 *Deep Web Source Repository* 141
    4.9.7 *Challenges* 142
    4.9.8 *Counter Measures to Overcome Challenges with Deep Web* 143

5. Introduction to Cyber Forensics 149

5.1 Interrelation among Cybercrime, Cyber Forensics, and Cyber Security 149
    5.1.1 *Security* 150
5.2 Cyber Forensics 151
    5.2.1 *Definition* 151
    5.2.2 *Need* 151
    5.2.3 *Objectives* 151
    5.2.4 *Computer Forensics Investigations* 152
    5.2.5 *Steps in Forensic Investigation* 152
    5.2.6 *Forensic Examination Process* 154
    5.2.7 *Methods Employed in Forensic Analysis* 155

	5.2.8	*Classification of Cyber Forensics* 155	
	5.2.9	*Benefits of Cyber Forensics*	155
5.3	Disk Forensics		155
	5.3.1	*Challenges* 156	
5.4	Network Forensics		156
	5.4.1	*Tools for Analysis* 156	
	5.4.2	*Challenges* 156	
5.5	Wireless Forensics		157
	5.5.1	*Forensic Tools* 157	
	5.5.2	*Challenges* 157	
5.6	Database Forensics		158
	5.6.1	*Forensic Approaches* 158	
	5.6.2	*Forensic Methodology* 159	
5.7	Malware Forensics		160
	5.7.1	*Malware Analysis* 160	
5.8	Mobile Forensics		160
	5.8.1	*Stages* 161	
	5.8.2	*Analysis Tools* 162	
5.9	GPS Forensics		163
5.10	Email Forensics		164
	5.10.1	*Client and Server in Email* 164	
	5.10.2	*Structure of Email* 164	
	5.10.3	*Working of Email* 165	
	5.10.4	*Email Protocols* 165	
	5.10.5	*Examining Email Messages* 166	
	5.10.6	*Viewing Email Headers* 166	
	5.10.7	*Examining Email headers* 167	
	5.10.8	*Examining Additional Email Files* 168	
	5.10.9	*Tracing Email Messages* 168	
	5.10.10	*Email Servers and their Examination* 170	
	5.10.11	*Email Forensics Tools* 170	
	5.10.12	*Tracking Emails* 171	
5.11	Memory Forensics		172
	5.11.1	*RAM Artifacts* 172	
	5.11.2	*RAM Analysis* 172	
	5.11.3	*Forensic Tools* 173	
5.12	Building Forensic Computing Lab		173
5.13	Incident and Incident Handling		174
	5.13.1	*Incident* 174	
	5.13.2	*Incident Handling* 175	
	5.13.3	*Incident Reporting* 176	
	5.13.4	*Incident Response* 176	
5.14	Computer Security Incident Response Team		177
	5.14.1	*Forensic Readiness* 178	

6. Digital Evidence   **184**

6.1	Introduction to Digital Evidence and Evidence Collection Procedure		184
	6.1.1	*Types of Digital Evidence*	184
	6.1.2	*Evidence Collection Procedure* 185	
	6.1.3	*Mechanisms Associated with Digital Evidence Collection* 187	
6.2	Sources of Evidence		187
6.3	Digital Evidence from Standalone Computers/Electronic Communication Devices		187
6.4	Operating Systems and their Boot Processes		188
6.5	Storage Medium		190
	6.5.1	*Disk Drive* 190	
	6.5.2	*Other Storage Media* 194	
6.6	File System		194
	6.6.1	*FAT File System and its Components* 195	
	6.6.2	*Extended File Allocation Table File System* 198	
	6.6.3	*New Technology File System* 202	
	6.6.4	*ext family of File Systems*	209
	6.6.5	*Hierarchical File System*	213
6.7	Windows Registry		215
6.8	Windows Artifacts		217
6.9	Browser Artifacts		219
6.10	Macintosh Artifacts		220
6.11	Linux Artifacts		221
6.12	Whole Disk Encryption or Full Disk Encryption		221
6.13	Evidence from Mobile Devices		222
6.14	Digital Evidence on the Internet		223
6.15	Digital Evidence as Alibi		224
6.16	Impediments to Collection of Digital Evidence		225
6.17	Challenges with Digital Evidence		226

7. Cyber Forensics—The Present and the Future   **232**

7.1	Forensic Tools		232
	7.1.1	*Types* 233	
	7.1.2	*Categories* 233	
7.2	Cyber Forensic Suite		235
	7.2.1	*Free and Open-source Forensic Suite* 235	
	7.2.2	*Proprietary Forensic Suites*	238

- 7.3 Drive Imaging and Validation Tools 239
- 7.4 Forensic Tool for Integrity Verification and Hashing 240
- 7.5 Forensic Tools for Data Recovery 241
- 7.6 Forensic Tools for RAM Analysis 242
- 7.7 Forensic Tools for Analysis of Registry 243
- 7.8 Forensic Tools for Encryption/Decryption 243
- 7.9 Forensic Tools for Password Recovery 244
- 7.10 Forensic Tools for Analysing Network 245
- 7.11 Forensic Utility for Metadata Processing 246
- 7.12 Miscellaneous Tools 247
- 7.13 Forensic Tools for UNIX System Analysis 250
- 7.14 Forensic Tools for Other media 251
- 7.15 Forensic Hardware 252
- 7.16 Forensic Analysis Tools for Mobile Devices 252
  - 7.16.1 *Free and Open-source Forensic Tools for Mobile Devices* 252
  - 7.16.2 *Proprietary Forensic Tools for Mobile Devices* 254
  - 7.16.3 *Forensic Hardware for Mobile Devices* 255
- 7.17 Forensic Tools for Email Analysis 256
- 7.18 Need for Computer Forensic Investigators 257
- 7.19 Career Prospects for Forensic Investigators 258
- 7.20 Forensic Training and Certifications 258

8. Acquisition and Handling of Digital Evidence 269

- 8.1 Preliminaries of Electronic or Digital Evidence 269
  - 8.1.1 *Categorization of Source of Digital Evidence* 270
  - 8.1.2 *Locality of Digital Evidence* 271
  - 8.1.3 *Roles played by Digital Evidence* 271
  - 8.1.4 *Characteristics of Digital Evidence* 271
  - 8.1.5 *Physical versus Digital Evidence* 271
  - 8.1.6 *Order of Volatility of Digital Evidence* 272
  - 8.1.7 *List of Crimes and Probable Location of Evidence* 273
- 8.2 Acquisition and Seizure of Evidence 274
  - 8.2.1 *Acquisition of Evidence* 274
  - 8.2.2 *Precautionary Measures before Acquisition* 274
  - 8.2.3 *Search and Seizure* 275
  - 8.2.4 *Seizure Memo* 276
- 8.3 Chain of Custody and Digital Evidence Collection Form 277
  - 8.3.1 *Chain of Custody* 277
  - 8.3.2 *Digital Evidence Collection Form* 278
- 8.4 Fourth Amendment and Seizure 279
  - 8.4.1 *Search and Seizure with Search Warrant* 279
  - 8.4.2 *Warrantless Searches* 279
- 8.5 Acquisition of Computer and Electronic Evidence 280
  - 8.5.1 *Acquisition of Configuration Information through Controlled Boots* 281
  - 8.5.2 *Acquisition of Evidence from Switched-off Systems* 281
  - 8.5.3 *Collection of Volatile Data* 283
  - 8.5.4 *Acquisition of Evidence from Live Systems* 283
  - 8.5.5 *Acquisition of Evidence from Standalone Hardware Device* 284
  - 8.5.6 *Acquisition of Evidence from Non-detachable Hard Disk Drive* 284
- 8.6 Acquisition Procedure using Target Disk Mode from Apple Macintosh Computer 286
  - 8.6.1 *Social Media* 288
- 8.7 Acquisition of Evidence from Mobile Phone and PDA 288
  - 8.7.1 *Procedure for Acquiring Evidence from Mobile Phones* 292
- 8.8 Acquisition of Evidence from Optical and Removable Media, Digital Cameras 292
  - 8.8.1 *Evidence from Optical Media* 292
  - 8.8.2 *Evidence from USB Drives* 292
  - 8.8.3 *Evidence from Digital Cameras* 293

8.9	Acquisition of Evidence from Third Party, External Agency, or Organization	293
8.10	Challenges to Acquisition of Digital Evidence	293
8.11	Handling of Digital Evidence	294
8.12	Precautions Involved in Handling Digital Evidence	296

9. Analysis of Digital Evidence — 301

- 9.1 Introduction to Analysis of Digital Evidence — 301
- 9.2 Capturing of Forensic Copy of Memory and Hard Drive with Toolkit Forensic Imager — 302
- 9.3 RAM Analysis with Volatility — 307
- 9.4 Analysing Hard Drive with WinHex — 312
  - 9.4.1 *Acquiring Forensic Copy of Drive* 312
  - 9.4.2 *Computing Hash* 314
  - 9.4.3 *Analysing Hard Disk* 316
  - 9.4.4 *Analysing Slack Space and Free Space* 327
  - 9.4.5 *File Carving* 328
- 9.5 Working with Autopsy — 328
  - 9.5.1 *Analysis Basics* 330
  - 9.5.2 *Timeline* 330
  - 9.5.3 *Example Use Cases* 330
  - 9.5.4 *Analysis of Deleted Files with Autopsy* 331
- 9.6 Email Tracking and Tracing — 336
  - 9.6.1 *Email Tracking* 336
  - 9.6.2 *Email Tracing* 338
- 9.7 Role of Forensic Analyst in Analysis — 343

10. Admissibility of Digital Evidence — 350

- 10.1 Introduction — 350
- 10.2 Digital Evidence—Electronic Record — 351
  - 10.2.1 *Prerequisites* 352
  - 10.2.2 *Retention of Electronic Records* 352
- 10.3 Section 5 of ITA 2000—Legal Recognition of Digital Signatures — 352
  - 10.3.1 *Rules of Admissibility of Electronic Evidence* 357
  - 10.3.2 *Categorization and Characteristics of Evidence with Respect to Law* 357
- 10.4 Pre-trial Preparation — 360
- 10.5 Presenting Digital Evidence — 360
  - 10.5.1 *Reporting—Expert Report* 360
  - 10.5.2 *Guidelines for Cyber Forensic Examiners* 361
  - 10.5.3 *Testimony* 362
  - 10.5.4 *Courtroom Presentation System* 363
  - 10.5.5 *Challenges with Admissibility and Presentation of Digital Evidence* 363
- 10.6 Summary of Investigation Process Involving Digital Evidence — 364

11. Cybercrime Case Studies — 371

- 11.1 Introduction — 371
- 11.2 Cybercrime against Individual — 371
  - 11.2.1 *Posting of Obscene, Defamatory, and Annoying Messages against Women Online (State of Tamil Nadu vs Suhas Katti)* 371
  - 11.2.2 *Phishing Fraud* 372
  - 11.2.3 *Hacking using Key Logger* 373
  - 11.2.4 *Impersonation for Purpose of Cheating* 375
  - 11.2.5 *Transmission of Sexually Explicit Material through Internet* 376
  - 11.2.6 *Criminal Intimidation and Sending Obscene Material through Internet* 377
  - 11.2.7 *Trolling on Social Media* 377
- 11.3 Cybercrime against Property — 379
  - 11.3.1 *Online Lottery Scam* 379
  - 11.3.2 *Theft in ATM* 380
  - 11.3.3 *Swindling of Money by Bank Employee* 381
  - 11.3.4 *Data Theft* 383
  - 11.3.5 *Hacking* 384
  - 11.3.6 *Data Theft by Ex-employee* 385
- 11.4 Cybercrime against Nation — 386
  - 11.4.1 *Preparation of Forged Counterfeits using Computers/ Printers/Scanners* 386
  - 11.4.2 *Blocking of Websites* 388

12. Introduction to Cyber Laws — 391

- 12.1 Cyber Laws — 391
- 12.2 Need for Cyber Laws — 391
- 12.3 Cyber Laws and Legal Issues — 392
- 12.4 Cyber Security — 393

12.5	Strategies Involved in Cyber Security	393	13.10	Summary of Cyber Laws in India 427
12.6	Minimizing Risk with Cyber Laws	393	13.11	Amendments to the Indian Evidence Act 1872 in View of Information Technology Act 2000 429
12.7	Initiatives Promoting Cyber Security	394		
12.8	Terms and Terminologies Associated with Cyber Laws	394	13.12	Amendments to the Banker's Book Evidence Act 1891 in View of Information Technology Act 2000 431

13. Cyber Laws in India and Case Studies   399

- 13.1 Cyber Laws, Cybercrime, and Cyber Security   399
- 13.2 Cyber Laws in India   399
- 13.3 Information Technology Act 2000   400
  - 13.3.1 *Scheme of IT Act 2000*   400
  - 13.3.2 *Salient Features of the Information Technology (Amendment) Act 2008*   401
- 13.4 Cybercrimes and Cyber Laws   402
- 13.5 Crime against Individual   404
  - 13.5.1 *Cyber Defamation*   404
  - 13.5.2 *Cyber Stalking*   405
  - 13.5.3 *Web Jacking*   406
  - 13.5.4 *Violation of Privacy*   407
- 13.6 Crime against Property   407
  - 13.6.1 *Theft of Data, Viral Attack, Hacking, Denial of Service Attack, and Cyber Bullying*   407
  - 13.6.2 *Forgery*   409
  - 13.6.3 *Data Diddling*   411
  - 13.6.4 *Email Bombing*   411
  - 13.6.5 *Possession of Stolen Electronic Communication Devices*   411
  - 13.6.6 *Identity Theft and Password Theft*   412
  - 13.6.7 *Financial Crime*   412
  - 13.6.8 *Email Spoofing*   414
  - 13.6.9 *Email Fraud*   414
  - 13.6.10 *Copyright Infringement Crimes*   414
  - 13.6.11 *Sale of Illegal Articles on the Internet*   415
- 13.7 Crime against Nation   416
  - 13.7.1 *Cyber Terrorism*   416
  - 13.7.2 *Website Defacement*   416
  - 13.7.3 *Pornography*   417
- 13.8 Cyber Laws for Cyber Security   421
- 13.9 Other Cyber Laws Associated with Cybercrime and Cyberspace   424
- 13.13 Indian Laws Related to Intellectual Property   432
- 13.14 Indian Case Laws   432

14. International Cyber Laws and Case Studies   439

- 14.1 Introduction   439
- 14.2 Cybercrime Legislation in the Netherlands   439
  - 14.2.1 *Specific Cybercrime Legislations*   439
  - 14.2.2 *Traditional Laws to Prosecute Cybercrimes*   442
  - 14.2.3 *Powers for Search and Seizure*   443
  - 14.2.4 *Other ICT-related Investigation Powers*   443
- 14.3 Cyber Laws in Malaysia   443
- 14.4 Cybercrime Laws in the UK   445
- 14.5 Cybercrime Laws of the United States   451
  - 14.5.1 *Computer Fraud and Abuse Act*   452
  - 14.5.2 *Provisions for Handling Cyber Stalking*   455
  - 14.5.3 *Provisions to Handle Cyber Terrorism*   456
  - 14.5.4 *Electronic Communications Privacy Act*   456
  - 14.5.5 *Cyber Security Enhancement Act*   457
  - 14.5.6 *Digital Millennium Copyright Act*   457
  - 14.5.7 *Traditional Laws to Prosecute Cybercrime*   458
  - 14.5.8 *Summary of US Federal Cyber Laws*   459
- 14.6 Australian Laws Related to Privacy and Cyber Security Domains   460
  - 14.6.1 *Legal, Legislative, and Regulatory Environment*   460
  - 14.6.2 *Other Federal Legislative Acts*   463

Appendix   474

Index   477

Networks and Network Security

Learning Objectives

This chapter provides an overview of computer networks and network security. The objective of this chapter is to provide a deeper insight into computer networks, networking architecture, networking technologies, and network models (the OSI and the Internet), networking devices, LAN technologies, and various network topologies. The chapter presents the network protocol, the TCP/IP protocol suite. It also explains the characteristics and functions of every layer in the OSI and discusses the associated protocols. Besides this, the security vulnerabilities in the TCP/IP suite and the security mechanism are elucidated. Protocols for network security at the network, transport, and the application layer are explained in detail. The establishment of network security with firewalls is also explained. The reader will be familiar with the following after studying the chapter:

- Networking architecture and technologies
- OSI model—characteristics, functions, and associated protocols
- Networking devices, LAN technologies, and network topologies
- TCP and IP suite
- Security vulnerabilities in the TCP/IP suite and the security mechanism developed for networking layers
- Security options and protocols at the network, transport layer, and application layer
- Firewall and its use in network security

1.1 NETWORK

A computer network is a collection of computers and devices interconnected by communication channels to facilitate communication and the sharing of information (data, messages, graphics) and resources (printers, fax machines, modems, and other hardware) among interconnected devices.

Thus, computer networks are primarily needed for information exchange and resource sharing.

1.2 NETWORKING ARCHITECTURE

The three common forms of networking architecture are (a) peer-to-peer network, (b) client/server architecture (or) server-based network, and (c) hybrid network. This is shown in Fig. 1.1.

Peer-to-peer network In this network, any computer can act as a server or a client and there is no designated server. All the computers are referred to as peers. There is no designated administrator in such a network. Only the users on a peer-to-peer network determine what data from their computers can be shared on the network.

Server-based network In this network, there is a dedicated server and all the other computers are clients. The dedicated server services requests from network clients. More servers are required as the size and traffic in a network scale up. The server may be a *file and print server* that manages user access and the use of file and printer services; an *application server* that makes client/server applications and data available to the clients; a *mail server* to manage electronic messaging between the network and the users; a *fax server* to manage incoming and outgoing data in the network by sharing one or more fax modem boards; or a *communication server* which handles data flow and email between its network and other networks as well as with remote users through a modem or telephone lines in a dial-up connection. The security aspect of servers is managed by the administrator, by setting up a policy that applies to every user on the network.

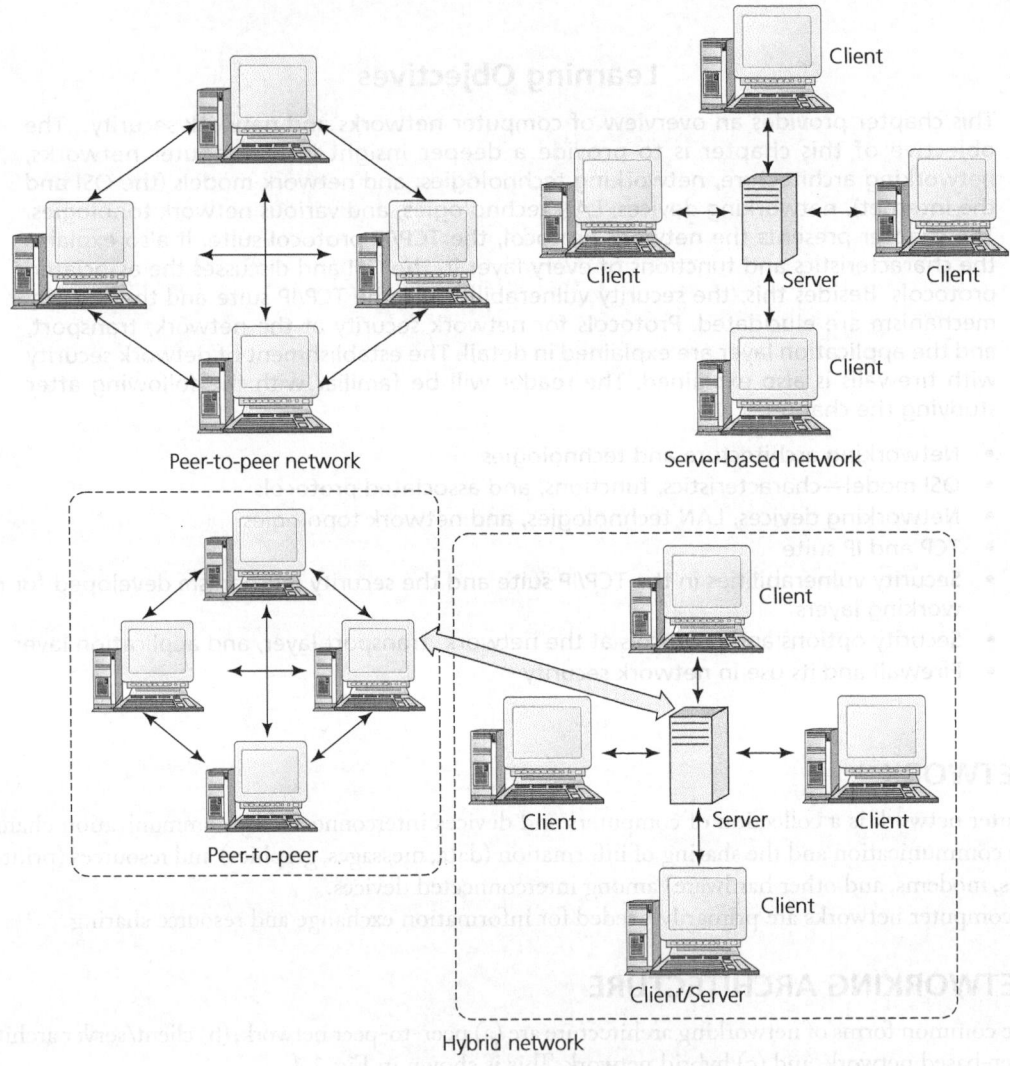

Fig. 1.1 Networking architecture

Hybrid network   This network is a combination of both, peer-to-peer network and server-based network.

1.3 NETWORKING TECHNOLOGIES

The four basic types of networking technologies based on geographical span are (a) local area network (LAN), metropolitan area network (MAN), wide area network (WAN), and Internetwork. This is shown in Fig. 1.2.

Networks and Network Security **3**

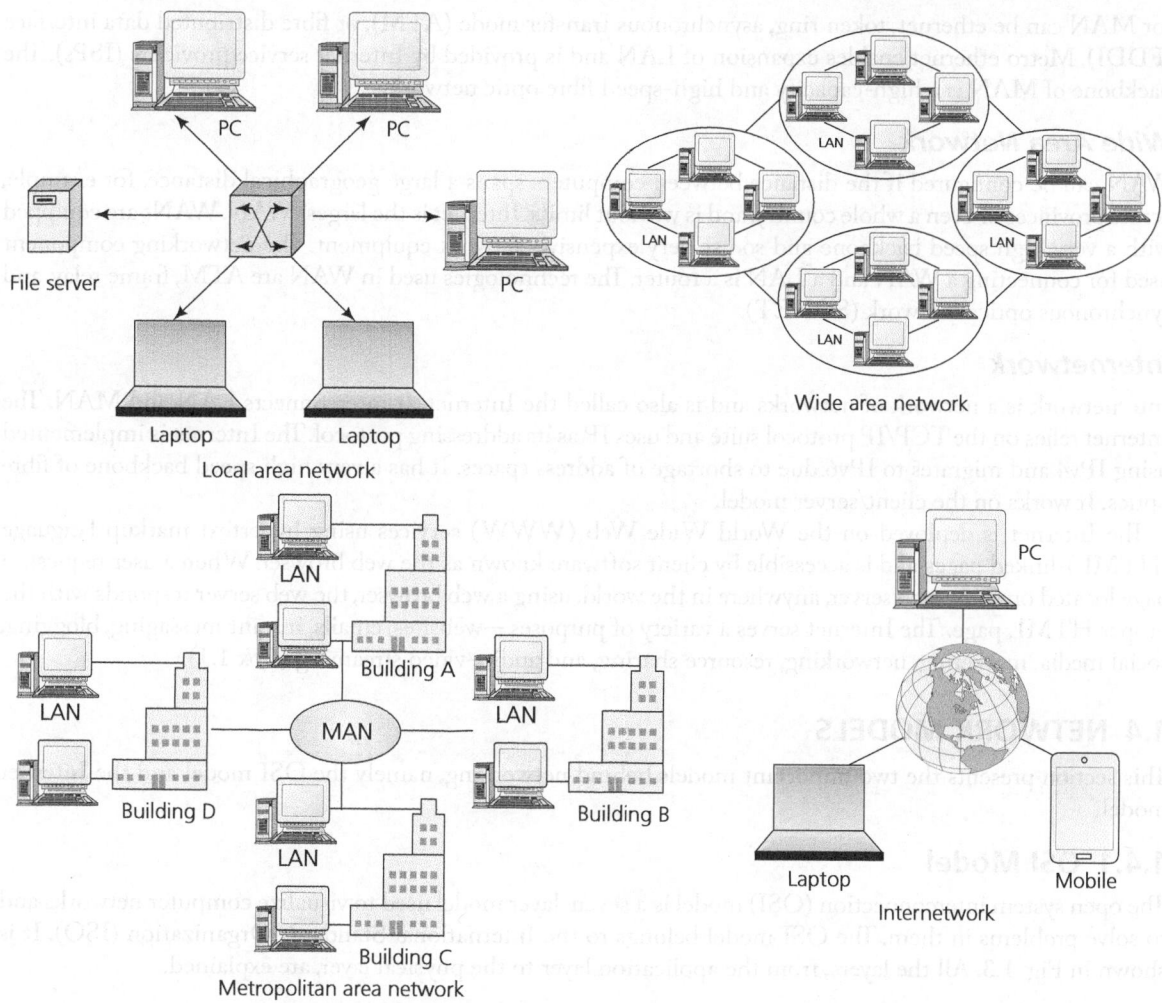

Fig. 1.2 Networking technologies

Local Area Network

LAN can be configured only if the distance between the computers is not much and they are spread over a small geographical area. LAN facilitates sharing of resources such as printers, scanners, file servers, and the Internet among computers. LAN is suitable for small organizations and for home-based networks. The networking components used for connectivity in a LAN include hubs, switches, and cables (Cat-5 and Cat-5e). LAN uses private IP addresses and does not involve any complicated routing. It works under its own domain and can be controlled centrally.

It uses either ethernet or token ring technology—the former is widely used, whereas the latter is only rarely used. Ethernet technology on LAN relies on star topology.

LAN can be either wired, wireless, or both.

Metropolitan Area Network

MAN can be configured when the distance between the computers is large and the computers are far apart from each other, spanning a city or a town. It is larger than a LAN and smaller than a WAN. The technology for MAN can be ethernet, token ring, asynchronous transfer mode (ATM), or fibre distributed data interface (FDDI). Metro ethernet enables expansion of LAN and is provided by Internet service providers (ISPs). The backbone of MAN is a high-capacity and high-speed fibre optic network.

Wide Area Network

WAN can be configured if the distance between computers spans a large geographical distance, for example, across provinces or even a whole country, and is without limits. Internet is the largest WAN. WANs are equipped with a very high speed backbone and so use very expensive network equipment. The networking component used for connecting a WAN and a LAN is a router. The technologies used in WAN are ATM, frame relay, and synchronous optical network (SONET).

Internetwork

Internetwork is a network of networks and is also called the Internet. It interconnects LAN and MAN. The Internet relies on the TCP/IP protocol suite and uses IP as its addressing protocol. The Internet is implemented using IPv4 and migrates to IPv6 due to shortage of address spaces. It has a very high speed backbone of fibre optics. It works on the client/server model.

The Internet is deployed on the World Wide Web (WWW) services using hypertext markup language (HTML)-linked pages and is accessible by client software known as the web browser. When a user requests a page located on some web server, anywhere in the world, using a web browser, the web server responds with the proper HTML page. The Internet serves a variety of purposes—websites, emails, instant messaging, blogging, social media, marketing, networking, resource sharing, and audio–video streaming (Box 1.1).

1.4 NETWORK MODELS

This section presents the two important models behind networking, namely the OSI model and the Internet model.

1.4.1 OSI Model

The open system interconnection (OSI) model is a seven-layer model used to visualize computer networks and to solve problems in them. The OSI model belongs to the International Standards Organization (ISO). It is shown in Fig. 1.3. All the layers, from the application layer to the physical layer, are explained.

Layer 7—Application Layer

It is responsible for user interaction in the form of input and output. The application is some software that runs on the local machine but depends on network architecture. The software could be either cloud-based, that is, run on a remote server from where data is transferred over the Internet or software that is run on a local server. Thus, the application layer is responsible for providing services for email, telnet, file transfer, etc. For instance, the application layer can be the Internet browser, an FTP client, or Microsoft Word.

> **Box 1.1 Internet and Intranet**
>
> The Internet is an interconnection of many combinations of networks, such as LAN, WAN, and MAN. It is an unlimited source of a wide variety of information, and enables sharing and access to its users. It can offer connectivity to an unlimited number of users. There is no limit on bandwidth. Traffic on the Internet is unrestricted.
>
> On the other hand, in an intranet, the network is restricted for use by a single corporate entity which has full control and management over the network. The functionality of the intranet is however the same as that of the Internet.

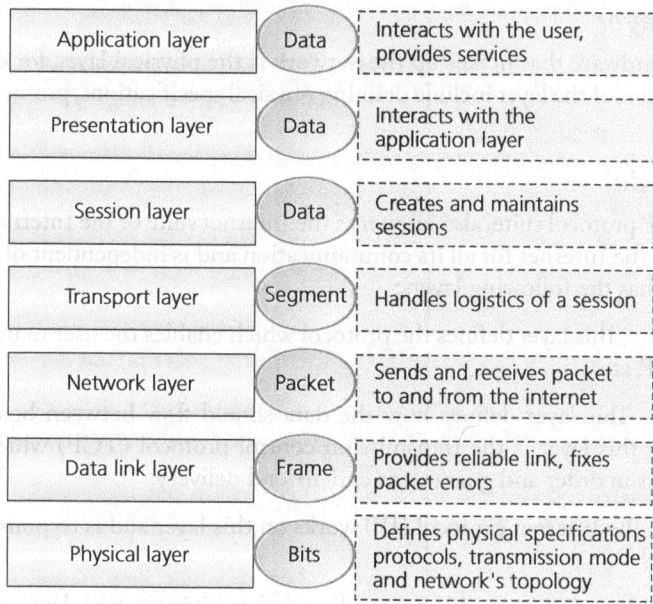

Fig. 1.3 OSI model

Layer 6—Presentation Layer
This layer directly interacts with the application layer above and this is where the operating system lies. Interaction happens either directly or through the Java runtime environment (JRE).

Layer 5—Session Layer
The session layer is responsible for creating and maintaining sessions between the operating system on the presentation layer and other third-party machines. For example, while browsing, the user interacts with the application layer which in turn interacts with the presentation layer, and the session layer facilitates interaction between the operating system and the web server.

Layer 4—Transport Layer
The logistics of a session are taken care of by the transport layer. For example, while browsing, the transport layer determines what and how much information should be exchanged between the operating system and the web server.

Layer 3—Network Layer or Internet Layer
This layer is responsible for sending and receiving packets to and from the Internet as governed by the IP address of the router. The router, the hardware device essential for forwarding packets between computers on a network, operates in this layer.

Layer 2—Data Link Layer
This layer is responsible for providing a reliable link between two directly connected nodes. It is also responsible for fixing packet errors which may arise in the physical layer below. Switches operate in this layer. On the basis of functionality, this layer is divided into two: the media access control (MAC) layer (which is responsible for the way in which the devices that are connected to the network gain access) and the logical link control (LC) layer (which is responsible for error checking and packet synchronization).

Layer 1—Physical Layer

Any physical device or hardware that makes up the network is the physical layer, for example, ethernet cables and bluetooth. The functions of the layer include defining physical specifications, protocols, transmission modes, and network topology.

1.4.2 Internet Model

The Internet uses TCP/IP protocol suite, also known as the Internet suite or the Internet model. It is a four-layered architecture used by the Internet for all its communication and is independent of the underlying network architecture. This model has the following layers:

Layer 4—application layer This layer defines the protocol which enables the user to interact with the network, for example, FTP, HTTP, etc.

Layer 3—transport layer This layer defines how the data should flow between hosts. The most important protocol that operates in this layer is the transmission control protocol (TCP) which ensures that the data delivered between hosts is in order and guarantees end-to-end delivery.

Layer 2—Internet layer The Internet protocol (IP) works on this layer and is responsible for host addressing, recognition, and routing.

Layer 1—link layer This layer is responsible for sending and receiving actual data, and is independent of the underlying network architecture and hardware.

1.5 NETWORKING DEVICES

Networking devices are used for connecting to a network, routing the packets, strengthening the signal, communicating with others, sharing files on the network, etc. Networking devices include repeaters, hubs, bridges, switches, gateways, and modems. They are explained here:

Repeater

A repeater is an electronic two-port device that operates at the physical layer. Its job is to regenerate the signal over the same network before the signal becomes too weak or corrupted so as to extend the length to which the signal can be transmitted over the same network. Repeaters do not amplify the signal. When the signal becomes weak, they copy the signal bit by bit and regenerate it to the original strength.

Hub

A hub is a networking device which is used to connect multiple network hosts. It is also called multiport repeater. Hubs cannot filter data, hence data packets are sent to all connected devices. They carry out data transfer in terms of packets. When a host sends a data packet to a hub, the hub copies the data packet to all its ports. This makes it slower and more congested. Further, it does not have the intelligence to determine the best path for these data packets; this leads to inefficiencies and wastages. Hubs are of two types:

Passive hub It forwards the data signal from all the ports except the port on which the signal arrived. It does not interfere with the data signal.

Active hub It also forwards the data signal from all ports except the port on which the signal arrived. However, before forwarding, it improves the quality of the data signal by amplifying it. Due to this, an active hub is also known as a repeater.

Bridge

A bridge connects two subnetworks (or two LANs) which are part of the same network. In other words, a bridge is used to divide a large network into smaller segments. It can join different media types (e.g., UTP with fibre optics) as well as different types of network architecture (e.g., ethernet with token ring). It has a single input and single output port, thus making it a two-port device. A bridge operates at the data link layer. A bridge is a repeater, with capability to filter out content by reading the MAC addresses of the source and destination.

There are three types of bridges, which are as follows:
1. *Local bridge*: This bridge connects two LAN segments directly.
2. *Remote bridge*: This bridge connects another bridge over the WAN link.
3. *Wireless bridge*: This bridge connects another bridge without any wiring between them.

Limitations with bridges include limited ports and a slowdown in the overall performance of the network, as in bridge forwarding decisions are made through software.

Switch

Switch is a data link layer device. A switch is more intelligent than a hub. While a hub does only data forwarding, a switch does 'filtering and forwarding'. When a packet is received at one of the interfaces of the switch, it filters the packet and sends it only to the interface of the intended receiver. For this purpose, a switch maintains a content addressable memory (CAM) table and has its own system configuration and memory.

A switch can perform error checking before forwarding data and processes the frame only if it is valid. When a switch receives a frame, it checks the frame checksum sequence (FCS) field in it. All invalid frames are automatically dropped. This makes it very efficient as it does not forward frames that have errors. All valid frames are processed and forwarded to their destination MAC address.

Switches support three methods of switching which are as follows:

Store and forward   This is the basic mode of switching where the switch buffers the entire frame into the memory and runs FCS to check if the frame is valid or not. Only valid frames are processed and all invalid frames are automatically dropped.

Cut and through   In this method, the switch reads only the first six bytes from the frame after the preamble. These six bytes refer to the destination address of the frame. This is the fastest method of switching.

Fragment free   This is a hybrid version of the store and forward method and cut and through method. It checks the first 64 bytes of the frame for error. It processes only those frames that have the first 64 bytes valid. Any frame less than 64 bytes is known as runt and is invalid. This method filters runt while maintaining the speed.

Routers

A router is a device like a switch that routes data packets based on their IP addresses. It is mainly a network layer device which is responsible for routing traffic from one network to another. Routers are used to connect different network segments, network protocols, media types (e.g., UTP and fibre optic cable), network architecture, and to group smaller networks to form larger networks and to break larger networks into smaller networks, among others.

Gateway

Gateways are also called protocol converters and can operate at the network layer. Gateways are generally more complex than switches or routers, and are used to forward the packets which are intended for the remote network from the local network. Until the host is configured with a default gateway address, every packet should have a default gateway address. A default gateway address is the address of the gateway device. If the packet does not find its destination address in the local network, it would take the help of the gateway device to find the destination address in the remote network.

Modem

Modem stands for *modulator* + *dem*odulator. It modulates and demodulates the signal between the digital data of a computer and the analog signal of a telephone line.

> **Box 1.2 NIC and MAC Address**
>
> MAC address is the hardware address tied to the key connection device in the computer called the NIC. The NIC is a computer circuit card that enables a computer to connect to a network. It also facilitates conversion of data into an electrical signal that can be transmitted over the network. A MAC address is given to a network adapter when it is manufactured. It is unique and hardwired or hard-coded onto the NIC. MAC address is also called a networking hardware address, the burned-in address (BIA), or the physical address.
>
> The MAC address is a string of usually six sets of two-digit numbers or characters, separated by colons. An example of a MAC address is '00-14-22-15-17-90' where the first three octets '00-14-22' represents an organizationally unique identifier (OUI) which is DELL in this case.
>
> All devices on the same network subnet have different MAC addresses. These addresses are very useful in diagnosing network issues like problems with IP addresses as they are not as dynamic as IP. The MAC address remains the most reliable way to identify senders and receivers of data on the network.

1.6 LAN TECHNOLOGIES

The various LAN technologies are summarized in this section.

Ethernet

Ethernet relies on shared media which has the highest probability of data collision. It uses carrier sense multi access/collision detection (CSMA/CD) technology to detect collisions. When a collision occurs in ethernet, all its hosts roll back, wait for some random amount of time, and then re-transmit the data. Ethernet connectivity is provided through a network interface card (NIC) equipped with a 48-bit medium access control (MAC) address which helps one ethernet device to identify and communicate with other remote devices over ethernet (Box 1.2).

Traditional ethernet uses a 10BASE-T specification, where the number 10 depicts speed of the order of 10Mbps, BASE stands for baseband, and T stands for thick ethernet. 10BASE-T ethernet can provide a transmission speed of upto 10Mbps and uses a coaxial cable or a Cat-5 twisted pair cable with an RJ-45 connector. An ethernet segment length can extend up to 100 metres.

Fast Ethernet

It is an extension of ethernet, and can run on optical fibre and in wireless mode with a speed of upto 100 Mbps. This standard is named 100BASE-T in IEEE 803.2 and uses Cat-5 twisted pair cable. Fast ethernet uses CSMA/CD technique while sharing the wired media among the ethernet hosts and CSMA/CA (CA stands for collision avoidance) technique for wireless ethernet LAN.

Fast ethernet on fibre is defined under the 100BASE-FX standard which provides speeds up to 100 Mbps on fibre. Ethernet over fibre can be extended upto 100 metres in half-duplex mode and can reach a maximum of 2000 metres in full-duplex over multimode fibres.

Giga Ethernet

Gigabit ethernet offers a speed of upto 1000 Mbps. IEEE802.3ab defines giga ethernet over UTP using Cat-5, Cat-5e, and Cat-6 cables. IEEE802.3ah defines giga ethernet over fibre.

Virtual LAN

LAN technology primarily relies on the ethernet which in turn works on shared media. Shared media in ethernet offers a single broadcast and a single collision domain. The introduction of switches to ethernet has resolved the single collision domain issue and each device connected to the switch works in separate collision domains. However, switches cannot divide a network into separate broadcast domains.

Virtual LAN enables division of a single broadcast domain into multiple broadcast domains, whereby the host in one VLAN cannot speak to a host in another. By default, all hosts are placed into the same VLAN. Hosts in one VLAN, even if connected on the same switch, cannot see or speak to other hosts in different VLANs. VLAN follows Layer-2 technology which works closely on ethernet. To route packets between two different VLANs, a Layer-3 device such as a router is required.

Wireless Fidelity

Wireless fidelity (Wi-Fi) is a wireless technology used in networking and communication. It is a cost-effective way to connect to the Internet and other electronic communication devices without the need for physical connection in the form of wires between them.

Wi-Fi technology offers the following advantages:

1. It allows flexible access to the Internet, file transfers, and print services, provided the electronic communication device is within a few metres of the Wi-Fi access point (AP).
2. It eliminates the need for wires, network cables, and sockets for interconnection which deteriorates over time.
3. Once a device is configured with the Wi-Fi AP, access to the Internet and the network at different locations is assured and there is no need to reconfigure the Internet settings every time.
4. It drastically reduces the IT set-up cost to offer Internet connectivity.

This technology has a few disadvantages as well, which are summarized here:
1. Any other Wi-Fi enabled electronic communication which is in close proximity to the Wi-Fi AP can make unauthorized access to the data and Internet connection unless the Wi-Fi is password-protected.
2. Since Wi-Fi networks are sensitive to signal strength, the electronic communication devices connected to it should have good signal strength at all times to ensure good connectivity.
3. Wi-Fi signals are sensitive to climatic conditions which make it less suitable during adverse weather conditions, for example, thunderstorms.

1.7 NETWORKING TOPOLOGIES

Network topology reflects the interconnection between computer systems and the networking devices in a network. Topology may define both the physical and logical aspect of the network, and both logical and physical topologies could be the same or different within the same network. The various network topologies are summarized here and shown in Fig. 1.4. See also Box 1.3.

Point-to-Point

Point-to-point networks contain exactly two hosts which may be a computer, switches, or routers and servers connected back-to-back using a single piece of cable.

In a logical point-to-point connection between the hosts, there may be multiple intermediate devices. However, the end hosts perceive each other as if they are directly connected and are unaware of the underlying network.

Bus Topology

In the case of bus topology, all devices share a single communication line or cable. The data is sent in the form of electronic signals to all the computers on the network, but is received only by the host whose address matches the address encoded in the signal. Both ends of the shared channel have a line terminator. The data is sent in only one direction and as soon as it reaches the other end, the terminator removes the data from the line.

In bus topology, only one host can send information at any point of time which inhibits network performance by slowdown and is proportional to the number of computers connected to the bus.

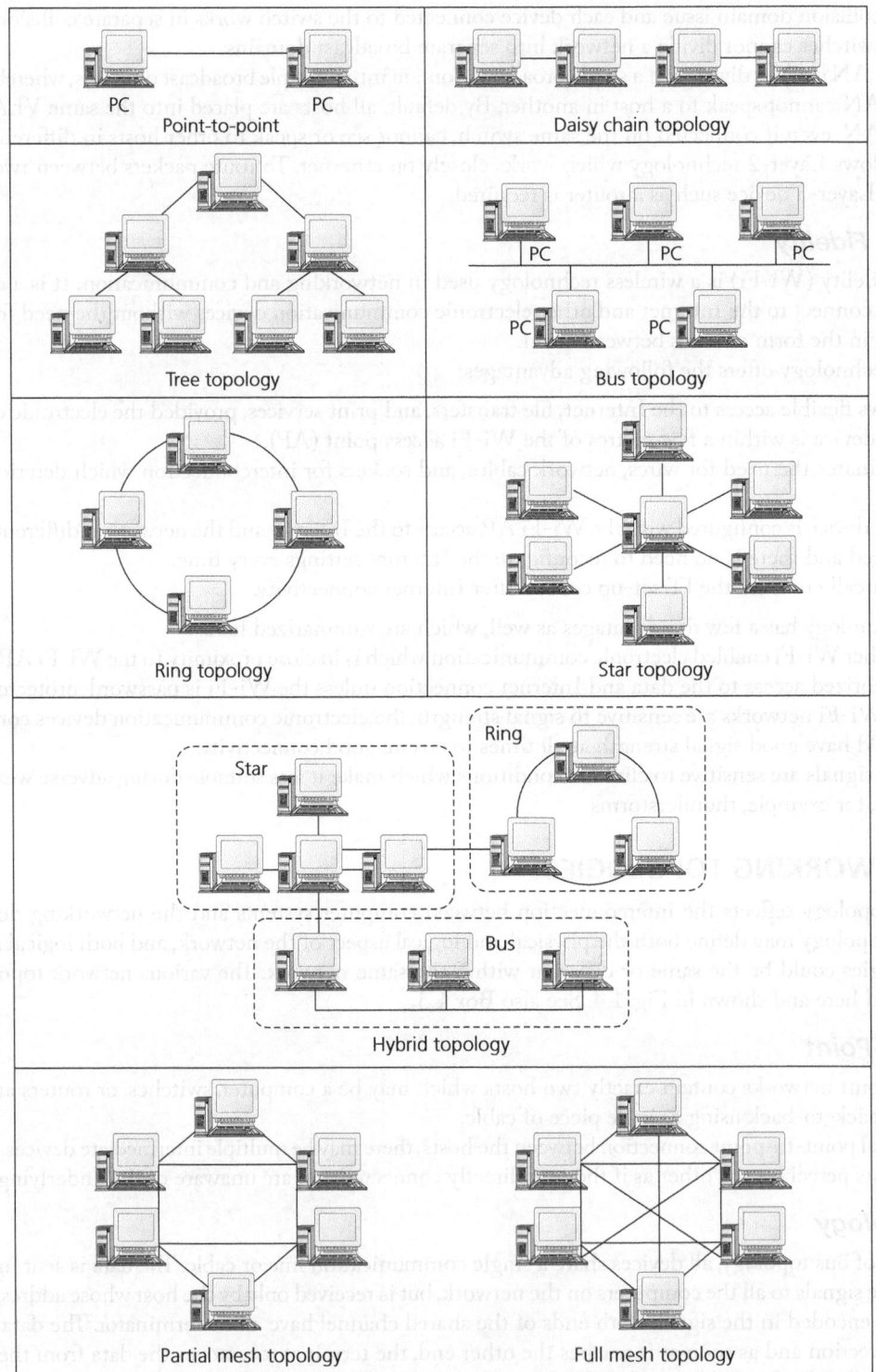

Fig. 1.4 Various LAN topologies

> **Box 1.3 WAN and the Internet**
> Most WANs are connected by means of dual-ring topology and the networks connected to them are mostly star topology networks. The Internet is the best example of the largest hybrid topology.

The advantage with bus topology is that it is simple, cheap, and easy to set up. The failure of one device does not affect other devices. The drawback with bus topology is that in the event of excess network traffic, multiple users may get affected and the network becomes difficult to troubleshoot. Further, in bus topology, problems arise while multiple hosts attempt to send data at the same time. In such a situation, the bus topology either uses CSMA/CD technology or designates one host as the bus master to solve the issue. It should also be noted that failure of the shared communication line can make all other devices stop functioning.

Star Topology

In star topology, all the hosts are connected to a central device known as hub device using a point-to-point connection. The hub device can be a Layer-1 device such as hub (passive hub) or a repeater (active hub), a Layer-2 device such as switch or bridge, or a Layer-3 device such as router or gateway.

An active hub regenerates and retransmits the signal just as repeaters do. They require electrical power to run. On the contrary, passive hubs act as mere connection points and do not amplify or regenerate the signal. It does not require electric power to run.

The advantage with star topology is that it is inexpensive to set up, easy to modify, and its configuration is simple. It facilitates centralized monitoring. The disadvantage is that just as in bus topology, the hub acts as a single point of failure where the failure of the hub leads to failure in the connectivity among all the hosts.

Ring Topology

In ring topology, every host is connected to exactly two other hosts, thus creating a circular network structure. When a host sends a message to another host not adjacent to it, the data travels through all the intermediate hosts. Connecting a new host to the existing structure requires an extra cable. This topology is in use in token ring networks.

The advantage of this topology is that it offers equal access privilege to any host. The disadvantage is that the failure of any host results in the failure of the whole ring. Thus, every connection in the ring is a point of failure. However, employing another backup ring would resolve this issue.

Mesh Topology

In mesh topology, a host is connected to one or more hosts by a point-to-point connection. Hosts in mesh topology act as a relay node for other hosts which do not have direct point-to-point links. In most cases, every host is connected with every other host by a point-to-point connection.

Mesh technology is of two types:

Full mesh All hosts have a point-to-point connection to every other host in the network. Thus for every new host $n(n - 1)/2$ connections are required. It provides the most reliable network structure among all the network topologies.

Partial mesh Not all hosts have a point-to-point connection to every other host. Hosts are connected to each other in some arbitrary fashion. This topology is suitable when only some nodes need to be reliable, and not necessarily the others.

Tree Topology

This is the most common network topology presently in use and is also called hierarchical topology. It imitates an extended star topology and inherits the properties of bus topology.

This topology divides the network into multiple levels/layers of network. The lowermost is the access layer where the computers are attached. The middle layer is known as the distribution layer, which works as the mediator between the upper layer and lower layer. The highest layer is known as the core layer, and is the central point of the network, that is, root of the tree from where all nodes fork. There is point-to-point connection between neighbouring hosts.

The drawback with this topology is the same as bus topology—if the root goes down, the entire network suffers even though it is not the single point of failure. Every connection serves as a point of failure and may lead to division of network into the unreachable segment.

Daisy Chain Topology

In this topology all the hosts are connected in a linear fashion. It is similar to ring topology where all the hosts are connected to only two hosts, except the end hosts. Every intermediate host works as a relay for its immediate hosts.

The drawback with this topology is that each link in the daisy chain topology represents a single point of failure and any link failure, therefore, will split the network into two segments.

Hybrid Topology

Hybrid topology integrates more than one topology and inherits the merits and demerits of all the incorporating topologies. The combining topologies may contain attributes of star, ring, bus, and daisy chain topologies.

1.8 NETWORK PROTOCOLS—TCP/IP PROTOCOL SUITE

Network protocol is a set of rules that govern communications between devices connected on a network. It includes mechanisms for establishing connections as well as formatting rules for data packaging and wrapping for every message exchanged over the network. TCP and IP are the two computer network protocols used in all operating systems of networked devices which operate in Layer-4 and Layer-3 of the OSI respectively. The various other protocols used in every layer are shown in Fig. 1.5.

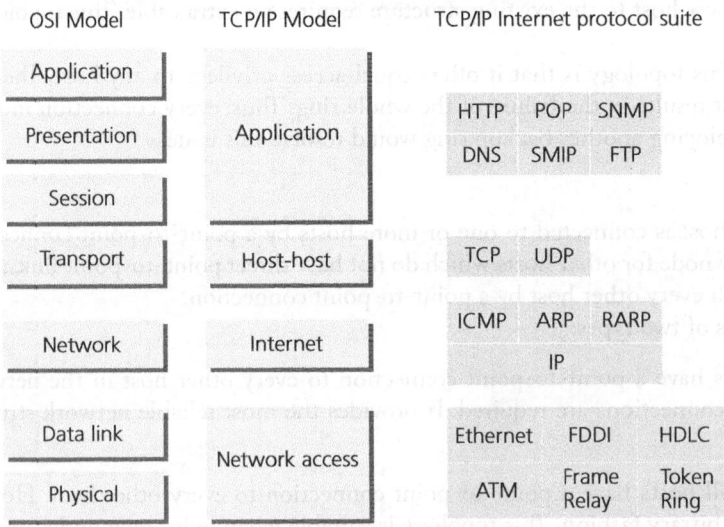

Fig. 1.5 TCP/IP protocol suite

The following sections briefly highlight the importance of every layer in the OSI model and the protocol attached to it, and as part of the TCP/IP protocol suite, briefly.

1.9 PHYSICAL LAYER

The physical layer deals with physical connectivity between any two hosts. It defines the hardware equipment, cabling, wiring, frequencies, and pulses used to represent binary signals, etc. Interaction with hardware and signalling are taken care of by the physical layer. Besides this, it offers services to the data link layer as it receives frames from it. The physical layer converts it into electrical pulses which represent binary data that is sent over either wired or wireless media.

Transmission Media

The transmission media over which data is exchanged between any two hosts may be either guided or unguided.
1. In *guided media*, the sender and the receiver are directly connected with wires and cables such as unshielded twisted pair (UTP) cable, coaxial cables, and fibre optics, and the information is exchanged or guided through it.
   (a) A twisted pair cable is made of two plastic insulated copper wires twisted together to form a single media of which one carries the actual signal and the other is used for ground reference. The twists between wires reduce noise due to electro-magnetic interference and crosstalk.
   Twisted pair cables may be either shielded twisted pair (STP) cable or unshielded twisted pair (UTP) cable. STP cables come with a twisted wire pair covered in metal foil which makes it more indifferent to noise and crosstalk. There are seven categories of UTP, each suitable for a specific use. In computer networks, Cat-5, Cat-5e, and Cat-6 cables are most widely used. UTP cables are connected by RJ45 connectors.
   (b) A coaxial cable has two wires of copper. The core wire at the centre is made of solid conductor and is enclosed in an insulating sheath. The second wire is wrapped around the sheath which in turn is encased by an insulator sheath. Both are covered by plastic. This makes it suitable for carrying high-frequency signals, compared to twisted pair cables. It also provides a good shield against noise and crosstalk. Coaxial cables can support a bandwidth of up to 450 Mbps.
   There are three categories of coaxial cables: RG-59 (cable TV), RG-58 (thin ethernet), and RG-11 (thick ethernet) where RG stands for radio government. Cables are connected using a BNC connector and BNC-T. BNC terminator is used to terminate the wire at the far ends.
   (c) Fibre optic cable has a core made of high-quality glass or plastic. From one end, light is emitted, it travels through it, and at the other end a light detector detects the light stream and converts it to electric data. Thus fibre optics offers the highest speed. Fibre optics come in two modes, namely single mode fibre and multimode fibre. Single mode fibre can carry a single ray of light, whereas multimode is capable of carrying multiple beams of light.
2. *Unguided media* refers to wireless or open air space wherein there is no physical connectivity between the sender and the receiver and the information is exchanged over air. Hence, anyone can intercept the actual communication and collect the information.

Channel Capacity

Channel capacity determines the speed of transmission of information and depends on factors such as *bandwidth* (which corresponds to the physical limitation of the underlying media), *error rate* (which is directly proportional to the noise in the channel, thus inhibiting the exact reception of information), and the *encoding mechanism* used which influences the number of levels used for signalling.

Multiplexing

Multiplexing is a technique by which different analog and digital streams of transmission can be simultaneously processed over a shared link. A multiplexer (MUX) refers to hardware that mixes multiple data streams and sends them over a single medium. A de-multiplexer (DMUX) takes information from the medium and distributes them to different destinations.

Switching

Switching is a mechanism by which data or information is sent from the source towards destinations that are not directly connected, but through interconnecting devices between them. Such interconnecting devices in the network receive data from the directly connected source, stores and analyses it, and then forwards it to the next interconnecting device closest to the destination.

Switching is categorized into three types: circuit switching, message switching, and packet switching.

Circuit switching The two nodes involved in communication establish a dedicated communication path (circuit) through which data will travel and no other data is transferred over that path. Circuits can be permanent or temporary. Once data transfer is completed, the circuit is disconnected. Circuit switching is best used by telephone networks.

Message switching Every switch in the transit path receives the whole message and buffers it until resources are available in the next hop node to transmit it. Hence every switch needs enough storage to accommodate the entire message. The store-and-forward and the wait-until-resources-are-available techniques make the process of message switching very slow. Moreover, it is not suitable for streaming media and real-time applications.

Packet switching The entire message is broken down into smaller chunks called packets wherein the switching information is added to the header of each packet and the packets are transmitted independently. The small size of packets, as opposed to message switching, makes it easy for intermediate networking devices to store them as they do not take up much resource, either on the carrier path or in the internal memory of switches. Packet switching enhances line efficiency as packets from multiple applications can be multiplexed over the carrier. The Internet uses the packet switching technique.

1.10 DATA LINK LAYER

The data link layer, the second layer of the OSI layered model, hides the details of the underlying hardware and represents itself to the upper layer as the medium to communicate.

This layer converts the data stream into signals bit by bit and sends them over the underlying hardware. At the receiving end, the data link layer picks up data from the hardware, which are in the form of electrical signals, assembles them in a recognizable frame format, and hands over to the upper layer.

The data link layer has two sub-layers:
1. *Logical link control* deals with protocols, flow-control, and error control.
2. *Media access control* deals with the actual control of media.

1.10.1 Functions

The functions of the data link layer are as follows:

Framing The data link layer receives packets from the network layer and encapsulates them into frames. Then every frame is sent bit-by-bit on the hardware. At the receiver's end, the data link layer picks up signals from the hardware and assembles them into frames.

Addressing The data link layer deals with a Layer-2 hardware addressing mechanism which is encoded into the hardware at the time of manufacturing and is unique on the link.

Synchronization When data frames are sent on the link, both the sender and the receiver must be synchronized so as to guarantee correct transfer.

Error control Data during transmission is prone to errors which may be single-bit, multiple-bit, or burst errors due to noise, crosstalk, etc., and the bits are flipped. Error control attempts to detect such errors and recover actual data bits. The error is reported to the sender as well.

Flow control The data link layer ensures flow control so that both the sender and the receiver (which otherwise may have different speeds or capacities) exchange data at the same speed. Flow control minimizes loss of data due to synchronization.

Multi-access The data link layer resolves collisions that arise while accessing shared media with a mechanism like CSMA/CD.

1.10.2 Error Control

The data link layer uses some error control mechanism to ensure that frames (data bit streams) are transmitted with a certain level of accuracy. Error control involves *error detection* and *error correction*.

Error detection is based on parity check or cyclic redundancy check.

Error detection can be carried out with *parity check* wherein one extra bit is sent along with the original data bits to mark whether the number of 1s is even (in the case of even parity) or odd (in the case of odd parity). The receiver at the other end checks for the said parity and can detect single-bit flips in transit. However, this tends to get tedious when more than one bit is erroneous.

In *cyclic redundancy check* (CRC) attempts are made to detect if the received frame contains valid data. This technique involves binary division of the data bits being sent. The divisor is generated using polynomials. The sender performs a division operation on the bits to be sent to the receiver and calculates the remainder. The sender adds the remainder at the end of the actual bits. The actual data bits plus the remainder is called a *codeword*. The sender transmits the data bits as codewords. At the other end, the receiver performs division operation on codewords using the same CRC divisor. If the remainder contains all zeros then the data bits are free from errors and are accepted. Otherwise it is obvious that some data has been corrupted in transit.

There are two types of error correction mechanisms: backward error correction and forward error correction.

Backward error correction The receiver detects an error in the data received and requests the sender to retransmit the data. This sort of error correction is used in a fibre optics medium as retransmission is cheaper.

Forward error correction When the receiver discovers an error in the received data, it executes error-correcting code to correct and auto-recover from errors in the data. This sort of error correction is used in wireless medium as retransmission is expensive.

To handle error control, the sender or the receiver should ascertain that there are errors during transit. The sender maintains a clock upon transmitting a data frame and expects an acknowledgement before the timer expires or a timeout has occurred. Failure to receive the acknowledgement necessitates the sender to retransmit the frame. The receiver sends a positive acknowledgement (ACK) upon receiving a correct data frame or a frame without errors and a negative acknowledgement (NACK) if either the frame has been lost during transit or has been received with errors. The three error control mechanisms or protocols are (a) stop-and-wait with automatic repeat request (ARQ), (b) go-back-N ARQ, and (c) selective repeat ARQ, and are shown in Fig. 1.6.

Stop-and-wait ARQ The sender, after transmitting a data frame, starts a timer and transmits the next frame if an acknowledgement is received before the timer expires. In the event of non-receipt of an acknowledgement, the sender retransmits the frame and resets the timer. The sender immediately retransmits the data frame in case a negative acknowledgement is received. However, stop-and-wait ARQ results in poor utilization of resources.

Go-Back-N-ARQ The sender and the receiver agree upon a window and this enables the sender to send multiple data frames rather than wait for acknowledgements for every data frame. This also enables the receiver to send acknowledgements for multiple frames rather than individual frames while keeping track of the sequence number of the frame. The sender, before initiating the next transmission, checks the sequence number of the frames that have received a positive acknowledgement. If all the frames have received a positive acknowledgement, the sender transmits the next set of frames. Otherwise, the sender retransmits those frames for which either a negative acknowledgement is received or an acknowledgement has not been received. However, Go-

Back-N-ARQ assumes that there is no buffer with the receiver and so a frame has to be processed as received and retransmitted in case of a negative acknowledgment.

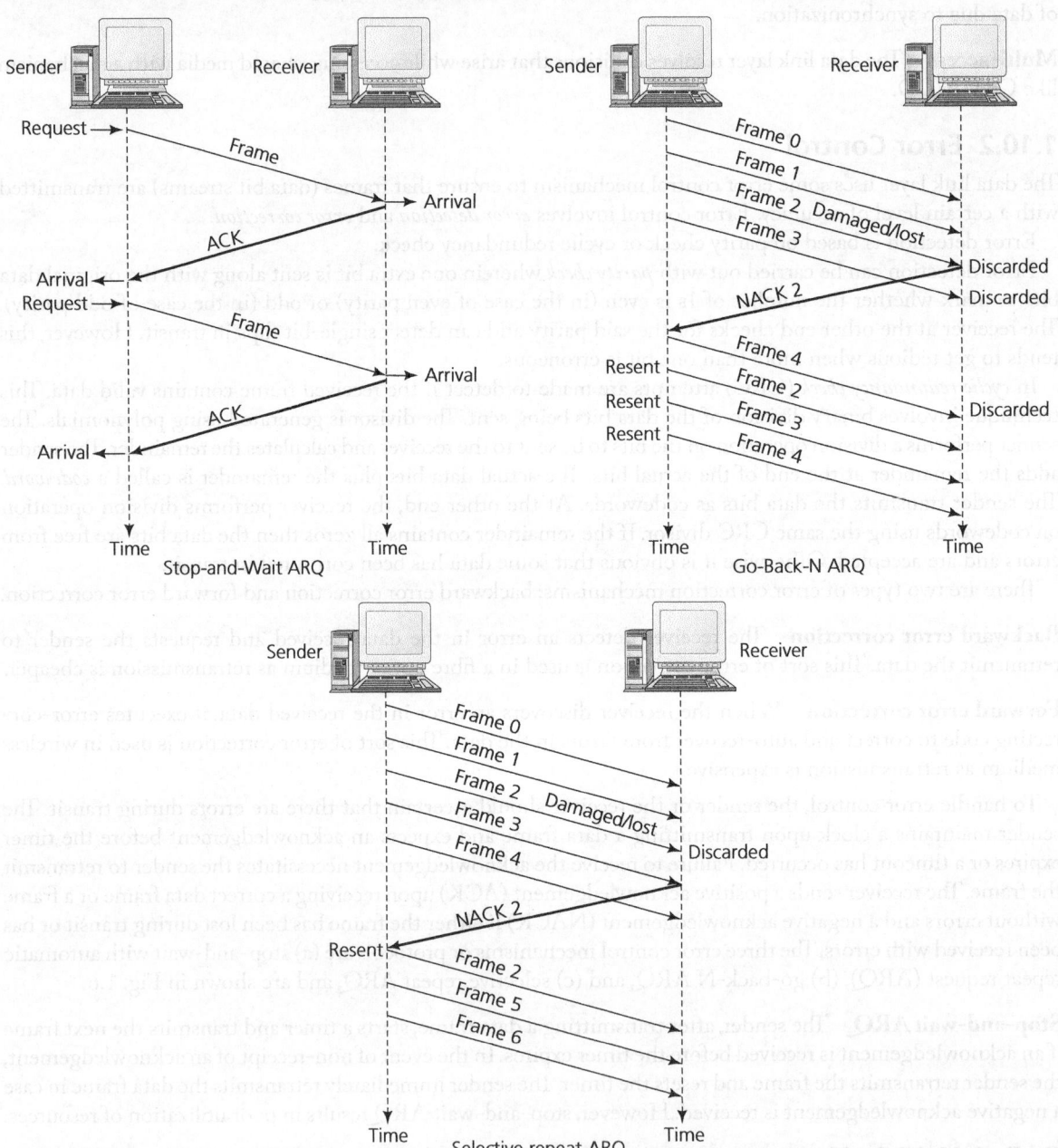

Fig. 1.6 Error control mechanisms in data link layer

Selective repeat ARQ The receiver buffers up the frame, tracks the sequence number of the frames, and sends NACK to those frames which are missing or damaged. This makes the sender retransmit only those frames for which a NACK is received.

1.10.3 Flow Control

When a frame is sent from one host (sender) to the other (receiver), it is necessary that they are synchronized and work at the same speed. Otherwise, it would cause the receiver to be overloaded and swamped resulting in data loss if the sender sends data frames too fast. The two popular flow control mechanisms are (a) stop and wait, and (b) sliding window.

Stop and wait This flow control mechanism forces the sender to stop transmitting a data frame and wait until an acknowledgement for the previously sent data frame is received.

Sliding window This flow control mechanism eliminates the problem of resource under utilization experienced with stop and wait protocol. In this mechanism, both the sender and the receiver agree on the number of data frames after which an acknowledgement has to be sent just before the initiation of every data transfer from the sender.

1.11 NETWORK LAYER

The network layer is responsible for routing packets from the source to the destination either within or outside a subnet irrespective of different, non-compatible addressing schemes and protocols.

The primary function of this layer is routing and it involves the following sub-tasks:
1. Addressing networking devices and the network as well as the internetworking of two different subnets
2. Maintaining the routing table and populating it
3. Queuing up incoming packets and forwarding them to devices or nodes closest to the destination without violating the quality-of-service (QoS)
4. Best effort in the delivery of packets to the destination

Besides this, the network layer is responsible for managing QoS and the link, load balancing, security, handling of different protocols and subnets, and end-to-end connectivity with VPN and tunnels.

The protocol that operates in the network layer is the Internet protocol (either IPv4 or IPv6) which is responsible for end-to-end communication between devices over the Internet.

1.11.1 Network Addressing

Every node/host in the network is uniquely identified with an IP address. This is the network address and is configured on the NIC. This address is mapped to the MAC address of the machine for Layer-2 communication. The network address is always logical in meaning; it is a software-based address and so can be changed by appropriate configurations.

Since IP addresses are assigned hierarchically, a host always resides under a specific network. The sending host can send a packet to the destination host, either inside or outside its subnet, provided it knows the destination network address. Hosts in different subnets can locate each other with the domain naming server (DNS), which is a server that maps the Layer-3 address of the remote host with its domain name. When such a host acquires the Layer-3 address (IP address) of the remote host, it forwards all its packets to its gateway. A gateway is a router equipped with routing tables to route packets to the destination host such that every packet gets forwarded to its next hop (adjacent router) towards the destination.

1.11.2 Routing in Network Layer

Routing is the process of selecting one among the multiple paths to reach a destination from the routing table and is performed by the network device, router. The decision to select a specific route is based on hop count, bandwidth, delay, prefix length, etc. Routing can also be performed with software but is limited in functionality and scope. Routes are configured in a router either statically or learnt dynamically. Every router has a default route which reveals where to forward a packet if no appropriate route to the specific destination is found in the routing table.

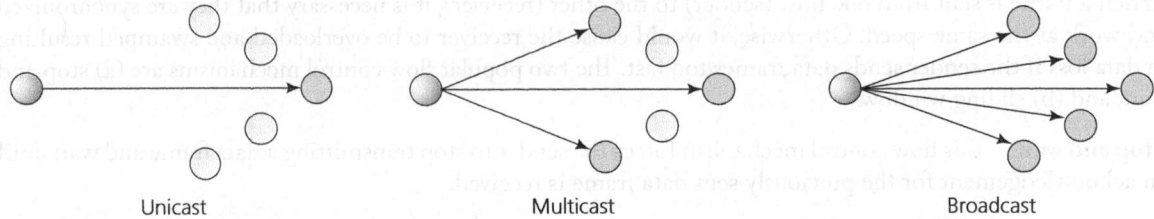

Fig. 1.7 Unicast, multicast, and broadcast communication

Unicast Routing

Routing unicast data (from a specific known destination) over the Internet is called unicast routing. This necessitates the router to look up the routing table and forward the packet to the next hop.

The routing protocols for unicast routing can be distance vector routing protocol or link state routing protocol.

Distance vector routing protocol The distance vector routing (DVR) protocol performs routing by taking into consideration a route with fewer hops between the source and the destination as the best route. Every router advertises its set best routes to other routers. Ultimately, all routers build their network topology based on the advertisements of their peer routers. The best example for DVR is routing information protocol (RIP). DVR is simple to implement.

Link state routing protocol The link state routing (LSR) protocol considers the states of links of all the routers in a network to build a graph of the entire network. The best path for routing is then calculated by all routers. Examples for LSR include open shortest path first (OSPF) and intermediate system to intermediate system (ISIS).

Broadcast Routing

Routing packets to every host and devices in the network is known as broadcast routing. This is done in one of the following two ways:
1. The router creates multiple copies of a single data packet with different destination addresses. All packets are sent as unicast to simulate router broadcasting. Since multiple unicast transfers are involved, it leads to larger consumption of bandwidth.
2. Second, when the router receives a packet that is to be broadcasted, it simply floods those packets out in all its interfaces.

Multicast Routing

Multicast routing is a special case of broadcast routing where the data packets are only sent to nodes which want to receive the packets as against broadcast routing, where the packets are sent to all nodes irrespective of whether they want them or not.

Multicast routing protocols build a tree rather than a graph with unicast routing protocols. They construct a minimum spanning tree so as to avoid loops. Example of multicast routing protocols include distance vector multicast routing protocol (DVMRP), multicast open shortest path first (MOSPF), protocol independent multicast (PIM), etc. PIM may be either *PIM dense mode* which uses source-based trees that are suitable for a dense environment like LAN or *PIM sparse mode* that uses shared trees and is suitable for a sparse environment like WAN.

Anycast Routing

Anycast packet forwarding is a mechanism where multiple hosts can have the same logical address. When a packet destined to this logical address is received, it is sent to the host which is nearest in the routing topology.

Anycast routing is done with the help of a DNS server which upon receiving a packet, destines it to the nearest IP address configured on it. Whenever an anycast packet is received, it enquires with DNS about where to send it. DNS provides the IP address which is the nearest IP configured on it.

Routing Algorithms

Routing algorithms may be based on either flooding or shortest path.

Flooding It is the simplest method of packet forwarding. When a packet reaches a router, it is forwarded through all the interfaces except the one through which it was received. Flooding results in lots of duplicate packets wandering in the network which later burdens the network. Every packet possesses time to live (TTL) which determines the lifetime of the packet and eliminates infinite looping of packets. Another variant of flooding is selective flooding wherein the router does not flood out on all the interfaces, but on selective ones which greatly reduces the network overhead.

Shortest path It performs routing decisions on the basis of cost between source and destination and considers a path in the network with minimum number of hops towards the destination to forward a packet until it reaches the destination.

Internetworking

Routing between two networks, either of the same kind or different kind, scattered geographically is called internetworking. Routers have each other's addresses and are either statically configured or learnt dynamically through the internetworking protocol. Routing protocols used within an organization are called interior gateway protocols (IGP). RIP and OSPF are examples of IGP. Routing between different organizations is called exterior gateway protocol (EGP). An example of EGP is the border gateway protocol (BGP).

Any two geographically separate networks that want to communicate with each other may deploy either a dedicated line between them or pass their data through intermediate networks. *Tunneling* is a mechanism by which two such networks communicate with each other, by passing intermediate networking complexities. Tunneling is configured at both ends. When the data enters from one end of the tunnel, it is tagged. This tagged data is then routed inside the intermediate or transit network to reach the other end of the tunnel. When data exits the tunnel, its tag is removed and delivered to the other part of the network. Tunneling gives the illusion that both the ends are directly connected. Tagging ensures data travel through the transit network without any modifications.

Internetworking refers to routing between different kinds of networks. Different networks are capable of handling data of different sizes. Most ethernet segments have their maximum transmission unit (MTU) fixed at 1500 bytes. Devices in the transit path have varying hardware and software capabilities which influence the amount of data that the device can handle and the size of packet it can process.

If the data packet size is less than or equal to the size of packet the transit network can handle, it is processed neutrally. If the packet is larger, it is broken into smaller pieces and then forwarded. This is called *packet fragmentation*. Each fragment contains the same destination and source address and is routed through the transit path. At the receiving end, it is assembled again.

If a packet with a don't fragment (DF) bit set to 1 arrives at a router, and if it cannot handle the packet because of its length, the packet is dropped. Similarly when a packet that arrives at a router has its more fragment (MF) bit set to 1, the router knows that it is a fragmented packet and parts of the original packet are on the way.

1.12 NETWORK LAYER PROTOCOLS

Every computer in a network has an IP address with which it can be uniquely identified and addressed. An IP address (Layer-3 address) is a logical address which may change every time a computer restarts. A computer can have one IP at one instant of time and another IP at a different time. Communication with the destination host requires its MAC address (Layer-2 address) as well, which is physically burnt into its NIC and never changes.

Four protocols operate at the network layer: address resolution protocol (ARP), reverse address resolution protocol (RARP), Internet control message protocol (ICMP), and the Internet protocol (IP). IP is available in two variants, namely IPv4 and IPv6.

1.12.1 Address Resolution Protocol and Reverse Address Resolution Protocol

To initiate a communication, the source should know the MAC address of the remote host on a broadcast domain. It sends out an address resolution protocol (ARP) broadcast message asking, 'Who has this IP address?' An ARP packet contains the IP address of the destination host with which the source wishes to communicate. Because it is a broadcast message, all hosts on the network segment (broadcast domain) receive this packet and process it. When a host receives an ARP packet destined to it, it replies with its own MAC address so that the host can communicate with the remote host using Layer-2 link protocol. This MAC-to-IP mapping is saved in the ARP cache of both the sending and receiving hosts for future reference during communication.

Reverse address resolution protocol (RARP) is a mechanism wherein a host that knows the MAC address of the remote host gets to know the IP address of the remote host so as to communicate. ARP and RARP are shown in Fig. 1.8.

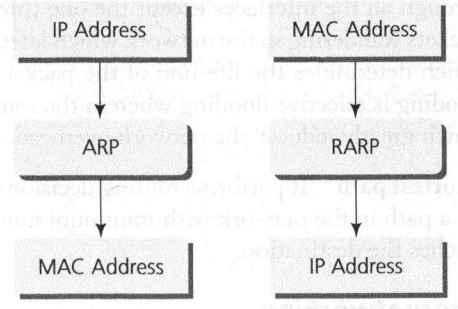

Fig. 1.8 Address resolution protocol (ARP) and reverse address resolution protocol (RARP)

1.12.2 Internet Control Message Protocol

Since IP does not have an inbuilt mechanism for sending error and control messages, it depends on the Internet control message protocol (ICMP), a network diagnostic and error reporting protocol. ICMP belongs to the IP protocol suite and uses IP as its carrier protocol. After constructing the ICMP packet, it is encapsulated in the IP packet.

ICMP contains dozens of diagnostic and error reporting messages. Any feedback about network as well as error in the network is sent back to the originating host. Some of the ICMP messages include the following:

Source quench message When the rate at which packets (traffic rate) sent by the source is very fast, the packets get dropped ultimately. When a receiving host or the router detects this, ICMP will take the source IP from the discarded packet and inform the source by sending a source quench message. As a result, ICMP slows down the pace so that no packet is lost.

Parameter problem Whenever packets come to the router, its header checksum calculated earlier should be equal to the received header checksum for the packet to be accepted by the router. In case of mismatch, the packet is dropped by the router. ICMP will take the source IP from the discarded packet and inform the source by sending a parameter problem message.

Time exceeded message When some fragments are lost in a network, the fragment held by the router will be dropped. ICMP will take the source IP from the discarded packet and inform the source by sending a time exceeded message stating that the TTL field has reached zero.

Destination unreachable A destination host in the network may become unreachable during any failure in the link, hardware, or port. This is informed to the source with ICMP messages. ICMP-echo and ICMP-echo-reply are the most commonly used ICMP messages to check the reachability of end-to-end hosts.

1.12.3 Internet Protocol Version 4

Internet protocol version 4 (IPv4) is the fourth revision of IP and a widely used protocol in data communication over different kinds of networks. IPv4 is a connectionless protocol used in packet-switched layer networks like ethernet, and can be configured either manually or automatically depending on the network type.

IPv4 is based on the best-effort model which means it guarantees neither delivery nor avoidance of duplicate delivery. These are handled by the upper layer transport.

IPv4 is a 32-bit numeric address that is written in decimal as four numbers separated by dots. This addressing scheme is used as a TCP/IP host addressing mechanism. IP addressing enables every host on the TCP/IP network to be uniquely identifiable.

IPv4 provides a hierarchical addressing scheme which enables the network to be divided into sub-networks, each with a well-defined number of hosts. IP addresses are divided into many categories:

Class A   It uses the first octet for network addresses and the last three octets for host addressing.

Class B   It uses the first two octets for network addresses and the last two octets for host addressing.

Class C   It uses the first three octets for network addresses and the last one for host addressing.

Class D   It is reserved for multicasting.

Class E   It is reserved for future use.

IPv4 also has well-defined address spaces to be used as private addresses (not routable on the Internet) and public addresses (provided by ISPs and routable on the Internet).

1.12.4 Internet Protocol Version 6

Internet protocol version 6 (IPv6) was developed to fulfil the need for more Internet addresses as the IPv4 address space was exhausted. IPv6 is also called Internet protocol next generation (IPng). IPv6 addresses are 128-bit wide and are written in hexadecimal separated by colon.

IPv6 has introduced anycast addressing but has removed the concept of broadcasting. It enables devices to self-acquire an IPv6 address and communicate within that subnet. This auto-configuration removes the dependability of dynamic host configuration protocol (DHCP) servers. This way, even if the DHCP server on that subnet is down, the hosts can communicate with each other.

IPv6 provides a new feature—that of IPv6 mobility. Mobile IPv6-equipped machines can roam around without the need for changing their IP addresses. Table 1.1 presents the differences between IPv4 and IPv6.

Table 1.1 Differences between IPv4 and IPv6

Characteristics	IPv4	IPv6
Number of bits in IP address	32	128
Format	Decimal	Hexadecimal
Address space	4.3 billion	Infinite
IPSec support	Optional	Inbuilt
Fragmentation	By routers and intermediate hosts	By sender
Support for mobile networks	Minimum support for mobile networks	Best compatibility with mobile networks
Payloads	Less	Supports bigger payloads

1.13 TRANSPORT LAYER

Transport layer is responsible for end-to-end connection between two processes on remote hosts. It takes data from the upper layer (application layer), breaks it into smaller-sized segments, numbers each byte, and hands it over to the lower layer (network layer) for delivery. Thus the functions of the transport layer include breaking down of data into segments, numbering of bytes in the segments, ensuring sequence during the receipt of data, and ensuring end-to-end data delivery. End-to-end communication is achieved when a host identifies its peer host through specified and agreed-upon ports called *transport service access points* (TSAPs) identified by port numbers. For example, a DHCP client communicates with a remote DHCP server on port number 67. A DNS client communicates with a remote DNS server on port number 53 (UDP). Port numbers range from 0–65535 where system ports correspond to (0–1023), user ports correspond to (1024–49151), and private/dynamic ports correspond to (49152–65535).

The two main transport layer protocols are TCP which provides reliable communication between two hosts and user datagram protocol which provides unreliable communication between two hosts, of which TCP is widely used in communication networks like the Internet.

1.13.1 Transmission Control Protocol

TCP is a transport layer protocol used by applications that require guaranteed delivery. TCP establishes a full duplex virtual connection between two endpoints. Each endpoint is defined by an IP address and a TCP port number. The byte stream is transferred in segments. It is a sliding window protocol that provides handling for both timeouts and retransmissions. The window size determines the number of bytes of data that can be sent before an acknowledgement from the receiver is necessary.

Characteristics

The following are the characteristic features of TCP:
1. It is a connection-oriented protocol which means that the connection is established between two end points (between the host and the remote host) before the actual transmission of data.
2. It is a reliable protocol which means that for every data packet sent from the sender, an acknowledgement, either positive or negative, is received from the receiver. Positive acknowledgement marks the successful delivery of a data packet, whereas negative acknowledgement necessitates retransmission of the packet by the sender.
3. It operates in point-to-point client/server mode.
4. It ensures the ordering of data sent from the sender to the receiver.
5. It guarantees end-to-end communication.
6. Besides this, TCP provides flow control, error checking, and a recovery mechanism, ensuring QoS.
7. It supports full duplex server which means that it can perform the roles of both the sender and the receiver.

TCP Header

A TCP header and the segment format are shown in Fig. 1.9. The length of a TCP header ranges from a minimum of 20 bytes to a maximum of 60 bytes. The fields and its purpose are mentioned here:

1. Source port (16-bits): It identifies the source port of the application process at the sender's end.
2. Destination port (16-bits): It identifies the destination port of the application process on the receiving device at the receiver's end.
3. Sequence number (32-bits): It reveals the sequence number of data bytes of a segment in a session.
4. Acknowledgement number (32-bits): When ACK flag is set, this number contains the next sequence number of the data byte expected by the receiver and serves as an acknowledgement of the previous data byte received.
5. Data offset (4-bits): This field implies both, the size of the TCP header (as a number of 32-bit words) and the offset of data in the current packet in the whole TCP segment.
6. Reserved (3-bits): Reserved for future use and all are set zero by default.
7. Flags (1-bit each)
   (a) NS: Nonce sum bit is used by an explicit congestion notification signalling process.
   (b) CWR: When a host receives a packet with an ECE (explicit congestion notification -echo) bit set, it sets the congestion windows reduced to acknowledge that the ECE has been received.
   (c) ECE: It has two meanings:

1. If SYN bit is cleared to 0, then ECE means that the IP packet has its congestion experience (CE) bit set.
2. If SYN bit is set to 1, ECE means that the device is ECT capable; ECT denotes ECN capable transport.
   (d) URG: It indicates that the urgent pointer field has significant data and should be processed.
   (e) ACK: It indicates that the acknowledgement field has significance. If ACK is cleared to 0, it indicates that the packet does not contain any acknowledgement.
   (f) PSH: When set, it is a request to the receiving station to PUSH data (as soon as it comes) to the receiving application without buffering it.
   (g) RST: Reset flag has the following features:
      1. It is used to refuse an incoming connection.
      2. It is used to reject a segment.
      3. It is used to restart a connection.
   (h) SYN: This flag is used to set up a connection between hosts and synchronize the sequence numbers.
   (i) FIN: This flag is used to release a connection and no more data is exchanged thereafter. Because packets with SYN and FIN flags have sequence numbers, they are processed in the correct order.
8. Windows size (16 bits): This field is used for flow control between two stations and indicates the amount of buffer (in bytes) the receiver has allocated for a segment, that is, how much data the receiver is expecting.
9. Checksum (16 bits): This field contains the checksum of header, data, and pseudo headers.
10. Urgent pointer (16 bits): It points to the urgent data byte if URG flag is set to 1.
11. Options: It facilitates additional options which are not covered by the regular header. Option field is always described in 32-bit words. If this field contains data less than 32 bits, padding is used to cover the remaining bits to reach the 32-bit boundary.

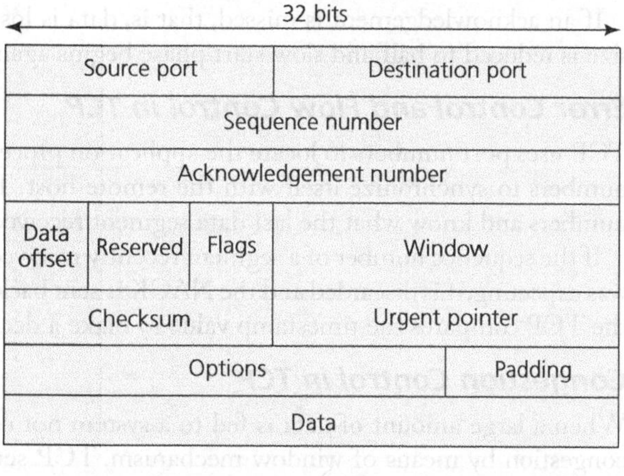

Fig. 1.9 Structure of TCP header and data

Connection Management by TCP

TCP communication works on a server–client model. Three-way handshaking is used for connection management. This is explained here.
1. The client initiates the connection and sends the segment with a sequence number. The server either accepts or rejects it.
2. The server acknowledges it with its own sequence number and ACK of the client's segment which is one more than the client's sequence number.
3. The client, after receiving the ACK of its segment, sends an acknowledgement of the server's response.
4. To terminate a connection, either the server or the client sends a TCP segment with FIN flag set to 1. When the receiving end responds by ACKnowledging FIN, the connection is said to have been closed and released.

Bandwidth Management by TCP

TCP uses the concept of window size to manage bandwidth. Window size tells the sender at the remote end, the number of data byte segments the receiver at this end can receive. TCP uses slow start phase by using window size 1 and increases the window size exponentially after each positive acknowledgement and successful communication.

If an acknowledgement is missed, that is, data is lost in transit network or NACK is received, the window size is reduced to half and slow start phase begins again.

Error Control and Flow Control in TCP

TCP uses port numbers to locate the application process it needs to hand over the data segment and sequence numbers to synchronize itself with the remote host. The sender and the receiver keep track of the sequence numbers and know what the last data segment received and sent respectively were.

If the sequence number of a segment recently received does not match the sequence number that the receiver was expecting, it is discarded and the NACK is sent back. If two segments arrive with the same sequence number, the TCP compares the timestamp value to make a decision.

Congestion Control in TCP

When a large amount of data is fed to a system not capable of handling it, congestion occurs. TCP controls congestion by means of window mechanism. TCP sets a window size telling the other end how much data segment it can send. TCP uses three algorithms for congestion control:

Slow start, exponential increase

1. This algorithm is based on the idea that the size of the congestion window (cwnd) starts with one maximum segment size (MSS).
2. The size of the window increases by one MSS each time an acknowledgment is received.
3. As the name implies, the window starts slowly, but grows exponentially.
4. Slow start cannot continue indefinitely. There must be a threshold to stop this phase. The sender keeps track of a variable named slow-start threshold (ssthresh), usually 65,535 bytes, to stop this phase.
5. When the size of the window in bytes reaches this threshold, slow start stops and the next phase begins.

Congestion avoidance, additive increase

1. Slow-start algorithm increases the size of the congestion window exponentially, which should then be slowed down to avoid congestion.
2. When the size of the congestion window reaches the slow-start threshold, the slow-start phase stops and the additive phase begins.
3. In congestion avoidance algorithm, each time the whole window of segments is acknowledged (one round), the size of the congestion window is increased by 1.

Congestion detection, multiplicative decrease

1. If congestion occurs, the congestion window size must be decreased. The only way for the sender to sense congestion is with retransmission. However, retransmission can occur in one of the two cases: when a timer times out or when three ACKs are received. If congestion is detected to be due to time-out, TCP starts a new slow-start phase. If detection is by three ACKs, it starts a new congestion avoidance phase. However, in both the cases, the size of the threshold is dropped to one-half, a multiplicative decrease.

Timer Management by TCP

TCP uses different types of timers to control and manage various tasks. The timers are listed here:

Keep-alive timer

1. This timer is used to check the integrity and validity of a connection.
2. When keep-alive time expires, the host sends a probe to check if the connection still exists.

Retransmission timer

1. This timer maintains a stateful session of the data sent.

2. If the acknowledgement of sent data is not received within the retransmission time, the data segment is sent again.

Persist timer
1. A TCP session can be paused by either host by sending window size 0.
2. To resume the session, a host needs to send a window size with some larger value.
3. If this segment never reaches the other end, both ends may wait for each other for infinite time.
4. When the persist timer expires, the host re-sends its window size to let the other end know.
5. Persist timer helps avoid deadlocks in communication.

Timed-wait
1. After releasing a connection, either of the hosts waits for a timed-wait time to terminate the connection completely.
2. This is to make sure that the other end has received the acknowledgement of its connection termination request.
3. The timed-out duration can be a maximum of 240 sec (4 min).

Crash Recovery in TCP

TCP is a very reliable protocol. When a TCP server crashes mid-way during communication and re-starts its process, it sends a transport protocol data unit (TPDU) broadcast to all its hosts. The hosts can then send the last data segment which has never been unacknowledged and carry on. This eliminates the need to retransmit all the segments which are part of the transmission but have been acknowledged positively by the receiver.

1.13.2 User Datagram Protocol

The user datagram protocol (UDP) is the simplest transport layer communication protocol which involves minimum amount of communication mechanism, compared to TCP. UDP is said to be an unreliable transport protocol but it uses IP services which provide the best effort delivery mechanism.

In UDP, the receiver does not generate an acknowledgement of packet received and in turn, the sender does not wait for any acknowledgement of packet sent. This shortcoming makes this protocol unreliable but easier on processing (Box 1.4).

Characteristics

The following are the characteristics of UDP:
- It is a stateless protocol.
- It is not connection oriented.
- It does not guarantee ordered delivery of data.
- It does not send any acknowledgement to the data sent.
- It is simple and suitable for data flowing in one direction, query-based communications, streaming applications like VoIP, and multimedia streaming.
- It does not provide a congestion control mechanism.

Box 1.4 Why UDP Is Preferred for Video Streaming

Video streaming involves thousands of packets to be forwarded over the network to the intended users. Acknowledging all the packets is troublesome and may result in wastage of bandwidth. The best delivery mechanism of the underlying IP protocol ensures best efforts to deliver its packets. Even if some packets in the course of video streaming are lost, the impact is not calamitous and goes unnoticed by the user.

UDP Header

The UDP structure is illustrated in Fig. 1.10. The fields in the UDP header are as follows:

1. *Source port*: These 16 bits of information is used to identify the source port of the packet.
2. *Destination port*: These 16 bits of information identifies application-level service on the destination machine.
3. *Length*: The length field specifies the entire length of the UDP packet (including header). It is a 16-bit field; the minimum value is 8-byte, that is, the size of the UDP header itself.
4. *Checksum*: This field stores the checksum value generated by the sender before sending. IPv4 has this field as optional. So when checksum field does not contain any value it is made 0 and all its bits are set to zero.

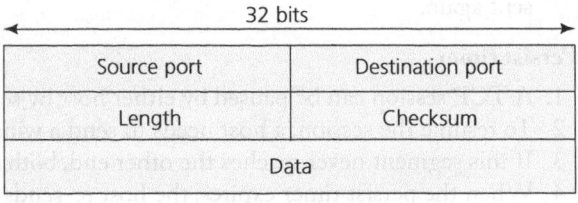

Fig. 1.10 Structure of UDP

1.14 APPLICATION LAYER

The application layer is the top-most layer in the OSI, and TCP/IP-layered model takes the help of transport and all the layers below it to communicate or transfer its data to the peer application layer protocol on a remote host. Applications which interact with the communication system are put up in this layer. For example, a web browser uses HTTP to interact with the network and so HTTP is an application layer protocol. Another example is file transfer protocol (FTP), which helps a user to transfer text-based or binary files across the network. Similarly, DNS is a protocol which helps HTTP to accomplish its work.

Any two processes can interact and communicate in any one of the following two ways:

Sockets

1. With this, the process acting as the server opens a socket using a well-known (or known by client) port and waits until a client makes a request.
2. The client also opens a socket but instead of waiting for an incoming request, the client processes 'requests first'.
3. When the request reaches the server, it is served.

Remote procedure call

1. In this, the interaction between processes happens through procedure calls. One process (client) calls the procedure lying on the remote host.
2. The process on the remote host is said to be the server. Both processes are allocated stubs.
3. This communication happens in the following way:
   (a) The client process calls the client stub. It passes all the parameters pertaining to the program local to it.
   (b) All parameters are then packed (marshalled) and a system call is made to send them to the other side of the network.
   (c) The kernel sends the data over the network and the other end receives it.
   (d) The remote host passes the data to the server stub where it is unmarshalled.
   (e) The parameters are passed to the procedure and the procedure is then executed.
   (f) The result is sent back to the client in the same manner.

1.14.1 Application Layer Protocols

Five of the most frequently used application protocols are listed here:

Domain name system

1. The DNS works on the client/server model.
2. It uses UDP protocol for transport layer communication.
3. It uses a hierarchical domain-based naming scheme.
4. The DNS server is configured with fully qualified domain names (FQDN) and email addresses mapped with their respective IP addresses.
5. When a DNS server is requested with FQDN, it responds with the IP address mapped to it. DNS uses UDP port 53.

Simple mail transfer protocol

1. The simple mail transfer protocol (SMTP) is used to transfer electronic mail from one user to another.
2. This task is done by means of email client software (user agents) that the user is using.
3. User agents help the user to type and format the email and store it until Internet is available.
4. When an email is submitted to send, the sending process is handled by a message transfer agent that normally comes inbuilt in email client software.
5. The message transfer agent uses SMTP to forward the email to another message transfer agent on the server side.
6. While SMTP is used by the end user to only send the emails, the servers normally use SMTP to send as well as receive emails.
7. SMTP uses TCP port numbers 25 and 587.
8. Client software uses Internet message access protocol (IMAP) or POP protocols to receive emails.

File transfer protocol

1. FTP is the most widely used protocol for file transfer over the network.
2. FTP uses TCP/IP for communication and works on TCP port 21.
3. FTP works on a client/server model where the client requests the server for a file. When the server receives a request for a file, it opens a TCP connection for the client and transfers the file. Once the transfer is complete, the server closes the connection. For a second file, the client requests again and the server reopens a new TCP connection.
4. FTP uses out-of-band controlling, that is, FTP uses TCP port 20 for exchanging and controlling information and the actual data is sent over TCP port 21.

Post office protocol

1. The post office protocol version 3 (POP 3) is a simple mail retrieval protocol used by user agents (client email software) to retrieve mails from the mail server.
2. When a client needs to retrieve mails from the server, it opens a connection with the server on TCP port 110.
3. The user can then access mails and download them to the local computer. POP3 works in two modes. In delete mode, the emails get deleted from the remote server after they are downloaded to local machines, whereas the keep mode does not delete the email from the mail server but gives the user an option to access mails later on the mail server.

Hyper text transfer protocol

1. It is the foundation of WWW.
2. It works on client/server model.
3. It is a stateless protocol, which means the server maintains no information about earlier requests by clients.
4. When a user wants to access any HTTP page on the Internet, the client machine at the user end initiates a TCP connection to the server on port 80.

5. When the server accepts the client request, the client is authorized to access the web pages.
6. To access the web pages, a client normally uses web browsers, which are responsible for initiating, maintaining, and closing TCP connections.

1.15 SECURITY VULNERABILITIES IN TCP/IP SUITE

Security vulnerabilities exist in the design and implementation of the TCP/IP protocol suite. Compromising the protocols in any of the layers can compromise the entire communication. This necessitated the employment of a layered security mechanism to ensure foolproof security. Some of the vulnerabilities which exist in the protocols part of the TCP/IP suite are listed here:

1. In HTTP, an application layer protocol in the TCP/IP suite file transfers are made in plain text and so it is easy for an intruder to read the data packets exchanged between the server and a client. Moreover, weak authentication during session initialization leads to a session hijacking attack where the HTTP session of the legitimate user is stolen by the attacker.
2. A TCP protocol relies on a three-way handshake for connection establishment. A SYN-flooding attack (a kind of DoS attack) can overload the server and lead to a crash.
3. IP spoofing can be launched by modification of the IP protocol header.
4. As an attack on DNS, a DNS record can be modified by the attacker to get resolved to an incorrect IP address.
5. ICMP can be exploited to discover all host IP addresses that are alive in a target network with ICMP sweep attack.

1.16 SECURITY MECHANISMS IN NETWORKING LAYERS

Security in a TCP/IP protocol-based network is implemented in physical and data link layers at the user terminal and NIC, in the TCP and IP layers at the operating system, and as user process for the layers above TCP/IP.

The goal of network security is to ensure that the entire network is secure in terms of confidentiality, availability, and integrity. The confidentiality aspect of network security ensures that the data in the network is available only to intended and authorized recipients. Availability attempts to ensure that the data, network resources, and services are available to intended recipients at all times. Integrity attempts to make sure that the data in the network is reliable and not altered by unauthorized persons. These three conflicting metrics are achieved with mechanisms such as encryption, digital signatures, and access control measures.

Any communication between two hosts can be secured by employing security mechanisms at the network layer. This is achieved with a security protocol like Internet protocol security (IPsec). Security measures at the transport layer can protect the data in a single communication session between two hosts. The transport layer security (TLS) and secure socket layer (SSL) are the most common protocols used for this purpose. An application-specific security measure can be offered at the application layer. Secure multipurpose Internet mail extensions (S/MIME) is the best example of an application layer security protocol meant to encrypt email messages.

1.17 NETWORK SECURITY AT NETWORK LAYER WITH INTERNET PROTOCOL SECURITY

IPSec is a framework for ensuring security at the network layer. The security functions offered by IPSec are as follows:

1. *Confidentiality*: IPSec enables communicating nodes to encrypt messages which prevent eavesdropping by third parties.
2. *Origin verification and data integrity*: IPSec verifies whether a received packet is actually transmitted by the source from the packet header and also confirms that the packet has not been altered.

3. *Key management*: IPSec allows secured exchange of keys which guarantee safe exchange of data between hosts besides offering protection against many attacks.

IPSec Operations

The two operations of IPSec which provide security services include IPsec communication and Internet key exchange (IKE).
1. IPSec Communication deals with packet processes such as encapsulation, encryption, and hashing the IP datagrams. It manages the communication according to the available security associations (SAs) established between communicating parties using security protocols such as authentication header (AH) and encapsulating security payload (ESP).
2. IKE is an automatic key management protocol used by IPsec. It is responsible for the creation of keys for IPsec and providing authentication during the key establishment process.

IPSec Communication Modes

IPsec communication can function in transport and tunnel modes, either individually or in combination.

Transport mode   In this mode, IPsec does not encapsulate a packet received from the upper layer. The original IP header is preserved and the data is forwarded based on the original attributes set by the upper layer protocol. No gateway services are provided in this mode and it is reserved for point-to-point communications.

Tunnel mode   Tunnel mode is typically associated with gateway activities. In tunnel mode, the entire packet from the upper layer is encapsulated before applying a security protocol. A new IP header is added. The encapsulation provides the ability to send several sessions through a single gateway.

1.17.1   IPSec Communication

IPSec is equipped with a flexible, powerful way of specification with security policies and security associations, and how different types of datagrams should be handled.

Security policies   A security policy is a rule that is programmed into the IPSec implementation telling it to process different datagrams received by the device. If security is required, the security policy provides general guidelines for how it should be provided, and if necessary, links to more specific details. Security policies are used to decide if a particular packet needs to be processed by IPSec or not. If packets are not to be processed by IPSec, they can bypass AH and ESP completely. Security policies for a device are stored in the device's *security policy database* (SPD).

Security associations   A security association (SA) is a set of security information that describes a particular kind of secured connection between one device and another. A device's security associations are contained in its *security association database* (SAD).

The main difference between SAP and SAD is that security policies are general, whereas security associations are more specific. To determine how a particular datagram must be handled, a device first checks the SPD. The security policies in the SPD may make reference to a particular security association in the SAD. If so, the device will look up that security association and use it for processing the datagram.

IPSec Authentication Header

One of the two core security protocols in IPSec is the *authentication header* (AH). AH is a protocol that provides authentication of either all or part of the contents of a datagram through the addition of a header that is calculated based on the values in the datagram. What parts of the datagram are used for the calculation, and the placement of the header, depend on the communication mode (tunnel or transport) and the version of IP (IPv4 or IPv6).

The operation of the AH protocol is as follows:
1. A security association between two devices is set up, specifying these particulars so that the source and destination know how to perform the computation but nobody else can.
2. On the source device, AH performs the computation and puts the result called the integrity check value (ICV) into a special header with other fields for transmission.
3. The destination device does the same calculation using the key the two devices share, which enables it to see immediately if any of the fields in the original datagram were modified (either due to error or malice).
4. The presence of the AH header helps only in the verification of the integrity of the message and not in encryption. Thus, AH provides authentication but not privacy.

IPSec Encapsulating Security Payload

Datagrams should be protected from intermediate devices against changes and should be protected from examining their contents. This is achieved with *encapsulating security payload* (ESP) protocol.

The main job of ESP is to provide privacy for IP datagrams by *encrypting* them. This is then repackaged using a special format and transmitted to the destination, which decrypts it using the same algorithm. ESP also supports its own authentication scheme like the one used in AH, or can be used in conjunction with AH.

Encapsulating Security Payload Fields

ESP has several fields that are the same as those used in AH, but packages its fields in a very different way. Instead of having just a header, it divides its fields into three components:

ESP header   This contains two fields, the SPI and sequence number, and comes before the encrypted data. Its placement depends on whether ESP is used in transport mode or tunnel mode.

ESP trailer   This section is placed after the encrypted data. It contains padding that is used to align the encrypted data, through a padding and pad length field. Interestingly, it also contains the next header field for ESP.

ESP authentication data   This field contains an integrity check value (ICV), computed in a manner similar to how the AH protocol works, for when ESP's optional authentication feature is used.

1.17.2 Internet Key Exchange

It is necessary for the two devices involved in communication to exchange the 'secret' that the security protocols themselves will use. The primary support protocol used for this purpose in IPSec is called Internet key exchange (IKE).

The purpose of IKE is to allow devices to exchange information required for secured communication. It includes cryptographic keys used for encoding authenticated information and performing payload encryption. Any two devices that securely exchange information encode and decode it using a piece of information known only to them. Anyone can intercept the information but is prevented either from reading it (if ESP is used to encrypt the payload) or from tampering (if AH is used). IKE works by allowing IPSec-capable devices to exchange SAs to populate their SADs. These are then used for the actual exchange of secured datagrams with the AH and ESP protocols.

1.18 NETWORK SECURITY AT TRANSPORT LAYER

Transport layer security (TLS) protocols operate above the TCP layer. The design of these protocols uses popular application program interfaces (API) to TCP, called 'sockets' for interfacing with the TCP layer. The family of protocols designed for TLS include SSL versions 2 and 3 and TLS protocol. Netscape developed SSLv2 and SSLv3. SSL used the patented RSA crypto.

The Internet Engineering Task Force (IETF) subsequently introduced a similar TLS protocol as an open standard. TLS protocol is non-interoperable with SSLv3. TLS modified the cryptographic algorithms for key expansion and authentication and used open crypto Diffie–Hellman (DH) and digital signature standard (DSS).

1.18.1 Secure Socket Layer

SSL protocol works in between the application and transport layer. It is a two-layer protocol with *SSL record protocol* in the lower layer and an upper layer comprising *SSL handshake protocol, change cipher spec protocol*, and *alert protocol* for message exchange and an *application protocol* for providing information transfer service between client/server interactions.

The functions of these protocols are as follows:
1. The *record protocol* formats the protocol messages from the upper layer, fragments it into blocks, compresses it (optionally), encrypts the data, adds a header to each message, adds a hash (preferably message authentication code, MAC) at the end and hands over the formatted block to the transport layer for transmission.
2. *SSL handshake protocol* and *change cipherspec protocol* creates SSL sessions between the client and the server. The handshaking mechanism is explained in detail in the following subsection.
3. *SSL alert protocol* is used to report errors.

Handshake Protocol of SSL and TLS

The handshake protocol is responsible for the authentication and key exchange necessary to establish or resume secure sessions. When establishing a secure *session*, the handshake protocol manages the following:

Cipher suite negotiation The client and the server contact each other and choose the cipher suite that will be used throughout their message exchange.

Authentication of server and optionally, client The server proves its identity to the client. The client might also need to prove its identity to the server. The use of *public/private key pairs* is the basis of this authentication. The exact method used for authentication is determined by the cipher suite negotiated.

Session key information exchange The client and the server exchange random numbers and a special number called the pre-master secret. These numbers are combined with additional data permitting the client and the server to create their shared secret, called the master secret. The master secret is used by the client and the server to generate the write MAC secret, which is the session key used for *hashing*, and the write key, which is the *session key* used for encryption.

The SSL/TLS handshake protocol is explained in detail in the following steps:
1. The client sends a 'Client hello' message to the server, along with cryptographic information such as the SSL or TLS version, the client's random value to be used in subsequent computations and supported cipher suites, and the data compression methods supported by the client.
2. The server responds by sending a 'Server hello' message to the client, the CipherSuite chosen by the server from the list provided by the client, the session ID, along with the server's random value.
3. The server sends its digital certificate to the client for authentication and may request a certificate from the client. The server sends the 'Server hello done' message.
4. The client verifies the server's digital certificate.
5. If the server requires a digital certificate for client authentication, the server sends a 'client certificate request' that includes a list of the types of certificates supported and the distinguished names of acceptable certification authorities (CAs).
6. If the server has requested a 'client certificate request', the client sends a random byte string encrypted with the client's private key, together with the client's digital certificate.
7. The server verifies the client's certificate.

8. The client creates a random pre-master secret, encrypts it with the *public key* from the server's certificate, and sends the encrypted pre-master secret to the server.
9. The server receives the pre-master secret. The server and the client each generates the master secret and *session keys* based on the pre-master secret.
10. The client sends a 'Change cipher spec' notification to the server to indicate that the client will start using the new *session keys* for *hashing* and encrypting messages. The client also sends a 'Client finished' message.
11. The server receives the 'Change cipher spec' and switches its record layer security state to *symmetric encryption* using the *session keys*. The server sends a 'Server finished' message to the client.
12. The client and server can now exchange application data over the secured channel they have established. All messages sent from the client to the server and from the server to the client are encrypted using the session key.

Table 1.2 Differences between SSL and TLS

Characteristic	TLS	SSL
Protocol version in segment header	Version number 3.1	Version number 3
Message authentication	Keyed-hash message authentication code (H-MAC) that can operate with any hash function	MD5 or SHA
Session key generation	Computation of master secret uses HMAC standard	Computation of master secret uses adhoc-MAC
Supported cipher suites	All suites except Fortezza	RSA, Diffie–Hellman and Fortezza cipher suites
Padding of data before encryption	Minimum to make the total data equal to a multiple of the cipher's block length	Padding can be any amount upto a maximum of 255 bytes
Alert protocol message	Supports more error messages	Supports fewer error messages than TLS

1.18.2 Transport Layer Security Protocol

The architecture of the TLS protocol is similar to the SSLv3 protocol. It has two sub-protocols: the TLS record protocol and the TLS handshake protocol. Though SSLv3 and TLS protocol have similar architecture, differences exist in the architecture and functioning, particularly for the handshake protocol, and are listed in Table 10.2.

1.18.3 HTTPS

HTTPS stands for HTTP over SSL. This protocol provides an encrypted and authenticated connection between the client web browser and the website server thereby ensuring 'secure' web browsing. HTTPS application protocol typically uses one of the two popular transport layer security protocols: SSL or TLS. When a web page is requested using a web browser by entering https:// followed by the URL in its address bar, connection to the web server is initiated with the use of SSL protocol. The browser uses system port 443 instead of port 80 reserved for http.

The handshaking mechanism is invoked by SSL for ensuring a secure connection wherein the website at the server sends its SSL digital certificate to the browser. Upon certificate verification, the SSL handshake involves the exchange of shared secrets for the session. When a trusted SSL digital certificate is used by the server, users get to see a padlock icon in the browser address bar meaning that a secured connection is established between the web server and the browser.

HTTPS offers confidentiality, server authentication, and message integrity to the user which enables safe browsing on the Internet. It also prevents the data exchanged during a session from eavesdropping and identity theft.

1.18.4 Secure Shell Protocol

The secure shell protocol (SSH protocol) is a method for secure remote login from one computer to another. It provides strong authentication, and protects the communications' security and integrity with strong encryption. It is a secure alternative to telnet and FTP, and is primarily used for file transfer and email.

SSH is organized as three sub-protocols, namely *SSH user authentication protocol*, *SSH connection protocol*, and *SSH transport layer protocol*.

SSH transport layer protocol   It focuses on server authentication, session key establishment, and ensuring data integrity.

SSH user authentication protocol   It authenticates users with passwords, Kerberos, or public-key authentication and gives access only to intended users.

SSH connection protocol   It attempts to provide multiple logical channels over a single underlying SSH connection.

The SSH protocol works in the client/server model as follows:

1. The SSH client initiates a connection by contacting the SSH server.
2. The SSH server sends it the public key.
3. The SSH client uses public key cryptography to verify the identity of the SSH server.
4. The SSH server and SSH client enter into a negotiation phase during which they agree on the symmetric encryption algorithm to be used and generate the encryption key that will be used. This is to ensure that the traffic between the communicating parties is protected with industry standard-strong encryption algorithms.
5. A secure channel is opened. Once a connection has been established between the SSH client and server, the data that is transmitted is encrypted using strong symmetric encryption according to the parameters negotiated in the set-up. Hashing algorithms are used to ensure the privacy and integrity of the data that is exchanged between the client and the server.

SSH service includes secure command shell which facilitates remote log on, secure file transfers with SSH file transfer protocol (SFTP), and port forwarding which facilitates data forwarding through a secured tunnel to the remote machine.

SSH protocol is predominantly used for secure access and file transfers, issuing remote commands, and to manage network infrastructure and other mission-critical system components.

1.19 NETWORK SECURITY AT APPLICATION LAYER

Email is a widely used application in the application layer, and relies on protocols such as simple mail transfer protocol (SMTP) used for forwarding e-mail messages, post office protocol (POP), and Internet message access protocol (IMAP) to retrieve the messages with the help of a mail client from the server. The process of securing emails ensures end-to-end security of the communication. It provides security services of confidentiality, sender authentication, message integrity, and non-repudiation.

Two schemes have been developed for email security: PGP and S/MIME. Both these schemes use secret-key and public-key cryptography and are presented in the following sub-section.

Besides this, standard DNS lookup is vulnerable to attacks such as DNS spoofing/cache poisoning. Securing DNS lookup is feasible through the use of DNSSEC which employs the public-key cryptography. It is also explained in the following sub-section.

1.19.1 Pretty Good Privacy

Pretty good privacy (PGP) is a public key encryption program. It is the most popular standard for email encryption. In addition to encrypting and decrypting an email, PGP is used to sign messages so that the receiver can verify both the identity of the sender and the integrity of the content. PGP uses a private key that must be kept secret and a public key that the sender and the receiver must share. PGP is a hybrid cryptosystem because it combines the features of both conventional and public key cryptography. Use of both the cryptographic approaches improves performance and key distribution without any compromise on security.

PGP works as follows:
1. When plaintext is encrypted with PGP, it is compressed to save transmission time and disk space, and strengthens cryptographic security.
2. PGP then creates a session key (random number) which is a one-time secret key generated from the random movements of the mouse and keystrokes by the user.
3. The session key is used with a conventional encryption algorithm to encrypt the plaintext.
4. Once the data is encrypted, the session key is encrypted using the recipient's public key.
5. The public key encrypted session key is transmitted along with cipher text to the recipient.
6. During decryption, the private key of the recipient is used by PGP to first recover the temporary session key, which in turn is used to decrypt the conventionally encrypted cipher text.

1.19.2 Secure MIME

Secure MIME (S/MIME) is an Internet standard for digitally signing MIME-based email data and its public key encryption. It was developed by RSA Security, Inc. and relies on the Rivest–Shamir–Adleman encryption system. S/MIME is a technology based on asymmetric cryptography that uses a pair of mathematically related keys to operate, namely a public key and a private key, to encrypt emails and protect it from unwanted access. It is computationally infeasible to figure out the private key based on the public key. S/MIME ensures that an email message is sent by a legitimate sender and provides encryption for incoming and outgoing messages. This makes it an effective weapon against phishing attacks.

S/MIME can work simultaneously with the following technologies but is not dependent on them (Box 1.5).
1. TLS to encrypt the tunnel or the route between email servers to help prevent snooping and eavesdropping.
2. SSL to encrypt the connection between email clients and servers.
3. BitLocker to encrypt the data on a hard drive in a data centre so as to prevent unauthorized access.

To enable S/MIME-based communication, the sender and the receiver must be integrated with a public key and signatures issued from a CA. A digital signature is used to validate a sender's identity, whereas a public key provides encryption and decryption services.

Box 1.5 Multipurpose Internet Mail Extensions—MIME

MIME is an extension of the original email standard, SMTP, which enables the sending of email containing different kinds of data files on the Internet, namely audio, video, images, application programs, and ASCII text. SMTP was extended so that Internet clients and servers could recognize and handle other kinds of data. As a result, new file types were added to mail as a supported IP file type.

Servers insert the MIME header at the beginning of any web transmission. Clients use this header to select an appropriate 'player' application for the type of data the header indicates. Some of these players are built into the browser (e.g., GIF and JPEG image players), whereas other players may be downloaded.

S/MIME works as follows:
1. When an email is created, it is signed and the private key applies the sender's unique digital signature into the message. With signing emails, S/MIME attempts to prove the identity as a sender or legitimate business.
2. Email is then encrypted with the recipient's public key.
3. The email can only be decrypted with the corresponding private key, which is supposed to be in sole possession of the recipient.
4. The recipient opens the email and uses the sender's public key to verify the signature. This satisfies the recipient that the emails really came from the sender.
5. Unless the private key is compromised, only the intended recipient will be able to access the sensitive data in emails.

Google encrypts the messages sent to Gmail.

1.19.3 DNSSec

The DNS is used to translate domain names (e.g., *example.com*) into numeric Internet addresses. Domain name information is stored and accessed on special servers known as domain name servers. The top level of the DNS resides in the root zone where all IP addresses and domain names are kept in databases and sorted by top-level domain name, such as .com, .net, .org, etc. Several vulnerabilities were discovered with DNS. Email servers use DNS to route their messages, which means they are vulnerable to security issues in the DNS infrastructure, for example, routing through rogue mail servers.

Domain name system security extensions (DNSSEC) are a set of protocols that add a layer of security to the DNS lookup and exchange processes. It helps to prevent malicious activities such as cache poisoning, pharming, and man-in-the-middle attacks.

DNSSec creates a secure domain name system by adding cryptographic signatures to existing DNS records. These digital signatures are stored in DNS name servers alongside common record types such as A, AAAA, MX, CNAME, etc. DNSSEC protects Internet clients from counterfeit DNS data by verifying digital signatures embedded in the data. These new records are used to digitally 'sign' a domain, using a method known as public key cryptography.

To facilitate signature verification, DNSSec has added a few new DNS record types, which are as follows:

1. RRSIG   : Contains cryptographic signature
2. DNSKEY   : Contains public signing key
3. DS   : Contains the hash of DNSKEY record
4. NSEC and NSEC3 : For explicit denial of existence of a DNS record
5. CDNSKEY and CDS : For a child zone requesting updates to DS record in the parent zone

DNSSEC uses a system of public keys and digital signatures to verify data. DNSSec can be visualized as follows:
1. All the records with the same type in a zone are grouped to form a resource record set (RRset).
2. Each Zone has a zone-signing key (ZSK) pair.
3. To enable DNSSec, the zone operator uses the private key to digitally sign each RRset in the zone and stores them in the name server as RRSIG records.
4. The public key is used for signature verification and is added by the zone operator to the name server in the DNSKEY record.
5. DNS servers have key-signing-keys (KSK) to validate the ZSK stored in the DNSKEY record. KSK signs the public ZSK creating an RRSIG for the DNSKEY. The name server publishes the public KSK in another DNSKEY record.
6. When a domain name is entered in the browser, the resolver verifies the digital signature as follows:
   (a) The desired RRset is requested, which also returns the corresponding RRSIG record.
   (b) The DNSKEY records containing the public ZSK and public KSK are requested which also return the RRSIG for the DNSKEY RRset.

(c) RRSIG of the requested RRset is verified with the public ZSK.
(d) RRSIG of the DNSKEY RRset is verified with the public KSK.
7. If the digital signatures in the data match those that are stored in the master DNS servers, the data is allowed to access the client computer making the request thus ensuring that the communication is to the intended Internet location.
8. When someone makes a request to the signed name server, it sends information signed with its private key; the recipient then unlocks it with the public key. If a third party tries to send untrustworthy information, it won't unlock properly with the public key. So the recipient will know that the information is bogus.

1.20 NETWORK SECURITY WITH FIREWALL

A *firewall* is a network security device that grants or rejects network access to traffic flows between an untrusted zone (e.g., the Internet) and a trusted zone (e.g., a private or corporate network). It acts as the demarcation point or 'traffic cop' in the network, as all communications should flow through it and this is where traffic is granted or rejected access.

Firewall may be hardware, software, or a combination of both and is categorized into four types: network level, application level gateway, circuit level gateway, and stateful multilayer gateways. They are explained here.

Network-level

This type of firewall examines packet headers and filters traffic based on the source and destination computer's IP address, the port used, and the service requested. They can also filter traffic based on different protocols. Most modern routers contain network-level firewalls.

Circuit-level Gateway

The circuit-level gateway firewall works at the session layer of the OSI model or at the TCP/IP. This class of firewall determines the genuineness of a requested session by monitoring the handshake between them. A circuit-level firewall can hide the network from the outside world and also restrict session rules to known computers. Typically, circuit-level gateways cost less than other forms of firewall protection.

Application-level Gateway

Application-level gateways, most commonly known as proxies, work in a manner similar to circuit-level gateways except that they work on specific applications. Application-level gateways configured as a web proxy do not allow FTP, telnet, or any other traffic through the firewall. These firewalls also block websites based on content. Because application-level gateways thoroughly examine packets of data, it takes longer for information to pass through these firewalls. Application-level gateways also require manual configuration on each user system and have zero transparency to the user. Application-level gateways protect the network from malicious attacks, spam and viruses.

Stateful Multilayer Gateways

Stateful multilayer (SML) gateways offer the best features than the other three firewall types, that is, they filter packets at the network layer, they determine packet legitimacy, and they evaluate packet contents at the application layer. SML gateways also provide a direct connection between the host and the client. This allows for transparency at the user level, unlike the application-level gateway. Because SML gateways do not use proxies, they work faster than their application-layer counterparts. However, their cost is more.

1.21 NETWORK SECURITY WITH INTRUSION DETECTION SYSTEM AND INTRUSION DETECTION AND PREVENTION SYSTEM

Intrusion detection system (IDS) is a device or a software application that monitors the network for any suspicious activity and notifies the network administrator (NA) or the system personnel when a suspicious activity is discovered. IDS can be configured to take preventive action to prevent any further access in which case it is termed intrusion detection and prevention system (IDPS). In other words, IDPS is a network security appliance that resets the connection to save the IP address from blockage or reprograms the firewall to block further network traffic from a suspicious source as its response to detecting a suspicious activity.

There are two approaches used by IDS to detect intruders: *profile-based detection* and *signature-based detection* wherein the former uses the profiles created by NA to distinguish between normal traffic/activity and anomaly (traffic that does not match configured profile) and thereby detect intruders. The latter relies on a preconfigured set of signatures that are compared with network traffic so as to detect an intrusion.

IDPS relies on three different approaches to detect intrusion: (a) *signature-based detection* where IDPS monitors the network traffic for preconfigured signatures and takes action in the event of a match, (b) *statistical anomaly-based detection* where baseline performance of network anomaly is defined and any activity that deviates from the baseline is termed as intrusion for the IDPS to take appropriate preventive action, (c) *stateful protocol analysis detection* where normal and benign activities are predetermined and predefined and an IDPS takes preventive action when a match occurs while comparing observed events with benign activity profiles.

IDS can be classified into three types:

Host-based Intrusion Detection System—HIDS

This type of IDS operates by installing software agents on all hosts on the network to monitor network traffic and all the activities, log files, operating system, system calls, error messages, etc. It is also called a passive system. It assumes that an attack can come through a network by generating network traffic or by gaining physical access. Host-based IDS are powerful enough to detect attacks that are performed from the console, but can fail to offer physical security in the event that an intruder with knowledge on IDS gains access to the host and disables the detection software. It can detect stealth attacks as well.

The drawbacks with host-based IDS are as follows:
1. Manageability becomes complicated and time-consuming and tedious during software maintenance as IDS software needs to be installed on all hosts
2. It can analyse only received traffic and not port scans and ping sweeps.
3. In the event that it gets compromised it fails to send an intrusion notification to the NA.
4. Witnesses operate on system limitations because of the need to support hosts running different platforms in a network.

The best examples for HIDS are Tripwire and OSSEC.

Network-based Intrusion Detection System—NIDS

This type of IDS uses sensors or probes installed on a network that runs IDS software and sniffs the network traffic possibly from a hub/switch by looking for a match with a defined signature or profile to detect intrusion. It is also called reactive system. It works on the basis of perceiving intrusions from a network perspective and so can detect port scans and ping sweeps. Sensors employed have command control interface (CCI) to send and receive management traffic and to communicate with a centralized computer over a highly secure management network and monitoring interface (MI) to monitor the network. Thus NIDS remains invisible to intruders who are unfamiliar with the security features in the network. The advantage of this class of IDS is that it does not consume CPU cycles of the host for its operation, and does not pose manageability issues and operating system limitations as host-based IDS.

The drawbacks with network-based IDS are as follows:
1. It consumes network bandwidth for its operation
2. It cannot detect intrusion if an intruder fragments packets that correspond to an intrusion, as the sensor is unable to reassemble the packets correctly.
3. It cannot handle packets that have their time-to-live (TTL) field manipulated to be low by an intruder.
4. It cannot handle packets that are encrypted.

The best example for NIDS is Snort.

Hybrid Intrusion Detection System or Hybrid IDS
It combines the advantages of both HIDS and NIDS while overcoming their drawbacks.

IDPS is classified into three which are as follows:
1. *Host-based intrusion detection and prevention system (HIDPS)* which is installed in a host as a software package to monitor the events that happen in the host and to detect any suspicious activity.
2. *Network-based intrusion detection and prevention system (NIDPS)* which monitors the network traffic for suspicious activity by protocol analysis.
3. *Wireless intrusion detection and prevention system (WIDPS)* which monitors the network traffic for suspicious activity with wireless networking protocols.

POINTS TO REMEMBER

- A computer network is a collection of computers and devices interconnected by communication channels to facilitate communication, sharing of information, and resources among interconnected devices.
- The three common forms of networking architecture are peer-to-peer network, client/server architecture (or) server-based network, and hybrid network.
- The four basic types of networking technologies based on geographical span are local area network (LAN), metropolitan area network (MAN), wide area network (WAN), and Internetwork.
- The Internet is an interconnection of many combinations of networks such as LAN, WAN, and MAN. On the other hand, in an intranet, the network is restricted for use by a single corporate entity which has full control and management over the network.
- The open system interconnection (OSI) model is a seven-layer model used to visualize computer networks and to solve problems in it.
- The layers of the OSI model are application layer, presentation layer, session layer, transport layer, network layer, data link layer, and physical layer.
- The Internet is a four-layered architecture which uses TCP/IP protocol suite, also known as Internet suite or the Internet model.
- The four layers of Internet model are application layer, transport layer, Internet layer, and link layer.
- Networking devices such as repeaters, hubs, bridge, switch, gateway, and modem are used for connecting to a network, routing the packets, strengthening the signal, communicating with others, sharing files on the network, etc.
- The various LAN technologies are ethernet, fast ethernet, giga ethernet, virtual LAN, and Wi-Fi.
- Network topology reflects the interconnection between computer systems and the networking devices in a network.
- The various network topologies are point-to-point, bus topology, star topology, ring topology, mesh topology, tree topology, daisy chain, and hybrid topology.
- Network protocol is a set of rules that govern communications between devices connected on a network.
- Transmission control protocol (TCP) and Internet protocol (IP) are the two computer network protocols used in all operating systems of networked devices.
- Physical layer defines the hardware equipment, cabling, wiring, frequencies, and pulses used to represent binary signals, etc.
- The transmission media over which the data is exchanged between any two hosts may be either guided or unguided.
- Channel capacity determines the speed of transmission of information and depends on factors such as bandwidth, error rate, and the encoding mechanism.
- Multiplexing is a technique by which different analog and digital streams of transmission can be simultaneously processed over a shared link.
- Switching is a mechanism by which data or information is sent from the source towards a destination that is not directly connected to it, but through interconnecting devices between them.
- The different types of switching are circuit

- switching, message switching, and packet switching.
- The data link layer hides the details of the underlying hardware and represents itself to the upper layer as the medium to communicate.
- The logical link control deals with protocols, flow-control, and error control.
- Media access control deals with the actual control of media.
- The functions of the data link layer are framing, addressing, synchronization, error control, flow control, and multi-access.
- Error detection mechanism can be based on either parity check or cyclic redundancy check.
- Error correction mechanism can be processed as either backward error correction or forward error correction.
- The error control mechanisms or protocols are stop-and-wait with automatic repeat request (ARQ), Go-Back-N ARQ, and selective repeat ARQ.
- The flow control mechanisms are stop and wait, and sliding window.
- The network layer is responsible for routing packets from the source to the destination either within or outside a subnet irrespective of different, non-compatible addressing schemes and protocols.
- Every node/host in the network is uniquely identified with an IP address called network address and is configured on the network interface card.
- Routing is the process of selecting one among the multiple paths to reach a destination from the routing table and is performed by the network device, router.
- Routing can be unicast, multicast, broadcast, and anycast.
- Routing between two networks either of the same kind or different kind scattered geographically is called internetworking.
- The protocols that operate at the network layer are address resolution protocol (ARP), reverse address resolution protocol (RARP), Internet control message protocol (ICMP), and the Internet protocol (IP). IP is available as two variants, namely IPv4 and IPv6.
- The transport layer is responsible for end-to-end connection between two processes on remote hosts.
- The two main transport layer protocols are transmission control protocol (which provides reliable communication between two hosts) and user datagram protocol (which provides unreliable communication between two hosts), of which TCP is widely used in a communication network like the Internet.
- The application layer is the topmost layer in OSI, and TCP/IP-layered model takes the help of transport and all the layers below it to communicate or transfer its data to the peer application layer protocol on a remote host.
- The application layer protocols are domain name system, simple mail transfer protocol, file transfer protocol, post office protocol, and hypertext transfer protocol.
- Compromising the protocols in the TCP/IP suite in any one of the layers can compromise the entire communication.
- The vulnerabilities in TCP/IP suite are session hijacking attack, SYN-flooding attack, IP spoofing, DNS attack, and ICMP sweep attack.
- The security functions offered by IPSec are confidentiality, origin verification and data integrity, and key management.
- IPSec is equipped with a flexible, powerful way of specification using security policies, security associations, and the handling of different types of datagrams.
- Internet key exchange (IKE) is useful for the two devices involved in communication to exchange the 'secret' that the security protocols themselves will use.
- A firewall is a network security device that grants or rejects network access to traffic flows between an untrusted zone (e.g., the Internet) and a trusted zone (e.g., a private or a corporate network).
- Intrusion detection system (IDS) is a device or a software application that monitors the network for any suspicious activity and notifies the network administrator (NA) or the system personnel when a suspicious activity is discovered.
- IDS can be configured to take preventive action to prevent any further access in which case it is termed intrusion detection and prevention system (IDPS).
- The approaches used by IDS to detect intruders are profile-based detection and signature-based detection.
- IDPS relies on three different approaches to detect intrusion: signature-based detection, statistical anomaly-based detection, and stateful protocol

analysis detection.
- IDS is classified into host-based IDS, network-based IDS, and hybrid IDS.
- IDPS is classified into host-based IDPS, network-based IDPS, and wireless IDPS.

KEY TERMS

Bridge This is defined as a networking device that connects two sub-networks (or interconnects two LANs) which are part of the same network.

Computer network This refers to a collection of computers and devices interconnected by communication channels to facilitate communication, sharing of information (data, messages, graphics) and resources (printers, fax machines, modems, and other hardware) among interconnected devices.

Firewall This is defined as a network security device that grants or rejects network access to traffic flows between an untrusted zone (e.g., the Internet) and a trusted zone (e.g., a private or corporate network).

Hub This means a networking device which is used to connect multiple network hosts. It is also called multiport repeater.

Multicast routing This refers to a special case of broadcast routing where the data packets are sent only to the nodes which want to receive the packets as against broadcast routing, where the packets are sent to all nodes irrespective of whether they want them or not.

Multiplexing This is a technique by which different analog and digital streams of transmission can be simultaneously processed over a shared link.

Network protocol This is defined as a set of rules that govern communications between devices connected on a network.

Open system interconnection (OSI) model This is defined as a seven-layer model used to visualize computer networks and to solve problems in them.

Repeater This refers to an electronic two-port device that operates at the physical layer whose job is to regenerate the signal over the same network before the signal becomes too weak or corrupted so as to extend the length to which the signal can be transmitted over the same network.

Router This refers to a device like a switch that routes data packets based on their IP addresses.

Routing This is defined as the process of selecting one among the multiple paths to reach a destination from the routing table and is performed by the network device, router.

Switching This refers to a mechanism by which data or information is sent from the source towards destinations that are not connected directly, but through interconnecting devices between them.

Transmission control protocol (TCP) This is defined as a transport layer protocol used by applications that require guaranteed delivery which establishes a full duplex virtual connection between two endpoints.

User datagram protocol (UDP) This is defined as the simplest transport layer communication protocol which involves only a minimum amount of communication mechanism, when compared to TCP.

MULTIPLE-CHOICE QUESTIONS

1. The common form(s) of network architecture is/are _____.
   (a) peer-to-peer network
   (b) client/server network
   (c) hybrid network
   (d) all of these
2. The backbone of MAN is _____.
   (a) high-capacity, high-speed fibre optics
   (b) low-capacity, low-speed fibre optics
   (c) high-capacity, low-speed fibre optics
   (d) low-capacity, high-speed fibre optics
3. IPv6 has introduced _____ addressing.
   (a) unicast   (c) broadcast
   (b) multicast   (d) anycast
4. The length of the UDP packet is _____.
   (a) 8 bits   (c) 32 bits
   (b) 16 bits   (d) 64 bits
5. The primary need of computer networks is for

(a) information exchange
   (b) resource sharing
   (c) resource planning
   (d) (a) and (b)
6. _____ handles data flow and email between its network and other networks as well as with remote users through a modem or telephone lines in a dialup connection.
   (a) File server
   (b) Communication server
   (c) Print server
   (d) Application server
7. The technologies used in WAN are _____.
   (a) ATM and frame relay
   (b) ATM and SONET
   (c) ATM, frame relay, and SONET
   (d) frame relay and SONET
8. The logical link control (LLC) layer is responsible for _____.
   (a) network gain access
   (b) network gain access and error checking
   (c) network gain access and packet synchronization
   (d) error checking and packet synchronization
9. The transport layer defines _____.
   (a) the protocol which enables the user to interact with the network
   (b) how the data should flow between hosts
   (c) host addressing, recognition, and routing
   (d) sending and receiving actual data
10. _____ does not interfere with the data signal.
    (a) Passive hub   (b) Active hub
    (c) Both (a) and (b) (d) None of these
11. _____ is the fastest method of switching.
    (a) Store and forward
    (b) Fragment free
    (c) Cut and through
    (d) Routers
12. Ethernet uses _____ technology to detect collisions.

    (a) carrier sense multi access (CSMA)
    (b) carrier sense multi access (CSMA)/collision detection (CD)
    (c) collision detection (CD)
    (d) None of these
13. IEEE802.3ab defines giga ethernet over UTP using _____ cables.
    (a) Cat-5   (c) Cat-6
    (b) Cat-5e   (d) All of these
14. _____ topology offers equal access privilege to any host.
    (a) Point-to-point (c) Star
    (b) Bus   (d) Ring
15. RG stands for _____.
    (a) radio government   (b) radio governance
    (c) radio generation   (d) radio generalization
16. _____ results in poor utilization of resources.
    (a) Go-Back-N-ARQ
    (b) Stop-and-wait ARQ
    (c) Selective repeat ARQ
    (d) None of these
17. An example of exterior gateway protocol (EGP) is the _____.
    (a) routing information protocol (RIP)
    (b) open shortest path first (OSPF)
    (c) border gateway protocol (BGP)
    (d) none of these
18. _____ addressing scheme uses the first two octets for network addresses and the last two for host addressing.
    (a) Class D   (c) Class B
    (b) Class C   (d) Class A
19. _____ remove(s) the dependability of dynamic host configuration protocol (DHCP) servers.
    (a) Auto-configuration (b) Configuration
    (c) Reconfiguration   (d) All of these
20. _____ indicates how much data the receiver is expecting.
    (a) Urgent pointer   (c) Data offset
    (b) Flags   (d) Windows size

REVIEW QUESTIONS

1. What is application layer? List out the most frequently used application layer protocols.
2. Define bridge. What are the types of bridge?
3. Compare and contrast Internet and intranet.
4. Explain the various network topologies.
5. What are IDS and IDPS?

6. Discuss data link layer in detail.
7. Explain network security in the transport layer and application layer.
8. Give a brief note on firewall.
9. Discuss in detail the following: (a) Networking architecture, (b) Networking technologies, (c) Network models, and (d) Networking devices.
10. Explain the following briefly: (a) TCP/IP protocol suite, (b) physical layer, (c) network layer and network layer protocols, (d) transmission control protocol (TCP), and (e) user datagram protocol (UDP)
11. Discuss the following briefly: (a) security vulnerabilities in TCP/IP suite, (b) security mechanisms in the networking layers, and (c) IPSec
12. Give a detailed account on achieving network security with IDS and IDPS.

APPLICATION EXERCISES

1. A company having its head office at location A plans to start another venture in a new city, at locations B, C, and D in the city where the distance between B and C is 150 km, B and D is 25 km, C and D is 15 km, and B is 1500 km away from the head office. It plans to install 50, 75, 100 and 125 computers in each of the locations, namely A, B, C, and D respectively.
   (a) Suggest the kind of networking technology to be used for connecting each of the offices at different locations and justify your answer.
   (b) Which networking device can be used to connect all the computers in the respective locations and why?
   (c) Which communication media can be procured so as to ensure a very effective high-speed communication?
2. Do you think that email security has to be taken seriously? How much effort must be invested in encryption and securing email? State the requirements for email security and vulnerabilities, if any. Suggest the components required for email security and justify your answer.

BIBLIOGRAPHY

1. Geeksforgeeks, *Internet Control Message Protocol*, available at: http://www.geeksforgeeks.org/internet-control-message-protocol-icmp/ (Accessed 06 December 2017)
2. Computernetworkingnotes, *Computer Networking Devices Explained with Function*, available at: https://www.computernetworkingnotes.com/networking-basic/computer-networking-devices-explained-with-function.html (Accessed 06 December 2017)
3. Codesandtutorials, *Modem–Definition, Working, Types*, available at: http://www.codesandtutorials.com/networking/networkdevices/modem-types-working.php (Accessed 06 December 2017)
4. Geeksforgeeks, *Network Device (Hub, Repeater, Bridge, Switch, Router and Gateways)*, available at: http://www.geeksforgeeks.org/network-devices-hub-repeater-bridge-switch-router-gateways/ (Accessed 06 December 2017)
5. Amar Shekar (30 March 2016), *Different Networking Devices and Hardware Types – Hub, Switch, Router, Modem, Bridge, Repeater*, available at: https://fossbytes.com/networking-devices-and-hardware-types/ (Accessed 06 December 2017)
6. TutorialsPoint, *DCN Application Layer Introduction*, available at: https://www.tutorialspoint.com/data_communication_computer_network/application_layer_introduction.htm (Accessed 06 December 2017)
7. Tcpipguide, *The TCP/IP Guide–IPSec Key Exchange (IKE)*, available at: http://www.tcpipguide.com/free/t_IPSecKeyExchangeIKE.htm (Accessed 06 December 2017)
8. Margaret Rouse (September 2005), *What is MIME*

(Multi-Purpose Internet Mail Extensions)?–Definition from WhatIs.com, available at: http://searchmicroservices.techtarget.com/definition/MIME-Multi-Purpose-Internet-Mail-Extensions (Accessed 07 December 2017)
9. Ssh, *SSH Protocol – Secure Remote Login and File Transfer*, available at: https://www.ssh.com/ssh/protocol/ (Accessed 07 December 2017)
10. IBM, *IBM Knowledge Center–An overview of the SSL or TLS handshake*, available at: https://www.ibm.com/support/knowledgecenter/en/SSFKSJ_7.1.0/com.ibm.mq.doc/sy10660_.htm (Accessed 07 December 2017)
11. Technet, *What is TLS/SSL?: Logon and Authentication*, available at: https://technet.microsoft.com/en-us/library/cc784450(v=ws.10).aspx (Accessed 07 December 2017)
12. MSDN, *TLS Handshake Protocol (Windows)*, available at: https://msdn.microsoft.com/en-us/library/windows/desktop/aa380513(v=vs.85).aspx (Accessed 07 December 2017)
13. Instant SSL, *What is SSL?, Definition and How SSL Works, Comodo SSL Wiki*, available at: https://www.instantssl.com/ssl.html (Accessed 07 December 2017)
14. Users, *How PGP Works*, available at: https://users.ece.cmu.edu/~adrian/630-f04/PGP-intro.html (Accessed 07 December 2017)
15. Kraken, *What is PGP encryption –Kraken*, available at: https://support.kraken.com/hc/en-us/articles/201648223-What-is-PGP-encryption- (Accessed 07 December 2017)
16. Ricky Publico, *What is S/MIME and How Does it Work?*, available at: https://www.globalsign.com/en/blog/what-is-s-mime/ (Accessed 07 December 2017)
17. Technet, *S/MIME for Message Signing and Encryption. Exchange Online Help*, available at: https://technet.microsoft.com/en-us/library/dn626158(v=exchg.150).aspx (Accessed 08 December 2017)
18. Margaret Rouse, *What is S/MIME (Secure Multi-Purpose Internet Mail Extensions)?–Definition from WhatIs.com*, available at: fromhttp://whatis.techtarget.com/definition/S-MIME-Secure-Multi-Purpose-Internet-Mail-Extensions (Accessed 08 December 2017)
19. Techopedia, *What is Secure MIME (S/MIME)?–Definition from Technopedia*, available at: https://www.techopedia.com/definition/9245/secure-mime-smime (Accessed 08 December 2017]
20. etutorials.org (2007), *Intrusion Detection Systems Overview*, available at: http://etutorials.org/Networking/Cisco+Certified+Security+Professional+Certification/Part+V+Intrusion+Detection+Systems+IDS/Chapter+23+Intrusion+Detection+System+Overview/Intrusion+Detection+Systems+Overview/ (Accessed 27 January 2018)
21. Vskills Govt Certifications, India's Largest Certification Body (2010) *Intrusion Detection and Prevention*, available at: https://www.vskills.in/certification/tutorial/basic-network-support/intrusion-detection-and-prevention/ (Accessed 27 January 2018)

Answers to Multiple-choice Questions

1. (d)	2. (a)	3. (d)	4. (b)	5. (d)
6. (b)	7. (c)	8. (d)	9. (b)	10. (a)
11. (c)	12. (b)	13. (d)	14. (d)	15. (a)
16. (b)	17. (c)	18. (c)	19. (a)	20. (d)

INTRODUCTION TO CYBERCRIME

2

Learning Objectives

This chapter provides an introduction to cybercrime. It presents the definitions, and the role of electronic communication devices (ECD) and information and communication technology (ICT) in cybercrime. It highlights the elements and types of cybercrime, and provides a deeper insight into cybercrimes against individuals, properties, and nations. A brief note on the crimes associated with mobile ECD is also presented. The chapter also focuses on the tools used to commit such crimes. This chapter includes factors that facilitate crime, challenges to cybercrime, and the cybercrime prevention strategies to be adopted by the government. Some of the incidents of cybercrime reported in the national and international arena are provided to make the reader understand the impact of cybercrime. Towards the end of the chapter, the terms and terminologies associated with cybercrime are presented. The reader will be familiar with the following after studying the chapter:

- Cybercrime and its associated concepts
- Role of ECD and ICT in cybercrime
- Types of cybercrime
- Cybercrimes against properties and persons
- Cybercrimes against individuals
- Cybercrimes against the nation
- Cybercrimes associated with mobile ECD
- Classification of cybercriminals
- Tools that facilitate cybercrime
- Challenges to cybercrime
- Strategies to prevent cybercrime
- Cybercrime incidents on the national and global arena

2.1 INTRODUCTION

'Cybercrime' encompasses acts committed in cyberspace. In other words, it refers to any activity in which electronic communication devices (ECDs) or networks are tools, targets, or places of criminal activities. The term cybercrime includes crimes against confidentiality, integrity and availability of data and computer system, computer related traditional crimes, content related offences, offences related to infringement of copyright and privacy. With the rapid growth in information and communication technologies (ICTs) in almost all frontiers of human activity and with the ever-increasing number of IT users, the possibility of assaults (cyberattacks) to information systems (targets) are increasing, resulting in damage and huge loss of revenue. The perpetrators of cybercrime indulge in such acts out of curiosity and for fun, or sometimes with the sole intention of making money.

Cyberattacks are executed fast and have the potential to affect thousands of devices across the world in a short span of time. Individuals and organizations should therefore take necessary precautions to ensure that their know-how and other confidential information (e.g., trade secrets) not fall into the hands of competitors or offenders. Besides this, any system is subject to attacks from even employees within an organization. It is a well-known fact that a secure computer is one that is not connected to any network and is not used by others. It is highly essential that anyone working in the IT department and in any organization with IT infrastructure be aware of the potential threats, and ensure there are adequate security policies in place. In reality, however, the complexity of technology and the global nature of the cyberspace make the detection of cybercrime difficult. Responding to cybercrime requires specialization and special training to gather evidence, follow appropriate procedures to analyse and prove the crime, and later present it in a court of law for punishing cybercriminals.

2.1.1 Definition

Cybercrime may be defined in a general way as *an unlawful act wherein the electronic communication device is either a tool, target, or both.*

More than one definition exists for cybercrime.

According to the US Department of Justice (DOJ):

Cybercrimes are any violations of criminal law that involve knowledge of computer technology for their perpetration, investigation or prosecution.

According to the EU Council (Justice and Home Affairs):

The communication addresses computer crime in its broadest sense as any crime involving the use of information technology. The terms "computer crime", "computer-related crime", "high-tech crime", and "Cybercrime" share the same meaning in that they describe (a) the use of information and communication networks that are free from geographical constraints and (b) the circulation of intangible and volatile data.

According to Computer Crime Research Center:

Cybercrime is defined as crimes committed on the internet using the computer as either a tool or a targeted victim.

Cybercrimes encompass any criminal act dealing with ECDs and networks. Additionally, cybercrimes also include traditional crimes conducted over the Internet. For example, online credit and debit card frauds, telemarketing, Internet fraud, and identity theft are also considered cybercrimes when such illegal activities are committed through the use of communication devices with or without the Internet.

2.2 ROLE OF ELECTRONIC COMMUNICATION DEVICES AND INFORMATION AND COMMUNICATION TECHNOLOGIES IN CYBERCRIME

The role of ICT and ECD in criminal offences is shown in Fig. 2.1 and is divided into the following:

ECD and ICT as tools to commit criminal offence This category may include the following: intrusion into a computer or other ECD, deleting, modifying, and/or entering data into the device under attack, disabling or obstructing network information, password and credit/debit card number theft, exchange of child pornography, participation in online chat rooms and the arrangement of meetings with children, online banking frauds, forgery of cheques and credit cards, fraudulent and deceptive advertising and sales of pyramid schemes, and intellectual property theft.

ECD and ICT as targets of criminal offence An ECD becomes a target if it contains confidential information or trade secrets, or if the offender wishes to obtain free services that are enabled by the device or if the device is used as an intermediary to attack other electronic devices.

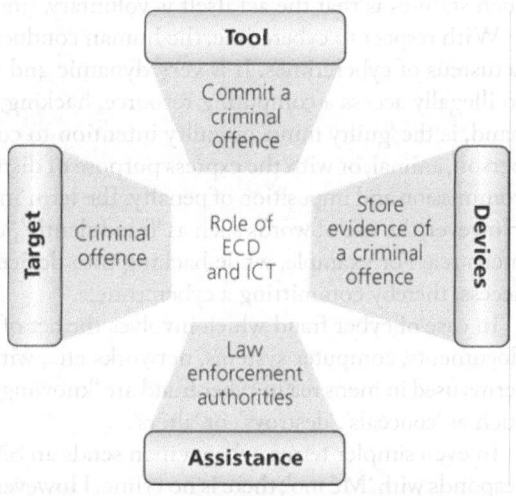

Fig. 2.1 Role of ECD and ICT in cybercrime

ECD and ICT as devices that store digital evidence of criminal offence The following are examples of digital evidence of criminal offences stored in ECDs: personal data, employee work calendars and project development data, databases, emails, letters, contracts, agreements, photographs, financial data, notebook, social media, chat messages, etc.

ECD and ICT as assistance to law enforcement authorities An ECD is used as a tool by law enforcement authorities to detect, investigate, and provide proof of the criminal offence. Proofs include databases of transactions and operations, tables of financial transactions, search tools on the Internet, various simulations, analytical software, etc.

2.3 MENS REA AND ACTUSREUS IN CYBERCRIME

A fundamental principle of criminal law is that a crime consists of both mental and physical elements. Mens rea, a person's awareness of the fact that his or her conduct is criminal, is the mental element, and actusreus, the act itself, is the physical element. Thus, actusreus unaccompanied with mens rea is not a crime. The elements of crime are shown in Fig. 2.2. With cybercrime, it is difficult to prove both the elements of crime.

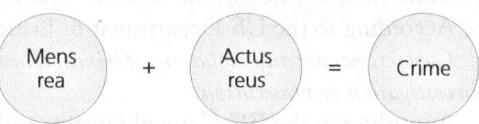

Fig. 2.2 Elements of crime

The concept of mens rea was developed in England during the latter part of the common-law era (about the year 1600) when judges began to hold that an act alone could not create criminal liability unless it was accompanied by a guilty state of mind. The degree of mens rea required for a particular common-law crime varied. Murder, for example, required a malicious state of mind, whereas larceny or accident required a felonious state of mind.

Today, most crimes, including common-law crimes, are defined by statutes that usually contain a word or phrase indicating the mens rea requirement. A typical statute, for example, may require that a person acts knowingly, purposely, or recklessly.

Sometimes a statute creates criminal liability for the commission or omission of a particular act without designating a mens rea. These are called strict liability statutes. If such a statute is construed to purposely omit criminal intent, a person who commits the crime may be guilty even though he/she had no knowledge that his/her act was criminal in nature and he/she had no intention of committing a crime. All that is required under such statutes is that the act itself is voluntary, since involuntary acts are not criminal.

With respect to cybercrime, the human conduct or actions in cyberspace, which laws seek to prevent, are the actusreus of cybercrimes. It is very dynamic and varied. Examples of such human conducts include attempts to illegally access a computing resource, hacking, sending viruses as attachments, etc. Mens rea, on the other hand, is the 'guilty mind' or guilty intention to commit a crime, with the objective of causing hurt to another person, animal, or with the express purpose of disturbing the peace. Mens rea is an important element for crime commission and imposition of penalty. The term 'mens rea' is not used and defined in the Indian Penal Code 1860. However, the use of words such as 'fraudulently', 'dishonestly', 'knowingly', 'recklessly', and 'intention' represent mens rea. For example, while hacking into devices, hackers do so with the intention of gaining unauthorized access, thereby committing a cybercrime.

In case of cyber fraud which involves the act of knowingly tampering with computer source codes or source documents, computer systems, networks etc., without the permission of the owner or authorized person, the terms used in mens rea in cyber fraud are 'knowingly' or 'intentionally' and the actusreus relate to human conduct such as 'conceals', 'destroys', or 'alters'.

In even simpler terms, when a man sends an SMS text to a lady carrying the text 'I love you', and if the lady responds with 'Me too', there is no crime. However, if she gets annoyed and is harassed, it becomes a cybercrime, where the actusreus is 'sending the SMS text' and the mens rea is 'the intention behind stalking'. When a morphed photograph of a lady, outraging her modesty is forwarded on Whatsapp, the actusreus is 'posting

Introduction to Cybercrime 47

the illegal content on social media' and the mens rea is 'to spoil her reputation'. Therefore, the intention is of significance in cybercrimes, like in any other crime.

2.4 TYPES OF CYBERCRIME

Generally, crimes are classified as crimes against the human body, crimes against property, and crimes against the nation. Cybercrime, like traditional crime, can be broadly classified into three categories, namely cybercrime against persons, cybercrime against properties, and cybercrime against the nation. This is shown in Fig. 2.3.

Cybercrimes exclusively committed against persons include crimes such as spreading of child pornography, harassment of people by using a computer (e.g., via email), and cyber stalking.

Cybercrimes against all forms of property include unauthorized computer trespassing through cyberspace, computer vandalism, transmission of harmful programs, and unauthorized possession of computerized information. However, some of these crimes can fall under all these categories. For example, illegal access can fall under both, cybercrime against person and cybercrime against the nation.

Cybercrime against person	Cybercrime against property	Cybercrime against the nation
• Internet grooming • Stalking • Harassment • Extortion • Pedophilia	• Illegal acces • Illegal data acquisition • Illegal interception • Data interference • System interference • Copyright and trademark-related offence • Computer-related offence	• Information warfare • Computer terrorism • Giving false propaganda against the nation • Content-related offence

Fig. 2.3 Types of cybercrime

The third category of cybercrimes is that of cybercrimes against the government which includes cyber terrorism, information warfare, etc.

Each of these crimes is briefly explained in the following section and discussed in detail in Chapter 3.

2.5 CYBERCRIME AGAINST INDIVIDUALS

Some forms of cybercrime may victimize individuals over the Internet. They are also termed *cyber nuisance* and are listed here:

Internet grooming Internet grooming is the process of befriending children to perpetrate sexual crimes, abuse, or exploitation over the Internet. Groomers persuade young children to send sexually explicit images, take part in sexual activities via webcam, or have sexual conversations online.

Stalking Stalking refers to a crime that relies upon electronic communication for sending threatening emails or messages to follow and harass the victim persistently.

Cyber stalking occurs in many forms such as harassing, embarrassing, humiliating, isolating or frightening the victim by following him/her online. It can happen via email (called email stalking), chat rooms, and discussion forums. Usually, people who are not aware of netiquette and rules of Internet safety, and inexperienced web users are prone to this attack. Social networking has opened the doors to cyber stalking by revealing the victim's online presence.

In *Internet stalking*, the victim is harassed via the Internet, by sending obscene content or a virus repeatedly through email.

Computer stalking refers to the usage of computer skills to gain unauthorized access to a victim's computer, possibly by exploiting the gaps prevailing on the Internet or the host operating system.

Harassment Harassment is a threat that uses online communication facilities such as email, social media, apps, and websites to cause emotional distress to targeted individuals. Sometimes, online harassments have even lead to death and suicide.

Extortion Extortion or digital blackmail is a crime which aims at damaging the reputation of an individual or an organization for exploiting money or any other benefit unlawfully.

Pedophilia Pedophilia is when sexual predators approach children. Online pedophiles exploit children over the Internet (through email, chat, and instant messages).

Facebook stalking Facebook stalking refers to the act of using a Facebook account to follow the actions of the targeted Facebook user. It may take the form of either excess viewing of the target user's profile or posting comments to the target user very often. However, Facebook stalking can be spotted with Stalker apps which provide information to the targeted user on who views the profile often and so on.

Internet troll Troll refers to a person who uses the Internet to post unwanted, provocative messages to an online community, especially on forums and chat rooms, with the intention to provoke the person engaged in a discussion to give an emotional response or to disrupt an ongoing healthy discussion. Trolling can happen in the form of YouTube video comments, blog comments, and through social networking sites such as Facebook, Twitter, Reddit, Instagram, and Tumblr.

Pyramid scheme fraud Multi-pyramid marketing risk or pyramid scheme fraud rewards people for enrolling others into an unsustainable business. The victim first has to pay to enter the investment scheme and then enrol more people into the scheme for which he/she will be paid. This continues and the chain continues to grow. The money paid is not invested in anything substantial and so in the end, the fraudster (at the top of the pyramid) makes more money and the ones who entered later in the chain end up with nothing.

Credit card fraud Credit card fraud occurs when the fraudster steals the card's number, pin, and security code so as to make purchases on behalf of the victim without his/her authorization. The fraudster can acquire the victim's credit card-related credentials using phishing or through online data breach. A hacker can manage to get that information by employing a key logger or by installing malware that captures the card information while the victim is making an online purchase.

2.6 CYBERCRIME AGAINST PROPERTY

Offences against confidentiality, integrity, and availability of computer data and systems include cybercrimes against both the property and the person. This may take any one or more of the following forms:

Illegal Access—Hacking and Cracking

Hacking is carried out by an intruder who gains access to a system without permission because of greed, or to get fame and power. Hacking when done with a destructive motive (such as stealing personal banking information or corporate financial information) is called *cracking*.

The intention of the offender in this case is primarily to break the security measures adopted so as to prove his/her skill. The factors which promote this kind of crime include failure on the part of organizations and individuals to adopt appropriate protection and security measures and the development of many automated tools to locate possible targets for the attack and to execute them. The offender could scale this attack to more targets with *botnets*. In general, any computer system that is connected to the Internet and left without protective measures is more prone to attack. Adoption of security measures can however lower the risk but cannot eliminate it once and for all. Investing in a firewall would help to protect the systems from hackers.

Illegal access may include cracking of the passwords of password-protected computer systems and websites. It also includes illegally gaining access to systems through loopholes in the software and hardware, spoofing systems and websites, and making users reveal their passwords and even employ keyloggers to capture sensitive information such as passwords of other applications and financial data. Besides these classes of attacks, illegal access is the primary step in perpetrating cybercrimes.

Hackers reach the victim via the Internet, employing one of the following ways:

SQL injections This is the most common web hacking technique that can destroy the SQL database if it is unprotected. It implies the placement of malicious code in SQL statements through web page input so that the hacker gains access to the backend. A hacker may gain access to the database if he/she is not prevented from giving a wrong input that runs on the database successfully. For example, an input given in the web form for Username as "or""=" and Password as "or""=" may be executed as follows:

SELECT * FROM Table WHERE UserName = "" or ""="" AND Password= " " or "" =""
and will return all rows from the table "Table" as OR "" ="" will always evaluate to TRUE.

Theft of FTP passwords Theft of FTP passwords happen when the hacker acquires the website login information retained or saved in an unprotected computer system by the webmaster. Using the stolen login credentials, the hacker logs into the website from a remote computer with the intention of tampering with the website.

Cross-site scripting or XSS attack This is meant to compromise the users (victims) of a website so that the attacker can gain access to the user's cookies, session IDs, and passwords. It is also an injection attack executed on web applications that accept the input but fail to separate the data, and is executable before it is delivered to the user's browser. Thus a compromised or vulnerable website is used to deliver a malicious script to the victim's browser. When the web page is loaded, the script is executed in the victim's browser and the attack is successful.

Web jacking This is a case of web hijacking where the hacker fraudulently takes control over the website and alters the content or posts obscene content. In another form, just clicking on the website may redirect the victim to a fake but similar website controlled by the attacker.

Exploit kits They enable attackers to gain control over the computer by exploiting the vulnerabilities in the software.

Illegal Data Acquisition—Data Espionage

This class of crime includes gaining access to sensitive information from any computer system across the globe via the Internet. This is possible by looking for any protected ports via some software or by *phishing*. The solution to this class of crime primarily involves educating computer users of the need for data protection. Usually, organizations adopt powerful security measures for their business data. However, the most prone targets of this class of cybercrime are the users of standalone computers, wherein sensitive information such as passwords to bank accounts and credit card particulars are preserved and they are less protected.

Corporate espionage is the act of spying on a business competitor or even a government with the intention of gaining secret information. Many corporate companies hire ex-military and government agents who occupy leading positions within the company and are trained to be spies to gather information about competitors. These spies steal information employing one of the following means, namely key loggers, USB drives, the Internet, hacking, password cracking, or even Wi-Fi phishing techniques. Corporate espionage could be avoided to some extent if the internal network is not exposed to outsiders, the data is encrypted, and access control policies are in force.

Illegal Interception

Any communication over the Internet involves exchange of data between the users; this may include email, Voice over Internet protocol (VoIP), chat, etc. With this class of crime, the offenders target the communication medium to gain access to the sensitive information being exchanged. Usually data exchanged to and from Internet service providers (ISPs) are secure. Besides this, offenders can gain access to the communication medium, irrespective of whether it is wired or wireless. In case of wired medium, offenders use appropriate devices to extract information from the medium as it is electromagnetic in nature while it is in transit in wires. In case of wireless medium, any wireless access point deployed by the offender may deceive the users, making them use this for their communication. This is however insecure as the information exchanged is recorded by the offender.

Spoofing This is an attack where the attacker masquerades data packets, IP addresses, MAC addresses, and even email addresses to make it seem as if they are originating from a genuine source.

Skimming This is a card-related fraud where a handheld device called skimmer captures the information—name of the card holder, the card number, and the date of expiry—from the card and transfers it into the attacker's computer so as to use it to commit a crime later.

ATM hacking This refers to the act of launching cyberattacks on ATM machines with the intention of looting money from it. Such attacks are also called jackpotting attacks. ATM hacking happens in one of the four ways:
1. The attacker disconnects the ATM from the bank's network and connects it to an appliance that can control cash trays within it. Hence, the attacker can use any card and PIN to make a rouge transaction that appears legitimate.
2. An insider in a bank lends the key of the ATM chassis which the attacker uses to disconnect the cable, thereby disconnecting the ATM from the bank. Thereafter, the attack proceeds as discussed earlier.
3. Alternately, after obtaining the key, the attacker can even set the ATM machine in maintenance mode and use the appliance that controls the cash trays and accesses the cash.
4. The attacker can infect the ATM with malware using a USB drive, or remotely by infecting the bank's network.

Data Interference

This class of cybercrime includes attacks on the integrity and availability of the data. Offenders who gain access to the data may either delete it or alter it. Loss of access to data is also possible. All these may happen in the form of viruses, worms, Trojans, and logic bombs if there are no adequate protection measures in the system.

Viruses These are distributed via the Internet or through removable storage devices that are plugged into the computer system. The situation could be worse if backdoors are installed.

A similar effect can also be introduced by worms which, instead of clinging on to the computer system, self-replicate until the available memory is consumed.

Trojan horse This differs from a virus in the way it propagates itself. It pretends to be a legitimate file and enters the system as an email attachment or as a download (when using the Internet for surfing web pages or playing online games). Trojans cause damage by way of stealing information or by disrupting the normal functioning of the computer system.

Logic bombs These are malicious code-like viruses and worms that do not replicate but remain dormant until triggered by a specific event at a specific time.

Ransomware This is a malware-based attack which encrypts the files of the system it attacks, using public-key encryption and makes the data unavailable. The data can be decrypted with the secret key or the encryption key available with the hacker which will be provided only after paying a huge ransom.

System Interference

This class of cybercrime includes attacks on the integrity and availability of computer systems and can result in huge financial loss to the victims. This can happen via worms and *denial of service* (DoS) *attacks*.

1. In a DoS attack, attackers flood a computer resource with requests that are beyond its capacity to handle and consume its available bandwidth and overload it. This will cause the system to either crash or slow down, thereby denying the intended user of the service. DoS attack is thus an explicit attack by attackers. Another variant of DoS is *distributed denial of service* (DDoS) *attack* where perpetrators from different geographic locations target web servers by flooding the network with traffic, eventually resulting in a crash.
2. In *email bombing*, the attacker sends huge volumes of email to a target in order to consume network resources, resulting in crashing of either the victim's mail box or the mail server.
3. *Email spamming* is a variant of email bombing where bulk messages are sent as attachments to many victims. Opening the attachments in the spam mail will lead to phishing websites hosting malware.
4. *Malvertising* is a type of crime where the attacker inserts malicious advertisements on websites from where the victims download the malicious code by clicking on some advertisement without being aware of the fact that the website is infected.
5. *Publicly unwanted programs* (PUPs) automatically install unwanted software in the computer system, for example, search agents and toolbars with the intention of harming the system.

Copyright and Trademark-related Offences

Ownership over the content distributed over the Internet may be claimed if it is covered under digital rights management (DRM) or copyright. Today's peer-to-peer systems enable users to share and distribute digital content to millions of users. Rapid growth in digital technology makes it possible to duplicate copyright-protected digital content without any loss of quality. Law enforcement agencies can act against this menace if the file sharing system depends on a central server. In the case of distributed systems, it is however complicated as offenders can be tracked with IP addresses. Further difficulty arises when the offenders use software tools to remove the DRM protection from the copyright-protected digital content and make copies of the material without any limitation.

1. Copyright infringing software or *software piracy* is the unauthorized use, copy, or distribution of software. The copying of proprietary software to multiple computers without multiple licences is considered illegal. The copyright holder has certain exclusive rights such as the right to display, distribute, or reproduce the copyright-protected digital content.
2. *Trademark-related offences* include using brand names to mimic legitimate companies and sending disguising emails and domain-related offences like registering a domain name which resembles the trademark of a product.
3. A *clone* is another type of attack where the idea behind the developed software is used to develop new software. This is possible as the idea is not copyright-protected.
4. *Software crack* is an illegally obtained version of the software by defeating the copyright protection. This is possible by generating a serial number which unlocks the evaluation version of the software. Using pirated software comes with risks such as viruses, worms, and other malware.

Computer-related Offences

This class of offences include computer-related fraud, computer-related forgery, phishing, identity theft, and misuse of devices.

Computer-related fraud refers to offences where offenders target either the computer or the data processing system by masking their identities using automation and software tools. Online auction fraud and advanced fee fraud are the common ones. In the former, offenders exploit the auction platforms to disguise either the buyer or the seller. It can involve either forcing the buyer to make the payment prior to delivery of non-existent goods, or buying goods and asking for delivery with no intention of paying. In the case of the latter, offenders send emails asking for recipients to help in the transfer of a large sum of money to third parties. Following this, the offenders may make use of bank account information for executing their fraudulent activities. Another popular fraud of this kind includes the Nigerian Letter or '419' fraud. It is an impersonation fraud that offers a large amount of money to a person on the condition that the recipient should help the criminal for transaction. 419 fraud is intended to gather bank account details of the recipient by email or messages.

Impersonation   This refers to the offender creating a web page, or sending emails or instant messages over the Internet using either the name or domain name of the victim with the intention of harming, defrauding, intimidating, and threatening the other person/s.

Computer-related forgery   This refers to the manipulation of digital documents, for example, altering electronic images and text in the documents. Phishing also falls under this category. In *phishing*, confidential information is extracted by deceiving users (by seeming to be a legitimate enterprise), typically by email spoofing.

The most common form of *video forgery*, copy-paste forgery attempts to remove objects and regions from a set of frames, thereby altering the message conveyed by the video. It is considered a threat as videos serve as a source of information and evidence in many criminal investigations.

Data diddling   This refers to altering the data to be stored in a computer system before or during entry and changing it after processing is done. This can affect the integrity of data and is difficult to track as the attacker modifies the expected output.

Identity theft   This refers to fraudulently acquiring the identity of an individual [e.g., social security number (SSN), date of birth, passport number, address, phone number, passwords, etc.] and the misuse of the same. The digital era we live in has made this possible. The best example of the term identity is personal identification number (PIN) received on mobile phones while an online banking transaction is being executed. Identity theft may happen when an offender uses malicious software or executes a phishing attack to acquire identity information and uses it in the future to commit other crimes. The offender may also attempt to steal storage devices, use search engines, and look for any gaps in storage systems to acquire identity information. Credit card fraud is the simplest form of identity theft where the criminal uses the victim's credit card to fund transactions.

Misuse of devices   This refers to devices being used to commit cybercrimes. Today any computer with Internet access can be used to commit such crimes. In addition to this, some specialized software (either commercial or obtained online free of cost) may be used.

Salami slicing attack   This is an attack where criminals steal either money or resources in an unnoticeable way, in small proportions that accumulate to a considerable amount in due course of time. The attacker may even program the system to carry out this task automatically.

Remote commands   These can also be used to acquire sensitive information from the system.

Pharming   This is an attack where the victim is deceived by a fake website to enter sensitive information—PIN number, password, etc. It is different from phishing in that the attacker, instead of relying on URL links, redirects the network traffic from the genuine website to the fake one.

2.7 CYBERCRIME AGAINST NATION

Offenders may instigate cybercrimes by exploiting, say, network attacks against mission-critical infrastructure. Cyber terrorism, cyber warfare, and cyber laundering fall under crimes against the nation. The 9/11 attack is the best example of the use of ICT by terrorists.

Cyber terrorism is also called Internet terrorism, electronic terrorism, or information war. It relies on Internet-based terror attacks which either disrupt or create disturbances to the functioning of the Internet. Some of the possible cyber terrorist targets include the banking industry, military installations, power plants, air traffic control centres, and water systems.

Cyber warfare is a computer- or network-based conflict between countries, wherein the countries attempt to disrupt the activities of the other especially for military reasons and cyber espionage. Cyber warfare may take the form of viruses, worms, malware, DoS attacks, hacking and theft of data, and ransomware.

Cyber laundering is the act of electronically transferring illegal money without revealing the source and possibly the destination.

Further, the Internet also serves as a source of information to terrorists for carrying out more disruptive attacks. For example, high-resolution satellite images of critical infrastructure such as dams and bridges, hospitals, telecommunication systems, and electrical power systems are available online. Sensitive and confidential information may be retrieved by search engines. Unwarranted information such as loading bullets in weapons and construction of nuclear weapons may provide potential sources of information to terrorists. They may also use the Internet as a means of communication—for example, email was used for carrying out the 9/11 attacks. It is also used as a medium for financing (by way of electronic payments) and executing attacks.

2.7.1 Content-related Offences

This class of cybercrime refers to dissemination of content that is illegal and may include pornography, xenophobic material, insults to religious symbols, false information, and spam. The scope of these crimes is restricted to the nation where it is executed as it is governed by cultural and legal principles. Offenders cannot be prosecuted if it is not considered a crime in the nation they live, whereas the content may be illegal in other parts of the globe. However, ISPs can maintain a blacklist to prevent the content from being distributed via the Internet in the nation they operate.

Pornography This is the commercial distribution of sex-related content over the Internet. However, different countries criminalize this to varying extents. Some countries permit the exchange of this sort of content among adults with the help of adult verification systems. Though there are different conceptions about adult pornography, child pornography is condemned worldwide as an offence. This criminal act is also associated with huge exchange of money and is a highly profitable business. Attempts against child pornography fail because of two main reasons: (a) the non-digital way of exchanging money for this sort of offence prevents law enforcing agents from tracing the flow of money and the offenders; (b) this sort of content is usually stored and exchanged in encrypted form which makes it harder to block or filter the content while it is distributed over the Internet.

Racist and xenophobic material This refers to any written material, image, or any other representation of theories or ideas that advocate, promote, or incite hatred, discrimination, or violence against any individual or group of individuals based on race, colour, descent, or national or ethnic origin as well as religion if used as a pretext for any of these factors [Craig Barker and John Grant, 2006]. The Internet seems to be the best medium for its propagation. Since views and opinions, and their applicability vary across the globe, disparities prevailed during the drafting of the Council of Europe Convention on Cybercrime. Hence it is addressed in a separate First Protocol.

Religious offences These include anti-religious written statements and publication of cartoons defaming any religion, and is normally committed by way of debate over the Internet, posting material or written articles, and leaving comments on discussion groups.

Spread of false and defamatory information The spread of 'fake news' and disclosure of secret information via web forums and social networking sites is also a criminal act. ISPs usually facilitate promotion of the information without any requirement on the identity of the offender. Once such information is published on the Internet, it is duplicated and the offender generally loses control over the spread of such information. This makes such information permanently reside on the Internet even if the offender corrects or removes it from the source.

Email spam This is the predominant form of content-related offences. Even though today's email providers use anti-spam filter technology, its distribution is still uncontrolled as offenders use botnets. This makes it difficult for law enforcing agents to track offenders.

2.8 CRIMES ASSOCIATED WITH MOBILE ELECTRONIC COMMUNICATION DEVICES

Mobile ECDs such as laptops, notebooks, tablet PCs, mobile Internet devices (MIDs), cell phones, digital cameras, smartphones, and PDAs facilitate work from anywhere. These devices are widely used as a tool in cybercrime as they are portable and a cheap alternative to computers. They serve as personal communication device and are difficult to locate. They have capabilities on par with computer systems such as processing speed, inbuilt camera, voice recorder, video recorder, memory for storage, and sufficient power back up. In addition, they have the ability to connect to the Internet and access valuable information, provide GPS facility, have provisions for multiple SIM slots, and are compatible with almost all software. While connected to the Internet, these devices behave in the same manner as that of ECDs, that is, they can be the target or used as tools to commit cybercrime. However, some forms of cybercrimes are exclusive to mobile ECDs and are listed here:

Handset theft This is mainly done to either get valuable information from it (e.g., contacts, SMS, etc.) or for financial gain from selling it.

SMS-related crimes This includes SMiShing, flashing or auto deletion of SMSes, tampering with SMS contents, altering the dates in an SMS, and SMS spoofing.

SMiShing This is SMS phishing wherein the attacker dupes the victims with messages in such a way as to reveal their personal data. On the other hand, with Nigerian or Lottery fraud, the attacker sends a message to the victim to update the details of the bank account or credit card within a short duration, failing which the account may be deactivated. He/She may even insist on visiting a website or calling a specific number to cancel, confirm, or reactivate the account.

Flashing SMS This sort of message flashes on the screen and automatically gets deleted once the user exits the application. These messages are usually threatening or rumour-spreading in nature.

Altering dates in SMS This refers to sending back-dated or post-dated SMSes by changing the message time-stamp. This is done in an attempt to create a fake scenario, resulting in chaos.

SMS spoofing This refers to masquerading the identity of an offender and sending messages to the victim as a genuine user.

Bluetooth mobile hacking If the Bluetooth capability in a mobile is enabled, this attack is possible and can enable an offender to access the phonebook, SMSes, call register, and can even make remote controlling of the mobile handset possible.

Crimes with calls Voice phishing (vishing) refers to crimes that involve social engineering, wherein either information or money is stolen from a victim using the telephone network. It is analogous to online scams that induce the victim into revealing personal information. Vishing comes in two forms: wardialing and dumpster diving. In wardialing, a call is made from a bank's number with a pre-recorded audio seeking the victim (customer) to enter the credit/debit card number, PIN, and CVV. This is recorded and is used by the offender to make unauthorized financial transactions. In case of dumpster diving, an offender attempts to look for a bank customers' phone numbers to programme them as targets for the attack.

SIM card cloning This refers to attempting to connect to a SIM card and making a duplicate copy of it or an image of it.

MMS crime This includes scandals and morphing. In case of scandals, the offender captures the personal activities of a victim with the use of camera in a mobile handset and spreads it through MMS service, either via Bluetooth or via the Internet. In case of morphing, genuine photos are morphed with obscene content with the intention to defame the victim.

2.9 CLASSIFICATION OF CYBERCRIMINALS

Cybercrime is referred to as white collar crime since the Internet is used as a tool to commit the crime in most cases. Cybercriminals do not mean hackers alone. They can include crackers and network attackers. Their involvement in cybercrime may be categorized as incidental, accidental, and situational. The general term used to denote people who commit cybercrime is 'offender'. They may harass individuals for fun, and cause damage to the economic assets of individuals, organizations, and the nation. Offenders are categorized according to their level of technical knowledge, area of interest, and expertise in using the associated hardware and software, as follows:

Mules These refer to people who do not know that they are part of a criminal gang involved in money laundering. They commit crime with online adverts who deceive them into making money by working from home and include them in other organized crimes.

Toolkit newbies or getaways These are novices who have limited technical skills and they use the readily available software following guidelines and documentation. They cannot be prosecuted. However, they are likely to transform into serial cybercriminals.

Activists These are people who make use of the Internet to promote religion, politics, or other causes but do not possess the skills of hackers. They can either steal data or cause damage to the IT infrastructure.

Cyberpunks These are people who have the capability to develop small programs and use them for defacing, spamming, or stealing credit cards.

Internals These refer to disgruntled employees who seek revenge with the goal of disturbing the security of information and the system of an organization, potentially damaging them.

Nation state actors These are people who work for one government to disrupt its enemies' IT infrastructure, steal sensitive information, and even create other intentional incidents.

Coders These refer to people who develop codes to damage other systems.

Professionals These are people who associate themselves with cybercrime as a profession and use technology to hide the occurrence of the crime. They generally take the form of technical support personnel and deceive victims so as to gain access to the IT infrastructure. They are dangerous perpetrators of cybercrime whose intention is of a criminal nature.

Depending on whether the intention of the offenders is good or bad, they may be categorized as follows:

Black hats commit illegal acts with the intention of causing harm to the information system, steal information, etc.

Gray hats may be either good or bad. They might have penetrated the information systems but have not caused any harm.

White hats are people who possess knowledge and skills as that of black hats but work together with authorities and companies to combat cybercrime (Box 2.1).

The classification of cybercriminals is shown in Fig. 2.4.

> **Box 2.1 Ethical Hacker and Ethical Hacking**
>
> An ethical hacker is also known as a white hat hacker who is the ultimate security professional. His/Her role is similar to that of a penetration tester who breaks into systems legally and ethically. The difference between hackers and ethical hackers is the legality with which they carry out the task.
>
> According to the EU Council, an ethical hacker is defined as an individual who is employed within an organization, and who can be trusted to undertake an attempt to penetrate networks or computer systems using the same methods and techniques as used by a malicious hacker.
>
> An ethical hacker knows how to find and exploit vulnerabilities and weaknesses in various systems, just as a black hat hacker, the malicious hacker. They both use the same skills but the difference lies in the usage of skills. The ethical hacker uses those skills in a legitimate, lawful manner so that vulnerabilities can be found and fixed before anti-social elements get there and attempt to break in.
>
> Some of the duties that are part of ethical hacking include the following:
>
> 1. Scanning ports using port scanning tools like Nmap to look for open ports and any other vulnerabilities
> 2. Examining patch installations to ensure that they cannot be exploited
> 3. To engage in social engineering concepts looking for crucial information like passwords, which can be used to generate an attack.

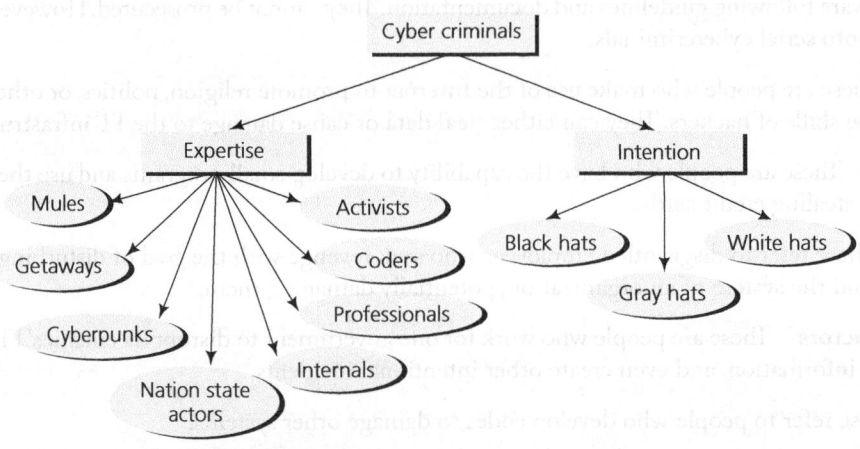

Fig. 2.4 Classification of cybercriminals

2.10 EXECUTION OF CYBERCRIME

The cybercrimes that are committed usually vary with the level of risk, associated cost, and complexity. Offenders intend to indulge in crimes that give them higher profits. Of course this depends on how they organize the crime and how much they have invested on assets.

In general, depending on the impact of the cybercrime executed (high or low), it is categorized as follows. This is shown in Fig. 2.5.

1. *Foreign intelligence services* that cause a major impact on the economy by being highly organized and using sophisticated techniques and extensive resources to commit crimes.
2. *Large organized crime networks* that focus on low-risk crimes with minimal investment.

Introduction to Cybercrime 57

Fig. 2.5 Intensity of cybercriminals

3. *Disreputable but legitimate organizations* that indulge in cybercrimes like IP theft to obtain sensitive information from a rival company.
4. *Individual or small group of opportunistic cybercriminals* who may target any particular person or any vulnerable organization.

2.11 TOOLS USED IN CYBERCRIME

Cybercriminals generally use various tools and techniques for executing crimes. They gain access to the victim's system in one of the following ways: (a) physical access, (b) relying on an intermediary system, (c) exploiting the vulnerabilities in the system and network, (d) deceiving victims into allowing access to his/her system, and (e) by gathering the victim's information.

The following are some of the common tools used for committing cybercrimes:

Proxy servers and anonymizers An anonymizer is a tool that attempts to make an activity on the Internet untraceable. It is a proxy server computer that acts as an intermediary and privacy shield between a client computer and the rest of the Internet. It accesses the Internet on the cybercriminal's behalf, protecting personal information by hiding the identifying information. There are hundreds of open proxies online. Such services are frequently used to break into foreign systems. Child pornographic material and other illegal content are usually exchanged through proxies.

Phishing Cybercriminals use spoof emails or direct people into using fake websites so as to deceive them into revealing personal financial details with which they access their accounts.

Malware It is a program that is inserted into a victim's computer, usually covertly, with the intent of compromising on the confidentiality, integrity, or availability of the victim's data, applications, and operating system, or otherwise annoying or disrupting the victim.

Keyloggers, password stealers, and spyware A keylogger is a software that has the ability to record every keystroke made by the victim onto a log file which is usually encrypted. A keylogger recorder can record instant messages, emails, and any information the victim types at any moment using his/her keyboard. The log file created by the keylogger can then be sent to the specified cybercriminal. Some keylogger programs also record the email addresses that are used as well as the URLs of the websites visited.

Password stealer is a type of malware that collects data such as account numbers and the associated passwords.

Spyware is a type of malware that is secretly installed in the victim's computer without his/her knowledge to gather information on individuals or the organization.

Virus and worm A virus is a self-replicating program that runs and spreads by modifying other programs or files. A worm is also a self-replicating program that is self-propagating and self-contained and it uses the network mechanism to spread itself.

Trojan and backdoors Trojan is a malicious program that masquerades as a benign application and can take complete control of the victim's computer.

A backdoor is a means of access to a computer program that bypasses security mechanisms. Programmers install a backdoor so that the program can be accessed for troubleshooting or other purposes. However, cybercriminals often use backdoors that they detect or install themselves, as part of an exploit. In some cases, a worm is designed to take advantage of a backdoor created by an earlier attack.

Steganography It is the art and science of writing hidden messages in such a way that no one besides the sender and the intended recipient suspects the existence of the message. An image file may contain hidden messages between terror groups which will be known only to the intended recipient and the sender.

DoS and DDoS attacks In a DoS attack, a cybercriminal uses a single Internet connection to either exploit a software vulnerability or flood the victim's computer with fake requests, usually in an attempt to exhaust server resources (e.g., RAM and CPU).

On the other hand, multiple connected devices that are distributed across the Internet can be used to launch DDoS attacks. These multi-person, multi-device barrages are generally harder to deflect, mostly due to the sheer volume of devices involved. Unlike single-source DoS attacks, DDoS assaults tend to attack the target network infrastructure and saturate it with huge volumes of traffic.

SQL injection The SQL injection attack is capable of targeting external websites or internal databases. It is used by cybercriminals to manipulate, steal, or destroy data. By taking advantage of vulnerabilities in the database layer of an application, hackers are able to inject malicious SQL queries into a website entry field, trick the application into executing unintended commands, and penetrate the backend database. An SQL injection attack may result in slowed application performance, data theft, loss or corruption, denial of access, or even complete takeover of the server.

Buffer overflow It is a condition wherein a program tries to store more information in a buffer than it was intended to hold. Since the buffers can hold only a finite amount of data, the extra information may overflow into adjacent buffers, thus corrupting or overwriting the valid data held in them.

Cracking It is the process of breaking into the victim's computer system, often on a network bypassing passwords or licences in computer programs, or in other ways intentionally breaching the victim's computer.

Data diddling It is adopted by cybercriminals to change the data before or during entry into a computer system or alter the raw data just before it is processed by a computer and changing it back after the processing is completed. Using this technique, criminals can manipulate the output so that it is not easy to identify.

Rootkit It is a set of tools that enable continuous privileged access to a computer while actively hiding its presence from the administrator. Usually a cybercriminal installs a rootkit on a computer after first gaining user-level access either by a known vulnerability or by cracking a password. Once the rootkit is installed, it allows the cybercriminal to mask the intrusion and gain root or privileged access to the computer and possibly other machines on the network.

Salami attack A salami attack is a small attack that can be repeated many times efficiently. Thus the combined output of the attack is massive. This attack involves making alteration so insignificant that it is easily concealed and would go completely unnoticed. These attacks are used for the commission of financial crimes.

The best example of this attack is the stealing of the round-off from interest in bank accounts. Even though it is less than Rs1 per account, when multiplied by millions of accounts over many months, the adversary can retrieve quite a large amount. It is also less likely to be noticeable since an average customer would assume that the amount is rounded down to the nearest rupee.

Sniffer   It is an eavesdropping program that is legally used by government surveillance agencies and telecom companies to identify network bottlenecks. Cybercriminals, however, use it for obtaining proprietary information.

Social engineering   It is a hacker term which involves non-technical intrusion for deceiving or manipulating unwitting people into giving out information about a network or how to access it.

Spoofing   In this case, cybercriminals try to get into the victim's computer by masquerading as a trusted source. Examples include email spoofing, IP spoofing, and address bar spoofing.

Rogue security software   It is a malware that is used by a cybercriminal and which pretends to be a malware removal software.

Pharming   It is when a cybercriminal redirects website traffic to a bogus website, usually an e-commerce or banking site.

Hijackware   It is a malware that changes the browser settings of the victim's computer to direct it to malicious sites.

Man-in-the-middle or MITM attack   This happens when a cybercriminal infiltrates a private conversation between two unsuspecting parties, eavesdropping on their entire conversation, and possibly altering the information the parties are trying to send each other. This attack will not be successful if the cybercriminal cannot establish a mutual authentication between the two parties. Generally, cybercriminals use a Wi-Fi router to intercept all of the user's communications. However, it can also be done via a rogue Wi-Fi network, with some malicious programs, so as to directly intercept the user's sessions on the router.

Watering hole   It is a *computer* attack strategy in which the victim is a particular group (organization, industry, or region). In this attack, the attacker guesses or observes which websites the group often uses and infects one or more of them with malware. Eventually, some members of the targeted group get infected.

2.12 FACTORS INFLUENCING CYBERCRIME

There are factors that influence the growth of criminals on the Internet and the rate of cybercrime. These are listed here.

Availability of tools to commit crime   The Internet is a boon to mankind and offers an array of services to almost all the domains. At the same time, it offers many tools free of cost to anyone—these are misused by criminals to commit cybercrime. Many crimes are committed every day without fear of being detected. Nowadays, free Wi-Fi facilities are available at airports, major railway stations, hotels, educational institutions, business establishments, etc., as a value added service, which also contributes towards commissioning of cybercrime by misuse of services.

No necessity of physical presence to commit crime   Today's world is technology-driven and even criminals are using technology to perpetrate hi-tech crimes. Technology has eliminated the need for criminals to be physically present to commit a crime.

Less investment to commit crime   Many service providers provide services at a subsidized price or even free of cost. This helps criminals execute cybercrimes with low investment. However, the damage caused by the criminal activity is huge.

Availability of forensic tools to mask crime   Masking and wiping tools available in the market help criminals hide their crimes and prevent themselves from being detected.

Jurisdictional concern of cybercrime   Cybercrimes span borders. Despite this, international cooperation and legislations supporting international jurisdiction do not exist, as a result of which most crimes remain undetected and unresolved.

Lack of awareness regarding usage of ECDs Many users of ECDs are either unprotected or unaware of the threats that ECDs are susceptible to while remaining connected to the Internet. This makes it possible for cybercriminals to execute attacks with ease.

Impact of social media Social media is a vital factor influencing cybercrimes, wherein without verifying the authenticity of the message, malafied or false messages are forwarded to many users of social media (Facebook, Whatsapp, etc.) either knowingly or unknowingly.

2.13 CHALLENGES TO CYBERCRIME

The following are some of the challenges from the general and legal perspectives to mitigate the impact of cybercrime:
1. ICT forms the basis for present-day communication infrastructure and the Internet. The dependence on ICT for systems and services is on the increase in developing countries than the developed countries. Since cybercrimes involve ICT, it is not possible to avoid ICT. Investing in proper protection measures will lower the incidence of crime and promote the use of ICT for critical infrastructure and services.
2. The tremendous growth of the Internet and allied technologies has raised the number of users. This poses difficulty for the law enforcement agencies to control and curb crimes. Automated online procedures for investigation have to be developed to ease the situation. For example, an automated search for illegal textual content can be made relatively easy with a keyword search. However, it is not that easy in the case of searching for an illegal image as this requires hash values of all existing images on the web.
3. Hardware, software, and Internet access are prone to cybercrime. The prices of hardware have come down but the computing power and the availability of software tools are on the rise. Even in developing countries, access to the Internet is not a big issue as a number of service providers have spawned, providing Internet access even free of cost. This has made it possible to commit crimes with ease. A proper balance has to be achieved by the law enforcement agencies to limit access to the Internet and to monitor the Internet activity of individuals without violating human rights.
Besides this, botnets also pose a risk, as offenders use software to gather systems connected to the Internet to execute an attack, say, DoS and it will be difficult to track the original offender.
4. The positive side of search engines is that it provides instant information once a search with any keyword is made. At the same time, cybercriminals make use of search engines to look for targets and possible victims. Satellite images from Google Earth have been used by terrorists in the past to gather information about other countries and plan attacks.
5. Lack of centralized authority to control the operability of ICT is one of the challenges that needs to be addressed to control cybercrime. This necessitates legal standards to be framed by a centralized controlling authority which will make investigation of cybercrime easy.
6. Cybercrime spans international borders and in most cases the victims and the targets are located in different countries. Therefore, investigations in such a situation require coordination and support of law enforcement agencies of both the countries involved. For example, in a crime against an individual that was reported in India where obscene content was associated, the pornographic content was hosted in a server in another country. When a request was made to the respective country to block access to that server, it was refused for the reason that pornography is an offence in India and not in that country.
7. Cybercrime can be executed from anywhere. Absence of centralized legal standards poses difficulties during investigations. Cybercriminals tend to avoid countries where there is strong cybercrime legislation and prefer countries where there is insufficient legislation. Such countries have to frame strict legislation.
8. Automated tools offer advantages if used for the right purpose. Offenders use it, for example, to send spam mails which consume a majority of the network resources and can even cause a crash of critical infrastructure. Law enforcement agencies have to be vigil over the use of automation tools.
Similarly, encryption tools and technologies, if used by offenders, pose challenges during investigation.

9. ICT facilitates data exchange in a fraction of time and the data exchanged can be deleted without any sign of the exchange having taken place. This poses hardship for agencies investigating cybercrime as against traditional crime where there are visible signs of the data exchange having taken place. However, forensic tools are employed to prove the occurrence of a crime.
   Besides this, investigators may have to intercept communication which is possible with traditional voice calls but not with VoIP services. In case of wireless access points (WAP) that do not require registration, it will be difficult to track offenders as an investigation can lead to the WAP but not the offender. This necessitates that with the improving trends in technology, new technical solutions and instruments for investigation are necessary, and investigators have to be technically sound to carry out the investigations successfully.
10. Anonymity is useful at times when a user does not want to reveal his/her identity in a public discussion forum. However, this could pose difficulty in the context of an investigation.
11. From a legal perspective, the available laws are insufficient under certain scenarios and for certain crimes. Hence, proper legislation should be framed to support investigation. Apart from this, on par with new crimes, relevant instruments for investigation have to be developed, along with appropriate procedures for acquiring digital evidence.
12. The problems faced by law enforcing authorities (LEAs) are crucial. Collection of digital evidence is very important to prove the cybercrime. Generally, LEAs do not get adequate cooperation while collecting digital evidence from mobile service providers (MSP) and social media service providers. For example, getting information from Facebook and Whatsapp is difficult for the LEA as the servers are located in western countries where the laws of the respective country are an impediment to the collection of digital evidence.

2.14 STRATEGIES TO PREVENT CYBERCRIMES

The following are some of the preventive measures that can help to control and curb cybercrime:
1. Turn off the systems when not in use.
2. Use of antivirus software and periodic updating of the same is essential to protect systems from viruses and other emerging threats on the Internet.
3. The firewall in the system should be turned on always.
4. All software in the system should be kept updated.
5. Social media accounts should be locked while not in use. Avoid invitations from social media and block unwanted invitations.
6. Usage of more than one email account and using each one of them exclusively for banking, shopping, etc., is desirable—if one of the accounts is hacked, not all the data is compromised.
7. Avoid clicking on pop-ups as they may contain malicious software.
8. Having two-step verification for email and social media accounts can prevent hackers from accessing accounts.
9. Avoid opening unknown attachments in emails.
10. Avoid shopping online and if necessary proceed only on secure sites.
11. As end users, one should be aware of the privacy policies of social media and websites. There could be issues when an individual does not want anyone to have access to photos posted earlier on social media. However, they could still be retained on websites as they are guarded by the website's privacy policies.
12. Usage of strong passwords can help to counter some forms of cybercrime. It is always advisable to change passwords often. Maintain different passwords for different sites rather than keeping the same one for all the accounts and websites.
13. Avoid maintaining credit card details and disclosing other valuable information on websites.
14. It is the responsibility of the individual to remain updated of major security breaches so as to avoid becoming a victim of cybercrime.

2.14.1 Indian Perspective

In India, the Department of Electronics and Information Technology (DeitY) has defined a cyber security strategy that includes the following [The Associated Chambers of Commerce and Industry of India (2015)]:

Framing Responsibilities

1. In India, to coordinate effective implementation of cyber security strategy, the government has set up an Inter Departmental Information Security Task Force (ISTF) with the National Security Council as the nodal agency. The Indian Computer Emergency Response Team (CERT-In) is the national nodal agency set up to respond to incidents of computer security. Its activities toward cyber security include coordination of responses to security incidents and major events, issuance of advisories and timely advice regarding imminent threats, analysis of product vulnerabilities, conducting of training sessions on specialized topics of cyber security, and development of security guidelines on major technology platforms.
2. Another major initiative is the creation of specialized teams at different departmental levels, such as the National Cyber Coordination Centre (NCCC), National Critical Information Infrastructure Protection Centre (NCIIPC), Grid Security Expert System (GSES), National Counter Terrorism Center (NCTC), Cyber Command for Armed Forces, Central Monitoring System (CMS), National Intelligence Grid (NATGRID), Network and Traffic Analysis System (NETRA), and Crime and Criminal Tracking Network & Systems (CCTNS) with specific roles. This includes screening online threats and coordinating with the intelligence agencies to handle issues related to national security by guarding against hackers and espionage, and tracking terrorist activity online.
3. The Information Technology (Procedure and Safeguards for Blocking Access of Information by Public) Rules 2009 read above have been notified by the Government of India under the provisions of clause (z) of sub-section (2) of Section 87, read with sub-section (2) of Section 69 A of the Information Technology Act 2000. Rules 3 to 6 read as follows:

Designated Officer

The Central Government shall designate by notification in Official Gazette, an officer of the Central Government not below the rank of a Joint Secretary, as the "Designated Officer", for the purpose of issuing direction for blocking access by the public any information generated, transmitted, received, stored or hosted in any computer resource under Sub-section (2) of Section 69A of the Act.

Nodal Officer of Organisation

Every organisation for the purpose of these rules, shall designate one of its officers as the Nodal Officer and shall intimate the same to the Central Government in the Department of Information Technology under the Ministry of Communications and Information Technology, Government of India and also publish the name of the said Nodal Officer on their website.

Direction by Designated Officer

The Designated Officer may, on receipt of any request from the Nodal Officer of an organisation or a competent court, by order direct any Agency of the Government or intermediary to block access by the public any information or part thereof generated, transmitted, received, stored or hosted in any computer resource for any of the reasons specified in Sub-section (1) of Section 69A of the Act.

Forwarding of Request by Organisation

1. Any person may send their complaint to the Nodal Officer of the concerned organisation for blocking of access by the public any information generated, transmitted, received, stored or hosted in any computer resource: Provided that any request, other than the one from the Nodal Officer of the organisation, shall be sent with the approval of the Chief Secretary of the concerned State or territory to the Designated Officer:

Provided further that in case a Union territory has no Chief Secretary, then, such request may be approved by the Adviser to the Administrator of that Union territory.

2. The organisation shall examine the complaint received under sub-rule (1) to satisfy themselves about the need for taking of action in relation to the reasons enumerated in Sub-section (1) of Section 69A of the Act and after being satisfied, it shall send the request through its Nodal Officer to the Designated Officer in the format specified in the Form appended to these rules.
3. The Designated Officer shall not entertain any complaint or request for blocking of information directly from any person.
4. The request shall be in writing on the letter head of the respective organisation, complete in all respects and may be sent either by mail or by fax or by e-mail signed with electronic signature of the Nodal Officer: Provided that in case the request is sent by fax or by e-mail which is not signed with electronic signature, the Nodal Officer shall provide a signed copy of the request so as to reach the Designated Officer within a period of three days of receipt of the request by such fax or e-mail.
5. On receipt, each request shall be assigned a number along with the date and time of its receipt by the Designated Officer and he shall acknowledge the receipt thereof to the Nodal Officer within a period of twenty four hours of its receipt.

Accordingly, any individual can send a complaint in case of defacement, abuse, misconduct, misuse of website, etc., to block access of information by the public.

Making Refinements to Law

1. The IT (Amendment) Act, 2008 has included the following: Electronic signatures, corporate responsibility, definitions of important terms such as intermediaries and communication devices, legal validity of electronic documents, role of adjudicating officers and requirements on data retention. It covers crimes like hacking, data theft, spreading malicious content, virus, identity theft and email spoofing. While such legal mechanisms are being developed, companies in India will need to increase investments to safeguard themselves against cybercrime.
2. The IT (Amendment) Act 2008, however, needs to be updated and has a wider scope to include a legal framework for cyber laws. It primarily deals with extraction, retention, communication and destruction of data. It does not cover majority of the crimes committed through mobiles.
3. A right balance needs to be found between blocking the privacy of the citizens and monitoring the crime levels. India has seen some region-specific violence where the reason was that improper information went viral on social networking websites. However, the recent scrapping of Section 66 A does not help the intermediaries to bring down such hatred information. The act needs to be expanded further and there is a strong need to introduce more clarity in terms of responsibilities of intermediaries and encryption mechanisms.

Providing Training and Awareness

1. DeitY as part of the training initiative developed internet security awareness materials in the form of presentations, posters, cartoons, guide books, security tools and parental controls for children.
2. A web portal 'www.infosecawareness.in' was also launched.
3. Government has formulated investigation manuals with procedures for search, seizure analysis and presentation of digital evidence in courts. The manuals had been circulated to law enforcement agencies in all the states of the country.

Developing Crime Reporting Mechanism

1. In India at present, reporting cybercrime involves registering complaints with the local police stations or cybercrime cells. Many of the Indian states have setup cybercrime cells, which monitor such crimes.
2. It is important to have provisions for online reporting of the crime so that the rate of reporting cybercrime increases. Using this system, an online cybercrime complaint can be made by the victims of cybercrime.

They will gain access to a convenient and easy-to-use reporting mechanism that alerts the law enforcement authorities of suspected criminal or civil violations.
3. Also, it will provide a central repository for reference to law enforcement and regulatory agencies at the national, state, and local levels.

Extending International Collaboration to Thwart Cybercrime

1. India, being one of the world's largest IT players and with the experience of working with almost all countries, is yet to form any substantial partnership on tackling cybercrime.
2. India signed a Convention on Mutual Assistance in Criminal Matters in 2008 to operationalize Mutual Legal Assistance Treaties (MLAT). MLATs play an important role in combating transnational organized crime and other serious offences, such as drug trafficking and money laundering.
3. However, India remains a non-signatory in the Budapest Convention, which is the international treaty seeking to address cybercrime by harmonizing national laws, improving investigative techniques, and increasing cooperation among nations. It is specifically designed to facilitate international cooperation to fight cybercrime.
4. It will be beneficial to have collaborations with International Cyber Security Protection Alliance, such as the Australian Cyber Security Centre (ACSC), National Crime Agency's National Cybercrime Unit (NCCM) and the UK's CEOP. This will help in not only adopting the best practices by other countries for prevention of cybercrime, but also in increasing the capability, knowledge, training, skills, capacity and expertise of cybersecurity task forces. Additionally, it will help to reduce the harm caused to businesses, customers and citizens due to international cyberattacks.

Developing Manpower for Cyber Security

1. In order to facilitate effective cybercrime investigation, Cybersecurity training facilities have been setup to provide training for law enforcement agencies. The state governments have been advised by the Ministry of Home Affairs to build adequate technical capacity in handling cybercrime, including technical infrastructure, cyber police stations and trained manpower for detection, registration, investigation and prosecution of cybercrime. Also, action has been taken to setup a national center of excellence, exclusively devoted to render cyber forensic services and to act as a national research and training center on cyber forensics.
2. CERT-In and CDAC are providing basic and advanced training to police officers, judicial officers and other law enforcement agencies on the procedures and methodology of collecting, analyzing and presenting digital evidences. Cybercrime investigation manuals had been circulated to all the state police stations.
3. A cyber forensics training lab was set up at the Training Academy of Central Bureau of Investigation (CBI) to impart basic and advanced training in cyber forensics and investigation of cybercrimes to police officers in India.
4. NASSCOM started an initiative of establishing cyber labs to train police officers. Such labs were set up in Mumbai, Thane, Pune, Bangalore, Chennai, Hyderabad and Haryana which are now under the control of Data Security Council of India (DSCI). The knowledge developed over a period of time by the police officers, instructors and experts in investigating cybercrime are summarized in *Cyber Crime Investigation Manual* to help police officers in handling crimes more effectively.

2.14.2 Global Best Practices

The following are some of the best practices adopted by developed countries to combat cybercrime:
1. A dedicated ministry is responsible for cyber security. The members have to devise a national strategy and foster local, national, and global cross-sector cooperation.
2. The national cyber security coordinator, who could be a department or an individual, oversees cyber security activities across the country
3. The National Cybersecurity Center Point, a multi-agency centre, exists which serves as a focal point for all activities dealing with the protection of a nation's cyberspace against all types of cyber threats.

4. Legal measures are taken after review of cyber laws and, if necessary, the existing laws are amended and new procedures and policies are created to deter, respond to, and prosecute cybercrime.
5. The national cyber security framework defines the minimum or mandatory security requirements on issues such as risk management and compliance.
6. Cybercrime reporting and analysis is performed where after the analysis of cyber threat trends, a response is coordinated and information is disseminated to all relevant stakeholders.
7. A national programme is convened to raise awareness about cyber threats and educate personnel on how to handle these issues on a continuous basis.
8. As part of capacity building, a programme is conducted to train cyber security professionals and adequate infrastructure for cybercrime prevention and investigation is developed.
9. International cooperation is extended to tackle the transnational nature of cyber threats.

2.15 EXTENT OF CYBERCRIME

There are numerous instances of cybercrime reported these days. Frauds include plastic card frauds, online shopping frauds, and offences from computer misuse. As many youngsters use gadgets, this group is an easy target for fraudsters. Cybercrimes may affect individuals, businesses, and the country. Threats today range from DDoS attacks to ransomware attacks. The following sub-sections present an account of cybercrimes reported across the globe and in India.

2.15.1 Cybercrime Statistics and World

According to Ecommerce Digest [eBusiness: Free ecommerce textbook: Complete guide to ebusiness and the digital economy], 73% of the Americans have suffered some form of cybercrime, against 68% globally, the worst

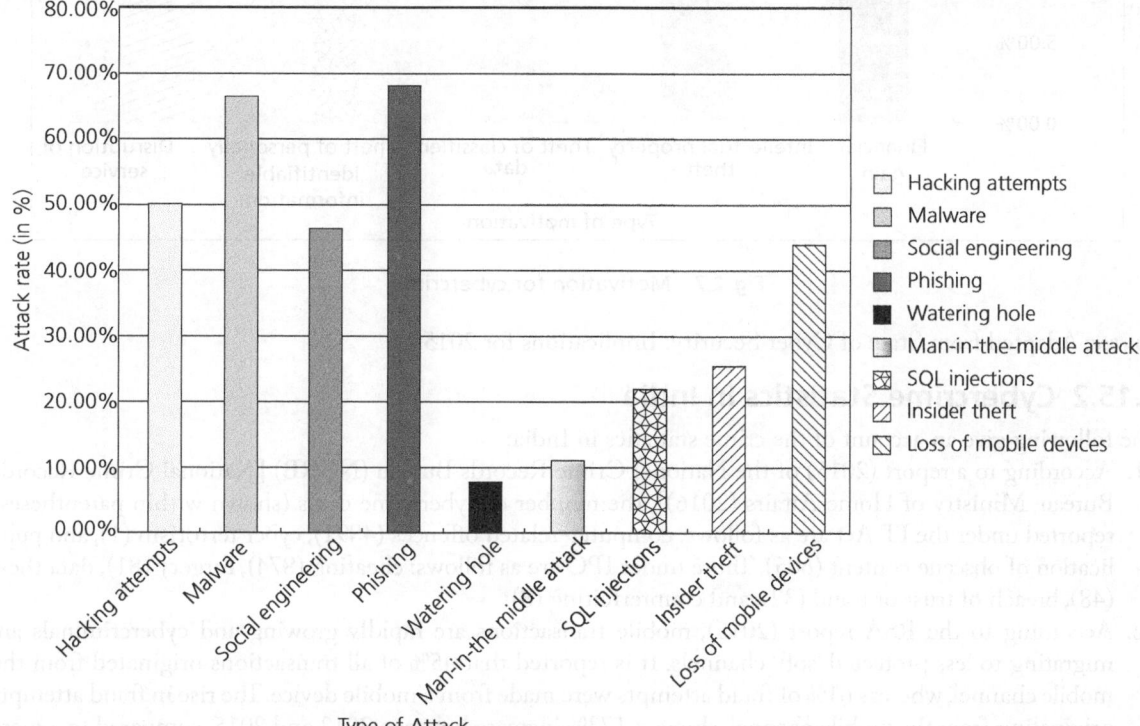

Fig. 2.6 Successful attack types in enterprise in 2014

Source: Adapted from State of Cyber Security: Implications for 2015

being Brazil at 83%. The average cost per fraud claim is US $223 for debit/credit card fraud, $610 for auction fraud, $800 for non-delivery payment of/for merchandise, and $1600 for Nigerian Letter fraud. The reasons for this seems to be that (a) 83% of the people do not use a separate email address for online shopping, (b) 69% of the people do not backup data regularly, (c) 62% of the people do not use strong passwords or change them regularly, and (d) 60% do not use a browser search adviser.

According to the report *State of Cyber Security: Implications for 2015* [ISACA, 2016], the reported cybercriminal attack types and the motivation behind them are shown in Figs 2.6 and 2.7 respectively.

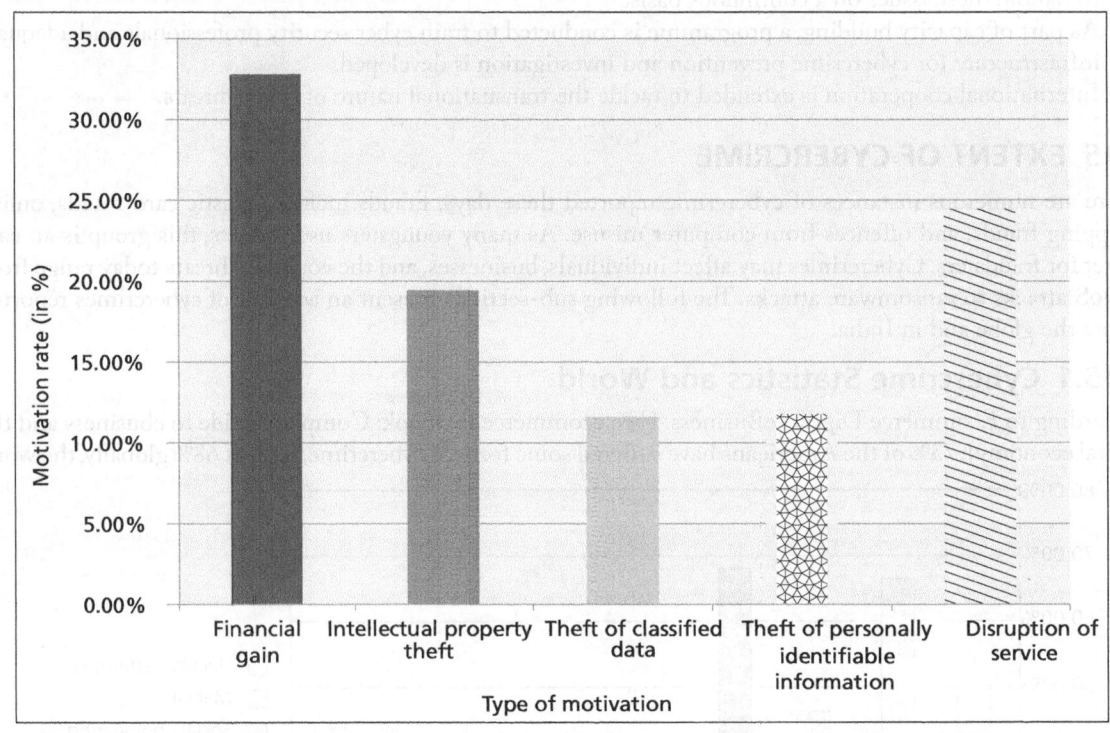

Fig. 2.7 Motivation for cybercrime

Source: Adapted from State of Cyber Security: Implications for 2015

2.15.2 Cybercrime Statistics in India

The following give an account of the crime statistics in India:
1. According to a report (2015) of the National Crime Records Bureau (NCRB) [National Crime Records Bureau, Ministry of Home Affairs (2016)], the number of cybercrime cases (shown within parentheses) reported under the IT Act are as follows: computer-related offences (4891), cyber terrorism (1), and publication of obscene content (605). Those under IPC are as follows: cheating (874), forgery (81), data theft (48), breach of trust or fraud (32), and counterfeiting (8).
2. According to the RSA report (2015), mobile transactions are rapidly growing and cybercriminals are migrating to less protected 'soft' channels. It is reported that 45% of all transactions originated from the mobile channel, whereas 61% of fraud attempts were made from a mobile device. The rise in fraud attempts originating from the mobile channel, shows a 173% increase between 2013 and 2015, compared to a mere one 1% in the web channel.

3. According to a report by software security firm Norton dated 19 November 2015, an estimated 113 million Indians lost about Rs. 16,558 on an average to cybercrime in addition to the 'emotional' stress caused by personal financial data breach.
4. According to the article *Over 55% Millennials in India Hit by Cybercrime: Report* published in the Indian Express dated 19 November 2016, (a) Over 55% of the millennials (those born between 1980 and 2000) in India have experienced cybercrime in the past year; (b) 39% of the Indian millennials have either experienced ransomware themselves or know someone who has; (c) 34% of millennials have admitted to sharing passwords of various accounts; (d) 30% of the millennials agreed to have used their neighbour's Wi-Fi network without their permissions; (e) putting their cyber security at risk, 60% of these millennials are willing to give into actions such as answering a survey, installing a third-party app (43%), providing access to files while online (25%), and turning off their security software (24%) to gain access to free public Wi-Fi.

2.15.3 Recent Sensitive Cybercrimes

This section presents some of the hypersensitive cybercrimes [National Crime Records Bureau, Ministry of Home Affairs (2016), Dan Elsom, 2017] reported worldwide. Their occurrence is highlighted in Fig. 2.8.

Ransomware

In 2017, WannaCry infected UK's NHS computer system and disabled it for a week. This was possible because of a major hole in the cyber security of healthcare systems and it left hospitals and practitioners in UK to completely operate offline.

Stolen Credit Card Data

In 2014, Home Depot's System was breached exposing the data of 50 million credit card holders. The offenders managed to gain entry into the company's network using a vendor's username and password, and installed a malware on its point-of-sale system. This resulted in the transfer of credit card information to the offenders as customers swiped their credit cards.

Data Breach

In 2015, Bank JPMorgan Chase disclosed that a massive breach had compromised the data of 76 million households and 7 million small businesses.

Similarly in 2017, Yahoo disclosed that a breach in security resulted in the theft of data of at least one billion users making it the biggest breaches of all time. However, now, Yahoo has admitted that all the three billion accounts were actually accessed. Earlier in 2013, a breach had compromised and exposed the data of 1 billion Yahoo users.

In 2015, the US Internal Revenue Service (IRS) faced a breach that disclosed more than 700,000 SSNs and other sensitive information.

The US Office of Personal Management (OPM) faced a data breach that exposed the records of 21.5 million people and is the largest breach of government data in the history of the US.

Consumer credit score company Equifax has revealed that hackers accessed up to 143 million customer account details in 2017.

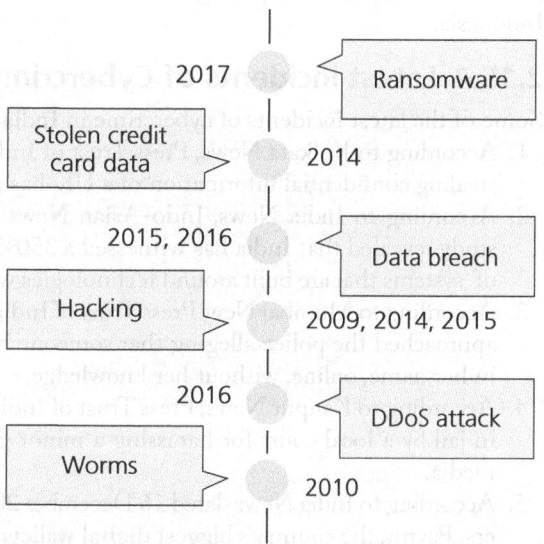

Fig. 2.8 Timeline of biggest cybercrimes in the world

Global consultancy firm Deloitte's systems were compromised through an unsecured administrator account, which allowed access to internal files. Details compromised include emails, usernames, passwords, health information, and details from Deloitte's clients.

Hacking
In 2009 Google's corporate servers in China were hacked and intellectual property and other information were stolen.

In 2014, the network of Sony Pictures was hacked and employee emails, information on executive salaries, and copies of unreleased movies were stolen. The North Korean Government was blamed by the US for this activity.

In 2015, offenders used a malware to infiltrate a bank's computer system and gather personal data. It was used to impersonate bank staff and authorize fraudulent transfers and dispense cash from ATM machines without a card. Around 650 million pounds were stolen in total.

In April 2017, American restaurant Chipotle announced that its payment systems had been hacked. The firm revealed that the malware which accessed payment card data was installed on point-of-sale terminals. The malware searched for track data (which sometimes has cardholder name in addition to card number, expiration date, and internal verification code) read from the magnetic strip of a payment card as it was being routed through the POS device.

In 2017, Canadian mobile phone, TV, and Internet service provider Bell was hit by an 'anonymous hacker' who swiped 1.9 million email addresses.

DDoS Attack
In 2016, a cyberattack targeted one of the companies that hosts the DNS. This took down many popular sites such as Twitter, Netflix, Paypal, and Spotify.

Worms
Stuxnet is a Microsoft Windows computer worm discovered in 2010 that targeted industrial software and equipment. It used USB drives for its propagation. The worm included a malware payload that targeted Siemens Supervisory Control And Data Acquisition (SCADA) configured to control and monitor specific industrial processes. It resulted in espionage and destruction which affected thousands of computers in Iran, India, and Indonesia.

2.15.4 Latest incidents of Cybercrime in India
Some of the latest incidents of cybercrime in India are presented here:
1. According to Kolkata News, Press Trust of India dated 22 January 2016, a man was arrested for allegedly stealing confidential information of a UK-based client from a company where he was employed earlier.
2. According to India News, Indo-Asian News Service dated 25 August 2016, an Assocham-PwC joint study revealed that India has witnessed a 350% rise in cybercrime cases in the period 2011–2014 because of systems that are built around technologies with weaker protocols which are inherently more vulnerable.
3. According to Mumbai New, Press Trust of India dated 1 October 2016, Bollywood actress Kareena Kapoor approached the police alleging that someone had filed income tax returns for the financial year 2015–16 in her name, online, without her knowledge.
4. According to Kanpur News, Press Trust of India dated 23 October 2016, a man was sentenced to two years in jail by a local court for harassing a minor girl and uploading her morphed, obscene images on social media.
5. According to India News dated 16 December 2016, Paytm was allegedly cheated of Rs6 lakhs by its customers. Paytm, the country's biggest digital wallets company said it had been cheated by a group of customers who colluded with executives at the company. The CBI, the country's premier investigating agency, was handling the case.

2.16 TERMS AND TERMINOLOGIES ASSOCIATED WITH CYBERCRIME

Backdoors These are secret (undocumented), hard-coded access codes or procedures for accessing information. Some backdoors exist in commercially provided software packages (e.g., consistent or canonical passwords for third-party software accounts). Alternatively, backdoors can be inserted into an existing program or system to provide unauthorized access later. Trojan Horse is an example of a program with an undocumented access method.

Blacklist This refers to a list of entities or people to whom access to a system, protocol, or service is denied.

Biometrics This refers to the use of a computer user's unique physical characteristics such as fingerprints, voice, and retina to identify that user.

Bot It is a program that is used for a specific function such as keeping a port open or launching a flood of packets in a DDoS attack especially for a robot.

Botnet It is a set of bots which are secretly installed on a number of victims' computers, referred to as zombies, to launch DDoS attacks or to send spam and virus.

Cipher text The data that has been encrypted is cipher text.

Computer emergency response team—CERT It is an IT security organization that collects and distributes information about computer-related security breaches and promotes effective security practices.

Cracking It is the act of unauthorized penetration into computer systems and networks, abuse of privilege, and unauthorized use of services.

Cybercriminal This is a person who commits cybercrime using a computer as a tool, target, or both.

Cyber extortion This refers to a growing threat that involves demanding money to minimize or stop attacks on public sector companies or corporates.

Cyber terrorism This is an attack that occurs over the Internet, computer systems, or networks against individuals, industries, or government organizations.

Cyber war It is the use of computer technology to disrupt the activities of a state or an organization especially any deliberate attack on information systems for strategic or military purposes.

Data diddling It is the act of modifying data before or during entry into computer systems for fun and profit (e.g., modifying grades, altering security clearance information, fixing salaries, or circumventing audit regulations).

Data leakage It is an uncontrolled, unauthorized transmission of information from a data centre or computer system to the outside world. Such leakage can be accomplished by physical means such as removal of data storage devices (diskettes, tapes), printouts, photographs, handwritten notes or by electronic means, for example, data hiding (steganography).

Denial-of-service or DoS attack It is a cyberwar technique designed for overwhelming or saturating resources on a target system to cause either unavailability or reduction of availability to legitimate users by spoofing packets or e-mail headers.

Digital signature It is an electronic form of a signature used to confirm the authenticity and integrity of digital document.

Distributed DoS or DDoS attack It is an Internet-mediated attack accomplished by enlisting the services of many target systems to launch a DoS.

DNS cache poisoning This type of attack modifies data in a DNS server so that calls to particular websites or even entire domains are misdirected for fraudulent purposes.

Easter egg This refers to any undocumented, unauthorized functions that are embedded in the production application. It is a kind of Trojan Horse.

Electronic mail—Email It is an application that allows the transmission of messages between computer users via a network.

Encryption It is the process of protecting sensitive information from unauthorized users or hiding its meaning by converting it into a code or unrecognizable form.

Exploit It is a method for utilizing a security hole in a computer or an application.

Firewall It is a security barrier designed as either a physical device, software, or both to enforce the boundary between systems or networks for secured communication.

Hacker This is a person who breaks into a computer system through skills for the purpose of stealing or destroying data.

Hacking It is the process of bypassing the security on a computer system or network for an illicit purpose.

Hacktivism (sometimes spelt hactivism) It is a means of politically or ideologically motivated vandalism. Defacing a website for no particular reason is vandalism; the same defacement to post political propaganda or to cause harm to an ideological opponent is hacktivism.

Identity theft This refers to the crime of creating a false identity using identity or key pieces of information of an individual such as name, social security number, and date of birth.

Impersonation It is a key element of social engineering that involves pretending to be authorized to enter a secure location.

Internet service provider This refers to any company that provides users with access to the Internet or other web services like the creation of websites.

Intranet It is a private network used within an enterprise or an organization for sharing information and resources among its employees.

Intrusion detection It is a security technique that is used for detecting breaches in a computer or a network.

Keystroke monitoring It is the non-destructive process of capturing and recording every keystroke of a computer user. It is often called key logging.

Leapfrog attack It is the process of using a password or a user ID obtained illegally as an initial attack to perform another attack.

Letter bomb This is an e-mail containing live data or virus intended to cause damage to the computer of the recipient.

Logic bomb It is a program designed to cause damage or payload when a particular logical condition occurs (e.g., not having the author's name in the payroll file). Logic bombs are a kind of Trojan Horse, time bombs are a type of logic bombs. Most viruses are logic bombs.

Malicious code It is a harmful computer code that is intentionally included in hardware or software for the purpose of creating potential damages.

Mail-bombing It is a malicious act of sending huge amounts of unwanted email messages to a single recipient or to a group of recipients. Mail-bombing is a form of DoS.

Malware It is a malicious program that causes damage to a computer or network. Examples include Trojan Horses, viruses, worms, logic bombs, exploits, and time bombs.

Master program It is a program that communicates with implanted zombies or slave programs on systems to launch DDoS attacks.

Offender He/She is an individual convicted of an online crime.

One-time password It is an automatic and randomly generated password that can be used only once for authenticating users during a login session or transaction.

Packet It refers to a block of data transmitted over a network.

Packet sniffer It is a program that monitors and captures data passing over a network. It is also known as packet analyser.

Password It is a secret data string used to validate the identity of a user.

Password sniffing It is the process of examining network traffic to capture passwords for performing masquerading attacks.

Payload It is a malicious program that performs unauthorized activities to execute desired tasks.

Penetration The unauthorized process of entering into a restricted system for gaining access.

Piggybacking It is the act of entering secured premises by following an authorized person through the security grid. Further, unauthorized access to information is gained by using a terminal that is already logged on with an authorized identification or ID.

Pharming It is the process of misdirecting traffic from one website to a website controlled by a hacker by changing the DNS (e.g., DNS cache poisoning) or by altering configuration files.

Phishing It is an attack that uses forged or spoofed emails or websites for duplicating an official communication or webpage to acquire logon or other confidential information of a victim that can be used for penetration, financial fraud, or identity theft.

Phreaker He/She is an individual who performs phone phreaking with computer hacking for making free calls.

Piracy It is the act of unauthorized copying, distributing, or use of right-protected software.

Pornography It is the representation of sexual behaviours that are erotic and are designed to arouse sexual interest (using books, magazines, photographs, films, and other media).

Pretty good privacy—PGP It is a freeware program designed for encryption and authentication.

Probe It refers to effort taken to collect information about the state of a computer or network for gaining unauthorized access.

Risk assessment It is the process of identifying vulnerabilities, controlling threats to resources, and analysing the likelihood of attacks on a computer system.

Root kit It is a script or set of scripts used for gaining unauthorized administrator access privileges on a system, for example, by script kiddies.

Salami theft It is an automatic technique of accumulating round-off errors or other small quantities in calculations such as withdrawal of small amounts of money from a large number of accounts.

Scavenging It is the process of using discarded storage media to determine sensitive information such as access codes and passwords. It is also known as dumpster diving.

Scripts It is a program or an automated system that is used for executing exploits especially using a scripting or macro language.

Simulation It is a program designed to mimic or reproduce the behaviour of a complex system using a computer or network of computers (e.g., inventing transactions to produce a pre-arranged bottom line in a financial report).

Smart card This is a plastic access card that contains encoded information used for data storage and user authentication.

Sniffer It is a program designed to monitor, capture, and analyse information across a network for legitimate or illegitimate purposes.

Spamming It is the process of sending junk email or an unsolicited commercial email (UCE) to many reluctant recipients in order to sell products or services or to cheat naïve customers.

Spam over instant messenger— Spim It is a type of spam delivered through instant messenger.

Spam over internet telephony—Spit It is unsolicited bulk messages broadcasted using voice over Internet protocol (VoIP) to phones using the Internet.

Spoofing It is a technique applied to gain unauthorized access to a computer system by pretending to be an authorized user using electronic misrepresentation or incorrect identification, for example, entering the wrong IP address on a TCP/IP packet to perform DoS and DDoS attacks.

Spyware It is a malicious software meant for transferring data from a victim's computer to an offender.

Superzapping It is utility software used to access secured information while bypassing security controls. Examples of superzapping include debug programs and disk editors.

Time bomb A Trojan Horse program or a type of logic bomb waits for a specific time to perform before causing damage. It is used by dishonest employees who find out that they are to be fired.

Trojan Horse or Trojan It is an innocent-looking program that has undocumented and nefarious functions such as altering data, recording passwords for later use, sending confidential information to unauthorized persons, or opening backdoors into compromised systems.

Vandalism It is a program that performs obvious, unauthorized, malicious modification or destruction of sensitive data such as information on websites.

Virus A viral code reproduces with the host code when loaded into the memory, thereby infecting executable code, boot sectors on disks, and macro programs or resides inside programs at all times or transforms a simple program into an unintended Trojan Horse.

Vulnerability It is a flaw permitting an attack on a computer system or network to reduce security controls.

Wiretapping It is an electronic surveillance that performs eavesdropping on data or voice transmissions by attaching unauthorized component or software to the communications medium or by intercepting and interpreting broadcast data.

Worm It is a destructive program that spreads through a system by replicating itself.

Zombie It is a program inserted into a vulnerable system that awaits remote instructions to perform malicious actions. It is usually a part of DDoS attacks.

POINTS TO REMEMBER

- Cybercrime refers to any activity in which computers or networks are tools, targets, or places of criminal activities.
- Cybercrime includes offences such as intrusion into a computer system, unlawful interception and interruption of data and systems, misuse of data and computer systems (hacking), forgeries and frauds committed using computer (phishing), dissemination of unauthorized content, and copyright infringement.
- ICT can be used as a tool to commit a criminal offence, a target of a criminal offence, a device to store the evidence of a criminal offence, and an assistance to law enforcement authorities.
- The two elements of crime are actusreus and mens rea.
- Cybercriminals can be classified as mules, toolkit newbies, activists, nation state actors, cyberpunks, internals, coders, and professionals according to their expertise.
- Cybercriminals can be classified based on their intention as black hats, gray hats, and white hats.
- Cybercrimes are categorized as cybercrime against individuals, cybercrime against properties, and cybercrime against the nation.
- The tools used to commit crime are proxy servers and anonymizers, phishing, malware, keyloggers, password stealers and spyware, viruses and worms, Trojans and backdoors, steganography, DoS and DDoS attacks, SQL injection, buffer overflow, cracking, data diddling, rootkit, salami attack, sniffer, social engineering, spoofing, rogue security software, pharming, hijackware, man-in-the-middle attack, and watering hole.
- Some crimes associated with mobile ECDs are handset theft, SMS crimes, flashing SMS, altering dates in SMS, SMS spoofing, Bluetooth mobile hacking, crimes with calls, SIM card cloning, and MMS crime.
- Factors influencing cybercrime include the availability of tools to commit crime, no necessity of physical presence to commit crime, less investment, availability of forensic tools, and jurisdictional concerns of cybercrime.

KEY TERMS

Activists This refers to people who make use of the Internet to promote religion or politics or other causes but do not possess the skills of hackers.

Coders This refers to people who develop codes to damage other systems.

Computer-related forgery This refers to the manipulation of digital documents, for example, altering electronic images and text in the documents.

Cracking This refers to breaking into the victim's computer system often on a network, bypassing passwords or licences in computer programs or in other ways intentionally breaching the victim's computer.

Cross-site scripting (XSS) attack This is meant to compromise the users (victims) of a website so that the attacker can gain access to the users' cookies, session IDs, and passwords.

Cyber laundering This refers to the act of electronically transferring illegal money without revealing the source and possibly the destination.

Cyber warfare This refers to a computer-based or network-based conflict between countries, wherein the countries attempt to disrupt the activities of the other especially for military purpose and cyber espionage.

Cyberpunks This refers to people who have the capability to develop small programs and use them for defacing, spamming, or stealing credit cards.

Data diddling This refers to altering the data to be stored in a computer system before or during entry and changing it after processing is done.

Extortion or digital blackmail This is defined as a crime which aims at damaging the reputation of an individual or an organization for exploiting money or any other benefit unlawfully.

Harassment This refers to a threat that uses online communication facilities such as email, social media, apps, and websites to cause emotional distress to the targeted individuals. Some online harassments have even led to death and suicide.

Identity theft This refers to fraudulently acquiring the identity of an individual (e.g., SSN, date of birth, passport number, address, phone number, passwords, etc.) and the misuse of the same.

Impersonation This refers to creating a web page or sending email or instant messages over the Internet by the offender using either the name or domain name of the victim with the intention to harm, defraud, intimidate, and threaten the other person/s.

Internals This refers to disgruntled employees who seek revenge with the goal of disturbing the security of information and the system of an organization potentially damaging them.

Internet grooming This is the process of befriending children for sexual reasons, such as abuse or exploitation, over the Internet. Groomers persuade young children to send sexually explicit images, take part in sexual activities via webcam, or have sexual conversations online.

Malvertising This refers to a type of crime where the attacker inserts malicious advertisements on websites from where the victims download the malicious code by clicking on some advertisement without being aware of the fact that the website is infected.

Misuse of devices This refers to the devices used to commit cybercrime. Today any computer with Internet access can be used to commit such crimes.

Mules This refers to people who do not know that they are part of a criminal gang involved in money laundering.

Nation state actors This refers to people who work for one government to disrupt its enemies' IT infrastructure, steal sensitive information, and create other incidents intentionally.

Professionals This refers to people who associate themselves with cybercrime as a profession and use technology to hide the occurrence of crime.

Ransomware This is a malware-based attack which, when entered in a system, encrypts the files using public-key encryption and makes the data unavailable. The data can be decrypted with the secret key or the encryption key available with the hacker, which will be provided after paying a huge ransom.

Skimming This refers to a card-related fraud where a handheld device called skimmer captures information (e.g., name of the card holder, the card number, and the date of expiry from the card) and transfers it into the attacker's computer later so as to use it to commit a crime.

SMS spoofing This refers to masquerading the identity of an offender and sending messages to the victim pretending to be a genuine user.

Sniffer This refers to an eavesdropping program that is legally used by government surveillance agencies and by telecom companies to identify network bottlenecks.

Social engineering This refers to a hacker term that involves non-technical intrusion for deceiving or manipulating unwitting people into giving out information about a network or how to access it.

Spoofing This is an attack where the attacker masquerades data packets, IP addresses, MAC addresses, and even email addresses to pretend that they are originating from a genuine source.

Stalking This refers to a crime that relies upon electronic communication for sending threatening emails or messages to follow and harass the victim persistently.

Toolkit newbies or getaways This refers to novices who have limited technical skills and they use the readily available software following the guidelines and the documentation of it.

Vishing Voice phishing (vishing) is used to refer to crimes involving a social engineering approach wherein either information or money is stolen from a victim using the telephone network.

MULTIPLE-CHOICE QUESTIONS

1. Cybercrime is also referred to as _____.
   (a) computer crime
   (b) computer-related crime
   (c) high-tech crime
   (d) all of these
2. An electronic communication device and ICT act as an assistance to _____.
   (a) store digital evidence
   (b) law enforcement authorities
   (c) commit criminal offence
   (d) (a) and (b)
3. _____ enables attackers to gain control over

the computer by exploiting the vulnerabilities in the software.
(a) Web jacking
(b) XSS attack
(c) Theft of FTP passwords
(d) Exploit kits

4. Which type of cybercriminals have the highest impact?
(a) Large organized cyber networks
(b) Disreputable but legitimate organizations
(c) Foreign intelligence services
(d) Small group of opportunistic cybercriminals

5. The factor(s) which influence(s) cybercrime is/are _____.
(a) availability of forensics tools to mask the crime
(b) impact of social media
(c) (a) and (b)
(d) high investment to commit a crime

6. Data espionage is an example of _____
(a) cybercrime against person
(b) cybercrime against property
(c) cybercrime against nation
(d) all of these

7. _____ is the unauthorized use, copy, or distribution of software.
(a) Software piracy
(b) Software crack
(c) Trademark-related offences
(d) Clone

8. _____ is an SMS-related crime wherein the attacker dupes the victims with messages in such a way as to reveal their personal data.
(a) SMS spoofing
(b) Altering dates in SMS
(c) Flashing SMS
(d) SMiShing

9. _____ is a tool that attempts to make an activity on the Internet untraceable.
(a) Malware
(b) Keylogger
(c) Phishing
(d) Anonymizer

10. _____ is an eavesdropping program that is legally used by government surveillance agencies and by telecom companies to identify network bottlenecks.
(a) Rootkit
(b) Data Diddling
(c) Sniffer
(d) Spoofing

11. _____ pose hardship for the law enforcement agencies to control and curb crime.
(a) Hardware, software, and Internet access
(b) Tremendous growth of the Internet and technologies
(c) Search engines
(d) All of these

12. _____ is the national nodal agency set up to respond to computer security incidents.
(a) Information Security Task Force (ISTF)
(b) National Security Council
(c) Indian Computer Emergency Response Team (CERT-In)
(d) Both (a) and (b)

13. A web portal _____ was launched for providing training and resources.
(a) www.infosecurityawareness.in
(b) www.infosecawareness.in
(c) www.informationsecawareness.in
(d) www.infosecaware.in

14. India signed a Convention on Mutual Assistance in Criminal Matters in 2008 to operationalize _____.
(a) Australian Cyber Security Centre (ACSC)
(b) National Crime Agency's National Cybercrime Unit (NCCM)
(c) International Crime Agency's National Cybercrime Unit (ICCM)
(d) Mutual Legal Assistance Treaties (MLAT)

15. A _____ lab was set up at the Training Academy of the Central Bureau of Investigation (CBI) to impart basic and advanced training in cyber forensics and investigation of cybercrimes to police officers in India.
(a) cyber forensics training
(b) cybercrime analysis training
(c) cyber investigation training
(d) none of these

REVIEW QUESTIONS

1. Define cybercrime.
2. Define and give the differences among hacker, phreaker, and sniffer.
3. Discuss the classes of cybercriminals.

4. What do mens rea and actusreus mean? How are they applicable to cybercrime?
5. Explain briefly the forms of cybercrime.
6. Discuss in detail the tools used in cybercrime.
7. What is a mobile crime? List its various forms.
8. Mention and explain the problems that influence the growth of criminals on the Internet and cybercrime.
9. Explain the challenges that mitigate the impact of cybercrime.
10. Briefly discuss the anti-cybercrime strategies.

APPLICATION EXERCISES

1. Having looked at the biggest cybercrimes across the globe, discuss the impact of any sort of cybercrime that persists even today but which has not been discussed in this chapter.
2. List some of the common cybercrimes and tabulate their mens rea and actusreus.

BIBLIOGRAPHY

1. Kebay, M.E. (2008), *Glossary*, available at: http://www.mekabay.com/overviews/glossary.pdf (Accessed 24 November 2017)
2. Dealers-insurance, *Glossary of Cyber Crime Terms*, available at: https://dealers-insurance.com/glossaryofcybercrimeterms.php (Accessed 24 November 2017)
3. Wikipedia, *Cybercrime*, available at: https://en.wikipedia.org/wiki/Cybercrime (Accessed 24 November 2017)
4. Chiesa R., Ducci S., Ciappi S. (2009), *Profiling Hackers. The Science of Criminal Profiling as Applied to the World of Hacking*. New York. Auerbach Publications
5. Detica (2011), *Detica and Office of Cyber Security and Information Assurance. The Cost of Cybercrime, Cybercrime 2011*, available at: https://www.gov.ik/government/uploads/system/uploads/attachment_data/file/60943/the-cost-of-cyber-crime-full-report.pdf (Accessed on 24 November 2017)
6. ITU (September 2012), *Understanding Cybercrime: Phenomena, Challenges and Legal Response, Telecommunication Development Sector*, available at: http://www.itu.int/ITU-D/cyb/cybersecurity/docs/Cybercrime%20legislation%20EV6.pdf (Accessed 24 November 2017)
7. RSA (2016), *2016: Current State of Cybercrime*, available at: https://www.rsa.com/content/dam/rsa/PDF/2016/05/2016-current-state-of-cybercrime.pdf (Accessed 24 November 2017)
8. IANS (19 November 2016), *Over 55% of Millennials in India hit by Cybercrime: Report*, available at: http://indianexpress.com/article/technology/tech-news-technology/over-55-millennials-in-india-hit-by-cybercrime-report-4384050/ (Accessed 24 November 2017)
9. Shruti Dhapola (18 November 2016), *Indian Cybercrime Victims Don't Learn from Past Experience: Norton Report*, available at: http://indianexpress.com/article/technology/tech-news-technology/indian-users-complacent-when-it-comes-to-cyber-security-norton-report/ (Accessed 25 November 2017)
10. NDTV, http://www.ndtv.com/topic/cyber-crime-in-india/news (Accessed 25 November 2017)
11. Craig Barker and John Grant (2006), *International Criminal Law Deskbook*, Cavendish Publishing Limited
12. eBusiness: Free ecommerce textbook—Complete guide to ebusiness and the digital economy, *Cyber Crime*, available at: http://www.ecommerce-digest.com/cyber-crime.html (Accessed 25 November 2017)
13. ISACA (2016), *State of Cybersecurity: Implications for 2015*, available at: https://www.isaca.org/cyber/Documents/State-of-Cybersecurity_Res_Eng_0415.pdf (Accessed 25 November 2017)
14. Digit, *The 12 Types of Cyber Crime*, available at: https://www.digit.in/technology-guides/fast-track-to-cyber-crime/the-12-types-of-cyber-crime.html (Accessed 25 November 2017)
15. Anand, K. (21 November 2014), *Types of Cybercrime Acts and Preventive Measures*, available at: http://www.thewindowsclub.com/types-cyber-

crime (Accessed 25 November 2017)
16. The Associated Chambers of Commerce and Industry of India (2015), *Strategic National Measures to Combat Cybercrime: Perspective and Learnings for India August 2015*, available at: http://www.ey.com/Publication/vwLUAssets/ey-strategic-national-measures-to-combat-cybercrime/$FILE/ey-strategic-national-measures-to-combat-cybercrime.pdf (Accessed 25 November 2017)
17. National Crime Records Bureau, Ministry of Home Affairs (2016), *Crime in India 2015 Statistics*, available at: http://ncrb.nic.in/StatPublications/CII/CII2015/FILES/Statistics-2015_rev1_1.pdf (Accessed 25 November 2017)
18. Dan Elsom (31 July 2017), *Attack of the Hack: Five of the Worst Causes of Cybercrime the World Has EverSseen – From Data Theft of One BILLION Yahoo Users to Crippling the NHS*, available at: https://www.thesun.co.uk/tech/4120942/five-of-the-worst-cases-of-cyber-crime-the-world-has-ever-seen-from-data-theft-of-one-billion-yahoo-users-to-crippling-the-nhs/ (Accessed 25 November 2017)
19. EC–Council (07 March 2017), *10 Biggest Cyber Crimes and Data Breaches Till Date*, available at: https://www.eccouncil.org/10-biggest-cyber-crimes-data-breaches-till-date/ (Accessed 25 November 2017)
20. Matt Cartoons (26 January 2016), What *Are the Six Types of Cybercriminals Identified by BAE?*, available at: http://www.telegraph.co.uk/news/2017/04/04/six-types-cybercriminals-identified-bae/ (Accessed 25 November 2017)
21. Shodhganga, available at: http://shodhganga.inflibnet.ac.in/bitstream/10603/7829/12/12_chapter%203.pdf (Accessed 25 November 2017)
22. Legal_dictionary, *Mens rea*, available at: https://legal_dictionary.thefreedictionary.com/mensrea (Accessed on 25 November 2017)
23. G.O. (D). No, 20 from Information Technology (D4) Department dt.11.07.2017.
24. Matt Burgess (04 October 2017), *That Yahoo Data Breach Actually Hit Three Billion Accounts*, available at: http://www.wired.co.uk/article/hacks-data-breaches-2017 (Accessed 25 November 2017)
25. Justine Bayod Espoz (2017), *Corporate Espionage Definition*, available at: htps://bizfluent.com/about-6742556-corporate-espionage-definition.html (Accessed 28 January 2018)
26. Techopedia, *Facebook Stalking*, available at: https://www.techopedia.com/definition/27873/facebook-stalking (Accessed 28 January 2018)
27. Elise Moreau (2018), *Internet Trolling: How Do You Spot a Real Troll? How Internet Trolling Affects us All Online*, available at: https://www.lifewire.com/what-is-internet-trolling-3485891(Accessed 28 January 2018)
28. Vaishali Lavekar (2018), *New Techniques Developed to Check Video Forgeries*, available at: http://www.thehindubusinessline.com/news/science/new-techniques-developed-to-check-video-forgeries/article9968569.ece (Accessed 28 January 2018)
29. Antivirus Protection & Internet Security Software | Kaspersky Lab UK (2016), *4 Ways to Hack an ATM*, available at: https://www.kaspersky.com/blog/4-ways-to-hack-atm/13126/(Accessed 28 January 2018)
30. Police UK-Action Fraud (National Fraud & Cyber Crime Reporting Centre), *Pyramid Scheme Fraud*, available at: https://www.actionfraud.police.uk/fraud_protection/ pyramid_schemes (Accessed 28 January 2018)
31. Dann Albright (2015), How *Credit Card Fraud Works, And How to Stay Safe*, available at: https://www.makeuseof.com/tag/credit-card-fraud-works-stay-safe/ (Accessed 28 January 2018)
32. MumbaiMirror (2013), *12 Ways To Protect Yourself From Cyber Crime*, available at: https://mumbaimirror.indiatimes.com/mumbai/other/12-ways-to-protect-yourself-from-cyber-crime/articleshow/20025280.cms?prtpage=1(Accessed 28 January 2018)

Answers to Multiple-choice Questions

1. (d)	2. (b)	3. (d)	4. (c)	5. (c)
6. (b)	7. (a)	8. (d)	9. (d)	10. (c)
11. (b)	12. (c)	13. (b)	14. (d)	15. (a)

CLASSIFICATION OF CYBERCRIME

3

Learning Objectives

This chapter provides an overview of the classification of cybercrime. The objective of this chapter is to offer a deeper insight into each of the common cybercrimes and its various forms. The chapter presents the methods to prevent and combat those cybercrimes. For certain cybercrimes, the best practices are provided. In addition, sample cases reported for each category of cybercrime are highlighted. The reader will be familiar with the following after studying the chapter:

- Classification of cybercrime
- In-depth insight of each cybercrime
- Preventive measures to be adopted for every cybercrime
- The best practices
- Cybercrime incidents

3.1 INTRODUCTION

This chapter presents the classification of cybercrime, as categorized in Chapter 2—cybercrime against individuals, cybercrime against property and organization, and cybercrime against nation and society—with applicable sample cases for almost all the cybercrimes.

Each cybercrime category uses a variety of methods and the methods used vary from one criminal to another.

Cybercrime against individuals This class of cybercrimes include cyber stalking, trafficking, grooming, etc. Today, the law enforcement agencies view this category of cybercrime very seriously and join forces internationally to reach and arrest the perpetrators.

Cybercrime against property This includes a wide variety of crimes, for example, stealing a person's bank details and siphoning off money, misusing credit cards to make numerous purchases online, running scams to get naive people into parting with their hard-earned money, using malicious software to gain access to an organization's website, or disrupting the systems of the organization. The malicious software can damage software and hardware, just as vandals damage property in the offline world.

Cybercrime against nation Cybercrimes against the government are referred to as cyber terrorism. If done or committed successfully, this category can wreak havoc and cause panic among the civilian population. In this category, criminals hack into government or military websites, or circulate propaganda. The perpetrators can be terrorist outfits or unfriendly governments of other nations.

For each of these crimes, appropriate measures to combat and prevent them are also highlighted.

3.2 CYBERCRIME AGAINST INDIVIDUALS

This section presents some of the cybercrimes against individuals, the ways to handle and prevent them. It can include harassment through emails, stalking, etc.

3.2.1 Internet Grooming

Internet grooming refers to the act of deliberately befriending and establishing an emotional connection with a child for sexual activities. This is usually done after establishing communication with a child privately by showing images of pornography so as to undermine his/her reluctance to participate in a sexual encounter or other activities and prevent the victim from seeking protection from his/her parents and teachers. This is then used in exploiting his/her popularity through social networking sites and chat rooms.

Handling The victim must do the following to book the perpetrators of these crimes:

1. All the evidence, such as screen captures, messages, and conversations, should be saved in order to prove the crime.
2. Support from law enforcement agencies should be obtained by lodging a complaint if harassment persists.

Preventive measures The following are some of the ways of preventing Internet grooming:

1. The computer should be protected with antivirus and strong passwords and kept malware free. Passwords should be changed often. All software should be updated. This is to avoid personal information from being stolen.
2. Personal information should not be uploaded on social networking sites, as it can be compromised.
3. Children should not yield to blackmail and seek the support of adults if information is stolen.

Case 1: Internet Grooming		
State : Tamil Nadu	City: Tirupur	Section of Law: Section 67B(c) of Information Technology Amendment Act, 2008
A thirteen-year-old girl was enticed, kidnapped, and raped by a twenty one-year-old man in Tamil Nadu. The rapist had first befriended the victim on Facebook and sexually assaulted her after having gained her trust online.		

3.2.2 Cyber Stalking

Cyber stalking is defined as a threatening activity or the repeated following of an individual in an effort to cause fear, using electronic communication. Cyber stalking is less dangerous than physical stalking. Examples of cyber stalking include tracking the online activity of an individual, sending a stream of threatening emails, etc. It is different from cyber harassment which focuses on activities like the assassination of one's character online. The characteristics of cyber stalking are false accusations, monitoring, threats, identity theft, data destruction, and data manipulation. Cyber stalking can be prevented by not revealing personally identifiable information on online profiles.

Types

The types of cyber stalking vary with the kind of offenders involved and can be classified as follows:

Delusional stalker These types of Internet stalkers have a history of mental illness and personality disorder. Their stalking and harassment characteristic reflect their pre-existing mental disorder.

Intimate stalker An intimate stalker cannot accept the end of his/her relationship and therefore begins to cyber-stalk the partner.

Vengeful stalker A vengeful stalker is the most dangerous among the cyber stalkers as they operate with the motive to cause misery to the victim with acts to publicly shame or create false stories to assassinate the character of the victim, his/her friends, and family.

Erotomanic stalker The erotomanic stalker believes he/she is in love with the individual and the individual loves him/her back. The victim is usually of a higher status than the stalker. Erotomanic stalkers are referred to as celebrity stalkers or obsessed fans.

Trolling stalker The trolling stalker posts violent statements which may be incomplete, irrelevant, or controversial. Trolling stalkers steal information, mimic behaviours, and copy actions.

Predatory stalker The predatory stalkers are sexual predators who are motivated by sexual gratification. Usually, these types of stalkers prey on females and children by engaging themselves in some activities such as obscene phone calls, monitoring Internet activities, and attempting to draw attention and admiration.

Handling The victim must do the following to book the perpetrators of these crimes:
1. All relevant documents related to cyber stalking must be documented in the form of screenshots.
2. A hard copy of the documents must be preserved so that there is a duplicate copy of the records.
3. Support from law enforcement agencies should be obtained by lodging a complaint if harassment persists.

Preventive measures The following are some of the ways of preventing cyber stalking:
1. The primary email account should only be used for communicating with trusted people.
2. Filtering options in email may be set up to prevent delivery of unwanted messages.
3. The online name used should not reveal one's identity and should be different from the original and gender neutral.
4. Investing in antispyware software and installing it on the electronic communication devices (ECDs) can minimize cyber stalking.
5. Security feature on wireless hubs or routers should be turned on and all privacy and security settings should be reviewed periodically.

Case 2: Cyber Stalking		
State: Haryana	City: Delhi	Section of Law: Section 509 of the Indian Penal Code (IPC) and Section 66 of Information Technology Amendment Act (ITAA) 2008.
An employee in New Delhi received a series of emails from an offender asking her to either pose in the nude for him or pay ₹1 lakh. The accused also threatened to put her morphed picture on display on sex websites along with her phone number and address. The accused also hacked her email account and accessed her pictures.		

3.2.3 Cyber Harassment

Cyber harassment is the use of information and communications technology (ICT) to harass, control, manipulate, or habitually disparage a child, adult, business, or group without direct or implied threat or physical harm. It takes the form of one or more of the following: verbal, sexual, emotional, or social abuse with the intention to exert power and control over the targeted victim(s).

An attacker performing cyber harassment on a victim will usually post undesired contents in the form of text, image, and graphics on social media, email, and instant messengers. It may even take the form of hate speech where the attacker uses his/her online persona to promote hate speech against a race, sex, religion, etc., or portray sexual or pornographic content about the victim on social media websites, or indulge in continuous abuse against the victim which eventually leads to suicide (Box 3.1).

Handling The victim must do the following to book the perpetrators of these crimes:
1. Avoid responding to the attacker.
2. Proof of the incident can be prepared by making a copy of the message, photo, or video, the URL of the website, or a screenshot of the webpage as applicable.
3. The website operators may be contacted by phone, email, or through any contact submission forms that are available on the site to take the content down immediately.
4. A case can be filed through the local law enforcement agency.

> **Box 3.1 Relationship among Cyber Harassment, Cyber Bullying, and Cyber Stalking**
>
> Cyber bullying is a term to describe cyber harassment when minors are involved. It refers to the use of ICT, primarily electronic devices, as the tool for information dissemination by minors to tease, deprecate and defame, humiliate, taunt, and disparage one another. It is usually done by the attacker who is a minor for peer acceptance and recognition and does not involve face-to-face contact.
>
> Cyber stalking is a term describing cyber harassment when direct or implied physical harm, habitual surveillance, and information gathering are done by the attacker on victim(s) who may be a child, adult, business, or group with the intention to control, manipulate, or threaten them.

Preventive measures   The following are some of the ways of preventing cyber harassment:
1. Avoiding contact initiation with unknown people and participating on forums or sites that encourage anonymous posts
2. Privatizing accounts so that only known people can access information, posts, and photographs
3. Avoiding personal information on social media accounts
4. Using Google Alerts to facilitate email notifications so that the name of the individual or relative appears online

3.2.4 Cyber Extortion

Cyber extortion is a cybercrime where cybercriminals take advantage of the vulnerability of intellectual property or threaten to release potentially embarrassing information publicly or encrypt data to render it useless to the right owners and hold it for ransom. In other words, it is the act of cybercriminals demanding payment through the use of or threat of some form of malicious activity against a victim, such as data compromise or denial of service (DoS) attack. Cyber extortion permeates actions such as ransomware, email ransom campaigns, and distributed denial of service (DDoS) attacks. Ransomware is a malware that typically enables cyber extortion for financial gain. More information on ransomware is presented in Chapter 4.

Handling   The victim must do the following to book the perpetrators of these crimes:
1. An incident response plan, if already available, can be activated at the onset of an attack.
2. The affected computer, server, or any networking equipment may be disconnected from the production environment rather than turning it off.
3. An attempt may be made to restore data from backup sources if one such is already in place.
4. Analysis of vulnerabilities in systems that were the cause for cyber extortion should be done to prevent any further attack and damage.

Preventive measures   The following are some of the ways of preventing cyber extortion:
1. Encryption of data can safeguard it from being exposed to cybercriminals. However, any unauthorized encryption could render the data inaccessible instead of protecting it.
2. Prompt backup of data can prevent tampering, for example, covet encryption. However, it cannot prevent data from being released from the originals stolen by the attackers.
3. Risk assessment and the mitigation policies in force can minimize cyber extortion.

Case 3: Extortion [Satheesh (2009)]		
State: Maharashtra	City: Mumbai	Section of Law: 420, 465, 471, 474 of the IPC
A former executive of Ambuja Cements (accused) posed as a woman and seduced an Abu Dhabi-based man (victim) online through a fake email ID. The accused trapped the victim in a cyber-relationship by sending emotional messages and indulging in online sex. Later, he sent an email stating that "she would commit suicide" if he (the victim) ended the relationship. The accused gave the victim another email ID claiming it to be his friend's ID but was actually his second fake ID. When the victim mailed the second fake ID, he was shocked to know that the accused (as implied by the first mail ID) had died. Then the accused started emotionally blackmailing the friend by calling up the victim to say that the police were searching for him (the victim). Then, the victim asked the accused to arrange for a good advocate and deposited a few lakhs in the account of the accused. The accused also sent emails posing as court and police officials to extort some more money. The situation ended when the victim finally lodged a complaint with an investigating agency.		

3.2.5 Online Pedophilia

Pedophilia is an offence which is committed by men who are more than 16 years of age. The offenders are known as pedophiles. The term 'online pedophile' refers to individuals whose interactions are focused essentially on child pornography in cyberspace. These pedophiles are generally sexually attracted towards children who have not yet attained puberty. The victims of pedophilia are generally girls. The age difference between the offender and the victims in pedophilia cases is generally more than 11. They lure the victims into sexual exploitation. These offenders are sometimes considered to be suffering from a psychiatric, mental disorder known as 'pedophilic disorder'. Child pornographers use the Internet's ease of distribution to sell their materials to pedophiles. In addition to purchasing child pornography, pedophiles also visit online chat rooms hoping to lure children into sex.

Online pedophiles can be categorized as *collector* who keeps in touch with other Internet users or pedophiles, *collector and distributor* who share child pornographic content and provides technical advice on how to avoid being penalized, *the predator or the abuser* who makes ads and distributes child pornographic content that he may even be part of, and *occasional user* who views but never downloads such pornographic content from cyberspace.

Handling The victim must do the following to book the perpetrators of these crimes:

1. Internet service providers (ISPs) and providers of social media should report incidents of child pornography to the police. Besides, steps should be taken to facilitate police surveillance and detection which will help address problems related to online pedophilia investigations. Collaboration between police forces governed by different laws is also required.
2. Promotion of civic action so that Internet users report websites, images, and speech deemed to be illegal and/or tendentious to the authority to take appropriate legal action.
3. Awareness and education campaigns targeting children, parents, and teachers to develop critical thinking about the use of the Internet and on educating young people about sex should be promoted, rather than promoting fear of online predators.

Preventive measures The following are some of the ways of preventing online pedophilia:

1. Children should be educated about online predators, the tactics they adopt, and the potential dangers.
2. Parental control software should be installed to monitor the usage of computer systems by children. Computers should be placed in the family room rather than in an isolated place.
3. Usage of computers by children should be limited. Internet usage, especially, should be monitored.
4. Children should not be allowed to have their own email account. They should be advised to not download images form unknown sources and not reveal their personal information online. They should be encouraged to keep a gender-neutral online name. Further, they should be taught to approach a well-known family member with whom they are comfortable with, in the event they are caught by online predators.

3.3 CYBERCRIME AGAINST PROPERTY

This section highlights cybercrime against property and the ways to handle and prevent them. This category of crime includes computer vandalism, unauthorized possession of information, and gaining access to unauthorized information.

3.3.1 Illegal Access—Hacking and Cracking

Hacking/Illegal access is the act of breaking into protected systems, networks, and files of an individual or enterprise. A hacker is an expert in a particular programming language or a geek who attempts to access someone's system. The term hacker and cracker can be used interchangeably. The two forms of hackers are knowledge seekers and digital intruders. Hackers also commit various cybercrimes such as data espionage and data manipulation. Conventional hackers enjoy learning the details of programming languages for creating innovation in the field of computer technology. The objective of hacking is to inform administrators about security holes. The following are the three main factors that contribute to hacking:

1. Lack of protection of computer systems
2. Automated tools for attacks
3. Growth of private computers

Hackers are classified into three types based on technical sophistication and motivation which are as follows:

1. *White hat hackers* identify system vulnerabilities.
2. *Black hat hackers* or *crackers* identify and exploit system vulnerabilities.
3. *Gray hat hackers* are a hybrid of white hat hackers and black hat hackers.

Modern hackers gain access to unauthorized systems for the purpose of earning money and destruction. Contemporary hacking has paved the way for non-technical experts to carry out hacking/cracking with the help of hacking software and remote access software. The major motivations of contemporary hacking are as follows:

1. Information theft (information voyeurism)
2. Intellectual challenge (knowledge mining)
3. Revenge (personal vengeance)
4. Sexual gratification (cyber stalking, cyber harassment)
5. Economical and political motives (cyber war, cyber terrorism)

Case 4: Hacking [Prateek Paranjpe]		
State: Karnataka	City: Bangalore	Section of Law: 66 and 67 of ITA Act 2008
A girl stated that she had been receiving obscene and pornographic materials on her email address and mobile phone from an individual who seemed to know a lot about her and her family and believed that her email account had been hacked.		

Preventive measures for hacking The following are the preventive measures for hacking:

- Systems should be patched and updated regularly.
- Access to system, database access should be restricted with adequate layers of security.
- Periodical auditing is essential to assess whether the adopted security measures are sufficient to withstand any attempt for hacking.

Handling hacking The following should be done in the event of hacking:

1. The victim's system should be isolated so as to prevent any further data migration from it.
2. All logs (corresponding to operating system, database, and network) on the victim's system should be preserved to characterize the attempt of hacking.
3. Any counter measures to prevent the occurrence of hacking should be enabled.

3.3.2 Illegal Data Acquisition—Data Espionage

This refers to gaining access to sensitive or confidential information by one way or the other as discussed in the following subsections.

Phishing

Phishing is a type of *social engineering attack* (the art of gaining access to systems by exploiting human psychology or tricking and manipulating people so as to gain confidential information such as passwords and financial data) causing an electronic communication to be made to a third party, often to steal user data, including login credentials and credit card numbers where the communication gives the impression that its source is some legitimate body. This happens when an attacker masquerades himself/herself as a trusted entity, and dupes a victim into opening an email, instant message, or text message. The recipient is then tricked into clicking a malicious link, which can lead to the installation of malware, the freezing of the system as part of a ransomware attack, or the revealing of sensitive information of a banking or financial nature such as credit card numbers or accounts and account-related details.

The most common form of phishing is an email, purportedly from a banking institution, lottery, or some overseas interest, which either warns of some problem with the recipient's bank account, notifies the recipient that he/she has apparently won some money, or makes some other gratuitous offer, requiring the recipient to click on a link and provide personal details to either fix the alleged problem or claim the prize or offer.

Phishing is also called carding or brand spoofing. It uses email messages that purport to come from legitimate businesses and online organizations such as eBay and PayPal, ISPs such as AOL, MSN, and Yahoo, and the messages may look official and authentic, featuring corporate logos and formats similar to the ones used for legitimate messages. It asks for verification of certain information, such as account numbers and passwords, allegedly for auditing purposes. When recipients respond to them unsuspectingly, it results in financial losses by way of unauthorized purchases, stealing of funds, identity theft, and other fraudulent activities.

Phishing is also used to gain entry into corporate or governmental networks as a part of a larger attack, such as an advanced persistent threat (APT) event. This is done by inducing employees into bypassing security perimeters, distributing malware inside a closed environment, or gaining privileged access to secured data.

Phishing attacks also rely on social networking techniques applied to email or other electronic communication methods, including direct messages sent over social networks, SMS text messages, and other instant messaging modes, social engineering and other public sources of information to gather background information about the victim's personal and work history, his/her interests and activities, uncover names, job titles, and email addresses of potential victims, as well as information about their colleagues and the names of key employees in their organizations. This information is then used to draft a believable email (Boxes 3.2 and 3.3).

Clues to Identify Phishing Email

Some of the clues which can be used to discover a phishing attack are as follows:
1. Any mail with an irrelevant company name and no affiliation can be a fake email.
2. Improper spelling and grammar are almost always a clue and so it is essential to look for obvious errors.
3. Any mail from a relevant source but without the user account details can be a fake email.
4. Email requesting an immediate response or a specific deadline can be a phishing attack.

Types

This subsection lists the various forms of phishing:

Deceptive phishing It is the most common type of phishing scam wherein an attacker attempts to steal people's personal information or login credentials, impersonating a legitimate company. For example, PayPal scammers sent out an attack email that instructed clicking on a link to rectify a discrepancy with the user account. In actuality, the link led to a fake PayPal login page that collected user's login credentials and delivered them to the attackers.

Box 3.2 Uttar Pradesh Special Task Force (UP STF) busts gang issuing fake biometric cards [23]

Damning details related to Aadhaar card security have emerged after the Uttar Pradesh Special Task Force arrested 10 members of a gang allegedly involved in issuing fake biometric cards. The gang had not only hacked into the secure source code to access the application but also cloned fingerprints of authorized issuing authorities using gelation gel, laser, and silicon.

Box 3.3 Lazarus

Lazarus is a notorious cyber espionage and sabotage group known for attacking manufacturing companies, media, and financial institutions in at least 18 countries around the world since 2009. The malware by the Lazarus group was behind the devastating cyber attacks, including the $81 million heist of Central Bank of Bangladesh in August 2016, and several other attacks against banks worldwide. The attack formula used by hackers to penetrate targeted systems and gain entry into the critical ones comprises an initial compromise, wherein a single system within a bank is breached, either with remotely accessible vulnerable code or through a watering hole attack. This is followed by establishment of a foothold, whereby persistent backdoors are deployed in other bank hosts. Then, internal reconnaissance begins wherein the group learns about the network and other valuable resources, for example, backup server, mail server, etc., over a period of time. As the last step, malware is deployed which is capable of bypassing the internal security features of financial software and issue rogue commands on behalf of the bank.

The Lazarus group had earlier attacked Sony Pictures in the United States in 2014 and is best known for catastrophically interrupting businesses in response to the movie 'The Interview' which depicted the assassination of North Korean Leader Kim Jong Un. It had also caused destructive attacks on the South Korean government systems and has been responsible for a series of digital bank robberies involving millions of dollars since 2015. Lazarus was tied to the outbreak of Wanna Cry malware in the year 2017, which took a heavy toll on international businesses, government networks, and hospitals in the United Kingdom.

Spear phishing attacks These are directed at specific individuals or companies with information specific to the victim that has been gathered to more successfully represent the message as being authentic. Spear phishing emails might include references to co-workers or executives at the victim's organization, as well as the use of the victim's name, location, or other personal information.

Whaling attacks These are a type of spear phishing attack that specifically targets senior executives within an organization, often with the objective of stealing their login credentials and thereby large sums of money. It is also called *CEO fraud*. A typical whaling attack targets an employee with the ability to authorize payments, with the phishing message appearing to be a command from an executive to authorize a large payment to a vendor when, in fact, the payment would be made to the attackers.

Pharming It is a type of phishing that depends on DNS cache poisoning, whereby a pharmer targets a DNS server and changes the IP address associated with a similar website name to redirect users from a legitimate site to the fraudulent one, and tricking users into using their login credentials to attempt to log in to the fraudulent site.

Clone phishing attacks These use previously delivered, but legitimate emails that contain either a link or an attachment. Attackers make a copy or clone of the legitimate email, replacing one or more links or attached files with malicious links or malware attachments. Because the message appears to be a duplicate of the original legitimate email, victims can often be tricked into clicking the malicious link or opening the malicious attachment. This technique is often used by attackers who have taken control of another victim's system. In this case, the attackers leverage their control of one system to pivot within an organization using email messages from a trusted sender known to the victims.

Evil twin Wi-fi attack Phishers initiate evil twin Wi-fi attack by introducing a Wi-fi access point and advertising it with a deceptive name that is similar to a legitimate access point. When victims connect to the evil twin Wi-fi network, the attackers gain access to all the transmissions sent to or from victim devices, including user IDs and passwords. Attackers can also use this vector to target victim devices with their own fraudulent prompts for system credentials that appear to originate from legitimate systems.

Voice phishing Also known as vishing, it is a form of phishing that occurs over voice communications media, including voice over IP (VoIP) or plain old telephone service (POTS). A typical vishing scam uses speech synthesis software to leave voicemails purporting to notify the victim of suspicious activity in a bank or credit account, and solicits the victim to respond to a malicious phone number to verify his/her identity, thus compromising the victim's account credentials.

SMS phishing Another mobile device-oriented phishing attack is SMS phishing which is also called SMishing or SMShing. This uses text messaging to convince victims to disclose account credentials or to install malware.

Service specific phishing This is a scam where phishers specialize their attack emails according to an individual company or service. For example, Dropbox, popularly used by users to back up, access, and share their files, has been used by attackers to target users with phishing mails. The disguised email convinces users to enter their login credentials on a fake Dropbox sign-in page hosted on Dropbox itself. Attackers have even used Google Drive to prey on Dropbox users.

Techniques

The most common technique used in a phishing attack is that of email to get the victim to follow a link that appears to be a legitimate web resource, but is actually a malicious web resource. Phishers use many tactics to trick victims, which include URL hiding, link shortening services to hide the link destination, homograph spoofing where URLS are created using different logical characters, for example, replacement of 'o' by zero in a domain name to read exactly like a trusted domain, etc. Attackers even bypass phishing defences and deliver phishing emails by rendering all or part of the message as a graphical image.

Handling The victim must do the following to book the perpetrators of these crimes:

1. Isolate the victim's ECD from the network and the Internet so as to avoid spreading of malware.
2. If antivirus software is already installed, a complete scan may be performed so as to quarantine or delete malicious files.
3. User credentials, if any, preserved in the system should be reset.
4. Information about the source of the phishing attack should be gathered from the email header in case of phishing emails and from log files. Such a host may be marked permanently in the blacklist.
5. The assistance of the incident response team and the police may be sought.

Preventive measures The following are some of the ways of preventing phishing:

1. Educating users to identify phishing messages and the tactics used by phishers can cut down successful attacks.
2. Employing a gateway email filter can trap phishing emails and reduce the number of phishing emails that reach the user's inbox.
3. Using an email authentication standard, for example, sender policy framework (SPF) protocol by enterprise mail servers to verify inbound email.

4. Employing a web security gateway provides another layer of defence by preventing users from reaching the target of a malicious link.
5. Two-factor authentication (2FA) is the most effective method for countering phishing attacks, as it adds an extra verification layer when logging into sensitive applications.
6. Avoiding email links to access sites and instead logging into the official website by opening it in a web browser.
7. Avoiding personal information from being sent through email.
8. Double checking hyperlinks to ensure if it is the actual URL, which otherwise would lead to malicious sites.
9. Avoid opening attachments that don't make sense and also from unknown senders.

> **Case 5: Phishing**
>
> A spoofed mail from myuniversity.edu was mass-distributed to many faculty members. The email claimed that the user's password was about to expire. Instructions were given to go to myuniversity.edu/renewal to renew the password within 24 hours. While attempting to renew, the users were redirected to myuniversity.edurenewal.com, a bogus page that appeared exactly like the real renewal page, where they typed both the new and existing passwords. While being redirected, a malicious script was activated in the background to hijack the user's session cookie. Thus, the attacker who monitored the page hijacked the original password to gain privileged access to secured areas of the university network.

Phreaking

Phreaking or telecommunication fraud is the process of gaining unauthorized access into secured telecommunication systems for exploiting the telecommunication services such as manipulating the phone networks and copying dialling tones. Phreak is a combination of phone and freak. Phreakers include customers, geeks, and communication service providers. Originally, the motivation of phreakers was to break the telecommunication system. The phreakers held illegal conference calls to discuss phreaking activities with their counterparts around the world and billed these to someone else.

Switch hooking is the first phreaking method which allows one to make calls by disabling the rotatory keypad. It is accomplished by pressing and releasing the switch hook to open and close the circuit quickly. War dialing is the standard phreaking method which involves random number generators to test a large number of access codes until one is positive. These codes are then stored in a database for making calls illegally. Due to the increased complexity of network security, war dialing has become outdated. The invention of the blue box is the most sophisticated method which enables free access to long-distance lines. Blue boxes are now obsolete because of digital switching technology.

Private branch exchange (PBX) is an internal telephone system that directs calls from one person to another within an enterprise. An act of breaking into the PBX system and selling long-distance calls to third parties around the world is the leading type of telecommunication fraud. Today, most PBXs are software driven, such as voice mail, maintenance port, and direct inward system access (DISA). Voice mail is used to make outbound voice calls. Phreakers make or forward voice calls to unauthorized users using the administrator's account. The PBX administrator manages the PBX system through the PBX maintenance port by connecting to the system remotely. Phreakers change the configuration and access codes of the PBX maintenance port to affect the operations of an enterprise. DISA enables remote users to access an outside line using PBX systems with authorization codes. Phreakers access the authorization code to make calls at the cost of an enterprise.

3.3.3 Illegal Interception

Illegal interception refers to listening to, monitoring, or recording any aspect of the contents of a communication originating from or destined to an ECD without the consent of either parties involved in the communication.

Spoofing

Spoofing is the act of assuming a fake identity, usually to trick another party into exchanging, submitting, or accepting either authentication-related data or other confidential data. Spoofing often includes attempts by crackers to create fake records or messages in a system, such as false bank accounts or forgeries. The variants of spoofing are given here:

Email spoofing It is the forgery of an email header, that is, the 'From' field so that the message appears to have originated from someone or somewhere other than the actual source. Email spoofing is a tactic used in phishing to get recipients to open the mail, and possibly even respond to it. Email spoofing is possible because the simple mail transfer protocol (SMTP) does not provide a mechanism for address authentication. However, solutions do exist to authenticate mail senders, for example, SPF protocol. SPF is an anti-spam approach that authenticates the sender of the mail by providing a mechanism for the receiving email exchanger to check if the incoming mail comes from a host authorized by domain administrators.

URL spoofing The website poses as another, by editing the details of the URL displayed in the web browser. URL spoofing is primarily used to conceal a pharming attack, so as to convince a victim with a false URL [Mark Johnson, (2016)].

IP spoofing It refers to the process of forging a computer's IP address and is used by an attacker to conceal his/her online identity and assume the identity of another in order to impersonate him/her [Pzimmerm, (2012)]. This is usually done by spammers as it is difficult to trace the actual owner. It is also used in DoS attacks, as the attacker is least bothered if the response messages are sent to the spoofed address. With IP spoofing, an attacker from outside the network pretends to be a trusted computer, either by using an IP address that is within the range or an external IP address that is trusted to provide access to specific resources on the network. Thus, IP spoofing attack is limited to injection of data or commands into the existing stream of data passed between a client/server application or a peer-to-peer network connection. Several tools and techniques exist to spoof a computer's IP address. For example, anonymous remailer is a service that hides the IP address of the original sender before forwarding the message to the intended recipient. However, investigators can get information about such email as the emailer service maintains a log of the emails that have been received and forwarded, but such logs are retained only for a short while.

Call spoofing Also called caller ID forging, it is the practice of causing the telephone network to display a number on the recipient's caller ID display which is not that of the actual originating station. Thus, caller ID forging can make a call appear to have come from any phone number the caller wishes.

SMS spoofing Here, the offender steals the identity of another person in the form of a mobile phone number and sends an SMS via the Internet so that the receiver gets the SMS from the mobile phone number of the victim.

Handling The victim must do the following to book the perpetrators of these crimes:
1. In case of email spoofing, a complete scan of the system with the installed antivirus software should be made. Further, the email password should be reset. This can eliminate any third-party connection that has gained control over the email account.
2. In case of call spoofing, avoid revealing personal and sensitive information. Authenticity of the source seeking information with call spoofing should be ensured.
3. For IP spoofing, it is sufficient to turn off the network components for a few minutes as this would kill the network traffic and the components can be resumed thereafter.

Preventive measures The following are some of the ways of preventing spoofing:
1. Implementing SPF can prevent email spoofing. IP *spoofing* can be *prevented* using source address verification on the router, by applying ingress and egress filtering. Ingress ensures if the data packets have come from

the original source and are not spoofed, whereas egress verifies if the data packets leaving the device are authorized to do so. Other steps that can be taken to *protect against spoofing* include the following: using encrypted authentication to prevent anyone from spoofing the IP and hijacking the data, configuring the router for traffic filtering to scan incoming and outgoing traffic, to look for suspicious signs such as malware and data leaks, and to reject any messages from outside that appear to come from an internal (local) address.
2. Protection from caller ID spoofing can be made as follows: the suspected phone number can be searched for association with any previous scams, the call can be trapped, if possible, and reported to the law enforcement agency.

Case 6: Spoofing

A branch of the erstwhile Global Trust Bank in India experienced a chaotic situation. Numerous customers decided to withdraw all their money and close their accounts. An investigation revealed that someone had sent out spoofed emails to many of the bank's customers stating that the bank was in very bad shape financially and could close operations any time. The spoofed email appeared to have originated from the bank itself.

Skimming

Skimming is a form of magnetic strip counterfeiting in which criminals are able to copy the magnetic strip and track information including card verification value (CVV) from a valid card. Information may then be encoded on a counterfeit card and used fraudulently. It is a method used by criminals to capture data from the magnetic strip on the back of an ATM card.

Skimming is done by a machine called skimmer. It is fitted in the opening of the ATM machine where the card holder inserts his/her card. When the card holder enters his/her secret PIN it is recorded in the machine and later, with the help of a fake card, the criminals withdraw the money.

Skimming uses certain other devices which are installed on machines and their parts. These include the following:

Skimming plates on ATMs These are installed over the card slot of the machine which reads the magnetic strip before the card is inserted.

PIN capturing This method employs positioning cameras or other imaging devices which are attached to ATMs to fraudulently capture PIN numbers. Once captured, the electronic data is put into a fraudulent card and the captured PIN is used to withdraw money from accounts. PIN capturing devices are normally fitted to the top of the ATM and these devices are usually difficult to detect.

Handling The victim must do the following to book the perpetrators of these crimes:
1. Immediately report this to concerned authorities.
2. The PIN should be changed.
3. The account should be monitored for suspicious activity and should be reported in case of any adverse activity.

Preventive measures The following are some of the ways of preventing skimming:
1. Inspecting the ATM machine and its area before using a card to make sure that any other devices are not installed on the machine.
2. Keeping track of the account for any undesirable activity regularly.
3. Shielding the number pad with one hand to prevent tracking of the PIN by a hidden camera.
4. Ensuring accuracy of the transaction, once done.
5. Changing the PIN often.

Besides this, old cards must be upgraded with new cards that are more secure because they have an embedded chip.

Case 7: Skimming	
State: Karnataka	City: Bengaluru
The accused (a gang of tech-savy thieves) stole at least ₹8.68 lakhs from the HDFC Bank accounts of 17 customers in Bengaluru between 31 July 2017 and 4 August 2017. The customers approached the bank, alleging the transactions happened without their knowledge or authorization. Prima facie, the accused appeared to have skimmed card details and later cloned the cards to withdraw money from ATMs.	

Computer Intrusion

Computer intrusion refers to any malicious activity that harms a computer, or causes a computer or a computer network to work in an unexpected manner. It refers to unauthorized access to a computer system, network, device, or data. The access may be either physical access or logical access. Physical access refers to breaking into the house and accessing the computer using the username and password. Logical access means that the attacker accesses the computer system, service, or data over the network. Computer intrusion is carried out with the following intentions:

1. Framing or executing a plan or artifice to defraud
2. Obtaining money, property, or services by means of fraudulent practices
3. Committing theft including but not limited to copyright-protected information

A computer virus is a tool that exploits the operating system or software vulnerability that causes denial of service.

3.3.4 Data Interference

Computer threat is the potential to cause serious damage, such as loss or corruption of data, physical damage to the hardware and to a computer, among others, which can lead to attacks and forms the basis for vulnerabilities. Computer threats may be intentional, accidental, or caused by natural disasters (Box 3.4).

Computer threat may be of two types:

Physical threat   A physical threat is a potential cause for an incident to occur and which may result in loss or physical damage to the computer system. The physical threat is classified into three types as follows:

Internal physical threat   Fire, power fluctuation, etc.

External physical threat   Natural disasters such as lightning, flood, and earthquake

Human physical threat   Stealing, vandalism, interruption and accidental or intentional errors

Logical threat   A logical threat or non-physical threat is a potential cause for an incident that may result in loss or corruption of data and sensitive information, disruption of computer system operations, and unauthorized monitoring of computer system activities. The logical threat can be any one of the following:

> ### Box 3.4 Threat, Attack, and Vulnerability
> *Threat* is the potential for the occurrence of a harmful event, for example, an attack.
>
> *Attack* is the execution of a harmful event on the target (victim), which may be either gaining unauthorized access or making unauthorized use of an asset with the intention to expose, alter, destroy, or steal data from the system or network.
>
> *Vulnerability* is a weakness that makes targets susceptible to an attack.

1. Worm
2. Trojan Horse
3. Logic bomb
4. Ransomware
5. Virus
6. Malware
7. Rootkit
8. Rogue software
9. Spyware
10. Botnet
11. Spam

Worm

A worm is a self-replicating malicious computer program that replicates itself from one computer to another in huge volume and with high speed with the intent of spreading malicious source code. For example, a worm can send copies of itself to all the contacts in the address book (email and phone contacts) without user interaction. The worm replicates itself through parts of an operating system, file attachments, file sharing networks, storage media, instant messages, and infected websites. Worms have the ability to infect computers across the globe overnight due to its high speed of infection. For example, Conficker Worm (Downadup) infected systems across the globe in no time by tripling the number of computers it infected to 8.9 million in just four days when the victims around the world switched on the system and opened their email. Morris Worm is considered the first computer worm. The most damaging computer worm was ILOVEYOU. It propagated through email attachments, IM chat, and copies of the executable file renamed with system files.

Types

Worms can be classified into four:

Bot worm A bot worm infects a computer and then changes the computer into zombies or bots for the purpose of using them in coordinated attacks through botnets.

Instant messaging worms It propagates itself through instant messaging services and exploits access to contact lists on the target computer.

Email worm Computer worms that have @m or @mm signify that the primary distribution method is through email or mass mail. Email worms are malicious executable files that spread by forcing the target system to forward the email worm to all the email addresses in the contact list but pretends to be a normal email. Email worms use social engineering methods to prompt the email recipients to open the file, thereby infecting the new computer.

Ethical worm An ethical worm is a computer worm that spreads to computers across the network by pretending to deliver security patches for fixing security vulnerabilities (Box 3.5).

Handling The victim must do the following once a worm is sensed:

1. The system should be isolated from the network.

Box 3.5 Virus, Worm, and Trojan Horse

A *virus* attaches itself to a program or file for spreading itself from one computer to another. Most viruses are attached to an executable file. Viruses cannot infect the system unless the user runs or opens the harmful program. It spreads through human actions, for example, sending emails with viruses as attachment, sharing infected files, etc.

A *worm* is a subtype of a virus. It spreads from one computer to another without human action. Worms have the capability to replicate itself from one copy to thousands of copies, leading to devastation.

A *Trojan Horse* appears as a useful software but it causes harm to the computer system once installed or run. It gives access to malicious users by creating backdoors. Unlike viruses and worms, a Trojan Horse neither spreads nor self-replicates.

2. A complete scan of the system should be performed to either disinfect or delete the worm.
3. A backup of data may be acquired, if not obtained earlier.

Preventive measures Since it is difficult to get rid of malware like worm once it infects a system, it is essential to prevent it from getting into the computer in the first place. The following are some of the ways of protection:

1. Avoid clicking on links shared on social media and email messages.
2. Install a good anti-malware protection on the computer.
3. Use personal firewall to block external access to network services.

Trojan Horse

Trojan horse, Trojan, or Trojan horse virus is a malicious software but appears harmless. It is used by cyber-criminals to gain unauthorized access into the system in the form of tricking users into installing malicious software. After the installation of Trojan Horse, it performs several actions on the user's computer, such as spying, stealing, deleting, blocking, altering, copying sensitive information, gaining backdoor access to system, and disturbing computer performance (Box 3.6).

Types

Trojan horse can be classified as follows:

Trojan-Banker Trojan-Banker steals the target user's account information through online banking, e-payment system, and credit or debit cards.

Trojan-DDoS Trojan-DDoS performs DoS attack against a targeted web address by sending multiple requests from the target computer or different infected systems.

Trojan-Downloader Trojan-Downloader downloads and installs the latest versions of malicious software such as Trojan and adware in the computer.

Trojan-Dropper Trojan-Dropper is used to install a Trojan or virus, or to prevent the identification of malicious software in the system. Not all anti-virus programs have the ability to scan this type of Trojan.

Trojan-FakeAV Trojan-FakeAV simulates all the activities of antivirus software. It is programmed to obtain money from the target user in return for the detection and removal of threats even though the threat is not present in the computer.

Trojan-GameThief Trojan-GameThief steals account information of the user through online games.

Trojan-IM Trojan-IM steals usernames and passwords for the purpose of instant messaging using Skype, ICQ, MSN Messenger, AOL Instant Messenger, Yahoo Pager, etc.

> **Box 3.6 New Malware Xafecopy Trojan Stealing Money through Phone**
>
> A new malware Xafecopy Trojan has been detected in India which steals money through the victims' mobile phones. Around 40 percent of the targets of this malware have been detected in India. Kaspersky Lab experts have uncovered a mobile malware which targets the WAP billing payment method, stealing money through victims' mobile accounts without their knowledge. Xafecopy Trojan is disguised as a useful app like BatteryMaster and operates normally. It does not require the mobile user to create a username or password, or register a debit or a credit card.

Trojan-Ransom Trojan-Ransom alters the data on the victim's system which leads to malfunctioning of the computer system. The performance of the computer or original data will be restored only after a ransom amount is paid to the criminal.

Trojan-SMS Trojan-SMS sends text messages from the target mobile phone to a premium rate phone number which can cost money.

Trojan-Spy Trojan-Spy can spy on the computer such as recording key strokes, taking snapshots, and obtaining a list of running programs.

Trojan-Mailfinder Trojan-Mailfinder gathers email addresses from the system.

Handling The following should be done when a system is infected by a Trojan:
1. The system should be isolated from the network.
2. A complete scan of the system should be performed to either quarantine or delete the Trojan.
3. A backup of data may be acquired, if not obtained earlier.

Preventive measures The following are some of the ways of preventing attacks by Trojan Horse:
1. Never open emails and attachments from unknown sources.
2. Install proper antivirus software.
3. Avoid autorun.
4. Back up data regularly.

Logic Bomb

A logic bomb is a malicious program timed to cause harm at a certain point of time, but is inactive until that point. It is triggered by a response to an event, such as launching an application or when a specific date/time is reached. Once activated, a logic bomb implements a malicious code that causes harm to a computer. Attackers can use logic bombs in a variety of ways. An arbitrary code may be embedded within a fake application or Trojan horse, and will be executed whenever the fraudulent software is launched. Attackers can also use a combination of spyware and logic bombs in an attempt to steal identity. A logic bomb is also known as slag code or malicious logic.

Time bomb is a variant of logic bomb that is programmed to execute when a specific date is reached. Disgruntled employees have made use of time bombs to execute within their organizations' networks and destroy as much data as possible in the event they are terminated.

Handling Since a logic bomb is mostly activated on a system date, changing the system date can deter its activation if its presence is sensed, before activation. Since every logic bomb is programmed with specific logic, understanding the logic can prevent any loss of data from the system.

Preventive measures Logic bombs are difficult to prevent because they can be deployed from almost anywhere. However, the following may be adopted to protect systems from logic bombs:
1. Logic bombs can be distributed by exploits that promote software piracy and so downloading of pirated software should be avoided.
2. Logic bombs can be embedded within Trojan horses. It is therefore essential to be careful while installing shareware or freeware applications as they may host one. It is advisable to ensure that applications are from a reputed source.
3. Email attachments may contain malware like logic bombs. Hence it is essential to be cautious when handling emails and attachments.
4. Most antivirus applications can detect malware like Trojan horses (which may contain logic bombs). However, antivirus software should be configured to routinely check for updates. It should possess the latest signature files to survive malware threats.

5. Installing the latest operating system patches can make the computer system less vulnerable to malware threats. The 'automatic updates' feature on Windows should always be turned on to automatically download and install Microsoft Security Updates. It is also advisable to ensure that the latest patches are installed on all software applications.

Case 8: Logical Bombs

Logic bomb sets off South Korea cyber attack in 2013
In a cyber attack, a logic bomb in the code was set to wipe the hard drives of computers and master boot record (MBR) belonging to at least three banks and two broadcasting media companies simultaneously. The logic bomb dictated the date and time the malware would begin erasing data from machines to coordinate destruction across multiple victims. The malware consisted of four files, including one called AgentBase. exe that triggered the wiping. Contained within that file was a hex string indicating the date and time the attack was to begin. As soon as the internal clock on the machine hit the specified time, the wiper was triggered to overwrite the hard drive and MBR on Microsoft Windows machines, and then reboot the system. Once the machine rebooted, users saw a message on the screens that read, 'Boot device not found. Please install an operating system on your hard disk.'
The malware also included a module for deleting data from remote Linux machines. The malware searched for remote connections and used stored credentials to access Linux servers and wipe their MBR. Prior to the cyber attack, a phishing email was sent to South Korean organizations that purported to come from a bank. It came with a malicious attachment that contained a Trojan.

Ransomware

Ransomware is a type of malware that blocks the ECD, encrypts data on it, and demands money for restoration of the facility. More information about ransomware and the prevention aspects are presented in Chapter 4.

Virus

A virus is the most common logical computer threat. A computer virus is a computer program that can infect other computer programs by modifying them in such a way as to include a copy of it. Viruses are dangerous as they spread faster than being stopped, and even the least harmful of viruses could be fatal. A computer virus might corrupt or delete data on a computer, use an email program to spread the virus to other computers, or even delete everything on the hard disk.

Computer viruses are frequently spread by attachments in email messages or by instant messaging messages. Computer viruses are also spread through downloads from the Internet. They can be hidden in pirated software or in other files or programs that are downloaded.

Types

A virus can be classified into two: *file infectors* and *boot record infectors*. File infectors attach themselves to ordinary program files. They usually infect .COM and .EXE extension programs. File infectors may work in either 'direct action' or 'resident' mode where the former selects one or more programs to infect each time an infected program is executed, whereas the latter hides itself in the memory the first time an infected program is executed and thereafter infects other programs when executed. Boot record infectors infect executable code found in certain system areas on the disk which are not ordinary files. Certain other types of viruses are listed here:

File system or cluster viruses These are types of file infectors that modify the directory table entries so that the virus is loaded and executed before the desired program gets executed.

Stealth virus It is a virus that resides in the memory and hides the modifications it made to the file or boot record. The virus monitors the system functions used by programs to read files or physical blocks from the storage media and forges the results so that programs which see the areas see the original uninfected form of the file instead of the actual infected form. Hence this virus goes undetected by antiviral programs.

Polymorphic virus This produces multiple copies of itself with the belief that the virus scanners cannot detect all instances of it.

Fast infectors These are a form of virus which when active in the memory infects not only programs that are executed but also those which are merely opened. Hence, such a virus, if present in memory, can infect even a scanner or an integrity checker.

Slow infector This refers to a virus which when active in the memory infects files as they are modified. This is done to fool integrity checkers and give an illusion that the modification is due to legitimate reasons.

Sparse infector This refers to a virus that infects only occasionally.

Companion virus It is a virus which, instead of modifying an existing file, creates a new program, and gets executed by the command-line interpreter instead of the intended program. On exit, the new program executes the original program to make things appear normal. This is done by creating an infected .COM file with the same name as an existing .EXE file.

Armored virus This is a virus that uses special tricks to make the tracing, disassembling, and understanding of its code more difficult.

Macro virus It is a virus that propagates through only one type of program, usually either Microsoft Word or Microsoft Excel. It can do this because these types of programs contain auto open macros, which automatically run when a document or a spreadsheet is opened. Along with infecting auto open macros, the macro virus infects the global macro template, which is executed anytime the program is run. Thus, once the global macro template is infected, any file that is opened later becomes infected and the virus spreads.

Handling The following should be done when a system is infected by a virus:

1. The system should be isolated from the network.
2. A complete scan of the system should be performed to either quarantine or delete the virus.
3. A backup of data may be acquired, if not obtained earlier, so as to avoid large volumes of files from getting infected.

Preventive measures The following are some of the ways of preventing viruses:

1. Unknown attachments in emails should not be opened as it may contain one or more viruses which may infect files in the computer.
2. Performing a virus scan after opening an email attachment is a must.
3. When a program or a data file is downloaded from the Internet or other shared networks, viruses might be transferred to the computer. Sometimes the free software programs on the Internet have viruses. It is therefore advisable to download files only from trusted sites.
4. It is a must to perform a virus scan immediately after a peripheral device is connected to the computer.
5. Antivirus software should be kept up to date as an outdated one can be as bad as no antivirus software.

Case 9: Virus [Prateek Paranjpe]

The VBS_LOVELETTER virus (better known as the Love Bug or the ILOVEYOU virus) was reportedly written by a Filipino undergraduate. In May 2000, this deadly virus became the world's most prevalent virus. Losses incurred during this virus attack were pegged at US $10 billion. VBS_LOVELETTER utilized the addresses in Microsoft Outlook and emailed itself to those addresses. The email which was sent out had 'ILOVEYOU' in its subject line. The attachment file was named 'LOVELETTER-FOR-YOU.TXT.vbs'. People wary of opening email attachments were influenced by the subject line and those who had some knowledge of viruses did not notice the tiny .vbs extension and believed the file to be a text file. The message in the email was 'Kindly check the attached LOVELETTER coming from me'.

Malware

Malware refers to any software that gets installed on the machine and performs unwanted tasks, often for some third party's benefit. It includes various forms of annoying software or programs such as viruses, Trojan horses, spyware, worms, and rootkits. Malware is used to do several operations such as encryption or modification of sensitive data, and unauthorized monitoring of computer activities of users. Malware seeks to destruct existing vulnerabilities in the system to make their entry easy. Malware programs can range from simple pop-up advertisements to serious computer invasion and damage, for example, stealing passwords and data or infecting other machines on the network. Some malware programs can even transmit information about web browsing habits to attackers.

Users are usually tricked into downloading malware. The primary source includes software that comes bundled with other software, that is, Trojan Horse. Peer-to-peer file sharing software also bundles various types of malware that are categorized as spyware or adware. Malware can spread through email attachments as well. Malware can exploit security holes in a browser as a way of invading computer systems.

Types

The different types of malware include spyware, adware, phishing, viruses, Trojan Horses, worms, rootkits, ransomware, and browser hijackers. Some of the categories are as follows:

Adware   Software that is financially supported or financially supports another program by displaying ads when connected to the Internet.

Spyware   Software that surreptitiously gathers information and transmits it to interested parties. Information gathered includes visited websites, browser/system information, and the computer's IP address.

Browser hijacking software   Advertising software that modifies browser settings, creates desktop shortcuts, and displays intermittent advertising pop-ups. Once a browser is hijacked, the software may also redirect links to other sites that advertise, or sites that collect web usage information.

Handling   The following should be done when a system is infected by malware:

1. The system should be isolated from the network.
2. A complete scan of the system should be performed to either quarantine or delete the malware.

Preventive measures   The following are some of the ways of preventing malware:

1. Use powerful antivirus and anti-malware software.
2. Email attachments from unknown or unexpected sources should not be opened.

Rootkit

A rootkit is a collection of programs that enable administrator-level access when installed in the computer system. After installation, rootkit gains user-level access of the target computer system, primarily by cracking a password or exploiting known vulnerabilities. Rootkits are associated with malware such as viruses, worms, and Trojan horses for concealing their presence and activities from the system user and system processes. Rootkits allow cybercriminals to mask certain objects or activities such as gaining access or intrusion and prevent the detection of malware in the system. The various activities performed by the rootkit are as follows:

1. Monitoring the network traffic of the computer
2. Monitoring the key strokes of the system
3. Creating a backdoor into the victim's computer
4. Altering log files
5. Attacking other computers on the network
6. Altering existing system files or tools

Sun and Linux operating systems were the initial targets of the rootkit. Nowadays, rootkits target a number of other operating systems. It is very difficult to detect the existence of rootkits in the system because it is activated before the operating system boots up. Once the rootkit is detected in the computer, it is essential to erase the hard drive of the affected system completely and reinstall the operating system. Some examples of rootkits are listed here:

1. NTRootkit
2. HackerDefender
3. Machiavaeli
4. Creek Wiretrapping
5. Zeus
6. Flame

Zeus and Flame are generally considered Trojan/bot. The latest variant has the capability of updating its bots/infected systems with updates that have the ability to drop a rootkit into infected systems and hide the Trojan to prevent the removal of malicious files and registry entries.

Handling Since rootkits can offer privileged access to an operating system, its presence should be curbed. Isolating the system and performing a system scan with appropriate tools can eliminate it from the system.

Preventive measures The following are some of the ways of preventing rootkits:
1. The system should be kept patched against known vulnerabilities.
2. Antivirus software should be updated and running.
3. Opening email file attachments from unknown sources should not be done.
4. When installing software before agreeing to end user licence agreements (EULAs), some may overtly state that a rootkit of some sort will be installed and this should not be encouraged.

3.3.5 System Interference—Computer Threats

System interference refers to the disruption of data or the asset (server or network) so that it is not available for use by the intended user.

DoS Attack and DDoS Attack

DoS attacks are the most costly attack listed under the 'real cost of cyberattacks'. A DoS attack tries to make a web resource unavailable to its users by flooding the target URL with more requests and beyond the limits the server can handle. That means that during the attack period, regular traffic on the website will either slow down or be interrupted completely. Cybercriminals use DoS attacks to extort money from companies that rely on their websites being accessible. Some legitimate businesses rely on underground elements of the Internet to help cripple rival websites. In addition, cybercriminals combine DoS attacks and phishing to target online bank customers and even take down the bank's website and then send out phishing emails to direct customers to a fake emergency site instead.

A distributed denial of service (DDoS) attack is a variant of DoS attack that comes from more than one source at the same time. A DDoS attack is typically generated using thousands (potentially hundreds of thousands) of unsuspecting zombie machines. The machines used in such attacks are collectively known as 'botnets' and would have previously been infected with malicious software. Hence, they can be remotely controlled by the attacker. These infected endpoints are usually computers and servers, but are increasingly IoT and mobile devices. The attackers will harvest the vulnerable systems that they can infect through phishing attacks, malvertising attacks, and other mass infection techniques. Increasingly, attackers will also rent these botnets from those who built them.

Generally, DDoS attacks work by drowning a system with requests for data. This could be sending a web server so many requests to serve a page that it crashes under the demand, or it could be a database being hit with a high volume of queries. The impact of DDoS attack can be witnessed from Internet bandwidth and from CPU and RAM capacity being overwhelmed.

Black-holing or sinkholing is the approach used to defend a DoS attack which blocks all traffic and diverts it to a black hole, where it is discarded.

Types

Typical types of DDoS attacks include bandwidth attacks and application attacks, wherein in the former, network resources or equipment are consumed by a high volume of packets, whereas in the latter, with an application attack, TCP or HTTP resources are prevented from processing transactions or requests [Paul Froutan, (2004)].

DDoS attack is executed using any one of the three approaches listed here with the primary goal of making online resources sluggish and unresponsive:

1. Generating massive amounts of bogus traffic using ICMP, UDP, and spoof-packet flood attacks to down a resource such as a website or server.
2. Using packets to target the network infrastructure and infrastructure management tools, for example, SYN floods and Smurf DDoS.
3. Targeting an organization's application layer by flooding applications with maliciously crafted requests.

Handling   The following can be done by a victim of a DDoS attack:

1. Wiping and reinstalling the operating system will help in bringing back the system to normal.
2. Turning off the networking components will also help. However, these can be performed only after acquiring logs which actually help to characterize the incident and the source of attack. This can help to avoid further attacks.

Preventive measures   A multi-layered cloud security developed and monitored by highly experienced and committed engineers offers the best protection from DDoS attack. The following can however prevent DDoS attacks [Paul Froutan, 2004]:

1. DDoS mitigation appliances are dedicated to sanitizing traffic or building DDoS mitigation functionality into devices used primarily for other functions such as load balancing or firewalling.
2. Proper configuration of server applications is critical in minimizing the effect of a DDoS attack. Combined with a DDoS mitigation appliance, optimized servers stand a chance of continued operations through a DDoS attack.
3. The network should be monitored with implementation of technology to know the amount of bandwidth pertaining to everyday activities. This would offer visual clues and predict the attacks if there is a deviation from the normal behaviour.
4. Firewalls and network security programs should be updated and patched. Firewalls can shut down a specific flow associated with an attack, but like routers, they can't perform antispoofing.
5. Routers can be configured to stop simple ping attacks by filtering non-essential protocols and can also stop invalid IP addresses.
6. It is necessary to ensure that the server capacity can handle heavy traffic spikes and has the mitigation tools needed to address security problems. Bandwidth may be added if possible. Such over-provisioning or buying excess bandwidth or redundant network devices to handle spikes in demand can be an effective approach to handling DDoS attacks.
7. Intrusion detection systems (IDS) solutions offer anomaly detection capabilities that can recognize when valid protocols are being used as an attack vehicle. They can be used in conjunction with firewalls to automatically block traffic.

Case 10: DDoS Attack

On 21 October 2016, the DDoS attack that disrupted the Internet was the largest of its kind in history. The cause of the outage was a distributed denial of service (DDoS) attack, in which a network of computers infected with special malware, known as a 'botnet', was coordinated into bombarding a server with traffic until it collapsed under the strain. Dyn, a company that controls much of the Internet's domain name system (DNS) infrastructure was the victim of the denial of service attack that was orchestrated using a weapon called the Mirai botnet [Symantec] as the primary source of malicious attack. Unlike other botnets, which are typically made up of computers, the Mirai botnet is largely made up of so-called 'Internet of Things' (IoT) devices such as digital cameras and DVR.

Email Bombing

A mail bomb is the sending of a massive amount of email to a specific person or system thus consuming system, storage, and network resources. A huge amount of mail may simply fill up the recipient's disk space on the server resulting in legitimate emails being rejected because the mailbox is out of space or, in some cases, may be too much for a server to handle and may cause the server to stop functioning.

An email bombing attack can lead to denial of service to legitimate users, loss of network connectivity, exhaustion of system and network resources, and financial loss due to disrupted online services.

Types

There are three methods of email bombing which are as follows:

Mass mailing This involves sending several duplicate emails to the same address but can be easily detected by spam filters.

List linking This involves subscribing the target email address to different email list subscriptions. The user would always receive spam mail from all these subscriptions and will have to manually unsubscribe from each list separately.

ZIP bombing This kind of email bombing is done using ZIP archived attachments containing millions and billions of characters.

Handling This can be handled similar to how viruses and worms are.

Preventive measures The following are some of the ways of preventing email bombs:

1. Using antivirus software and installing firewall to restrict traffic can reduce the chances of attack.
2. Email filter applications can be used to manage unsolicited emails by filtering emails according to the source address.
3. Proxy servers that are configured with certain rules for filtering the messages it receives helps in filtering malicious requests and messages from suspicious IP addresses before they are sent to the clients of the proxy server, thereby avoiding the difficulty to spam and filter each and every email bomb coming from different IP addresses.
4. Use of SMTP can minimize the attack as the message transfer agent (MTA) analyses the mail exchange record and IP address of the sender and rejects the message if it is found suspicious. Security mechanisms such as authentication and negotiation are processed during the exchange of data.

Case 11: Email Bombing [Prateek Paranjpe]

A foreigner who had been residing in Shimla (India) for almost 30 years wanted to avail of a scheme introduced by the Shimla Housing Board to buy land at lower rates. When he made an application it was rejected on the grounds that the scheme was available only for citizens of India. He decided to take revenge. Consequently, he sent thousands of emails to the Shimla Housing Board and repeatedly kept sending emails till their servers crashed.

Email Spamming

Spam or electronic junk mail refers to unwanted email that contains URLs of a website which when clicked installs harmful software in the system. Spam occupies some space in the email box and email server. The goal of spam is to gather sensitive information of the victim. Email that does not include a 'To' address or 'CC' is a general form of spam. Spamming is done in two ways:

1. Adding noise to the email message
   (a) Small letters and/or white text
   (b) Text not displayed

(c) Placing text outside the screen size
   (d) Disabling text visibility to users using tags
2. Masking hyperlinks
   (a) Obfuscation of domains using UTF
   (b) Mixing encoding
   (c) URL shortening with noise
   (d) Adding prefixes
   (e) Masking redirect

Handling The following can be done by a victim of spamming:
1. Spam mail can be deleted.
2. Employing multilevel authentication can serve as a defence.
3. If clicking on the spamming mail accidentally has resulted in a phishing attack, then the access granted earlier can be revoked.
4. Passwords to the accounts can be reset.
5. The spamming incident can be reported.

Preventive measures The following are some of the ways of preventing spam:
1. Setting up of multiple email addresses
2. Never responding to any spam
3. Never clicking on 'unsubscribe' links from unknown sources
4. Updating browsers
5. Using anti-spam filters

Malvertising

Malvertising (malicious advertising) is used by cybercriminals to victimize unsuspecting users. Malvertising can infect a user in two ways. In the first scenario, the user has to click on the ad for the infection to spread. The malicious ads appear as pop-ups or alert warnings. These social engineering tactics prompt users to install malware themselves by clicking on the ads. The second scenario involves drive-by download methods wherein the user becomes infected by simply loading a web page with malicious ads on it. The ads contain a script that looks for vulnerabilities to download and executes a file on the victim's system. This ultimately leads to the installation of information stealing malware.

Handling The following can be done if a victim has witnessed a malvertisng attack:
1. The system should be isolated from the network.
2. Security software such as smart sandboxes can be used to find and detect malicious behaviour.
3. Ad-blocking browser plugins can be enabled to lessen the risks. However, ad and script blockers kill legitimate ads. Alternatively, browsers can be set to flag malicious content. For example, Google Chrome has an option under Privacy Settings 'Enable phishing and malware protection'.
4. Antivirus software may be installed to prevent any further attack.

Preventive measures The following are some of the ways of preventing malvertising:
1. Web browsers and plugins such as Adobe Flash or Java should be kept up to date to alleviate risks.
2. Pop-up ads should be avoided in addition to random messages and unverified links so as to lessen the risks posed by this type of threat.

Botnet

Botnet is a network of infected systems. A bot is a form of malware that allows the attacker to take control of the infected systems. Generally, a bot is a part of the botnet. Botnet is used for data theft, sending spam email, performing DoS attacks, click fraud, mining bitcoins, and distributing other malware like Trojan Horse. The performance of a bot is based on the command of a bot master. A bot spreads itself across the Internet to search for any computer to infect. When an unprotected system is found, it is infected immediately and reported to the master. For example, ZeroAccess Botnet has been built with 1.9 million systems to generate a profit for the owner of ZeroAccess Botnet through bitcoin mining and click fraud.

Some of the botnet types include DoSBot, SpamBot, BrowseBot, AdSenseBot, ChatBot, idBot, CCBot, PollBot, BruteForceBot, and NetBot.

To get the malicious bot code on the victim's computer, cybercriminals use the following ways: sending emails that contain malicious attachments or links to websites they control where malicious code is hosted, sending messages through social networks/messaging apps which contain links to trick the victim and drive-by downloads which work by exploiting vulnerabilities from the web browser, browser plug-ins, or add-ons.

Handling A victim may become part of a botnet by any of these attacks—hacking, DDoS attack, or spam. Hence, appropriate measures for handling, discussed earlier, may be enabled to handle a botnet attack.

Preventive measures The following are some of the ways of preventing botnets:

1. Avoid clicking on suspicious links and downloading attachments that were never requested.
2. Avoid relying on online ads which disguise themselves and con users into believing that their computer systems are infected.
3. All the software required should be downloaded from a reputable source and should be up to date.
4. Antivirus and antispyware software in use should be activated, patched, and up to date.
5. The firewall should be turned on and set to the maximum security level. The user will be notified of all applications requiring Internet access so that incoming and outgoing traffic can be tracked.

Potentially Unwanted Programs

A potentially unwanted program (PUP) is a program that may be unwanted, despite the possibility that users consented to download it. PUPs include spyware, adware, and dialers, and are often downloaded in conjunction with a program that the user wants.

Generally third-party providers take popular free programs and pack them into a new executable file along with a PUP (bundling). This is then offered as a download on third-party provider sites, where the popular freeware acts as a bait. The websites with the bundled programs are positioned in search engines to increase the number of potential customers so that providers can earn money, not only per download but for displaying advertisements in the installed PUP.

Handling The following can be done if a victim has come across a PUP:

1. Such software can be uninstalled.
2. A complete scan of the system may be performed to quarantine or delete any unwanted adware/spyware, etc.
3. Further sites that were the source for such PUPs may be reported.

Preventive measures The following are some of the ways of preventing PUP installations:

1. All software should be downloaded from the original provider's websites.
2. Users should check for any pre-set and unwanted options set when downloading software.
3. While installing software, 'user-defined' or 'advanced' installation options may be preferred rather than 'quick' and 'default' options so that the user can have a look at the options displayed.
4. Users should read the information in the installation dialogue window carefully and check the boxes that discuss the data protection terms and terms of use.

> **Box 3.7 How is Scareware Different from Rogue Security Software?**
>
> Scareware relies on fear tactics like rogue security software. Scareware programs such as fake registry cleaners and system optimizers are designed to look authentic, but do very little or absolutely nothing of what it claims and so no or little harm. Scareware usually uses scary advertising and scary scan results to try to get users to buy something that's essentially useless. Scareware programs can be uninstalled from the system easily.

Rogue Security Software

Rogue security software usually appears as legitimate or useful software that prompts the user to click an erroneous or misleading link to install, update, or delete malicious software. It results in a variety of actions such as downloading malicious software, fraudulent transactions, stealing personal data, corrupting files or folders, and slows down computers.

Rogue security software, also known as rogue anti-malware, is a counterfeit software program that appears to be beneficial from a security perspective, but in reality, is not. Rogue security software masquerades as genuine security software, stays in the system, and constantly gives alerts about being infected, generates erroneous or misleading alerts, claims to help the computer to get rid of malware to scare victims, forces the user into buying the solution, and partakes in fraudulent transactions. Rogue security software lodges itself deep in the system and cannot be easily removed or uninstalled.

This software is delivered to the computer through advertisement, spam email, and manipulated search engine optimization (SEO) rankings. The risks associated with this software include giving a false sense of security to the user, inducing the user into perform fraudulent transactions, interrupting legitimate program security operations, and preventing the users from visiting legitimate security software vendor sites (Box 3.7).

Handling A victim of rogue security software can perform the following:

1. Isolate the system from the network.
2. Uninstall such software.
3. A complete scan of the system should be performed, installing legitimate security programs such as antivirus and antispyware.

Preventive measures The following are some of the ways of preventing these threats:

1. The computer should be installed with legitimate security programs such as antivirus, antispyware, and firewall, and the security programs have to be up to date and turned on always.
2. Clicking on links in website and emails and opening attachments in emails should be avoided.
3. Clicking on scary ads should be avoided.
4. It is necessary to be aware of common phishing scams and attacks.
5. Care should be taken while searching for security tools and such tools should be downloaded only from the official source.

Spyware

Spyware is a serious computer threat that monitors online activities of the user unlawfully and installs applications without the consent of the user to capture personal information. The spyware periodically sends all the recorded information to cybercriminals. The recorded information may be browsing habits, keyword strokes, login credentials, confidential information such as PIN, credit card information, and passwords. An example of spyware would be the keylogger software which records passwords. This software is widely used in houses and industries to monitor the usage of the Internet.

Types

Spyware has various forms:

Adware Adware or malvertising is a malicious program that displays advertisements in the computer system, sends search requests to advertising websites, and collects personal information without the knowledge of the user. In other words, adware is a malware that uses the activities of both Trojan Horse and spyware. Adware enters the system by means of freeware or shareware and infected websites.

Pornware Pornware is a program that displays pornographic content on the machine. The programs are installed either deliberately by the user or maliciously without the knowledge of the user. Trojan Downloader and Trojan Dropper are also used to infect the machine with pornware. Pornware is easy to install because antivirus software does not recognize the pornware program. The variants of pornware are as follows:

Porn-Dialer Porn-Dialer is a program that dials adult content-based phone sex lines and runs up a huge bill. It is similar to SMS hacking.

Porn-Downloader Porn-Downloader downloads pornographic content to the target user's computer system from the Internet.

Porn-Tool Porn-Tool acts as a toolbar for browsers to search and view porn. It also includes specific video players.

Riskware Riskware is a program that causes harm to the computer system when it is exploited by malicious users for deleting, blocking, and altering data, thereby disturbing the performance of a system or network. The various types of programs that are used by Riskware for malicious purposes include the following:

1. Remote administration utilities
2. IRC clients
3. Dialler programs
4. File downloaders
5. Computer activity monitoring software
6. Password management utilities
7. Internet server services

Handling The following may be performed by a victim of spyware:

1. The system should be isolated from the network.
2. Utilties that remove spyware, for example, Ad-ware, Spybot, etc., can be installed and run which will list down the suspected spyware entries which can either be quarantined or deleted.

Preventive measures The following are some of the ways of preventing spyware:

1. Anti-spyware software may be downloaded and installed.
2. Care should be taken while browsing so as to avoid links to websites from unknown sites, downloading pirated software, random pop-up alerts, and exploits which promote software piracy.
3. The operating system should be updated and all the installed software should be patched. Further, the web browser should be kept updated to prevent exploitation by taking defensive steps against spyware.
4. Firewalls should be employed to monitor the network and to block suspicious traffic which can prevent spyware from infecting the system.

3.3.6 Copyright- and Trademark-related Offences

Infringing on intellectual property rights is the process of copying, spreading, and using proprietary materials such as software, books, music, and videos protected by intellectual property laws illegally.

Software Piracy

Software piracy is the illegal copying, distribution, or use of software. Software piracy also refers to the use of computer application by violating an end user licence agreement. Software piracy causes loss of revenue to authorized developers and the software industry. Most software licence agreements allow placing one copy on a single computer and make a second copy for backup. The violation of software licence involves installing software with a single licence into multiple systems (end user piracy), downloading software from the bulletin (online piracy), duplicating proprietary software and distributing as legal software (reseller piracy).

Software piracy applies mainly to full-function commercial software. The time-limited or function-restricted versions of commercial software, called *shareware* and *freeware*, a type of software that is copyrighted but is freely distributed, are less likely to be pirated since they are freely available. The types of software piracy include the following:

Softlifting Borrowing and installing a copy of a software application from a colleague

Client–server overuse Installing more copies of the software than the number of licences purchased

Hard-disk loading Installing and selling unauthorized copies of software on refurbished or new computers

Counterfeiting Duplicating and selling copyrighted programs

Online piracy Typically involves downloading illegal software from peer-to-peer network, blog, etc.

Case 12: Software Piracy [Prateek Paranjpe]		
State: Karnataka	City: Bengaluru	Section of Law: 65 and 66 of the Information Technology Amendment Act 2008
A software company stated that some of its former employees had accessed the IT system of the company and tampered with the source code of the software under development.		

Trademark-related Offences

Trademarks are signs that can be represented graphically, and gives a product or service an identity. They are usually registered by businesses and are used to give an identity that resonates with both the clients and the customers. Once it has been registered, it cannot be used by anyone else as a brand symbol.

There are several criminal offences relating to trademarks. The first is counterfeiting a trademark. This includes making a sign identical to a registered trademark so as to deceive and falsify a genuine registered trademark, without the consent of the proprietor of the registered trademark. Another possible criminal offence is falsely applying a registered trademark to goods or services. This means applying a sign likely to be mistaken for that trademark to the goods or services.

Dealing with trademark infringement Trademark infringement is a violation of rights attached to a registered trademark without the authorization of the trademark owner. Infringement may be that the infringer's trademark is identical or confusingly similar to a registered trademark owned by a person and the product or service is identical to the service or product the registration covers. In such a case, the owner of the registered trademark can initiate legal proceedings against the infringer. A criminal complaint can also be filed. Under the provisions of TradeMarks Act 1999, the offences under the act are cognizable which means that the police can register an FIR and prosecute the offenders directly.

Data Piracy

Data piracy is the illegal printing, reproduction, and distribution or unauthorized use of copyright-protected materials such as books and journals.

Case 13: Data Piracy [Prateek Paranjpe]		
State: Haryana	City: Delhi	Section of Law: 420 / 408 / 120B IPC R/W 66 ITA ACT 2008
The complainant filed a case of fraud and cheating, alleging theft and sale of proprietary data. The complainant had a subsidiary company in the United States which did business with its US partner. The US partner provided mortgage loans to US residents for residential premises. The business of the complainant was providing leads to their US partner. The data included the details of loan seekers along with their telephone numbers. The complainant generated leads through arrangements with call centres in India who called from their database and shortlisted home owners who were interested in availing re-finance facility on their existing mortgage loans. The complainant realised that there was a sudden drop in the productivity of the call centres and therefore the production of leads, although the inputs meant to be given to various call centres by the employees of the company had remained the same as before. The concerned officials of the company got alarmed and made an in-house enquiry. On careful and meticulous scrutiny, it was revealed that one of the employees of the complainant (company), in connivance with some other officers, had been deceiving and causing wrongful loss to the company by selling the data purchased by the company and in effect wrongful gain for themselves.		

Audio and Video Piracy

Audio and video piracy is the illegal act of copying and distributing audio such as music, or videos such as movies and television programmes that are protected by the copyright holder. Bootlegging and counterfeiting are the two common forms of conventional audio and video piracy. Bootlegging is an illegal recording and copying of a live broadcast. Counterfeiting is an unauthorized copying of labels, artistic works, and packages. Video cassettes and optical discs are the main storage media of conventional audio and video piracy.

Preventive measures The Three E's—Engineering, Education, and Enforcement—help to combat any piracy.

1. Engineering refers to making software/data/digital object safer and more difficult to copy.
2. Education refers to making consumers aware as to where they can buy genuine copies of the software/data/digital object. They also need to be made aware of the risk involved with installing illegal copies. This is because, many consumers who are caught do not even know that the software/data/digital object that they are using is pirated or from illegal copies.
3. Enforcement refers to working together with the private sector, rights holders, and law enforcement officials to clamp down on the people who distribute the software/data/digital object to consumers.

Case 14: Audio Piracy	
State: Tamil Nadu	City: Chennai
Maestro Illayaraj served a legal notice on playback singer S.P. Balasubramanyam for rendering his songs in a concert in the US in August 2016 without his consent.	

Case 15: Video Piracy [Manish Raji (2016)]	
State: Tamil Nadu	City: Chennai
The Central Crime Branch (CCB), Egmore, Chennai, received a tip-off that a man was making pirated copies of recently released Tamil and English movies.	

3.3.7 Computer-related Offences

This class of offences refer to illegal action and crimes committed with the use of electronic communication devices.

Impersonation

Impersonation is one of the several social engineering tools used to gain access to a system or network in order to commit fraud, industrial espionage, or identity theft. The social engineer impersonates or plays the role of someone likely to be trusted to allow access to information, or to information systems. Impersonators obtain information about the target by stalking employees on social networking sites, email phishing, phone pretexting and dumpster diving, eavesdropping on employee conversations, and from black market websites or other social engineers and company websites.

Handling If any user credentials were subject to impersonation, it can be reset.

Preventive measures The following are some of the ways of preventing impersonation:

1. Avoid revealing passwords and sensitive information to anyone.
2. The premises should be physically secure. The surroundings should be checked to make sure no one is overhearing conversations and is involved in shoulder surfing.
3. Policies and procedures have to be framed within an organization to manage social engineering attacks, and these should be adhered to by everyone.

Computer-related Forgery

Computer forgery is the act of unlawfully inserting, altering, or deleting computer data or restricting the access to these data and supplying false data. Thus, in computer forgery, the computer is used to create a fraudulent document or illegally altering an otherwise legal document, for example, using a computer to create a fake certificate.

Computer-related forgery cannot be prevented completely. However, the use of message authentication code (MAC) exchanged between the parties can help in detecting forgery when a deviation is observed.

Data Diddling

Data diddling occurs when someone with access to information of some sort changes this information before it is entered into a computer. This is done to provide some sort of benefit to the data diddler, generally financial, and is a common method of computer-related crime.

Using this technique, cybercriminals can manipulate the output, which is not easy to identify. However, the use of cyber forensic tools can ascertain when the data was changed and can also change it back to the original form. It can be combated by ensuring that all the information is identical, whether it is a hard copy or the data within a digital system.

Case 16: Data Diddling [Prateek Paranjpe]

The New Delhi Municipal Council (NDMC) Electricity Billing Fraud Case took place in 1996. The computer network was used for receipt and accounting of electricity bills by the NDMC. Collection of money, computerized accounting, record maintenance and remittance in the bank were exclusively left to a private contractor who was a computer professional. He misappropriated huge amount of funds by manipulating data files to show less receipt and bank remittance.

Identity Theft

Identity theft is a cybercrime in which a person obtains and uses someone else's key piece of information, for example, identity information, deliberately without authorization, to commit a crime.

Types

Identity theft can be categorized as follows:

Financial identity theft   The most common type of identity theft is the financial identity theft. The criminal gains financial benefits such as loans, goods, and services by using someone else's financial identity information. The financial identity theft includes credit card and debit card fraud, tax fraud, savings account fraud, investment fraud, pension fraud, etc. Cybercriminals use various techniques such as phishing and skimming to obtain information from victims.

Child identity theft   Child identity theft is a cybercrime in which criminals target children under the age of 18 for personal data such as identity number, name, and date of birth. The criminals create a synthetic identity by using a child's identity number with a different date of birth to open bank accounts, apply for credit cards and debit cards, or to get employment, government, and health benefits. Child identity theft remains undetected for years until the child uses his/her personal information to apply for a bank account. It can be extremely shocking when the child reaches the age of 18 and comes to know that his/her identity has been stolen.

Medical identity theft   Cybercriminals make unauthorized use of names and health insurance numbers of the victims to file fraudulent claims with the insurance provider or get other medical services. Medical identity theft includes getting fake treatment, buying addictive drugs, obtaining free treatment which results in serious damages such as loss of health coverage, inaccurate health records, etc. It affects not only the victim but also the health and insurance provider.

Handling   The victim must do the following to book the perpetrators of these crimes:

1. Identity theft incident can be reported to the law enforcing agency and the respective financial institution.
2. User credentials such as passwords and PIN can be reset.
3. The associated account can either be blocked or closed permanently.

Preventive measures   The following are some of the ways of preventing identity theft:

1. Disposing sensitive information safely
2. Examining bank account statements monthly
3. Setting difficult passwords
4. Protecting computers with antivirus software
5. Monitoring credits

Case 17: Identity Theft [Prateek Paranjpe]		
State: Maharashtra	City: Pune	Section of Law: 467, 468, 471, 379, 419, 420, 34 of IPC and 66 of ITA ACT 2008
The accused in the case was working in a BPO which was handling the business of a multinational bank. During the course of work, the accused had obtained the personal identification number (PIN) and other confidential information of the bank's customers. Using this information, the accused and his accomplices transferred huge sums of money from the accounts of different customers to fake accounts through different cyber cafes.		

Misuse of Devices

Misuse of mobile devices is usually done by children and teens and is referred to as cyber bullying. It occurs in cyberspace by using technology to harass, embarrass, or target another person.

Preventive measures   The following are some of the ways of preventing misuse of devices:

1. Guidelines and rules should be framed while using technology.
2. Users should be educated on how to balance technology usage to their expectations. Further, they should be educated about how the misuse of devices affects other people.
3. In the case of cyber bullying, monitoring electronic devices and keeping communication lines open can help in the detection of misuse of ECDs.

Salami Slicing Attack

A Salami attack is a series of minor attacks that together results in a larger attack. Computers are ideally suited for automating this type of attack. It is also known as salami slicing/penny shaving where the attacker uses an online database to seize the information of customers, that is, bank/credit card details, deducting minuscule amounts from every account over a period of time. These amounts naturally add up to large sums of money that is taken without anybody noticing this, from the collective accounts. Victims who fall for such acts are usually bank account holders, and websites that store account information.

Handling The victim must do the following to book the perpetrators of these crimes:

1. The victims of Salami attacks can report the incident to authorized personnel.
2. User credentials can be reset.
3. If necessary, such accounts can either be blocked or closed permanently.

Preventive measures The following are some of the ways of preventing a Salami attack:

1. Banks have to update their security policies so that the attacker doesn't familiarize himself/herself with the way the framework is designed.
2. Be it a small or big amount, banks should encourage customers to come forward and openly notify them about this attack.

Case 18: Salami Slicing Attack

In 2008, a man was arrested for fraudulently creating 58,000 accounts which he used to collect money through verification of deposits from online brokerage firms, a few cents at a time. While opening the accounts and retaining the funds may not have been illegal by themselves, the authorities charged that the individual opened the account using false names (cartoon characters), addresses, and social security numbers, thus violating the laws against mail fraud, wire fraud, and bank fraud.

Pharming

Pharming is a form of online fraud. It is very similar to phishing as pharmers rely upon the same bogus websites and theft of confidential information. However, where phishing must entice a user to the website through 'bait' in the form of a phony email or link, pharming redirects victims to the bogus site even if the victim has typed the correct web address. This is often applied to the websites of a bank's core commerce site.

Handling A victim of pharming must do the following:

1. The victim of a pharming attack can isolate his/her system from the network.
2. Run antivirus software to perform a complete scan of the system.
3. Opt for a more reliable ISP.

Preventive measures The following are some of the ways of preventing pharming:

1. A trustworthy ISP may be used from among the vast majority of ISPs.
2. The first line of prevention against pharming is performed by the ISPs, as they filter out as many of the bogus redirects as possible.
3. While payment option is enabled, it is necessary to ensure if HTTPS is active.
4. Antivirus software can also help in protecting against instances of pharming, especially when entering an unsecured site without realising this. Such software should be kept updated.

Case 19: Pharming

In 2004, a German teenager hijacked the country's eBay domain name leaving thousands of users redirected to a bogus site.
In 2005, the domain name for Panix, a New York-based ISP was redirected to a bogus site in Australia. In the same year, a secured email service was attacked by redirecting users to a defaced website.

3.4 CYBERCRIME AGAINST NATION

This section presents cybercrimes against the nation and the ways to handle and prevent them.

3.4.1 Cyber Terrorism

The terrorist of the future may be able to do more damage with a keyboard than with a bomb.

Cyber terrorism is the anonymous use of cyberspace for terrorist activities, for threatening citizens, or a particular group, community, and country by an individual or a group of individuals known as cyber terrorist(s). Cyber Terrorism [UCLA] may be of three types:

Simple unstructured   Performs basic computer attacks against individuals using existing computer tools

Advanced structured   Performs complex computer attacks against groups by creating some tools

Complex coordinated   Performs coordinated attacks which cause massive disruption

Cyber terrorists hack into systems, create viruses, send threatening emails, and deface websites to cause disruption of networks on a large scale. The most common Internet-based attacks for cyber terrorism are IP spoofing, password cracking, and DoS attack (Box 3.8).

Types

The various forms of cyber terrorism are as follows:

1. Privacy violation
2. Data theft
3. Demolition of government database
4. DoS attack
5. Damage and disruption of networks
6. Packet sniffing

Handling   Cyber terrorist attacks are usually targeted towards particular victims for specific reasons. They exploit victims, usually with attacks such as the release of viruses and worms. Hence the ways to handle viruses and worms discussed earlier may be adopted.

Preventive measures   The following are some of the ways of preventing cyber terrorism:

1. An intrusion detection system (IDS) can be installed to immediately respond to any intrusion.

Box 3.8 Examples of Cyber Terrorism

9/11 Attack

The Al-Qaeda researched publicly available infrastructure information posted on websites and hit the websites that contained 'Sabotage Handbook'.

Ahmedabad Bomb Blast

A person named Kenneth Haywood's unsecured Wi-fi router was misused by terrorists to send an email with the ID alarbi_gujarat@yahoo.com. After the blast, three more mails were sent using the same unsecured Wi-fi routers.

26/11 Mumbai Attack

The terrorists communicated with handlers in Pakistan through CallPhonex using VoIP with an email ID kharak_telco@yahoo.com which was accessed from ten different IP addresses.

2. All patched programs should be kept up to date with the latest security updates.
3. Logs of activity should be preserved to detect any unusual event.
4. The system or network should be secured with strong passwords and effective firewalls. Such passwords should be changed regularly.
5. Employees should be trained to not respond to messages from unknown sources and to not open email attachments. Filters should be employed to screen out suspicious materials and messages.
6. Regular checks should be carried out to make sure that security precautions are followed.
7. News and computer information reports should be followed to know about new threats being circulated.
8. The defence mechanisms employed should be regularly tested by employing a testing or security service. This is just to report any deficiencies.

Case 20: Cyber Terrorism

In the year 2011, the Indian Parliament was attacked using information technology. The accused forged an official gate pass with the logo of the Ministry of Home Affairs and other information along with the layout of the Indian Parliament. Police found out a laptop from the main accused, Md. Afzal and S. Hussein Guru, and also learnt that they did it through a Pakistani ISP. They controlled the identity and email system of the Indian Army.

The 1 September 2011 attack on the World Trade Center (WTC) can also be categorized under cyber terrorism. The terrorist's unauthorized access over the network of one airline and the hijacking of two airlines resulted in the crash of those airlines into the WTC twin towers and Pentagon.

3.4.2 Cyber Warfare

Cyber warfare is the computer- or network-based conflict by a nation or a state to penetrate into another nation/state for political motivation. The nation/state actors attempt to disrupt the activities of an organization or nation/state mainly for military purposes and cyber espionage.

Case 21: Cyber Warfare [Maragaret Rouse]

The earliest instance of a nation waging cyber war was the Stuxnet worm which was used to attack Iran's nuclear program in 2010. The malware targeted the Supervisory Control and Data Acquisition (SCADA) systems and spread through infected USB devices. While the United States and Israel were both linked to the development of Stuxnet, neither formally acknowledged its role.

Handling The cause and the nature of attack has to be ascertained from the logs and appropriate counter measures have to be adopted.

Preventive measures The following are some of the ways of preventing cyber warfare:

1. An agency that solely focuses on cyberspace and cyber attacks should be formed by the government.
2. Since a major portion of the Internet is not managed by the governments but by large technological companies or universities, governments should support these companies and organizations financially to protect their network. If these gateways are protected well, a large number of cyber attacks can be stopped.
3. A proper security framework and policies have to be framed against probable cyber warfare targets.
4. Risk management towards critical assets such as intellectual property, national defence technology, etc., should be available as a roadmap to good defence.
5. Any change in the network or the network device should be observed carefully so as to ascertain whether security measures are in place. Monitoring helps to detect unwanted behaviour.
6. An effective response to probable attacks, by way of incident response, greatly reduces the impact of an attack. Hence, an effective response system should always be readily available.

3.4.3 Cyber Laundering

Cyber laundering is the act of electronic transfer of illegally obtained monies with the goal of hiding its source and possibly its destination. The laundering of funds over the Internet and through mobile phones is easy with the growth of peer-to-peer transactions. In this, the funds are directly transferred from one individual to another without the interaction of a third party, for example, a financial institution, thus avoiding financial oversight and potential detection.

Handling   The victim of cyber laundering can do the following:

1. Report the incident to the law enforcing agency and the respective financial institution.
2. Reset user credentials.
3. If necessary, such an account can either be blocked or closed permanently.

Preventive measures   The '4-I' approach has been used to combat cyber laundering and is explained here:

Better IT knowledge   Organizations and IT industries, and law enforcement agencies should strengthen their IT knowledge to keep pace with criminals across the world. This can include increased training and hiring former hackers.

Better ID checks online   Better ID checks with new financial instruments should be introduced to help reduce the use of anonymous payments.

Better IP tracking   Financial operations should never be conducted anonymously. Hence better IP tracking of payments is essential to prevent criminals from hiding their online identities.

Better international cooperation   Better international cooperation and coordination apart from strengthening national and international efforts and instruments aimed at combating online money laundering are needed.

3.4.4 Content-related Offences

Pornography

Cyberpornography is the act of using cyberspace to create, display, distribute, import, or publish pornography or obscene materials, especially materials depicting children engaged in sexual acts with adults. Cyberpornography is a criminal offence, classified as causing harm to persons. Lack of consistent and appropriate legislation across countries globally, however, remains a major impediment to successful investigations and prosecutions.

Case 22: Pornography

One of the biggest publicized plans to get the perpetrators of child pornography in the net was launched in May 2002 and was called Operation Ore. After the FBI accessed the credit card details, email addresses, and home addresses of thousands of pornographers accessing a British child pornography site, the particulars were given to the British police for investigation. The arrest of a computer consultant in Texas led to an international investigation that jailed Thomas Ready for 1335 years for running the pornography ring. About 1300 other perpetrators were also arrested including teachers, child-care workers, social workers, soldiers, surgeons, and 50 police officers.

Spread of False and Defamatory Information

Defamation refers to any statement that hurts someone's reputation (a civil wrong, rather than a criminal wrong). A person who has been defamed can sue the person who made defamatory comments. To prove it, the victim should file statements to attest that it is false, injurious, and unprivileged.

Email Spam

Email spam, the electronic junk mail, includes unsolicited messages sent by email, text messages, or instant messages sent without the recipient's consent. Spam messages often contain offers of free goods or prizes, cheap products, promises of wealth or other similar offers.

Handling   The victim must do the following:

1. When spam email is received, it is best to delete it.

Preventive measures The following are some of the ways of preventing email spamming:

1. Responding or attempting to unsubscribe, or call any telephone number listed in the email should never be done.
2. Money, credit card details, or other personal details should not be sent to the scammers.

Case 23: Email Spamming

Stephanie, a university student living in Cairns, received an email from an airline saying that she had won a $999 credit towards her next holiday. To redeem the credits, the email requested Stephanie to respond within 12 hours with her credit card details. She responded straight away, including her full name and credit card details. The next day, Stephanie noticed that $1000 had been taken from her bank account.

POINTS TO REMEMBER

- Internet grooming refers to the act of deliberately befriending and establishing an emotional connection with a child for sexual activities.
- Cyber stalking is defined as the threatening activity or repeated following of an individual in an effort to cause fear, using electronic communication.
- Cyber harassment is the use of information and communication technology (ICT) to harass, control, manipulate, or habitually disparage a child, adult, business, or group without a direct or implied threat of physical harm.
- Cyber extortion is a cybercrime where cybercriminals take advantage of the vulnerability of intellectual property, threaten to release potentially embarrassing information publicly, or encrypt data to render it useless to the right owners and hold it for ransom.
- Pedophilia is an offence committed by men who are more than 16 years of age.
- Hacking/Illegal access is an act of breaking into protected systems, networks, and files of an individual or enterprise.
- Phishing is a type of social engineering attack causing an electronic communication to be made to a third party, often to steal user data, including login credentials and credit card numbers, where the communication gives the impression that its source is a legitimate body.
- Phreaking or telecommunication fraud is the process of gaining unauthorized access into secured telecommunication systems for exploiting the telecommunication services, such as manipulating the phone networks, copying dialling tones, etc.
- Spoofing is the act of assuming a fake identity, usually to trick another party into exchanging, submitting, or accepting either authentication-related data or other confidential data.
- Skimming is a form of magnetic strip counterfeiting in which criminals are able to copy the magnetic strip, track information (including card verification value, CVV) from a valid card.
- Computer intrusion is any malicious activity that harms a computer or causes a computer or a computer network to work in an unexpected manner.
- A computer threat is the potential to cause serious damage such as loss or corruption of data or physical damage to the hardware or to a computer which can lead to attacks and forms the basis for vulnerabilities.
- A worm is a self-replicating malicious computer program that replicates itself from one computer to another in huge volume and with high speed with the intent of spreading malicious source code.
- Trojan Horse, Trojan, or Trojan Horse virus is a malicious software that appears harmless.
- A logic bomb is a malicious program timed to cause harm at a certain point of time, but is inactive until that point.
- Ransomware is a type of malware that blocks the ECD and encrypts the data on it, further demanding money for restoration of the facility.
- A computer virus is a computer program that can infect other computer programs by modifying them in such a way as to include a copy of it.
- Malware or malicious software refers to various forms of annoying software or programs such as viruses, Trojan Horses, spyware, worms, and rootkits.

- A rootkit is a collection of programs that enable administrator-level access by installing it in the computer system.
- DoS attack tries to make a web resource unavailable to its users by flooding the target URL with more requests than the server can handle.
- A distributed denial of service (DDoS) attack is a DoS attack that comes from more than one source simultaneously.
- A mail bomb is the sending of a massive amount of email to a specific person or system, thus consuming system, storage, and network resources.
- Spam or electronic junk mail refers to unwanted email that contains the URL of a website which when clicked installs harmful software in the system.
- Malvertising is used to victimize unsuspecting users by cybercriminals.
- A bot is a form of malware that allows the attacker to take control of the infected systems.
- Botnet is a network of infected systems.
- A potentially unwanted program (PUP) is a program that may be unwanted, despite the possibility that users consented to download it.
- Rogue security software usually appears as legitimate or useful software that prompts the user to click erroneous or misleading links to install, update, or delete malicious software.
- Spyware is a serious computer threat that monitors the online activities of the user unlawfully and installs applications without the consent of the user to capture personal information.
- Pornware is a program that displays pornographic content on the machine.
- Infringing on intellectual property rights is the process of copying, spreading, and illegally using proprietary materials such as software, books, music, and videos protected by intellectual property laws.
- Impersonation is one of several social engineering tools used to gain access to a system or network to commit fraud, industrial espionage, or identity theft.
- Computer forgery is the act of unlawfully inserting, altering, or deleting computer data or restricting the access to these data, resulting in false data.
- Data diddling occurs when someone with access to information of some sort changes this information before it is entered into a computer.
- Identity theft or identity fraud is a cybercrime in which a pretender obtains and uses someone else's key piece of information such as identity information deliberately.
- Misuse of mobile devices is usually done by children and teens and is referred to as cyberbullying.
- A Salami attack is a series of minor attacks that together result in a large attack. Computers are ideally suited for automating this type of attack.
- Pharming (pronounced 'farming') is very similar to phishing as pharmers rely upon the same bogus websites and theft of confidential information.
- Cyber terrorism is the anonymous use of cyberspace for terrorist activities, for threatening citizens, a particular group, or a community and country by an individual or a group of individuals known as cyber terrorist(s).
- Cyber warfare is the computer- or network-based conflict by a nation or a state to penetrate into another nation/state for political motivation.
- *Cyber laundering* is the electronic transfer of illegally obtained monies with the goal of hiding its source and possibly its destination.
- Cyberpornography is the act of using cyberspace to create, display, distribute, import, or publish pornography or obscene materials, especially materials depicting children engaged in sexual acts with adults.

KEY TERMS

Audio and video piracy This is the illegal act of copying and distributing audio (music) or video (movies and television programmes) that are protected by the copyright holder.

Bot This is defined as a form of malware that allows the attacker to take control of the infected systems.

Computer forgery This is defined as the act of unlawfully inserting, altering, or deleting computer data or restricting the access to these data, resulting in the dispersion of false data.

Computer intrusion This refers to unauthorized access to computer systems, networks, device, or data.

Computer threat This is defined as the potential to cause serious damage such as loss or corruption of data, physical damage to the hardware, to a computer, etc., which can lead to attacks and forms the basis for vulnerabilities.

Computer virus This is a computer program that can infect other computer programs by modifying them in such a way as to include a copy of it.

Cyber extortion This refers to a cybercrime where cybercriminals take advantage of the vulnerability of intellectual property, threaten to release potentially embarrassing information publicly, or encrypt data to render it useless to the right owners and hold it for ransom.

Cyber harassment This refers to the use of information and communications technology (ICT) to harass, control, manipulate, or habitually disparage a child, adult, business, or group without direct or implied threat or physical harm.

Cyber laundering This refers to the electronic transfer of illegally obtained monies with the goal of hiding its source and possibly its destination.

Cyber stalking This is defined as the threatening activity or repeated following of an individual in an effort to cause fear, using electronic communication.

Cyber terrorism This refers to the anonymous use of cyberspace for terrorist activities, for threatening citizens, a particular group or community and country by an individual or a group of individuals known as cyber terrorist(s).

Cyber warfare This is defined as the computer- or network-based conflict by a nation or a state to penetrate into another nation/state for political motivation.

Data piracy This is defined as the illegal printing, reproduction, and distribution or unauthorized use of copyright-protected materials such as books and journals.

Hacking/Illegal access This is an act of breaking into protected systems, networks, and files of an individual or enterprise.

Identity theft or identity fraud This is a cybercrime in which a person obtains and uses someone else's key piece of information like identity information, deliberately.

Illegal interception This refers to listening to, monitoring, or recording of any aspect of the contents of a communication originating from or destined to an ECD without the consent of either parties involved in the communication.

Internet grooming This refers to the act of deliberately befriending and establishing an emotional connection with a child for sexual activities.

Logic bomb This is defined as a malicious program timed to cause harm at a certain point of time, but is inactive until that point.

Malware This refers to any software that gets installed on a machine and performs unwanted tasks, often for some third party's benefit.

Online pedophile This refers to individuals whose interactions are focused essentially on child pornography in cyberspace.

Phishing This is defined as a type of social engineering attack causing an electronic communication to be made to a third party, often to steal user data, including login credentials and credit card numbers, where the communication gives the impression that its source is some legitimate body.

Phreaking or telecommunication fraud This is the process of gaining unauthorized access into secured telecommunication systems for exploiting the telecommunication services such as manipulating the phone networks and copying dialling tones.

Ransomware This refers to a type of malware that blocks the ECD and encrypts data on it, further demanding money for restoration of the facility.

Riskware This is defined as a program that causes harm to the computer system when it is exploited by a malicious user for deleting, blocking, and altering data, thereby disturbing the performance of a system or network.

Rogue security software This usually appears as legitimate or useful software that prompts the user to click an erroneous or misleading link to install, update, or delete malicious software.

Rootkit This is a collection of programs that enable administrator-level access when installed in the computer system.

Skimming This is defined as a form of magnetic strip counterfeiting in which criminals are able to copy the magnetic strip and track information including card verification value (CVV) from a valid card.

Software piracy This is the illegal copying, distribution, or use of software. This also refers to the use of computer applications by violating an end user licence agreement.

Spam or electronic junk mail This refers to unwanted email that contains the URL of a website which when clicked installs harmful software in the system.

Spoofing This is defined as the act of assuming a fake identity, usually to trick another party into exchanging, submitting, or accepting either authentication-related data or other confidential data.

Spyware This is a serious computer threat that monitors online activities of the user unlawfully, and installs applications without the consent of the user to capture personal information.

Trademark infringement This refers to a violation of rights attached to a registered trademark without the authorization of the trademark owner.

Trojan Horse This is a malicious software that appears harmless.

Worm This is a self-replicating malicious computer program that replicates itself from one computer to another in huge volume and with high speed with the intent of spreading malicious source code.

MULTIPLE-CHOICE QUESTIONS

1. A type of cybercrime against individuals is _____.
   (a) cyber extortion
   (b) hacking
   (c) phishing
   (d) computer intrusion
2. The electronic junk mail that includes unsolicited messages sent by email, text messages, or instant messages which is sent without the recipient's consent is called _____.
   (a) spread of false and defamatory information
   (b) email spam
   (c) pornography
   (d) Internet grooming
   A Salami attack is also referred to as _____.
   (a) Salami slicing
   (b) penny shaving
   (c) both (a) and (b)
   (d) none of these
4. The variants of pornware are _____.
   (a) Porn-Dialer
   (b) Porn-Downloader
   (c) Porn-Tool
   (d) (a), (b), and (c)
5. _____ produces multiple copies of itself with the belief that the virus scanners cannot detect all instances of it.
   (a) Slow infector
   (b) Stealth virus
   (c) Fast infectors
   (d) Polymorphic virus
6. Skimming is performed using _____.
   (a) skimming plates on ATM
   (b) PIN capturing
   (c) both (a) and (b)
   (d) none of these
7. A malware that modifies the browser settings, creates desktop shortcuts, and displays advertising pop-ups is called _____.
   (a) adware
   (b) spyware
   (c) browser hijacking software
   (d) rootkit
8. Pharming is a form of _____.
   (a) online fraud
   (b) copyright- and trademark-related offence
   (c) system interference
   (d) data inference
9. Cyber terrorism may be of three types such as _____.
   (a) simple structured, advanced unstructured, complex coordinated
   (b) simple unstructured, advanced structured, complex coordinated
   (c) complex structured, advanced unstructured, simple coordinated
   (d) complex unstructured, advanced structured, simple coordinated
10. Better IP tracking is a preventive measure for _____.
    (a) cyber stalking
    (b) cyber laundering
    (c) virus
    (d) worm

11. _____ has a history of mental illness and personality disorder.
    (a) A delusional stalker
    (b) An intimate stalker
    (c) A vengeful stalker
    (d) An erotomanic stalker
12. _____ identify and exploit system vulnerabilities.
    (a) White hat hackers
    (b) Black hat hackers
    (c) Grey hat hackers
    (d) Red hat hackers
13. Phishing is also used to gain entry into a corporate or governmental network as part of a larger attack such as _____ event.
    (a) an aligned persistent event
    (b) an accurate proper event
    (c) an advanced persistent event
    (d) an aligned proper event
14. _____ is also called CEO fraud.
    (a) Pharming
    (b) Spear phishing attack
    (c) Deceptive phishing
    (d) Whaling attack
15. _____ is the first phreaking method which allows making calls by disabling the rotator keypad.
    (a) Network hooking
    (b) Router hooking
    (c) Switch hooking
    (d) None of these

REVIEW QUESTIONS

1. Define cyber laundering.
2. What is infringing on property rights? List its types.
3. Explain cybercrime against property in detail.
4. Discuss briefly cybercrime against nation with a case study.
5. List out case studies and preventive measures for cybercrime against individuals.
6. What is cybercrime? Mention and define the types of cybercrimes.
7. Define cyber harassment, cyber bullying, and cyber stalking. Explain the relationship among them.
8. What are the clues used to identify phishing email?
9. List the types and subtypes of computer threats.
10. Compare virus, worm, and Trojan Horse.
11. Describe the methods of email bombing.
12. What are the ways of email spamming?

APPLICATION EXERCISES

1. Do you think using a public Wi-fi is a crime? What are the risks associated with it?
2. Give some safety measures to protect oneself from identity theft.
3. What do you think has been the most important cyber security incident in the recent past? Ascertain the threat and suggest appropriate measures for policy makers.
4. Categorize crimes according to their impact as high, medium, or low financial loss with justification for each.
5. Having learnt about all sorts of crimes, suggest the measures an individual should adopt to avoid being a victim to any of these crimes.

BIBLIOGRAPHY

1. Marjie Britz, T. (2013), *Computer Forensics and Cyber Crime*, 3rd edn, Pearson Education, Inc.
2. Neustar, *Telecom Fraud Impact Whitepaper*, available at: https://www.neustar.biz/resources/whitepapers/telecom-fraud-impacts-whitepaper (Accessed 09 December 2017)
3. Literacy, *Ohio Literacy Resource Center sprsum20011*, available at: http://literacy.kent.edu/

Oasis/Pubs/techtalk6-1.pdf (Accessed 09 December 2017)
4. Portal, 'Building Peace in the Minds of Men and Women, available at: http://portal.unesco.org/culture/en/ev.php-URL_ID=39412&URL_DO=DO_TOPIC&URL_SECTION=201.html (Accessed 09 December 2017)
5. Prateek Paranjpe, 'Cyber Forensics: Case Studies from India', available at: http://prateek-paranjpe.blogspot.in/p/cyber-forensics-case-studies.html (Accessed 09 December 2017)
6. Manish Raji (2016), *Video Piracy Case: Chennai Court Rejects Man's Bail Plea*, available at: http://timesofindia.indiatimes.com/city/chennai/Video-piracy-case-Chennai-court-rejects-mans-bail-plea/articleshow/55622444.cms (Accessed 09 December 2017)
7. Express News Service (2012), *Hard Disk Stolen from Home Ministry Computer*, available at: http://indianexpress.com/article/delhi/hard-disk-stolen-from-home-ministry-computer/ (Accessed 09 December 2017)
8. Telecommunication Development Sector (2013), *Understanding Cybercrime: Phenomena, Challenges and Legal Response*. ITU Publication
9. Socialmediacast, *Six Types of Cyber Stalkers*, available at: https://socialmediacast.wordpress.com/tag/six-types-of-cyber-stalkers/ (Accessed 10 December 2017)
10. Brett Singer, *What is Child Identity Theft*, available at: http://www.parents.com/kids/safety/tips/what-is-child-identity-theft/ (Accessed 10 December 2017)
11. Raj Singh (2010), PowerPoint Presentation On Cyber Terrorism, available at: http://powerpointpresentationon.blogspot.in/2010/07/powerpoint-presentation-on-cyber.html (Accessed 10 December 2017)
12. Indiaforensic, *Cyber Stalking Frauds in India*, available at: http://indiaforensic.com/cyberstalking.htm [Accessed on 10 December 2017].
13. Maragaret Rouse, *What is Cyberwarfare?*, available at: http://searchsecurity.techtarget.com/definition/cyberwarfare (Accessed 10 December 2017)
14. Guru99, *Potential Security Threats to Your Computer Systems*, available at: https://www.guru99.com/potential-security-threats-to-your-computer-systems.html (Accessed 10 December 2017)
15. Michael Sanchez (2010), 'Potential Security Threats to your Computer Systems', available at: https://blogs.cisco.com/smallbusiness/the-10-most-common-security-threats-explained (Accessed 10 December 2017)
16. Kaspersky, *What is Trojan Virus*, available at: https://usa.kaspersky.com/resource-center/threats/trojans (Accessed 10 December 2017)
17. Kaspersky, *Pornware*, available at: https://www.kaspersky.co.in/resource-center/threats/pornware [Accessed on 10 December 2017].
18. Searchsecurity, *What is Computer Worm?*, available at http://searchsecurity.techtarget.com/definition/worm (Accessed 10 December 2017) Norton*Bot and Botnets – A Growing Threat*, available at https://us.norton.com/botnet/ (Accessed 10 December 2017)
19. Darys Gudkova, Maria Vergelis, Nadezhda Demidova, and Tatyana Shcherbakova (2017), *Spam and Phishing in 2016 – Securelist*, available at: https://securelist.com/kaspersky-security-bulletin-spam-and-phishing-in-2016/77483/ (Accessed 10 December 2017)
20. Searchmidmarketsecurity, *What is rootkit?*, available at http://searchmidmarketsecurity.techtarget.com/definition/rootkit (Accessed December 2017)
21. Veracode, *Rootkit: What is a Rootkit and How to Drtect it*, available at https://www.veracode.com/security/rootkit (Accessed 10 December 2017)
22. Shashanik Shekhar (2017), *UPSTF Busts Gang which Hacked Secure Source Code to Crack Aadhaar, Issued Fake Biometric Cards*, available at http://indiatoday.intoday.in/story/aadhaar-card-fake-biometric-id-up-stf-uidai/1/1045133.html (Accessed on 10 December 2017)
23. Business Standard, *Beware New Malware Safecopy Trojan Horse is Stealing through your Phone*, available at http://www.business-standard.com/article/technology/beware-new-malware-xafecopy-trojan-is-stealing-money-through-your-phone-117091000343_1.html [Accessed on 10 December 2017].
24. Satheesh (2009), 'Selected Case Studies on Cyber Crime', available at http://satheeshgnair.blogspot.in/2009/06/selected-case-studies-on-cyber-crime.html (Accessed 11 December 2017)
25. Valencynetworks, *Cyber Security: Botnet Attack Explained*, available at: http://www.valencynet-

26. Margaret Rouse, *What is Phishing?*, available at http://searchsecurity.techtarget.com/definition/phishing (Accessed 11 December 2017)
27. Computer Hope (2017), *What is Phishing?*, available at: https://www.computerhope.com/jargon/p/phishing.htm [Accessed on December 2017]
28. *Phishing, What is Phishing?*, available at: http://www.phishing.org/what-is-phishing (Accessed 11 December 2017)
29. Tripwire, *6 Common Phishing Attacks and How to Protect Against Them*, available at: https://www.tripwire.com/state-of-security/security-awareness/6-common-phishing-attacks-and-how-to-protect-against-them/ (Accessed 11 December 2017)
30. Incapsula, *What is Phishing*, available at: https://www.incapsula.com/web-application-security/phishing-attack-scam.html (Accessed 11 December 2017)
31. Navado, *Hacking, Phishing & Phreaking*, available at: http://www.navado.com.au/sydney/lawyers-solicitors/cyberlaw-and-internet-law/hacking-phishing-phreaking.html (Accessed 11 December 2017)
32. Mark Kay Hoal (2012), *5 Ways to Handle and Prevent Cyber Harassment*, available at: http://abcnews.go.com/Technology/We_Find_Them/ways-handle-prevent-cyber-harassment/story?id=15973742 (Accessed 11 December 2017)
33. Nobullying (2015), *What is Cyber Harassment – NoBullying*, available at https://nobullying.com/what-is-cyber-harassment/ (Accessed 11 December 2017)
34. Ipredator, *Cyber Harassment Internet Defammation & Internet Trolls*, available at: https://www.ipredator.co/cyber-harassment/ (Accessed 11 December 2017)
35. Fireeye, *Ransomware: The Tool of Choice for Cyber Extortion*, available at: https://www.fireeye.com/current-threats/what-is-cyber-security/ransomware.html (Accessed 11 December 2017)
36. Pierluigi Paganini (2014), *Extortion Is a Common Practice*, available at: http://securityaffairs.co/wordpress/23849/cyber-crime/extortion-cyber-crime.html (Accessed 11 December 2017)
37. Cisecurity, *Cyber Extortion: An Industry Hot Topic*, available at: https://www.cisecurity.org/cyber-extortion-an-industry-hot-topic/ (Accessed 11 December 2017)
38. Inspq, *Online Pedophilia and Cyberspace*, available at: https://www.inspq.qc.ca/en/sexual-assault/fact-sheets/online-pedophilia-and-cyberspace (Accessed 11 December 2017)
39. Cybercrimechambers, *Cyber Crime Chambers*, available at: https://www.cybercrimechambers.com/blog-child-pornography-54.php (Accessed 11 December 2017)
40. Jrank, *Cyber Crime – Online Child Pornography – Jrank Articles*, available at: http://law.jrank.org/pages/11986/Cyber-Crime-Online-child-pornography.html (Accessed 11 December 2017)
41. Legal Service India, *The Menace of Cyber Crime – Hacking – IP Spoofing – Password Attacks - Email Scams – Pornography on the net*, available at: http://www.legalserviceindia.com/articles/article+2302682a.htm (Accessed 11 December 2017)
42. Margaret Rouse, available at: http://searchsecurity.techtarget.com/definition/email-spoofing (Accessed 11 December 2017)
43. Cybercrime, *Cyber Crime, E-mail related Crimes*, available at: http://cybercrime.planetindia.net/email_crimes.htm (Accessed 11 December 2017)
44. Cybercrimehelpline, *Types of Cyber Crime – Cyber Crime Helpline*, available at: https://www.cybercrimehelpline.com/cyber-crime/types-of-cyber-crime/ (Accessed 11 December 2017)
45. Heimdalsecurity, *These Counter Spoofing Measure Will Keep you Safe*, available at: https://heimdalsecurity.com/blog/ip-email-phone-spoofing/ (Accessed 11 December 2017)
46. Cybercrimechambers, *ATM Card Skimming & PIN Capturing*, available at: https://www.cybercrimechambers.com/atm-card-skimming-and-pin-capturing.php (Accessed 11 December 2017)
47. Technology Depot (2013), *Credit Card Skimming !!! A way of Cyber Crime*, available at: https://volumeoftech.wordpress.com/2013/02/15/credit-card-skimming-a-way-of-cyber-crime/ (Accessed 11 December 2017)
48. Kim Zetter (2013), *Logical Bomb Set Off south Korea Cyberattack*, available at: https://www.wired.com/2013/03/logic-bomb-south-korea-attack/ (Accessed 11 December 2017)

49. Tommy Armendariz (2017), *What is a Logic Bomb? Explanation & Prevention*, available at: https://www.lifewire.com/what-is-a-logic-bomb-153072 (Accessed 11 December 2017)
50. Nick Lewis (2013), *Understanding logic bomb attacks: Examples and countermeasures*, available at: http://searchsecurity.techtarget.com/tip/Understanding-logic-bomb-attacks-Examples-and-countermeasures (Accessed 11 December 2017)
51. Technopedia, *What is a Logic Bomb? – Definition from Technopedia*, available at: https://www.techopedia.com/definition/4010/logic-bomb (Accessed 12 December 2017)
52. Cyber Crime, *Cyber Crime, Viruses*, available at: http://cybercrime.planetindia.net/viruses3.htm [Accessed on 12 December 2017].
53. WikiHow, *How to Prevent Computer Virus Infection 3 Steps (with Pictures)*, available at: https://www.wikihow.com/Prevent-Computer-Virus-Infection (Accessed 12 December 2017)
54. Avast, *Malware – What it Is and How to Remove it with Anti-malware*, available at: https://www.avast.com/c-malware [Accessed 12 December 2017]
55. Paul Froutan (2004), *How to Defend against DDoS Attacks*, available at: https://www.computerworld.com/article/2564424/security0/how-to-defend-against-ddos-attacks.html (Accessed 12 December 2017)
56. Nicky Woolf (2016), *DDos Attack that Disrupted Internet was Largest of its Kind in History Experts Say*, available at: https://www.theguardian.com/technology/2016/oct/26/ddos-attack-dyn-mirai-botnet (Accessed 12 December 2017)
57. George Hulme, V., *DDos Explained: How Distributed Denial of Service Attacks Are Evolving*, available at: https://www.csoonline.com/article/3222095/network-security/ddos-explained-how-denial-of-service-attacks-are-evolving.html (Accessed 12 December 2017)
58. Bullguard, *What are DoS and DDoS Attacks?*, available at: https://www.bullguard.com/bullguard-security-center/internet-security/internet-threats/what-are-dos-and-ddos-attacks.aspx (Accessed 12 December 2017)
59. Margaret Rouse, *What is Mail Bomb? – Definition from WhatIs.com?*, available at: http://searchsecurity.techtarget.com/definition/mail-bomb (Accessed 12 December 2017)
60. Thewindowsclub, *Email Bombing and ways to protect yourself*, available at: http://www.thewindowsclub.com/email-bombing (Accessed 12 December 2017)
61. Trendmicro, *Malvertising: When online ads attack*, available at: https://www.trendmicro.com/vinfo/us/security/news/cybercrime-and-digital-threats/malvertising-when-online-ads-attack (Accessed 12 December 2017)
62. Christina Chipurici, *What is a Botnet & How to Prevent your PC from Being Enslaved*, available at: https://heimdalsecurity.com/blog/all-about-botnets/ (Accessed 12 December 2017)
63. Sabrina Berkenopf (2013), *Potentially Unwanted Programs: Much More Than Just Annoying*, available at: https://www.gdatasoftware.com/blog/2013/10/23983-potentially-unwanted-programs-much-more-than-just-annoying (Accessed 12 December 2017)
64. Superantispyware (2017), *SUPERAntiSpyware Blog – Remove spyware, Not just the easy ones!*, available at: https://www.superantispyware.com/blog/ (Accessed 12 December 2017)
65. Tommy Armendariz (2017), *How to Prevent Spyware from Infecting your Computer?*, available at: https://www.lifewire.com/spyware-prevention-tips-153401 (Accessed 12 December 2017)
66. SingaporeLegalAdvice 2015), *What Are the Criminal Offences Relating to Trademarks?*, available at: https://singaporelegaladvice.com/law-articles/criminal-offences-relating-trade-marks/ (Accessed 12 December 2017)
67. Radhika Shukla (2014), *Trademark Infringement and Remedies*, available at: http://www.legalservicesindia.com/article/article/trademark-infringement-and-remedies-1740-1.html (Accessed on 12 December 2017)
68. Margaret Rouse, *What is Piracy? Definition from WhatIs.com*, available at: http://whatis.techtarget.com/definition/piracy (Accessed 12 December 2017)
69. Malvastyle (2015), *Preventing Software Piracy vs Making Money*, available at: http://malvastyle.com/preventing-software-piracy/ (Accessed 12 December 2017) MySecurityAwareness, *Impersonation: Hacking Humans*, available at: http://

www.mysecurityawareness.com/article.php?article=390&title=impersonation-hacking-humans#.Wi_hPN-WZPY (Accessed 12 December 2017)

70. Aj Maurya, *What is a Salami Attack? – Aj Maurya. An Engineer*, available at: https://ajmaurya.wordpress.com/2014/03/27/what-is-a-salami-attack/ (Accessed 12 December 2017)Norton India, *Pharming - Online Fraud | Cybercrime Pharming*, available at: https://in.norton.com/cybercrime-pharming (Accessed 12 December 2017)

71. Bob Haring, *How can an organization prevent Cyber Terrorism?*, available at: http://smallbusiness.chron.com/can-organization-prevent-cyberterrorism-48558.html (Accessed 12 December 2017]

72. Killian Strauss, *Cyber Laundering – How Can We Combat Money Laundering over the Internet?*, available at http://www.academia.edu/1369342/Cyber-laundering_How_can_we_combat_money_laundering_over_the_internet (Accessed 12 December 2017)

73. Mitch Ratcliffe, *How to Prevent against Online Libel and Defamation?*, available at: http://www.socialbrite.org/2009/08/08/preventing-against-online-libel-and-defamation/ (Accessed 12 December 2017)

74. Acorn, *Email Spam and Phishing*, available at: https://www.acorn.gov.au/learn-about-cybercrime/email-spam-and-phishing (Accessed 12 December 2017)

75. Grupo Deidev, *What is Internet Grooming? – SecureKids*, available at: https://securekids.es/en/tag/what-is-internet-grooming/ (Accessed 12 December 2017)

76. Kaspersky (2012), *Antivirus Protection & Internet Security Software*, available at: https://www.kaspersky.com/about/press-releases/2017_chasing-lazarus-a-hunt-for-the-infamous-hackers-to-prevent-large-bank-robberies (Accessed 23 December 2017)

77. Thehill (2016), *The Hill - Covering Congress, Politics, Political Campaigns and Capitol Hill*, available at: http://thehill.com/policy/cybersecurity/361166-north-korea-aligned-hackers-branching-out-into-mobile-phones-report (Accessed 23 December 2017)

78. Symantec, *What Do you Need to Know about Botnet behind Recent Major DDoS Attacks*, available at: https://www.symantec.com/connect/blogs/mirai-what-you-need-know-about-botnet-behind-recent-major-ddos-attacks (Accessed 23 December 2017)

79. UCLA, *Cyberterrorism*, available at: http://classes.dma.ucla.edu/Fall13/161/projects/adam-lai-fatt/5-style/html/ cyberterrorism.html (Accessed 23 December 2017)

80. Mark Johnson (2016), *Cyber Crime, Security and Digital Intelligence*, Second Edition, Routledge (Taylor & Francis Group), New York

81. Pzimmerm, (2012), Security Technologies, available at: http://docwiki.cisco.com/w/index.php?title=Security_Technologies&oldid=49118 (Accessed 23 December 2017).

Answers to Multiple-choice Questions

1. (a)	2. (b)	3. (a)	4. (d)	5. (d)
6. (c)	7. (c)	8. (a)	9. (b)	10. (b)
11. (a)	12. (b)	13. (c)	14. (d)	15. (c)

Cybercrime—The Present and the Future

4

Learning Objectives

This chapter provides an overview of the present and the future of cybercrime. The objective of this chapter is to provide the basics of cyber war, the A to Z of ransomware, an overview of the cryptocurrencies that facilitate a ransomware attack, its associated technology called blockchain, a brief outline of the deep and dark web, and cyber space, the platform for cybercrime in the future. Further, the challenges of the Internet are presented. The reader will be familiar with the following after studying the chapter:

- Cyber war
- Cryptocurrency
- Bitcoin
- Ethereum
- Blockchain
- Ransomware
- Deep web and dark web and the challenges

4.1 INTRODUCTION TO CYBER WAR—THE PRESENT AND THE FUTURE OF CYBERCRIME

Since the 1980s, the global economy has been evolving and so is the cyber space. In the 21st century, global economy and cyberspace have interpenetrated each other. They both occupy the same space and share the risks. Cybercrime evolved from viruses and worms in the 1990s, extended to credit card frauds from early 2005 to late 2007, and is now targeting business and government organizations. Besides this, with technological advancements, the number of connected devices on the cyber space is rising and is expected to touch 26 billion units in 2020. This will invite cybercriminals to capitalize on their new-found access to consumers handling businesses and in the government. The four categories of cybercrime that are predominant are electronic commerce crime, economic espionage, infrastructure attacks, and personal cyber insecurity, of which electronic commerce crimes are prevalent at present. Ransomware attacks fall under this category. Besides these, there are risks and threats associated with cyberspace—the dark side of the Internet. Cyberattacks will become a growing concern if nations attempt to wage cyber warfare in a bid to disrupt competing states and gain a competitive advantage.

'2017 is the year of cyber warfare' —Forbes

Cyber war is a global web-based battle that is intended to attack computer systems and networks for underscoring patriotism, political motivation or revenge. Cyber war disables official websites or networks, disrupts services and steals the data and information of a state, a nation or an organization. Generally, cyber war targets military forces, government organizations, financial institutions, and IT industries as they are purely critical infrastructure network or Internet-centric. Cyberattacks can be prevented by installing security updates in the computer.

Cyber war tool is a logical tool (cyber war weapon) that is used to conduct cyber war. Most of the cyber war tools are freely available (open source). Cyber war tools are categorized as reconnaissance tools, scanning tools, access and escalation tools, exfiltration tools, sustainment tools, assault tools, and obfuscation tools.

1. *Reconnaissance tools* are used for gathering information about computer systems and networks such as official website contents, domain name server (DNS), and metadata.
2. *Scanning tools* are used for gathering detailed information about target systems such as scanning ports and enumerating users.
3. *Access and escalation tools* are used for gaining access to systems and escalating access privileges.
4. *Exfiltration tools* are used for copying, transferring, or retrieving data illegally, such as physical exfiltration, encryption, and steganography.
5. *Sustainment tools* are used for ensuring access into the system in the near future such as adding authorized access to ourselves and adding backdoors.
6. *Assault tools* are used to assault a compromised machine, for example, by making simple changes to system configuration and performing denial of service (DoS) attacks using botnets.
7. *Obfuscation tools* are used for spreading obscure content in the computer system such as location obscuration, log manipulation and file manipulation [US Department of Justice].

Thus, obviously, ransomware is the predominant tool of cyber war that involves the use of cryptocurrency. This chapter deals with cryptocurrencies, ransomware, and the deep and dark web.

4.2 CRYPTOCURRENCY

Cryptocurrency is an encrypted digital or virtual currency which is not issued by a central authority. It is transferred between peers and confirmed in a public ledger via mining, which is a process of confirming transactions and adding them to the ledger. Cryptocurrency is mainly used for international transactions and smart contracts (Box 4.1).

> ### Box 4.1 Smart Contracts
>
> The term *smart contract* was conceived by Nick Szabo a computer scientist, law scholar, and cryptographer, in 1997, long before the arrival of the bitcoin. He wanted to use a distributed ledger to store contracts. Smart contracts are analogous to contracts in the real world except that they are digital, self-executing contracts with the terms of agreement between the buyer and the seller directly written into lines of code. It is actually a small computer program that is stored across the distributed and decentralized blockchain network and does not require a third party to ensure trust.
>
> They are actually programmed to hold all received funds transferred from a buyer until a certain goal is reached. The money is passed to the seller if the goals are met by the contract automatically. In the event that the goals are not met, the money is automatically sent to the buyer. Since smart contracts are stored on a blockchain, everything is completely distributed. Thus no one is in control of the money. Smart contracts permit trusted transactions and agreements to be carried out among disparate, anonymous parties without the need for a central authority, legal system, or an external enforcement mechanism, thus rendering transactions traceable, transparent, and irreversible.
>
> Such contracts can be trusted as they inherit the properties of the blockchain, namely immutable (once a smart contract is created, it can never be changed) and distributed (the output of the contract is validated by everyone on the network). Hence the code of the contract can never be tampered with. Also there is no chance for anyone to force the contract to release the funds as other people on the network can spot this attempt and mark it as invalid.
>
> Smart contracts find applications in many fields: banks can use them to issue loans or to offer automatic payments, insurance companies can use them to process claims, etc.

(*Contd*)

Box 4.1 (Contd)

Blockchain supports smart contracts; the biggest one is ethereum as it was specifically designed to support smart contracts, which are programmed using a special programming language called Solidity. Solidity was specifically created for ethereum and its syntax resembles Javascript. Bitcoin also supports smart contracts but is limited when compared to ethereum.

Cryptocurrency can be bought from cryptocurrency exchanges and peer-to-peer means, through cryptocurrency wallets and ATMs, with cash and credit cards, and via bank transfer and PayPal. Cryptocurrencies can be stored, sent, and received through a cryptocurrency wallet. Based on the type of media, crytpocurrency wallets are classified as desktop wallet, online wallet, mobile wallet, hardware wallet, and paper wallet.

Desktop wallet   It is a commonly used application that directly connects to the cryptocurrency's client.

Online wallet   It is a web-based wallet but data is stored on the virtual server.

Mobile wallet   It runs from a smartphone application.

Hardware wallet   It is used to hold cryptocurrency which is generally a USB device.

Paper wallet   It is used to send and receive cryptocurrency by printing out a QR code for both public and private keys.

Cyptocurrency makes transferring of funds between two parties easy during a transaction that is facilitated through public and private keys for security, and requires minimum processing fees for fund transfer. However, the cryptocurrency balance can be wiped out from a computer crash if a backup of the holding does not exist because it has no central repository. The rate of cryptocurrency fluctuates since prices are based on supply and demand.

4.2.1 Characteristics

Cryptocurrency is unique due to its characteristics, shown in Fig. 4.1.

Adaptive scaling   It works well on both large-scale and small-scale level.

Virtual   It is created from computer code. Unlike real-world currency, it has no central bank and is not backed by any government.

Vulnerable   It is vulnerable to theft as it is stored in digital wallets. It is also prone to attacks.

Cryptographic   It uses an AKA encryption algorithm to create currencies and verify transactions (Box 4.2).

Box 4.2 Authentication and Key Agreement

Authentication and Key Agreement (AKA) is a security protocol used in 3G networks for any secure communication system, for example, electronic commerce. AKA provides ways to verify the communicating parties' authenticity and to establish a common secret session key between them for subsequent use.

The purpose of authentication is to identify the user to the network and vice versa. It uses a one-time password generation mechanism for direct access authentication. The purpose of key agreement is to generate a cipher key for confidentiality and an integrity key for data integrity. It is a challenge–response-based mechanism that uses symmetric cryptography. When AKA is performed, integrity protection of messages, and confidentiality protection of signalling data and user data are ensured.

Decentralized It is not controlled or regulated by central bodies, and relies on peer-to-peer networks.

Digital It is not a physical object, and is created, stored, transferred, verified, and validated digitally.

Open source Developers can create cryptocurrency without paying any amount. All cryptographic transactions are open source.

Proof-of-work It uses a proof-of-work scheme, which is a method of adding value to cryptocurrency. Proof-of-work is hard to compute, but easy when it comes to verification of computational problems to restrict the exploitation of cryptocurrency. The method of validating cryptocurrency is called proof-of-stake.

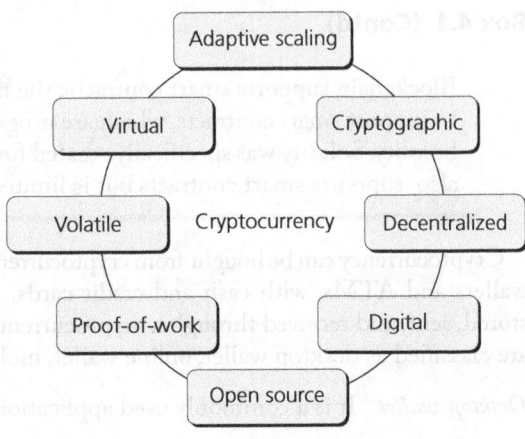

Fig. 4.1 Characteristics of cryptocurrency

4.2.2 Types

The following is the list of cryptocurrencies that exist. Figure 4.2 shows them along with their symbols.

Bitcoin Bitcoin is the first, highest priced, and the most used cryptocurrency. Outdoing all other cryptocurrencies, it is the leading cryptocurrency in the market.

Ether Ether is the ethereum currency created by Vitalik Buterin in 2015. It is known for decentralized peer-to-peer smart contracts. It makes coding and executing contracts without any third parties possible.

Monero Monero is known for its privacy and security. Individuals who remain incognito on the net use this type of cryptocurrency.

Lite Coin Lite Coin was created by Charles Lee in October 2011. It is very similar to Bitcoin and used in major exchanges.

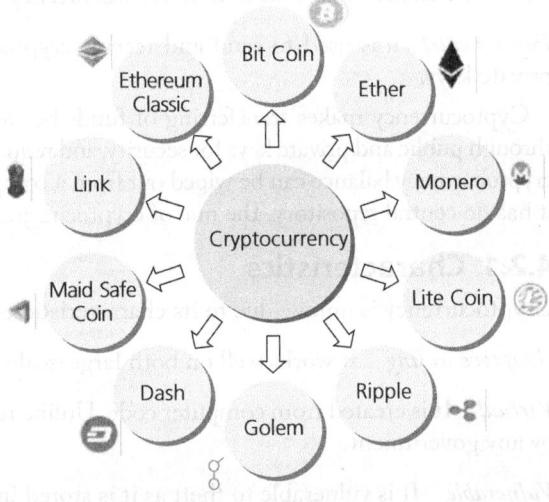

Fig. 4.2 Types of cryptocurrency and symbols

Ripple Ripple is a real-time solution system and a remittance network which mainly focuses on banking systems. Some of the banks have integrated with the ripple system to reduce costs.

Golem Golem is a decentralized, distributed computation network which is based on ethereum. Users can buy or sell computation power using Golem. It is considered the most powerful supercomputer in the world.

Dash Dash mainly focuses on privacy and transaction speed, and uses strong anonymization technology. Dash was previously known as Dark Coin. The reason behind the brand renaming was to stop its association with dark web. It is the second most exchanged cryptocurrency, and is known for being fast and anonymous.

Maid Safe Coin Maid Safe Coin is a security-centric system which provides space on an individual's computer during coin exchange. Most of the decentralized apps use the secure access for everyone (SAFE) network to store data securely.

> **Box 4.3   Decentralized Autonomous Organization Attack**
>
> The decentralized autonomous organization (DAO) attack is the most important case in cryptoeconomics since the birth of the bitcoin, as reported by Gravin Wood, the co-founder of Ethereum. DAO is digital. Launched on 30 April 2016, it is a form of investor-directed venture capital fund that aims to provide a new decentralized business model for organizing both commercial and non-profit enterprises. Ethereum is the blockchain platform that the DAO is built on and is independent from the applications based on it, that is, it is not responsible for DAO in anyway. On 28 May 2016, after a 28-day funding period, DAO raised $150 million which was the biggest crowdfund ever. On 17 June 2016, an unknown hacker spotted a flaw in the DAO's code and transferred 3.6 million Ether (the value token of Ethereum blockchain and a cryptocurrency) into his personal account. As a result, the price of Ether fell from $20.50 to $11.20. It is unlikely that the DAO will recover from the attack, but Ethereum and blockchain technology can learn from the attack to prevent future attacks.

Lisk   Lisk is a unique cryptocurrency with a modular system that allows users to create decentralized apps 'dapps' on their own in Javascript language. Lisk paves the way to create many apps, which in turn increases its value. It is best known for the making of dapps.

Ethereum Classic   Following DAO attacks (Box 4.3), the ethereum community decided to conduct a *hard fork* (splitting a blockchain path by invalidating transactions confirmed by nodes that have not been upgraded to the new version of the protocol software) of ethereum's blockchain so as to code the stolen money back to their owners. To accomplish this, a majority of the users have to take all the transaction records, up to the point of hacking, and start newly from there, while discarding all transactions that took place after hacking. The hard fork is responsible for two versions existing simultaneously, ethereum being the new one. The old one was named Ethereum Classic.

4.3   BITCOIN

Bitcoin is a virtual (digital) currency created in 2009 which facilitates instant payment using peer-to-peer (P2P) technology. It was worth $708 as of 10 November 2016. Bitcoins are used for online purchases and transfers, and are considered secure because each bitcoin is digitally signed during each transfer. According to Nakamoto (2009), bitcoin is a software-based online payment system that was introduced as open-source software in 2009.

It does not have a central issuing authority or a regulatory body and is not attached to any state or government. Though there is no need to *manufacture* bitcoins, it becomes difficult to track during investigation of frauds. Bitcoins that are exchanged are noted in a ledger called blockchain. This ledger held about 107 GB of data by 2016.

Tracking of the ledger is completely decentralized, meaning that anybody can volunteer to keep the blockchain up to date of all new transactions, thus ensuring it is accurate. Updating the ledger with information is analogous to writing in a notebook, where every page corresponds to a block of information and a set of pages correspond to a chain of blocks. That is why the ledger is called a blockchain.

The bitcoin blockchain is maintained by thousands of people around the globe which raises concerns on the synchronous maintenance of the ledgers. In a P2P network, some may exchange bitcoins while others volunteer to maintain the ledgers. Whenever bitcoins are exchanged, the account numbers and addresses of the sender and the receiver, and the number of bitcoins exchanged should be announced for volunteers to update in their ledgers.

Individuals can own bitcoins by creating an account called a *wallet*. When a wallet is created in the bitcoin network, two unique keys are linked to it—private key and public key. Wallets are secured using passwords. For every bitcoin payment, a bit of code is generated using the private key of the customer's bitcoin wallet which is associated with the transaction publicly. The private keys take data and sign them so that others can verify the signatures. This eliminates replication of data. When a signed message is sent out to the bitcoin network, everyone can make use of the corresponding public key to ensure the legitimacy of the sender. Bitcoins do not record any personal information during transaction. Usually, cyber criminals ask for payment of money using PayPal. Due to 'know your customer' (KYC) norms in PayPal, cyber criminals have now shifted to Bitcoin payment. To protect their identity, cyber criminals demand bitcoin payment for ransomware. Figure 4.3 illustrates a bitcoin transaction.

The bitcoin network and the wallet check if there are enough bitcoins to execute the current transaction. Since multiple copies of blockchain exist all over the world, the network delay associated with it contributes to the delay in receiving transaction requests in the same order.

To add a block of transaction to the chain, the person maintaining a ledger should solve a special kind of mathematical problem created using a cryptographic hash function. The hash function used by bitcoin is SHA256 and it is worth noting that it takes a special computer that is specifically designed to solve the SHA problem in ten minutes to guess the solution. Whoever solves the hash function first gets a chance to add the next block of transactions to the blockchain, and generate a new mathematical problem to be solved. If more than one person succeeds in solving the hash at the same time, then the network picks one arbitrarily and adds the block to the longest and the most trusted blockchain.

From a user's perspective, bitcoins appear much like a mobile app or a computer program that provides a personal bitcoin wallet. In addition, the bitcoin has a built-in system to reward volunteers maintaining ledgers. Upon successful solving of the hash function and gaining a chance to add a block to the blockchain, 12.5 new bitcoins (approximately $8550 by 2016) are awarded to the blockchain ledger keeper's account, who are otherwise called *miners*.

Consider Fig. 4.3, where the term 'address' is like a bank account into which a user can receive, store, and send bitcoins. Bitcoins are secured with public-key cryptography. Each address consists of a public key, which is published, and a private key, which the owner must keep secret. Anyone can send bitcoins to any public key, but only the person with the private key can use them. Bitcoin addresses are pseudonymous which means that the addresses are public, but nobody knows which addresses belong to whom. After depositing bitcoins into a wallet, the wallet broadcasts over the bitcoin network that it contains bitcoins. This information is incorporated into the block chain. In the figure, the buyer contacts the seller and creates a user account and then tells the

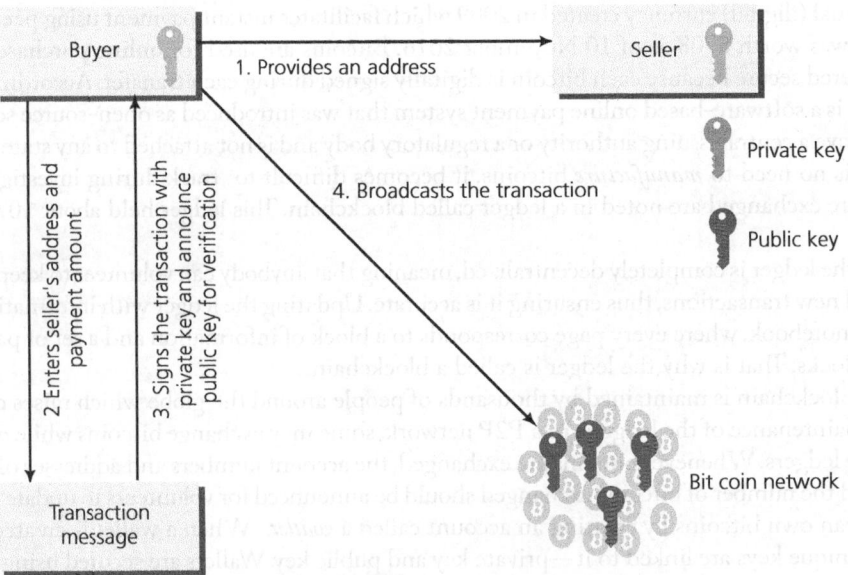

Fig. 4.3 Bitcoin transaction

seller that the payment will come from the buyer's public key. The buyer uses his/her wallet to exchange and broadcast the amount from his/her public key to the seller's public key. This is published on a peer-to-peer network which validates it against the buyer's public key, checks that the balance in the buyer's address is sufficient, and propagates it to all the other nodes on the network. The transaction is eventually received by a miner, who incorporates the transaction into a block. This block is then flooded into the network and is incorporated into the global block chain. The bitcoins now belong to the seller's public key.

Every bitcoin that exists was created to reward a bitcoin miner. Upon reaching 210,000 blocks, the number of coins generated when a new block is added goes down by half. The count, which started with 50, has now reached 12.5. It is soon expected to reach 6, and will further decrease in the years to come. It is expected that the last bitcoin will be mined in the year 2140. The idea here is that reducing the supply of bitcoins and keeping it limited will raise its value over a period of time.

However in India, the use of bitcoin and virtual currencies has not been authorized by the Reserve Bank of India yet. Further, countries such as Russia and China have heavily restricted the use of bitcoins.

4.3.1 Bitcoin Cash

Bitcoin is the world's biggest digital currency. Bitcoin's features as a *public* (anybody can use it), *permissionless* (anybody can build on top of it), *highly censor-resistant* (nobody can block your transactions), and *unseizable* form of money all stem from its decentralized architecture.

The idea behind creating bitcoin was to have a decentralized 'Internet currency' that was not bound to or controlled by banks and governments. Its limitation is that at any given moment in time there can only be 21 million bitcoins in existence—a rule which is hardwired into the currency's protocol.

Bitcoin's technology raises concern about the problem of scaling and increasing the speed of the transaction verification process. The bitcoin network can process seven transactions per second where every transaction takes about 10 minutes to process. And as the network of bitcoin users grows, waiting times will get longer, because there are more transactions to process without a change in the underlying technology that processes them. There are two major solutions to this problem, either to make the amount of data that need to be verified in each block smaller, making transactions faster and cheaper, or to make the blocks of data bigger, so that more information can be processed at a time.

Bitcoin cash (BCH) is the method adopted by the bitcoin community (BTC) to scale bitcoins to more users. A small group of miners, who were growing unsatisfied with bitcoin's technical limitations and lack of software updates, forced a fork in the bitcoin blockchain, resulting in a new currency known as bitcoin cash.

Bitcoin cash offers lower fees and a more reliable rate of transaction than bitcoins.

The BTC community is focused on keeping the bitcoin decentralized. Fast, cheap payments are a secondary priority. In contrast, the BCH community is focused on enabling fast, cheap payments over the network. The secondary priority is that of decentralization.

Bitcoin cash has increased the maximum blocksize limit parameter of the bitcoin codebase to 8MB, to accelerate the verification process, with an adjustable level of difficulty to ensure the chain's survival and transaction verification speed, regardless of the number of miners supporting it. Besides this, it allows for around two million transactions to be processed per day, whereas the bitcoin's block size limit remains at one megabyte thus allowing for approximately 250,000 transactions per day.

Most of the code for bitcoin cash is exactly that of the bitcoin. This makes developing software or altering existing software to support bitcoin cash quite simple. However, there are still concerns regarding the security of bitcoin cash. The value of BCH/USD as of January 2018 is $2718 and that of BTC/USD is $14422.

4.4 ETHEREUM

Ethereum is a result of convergence of technologies, namely peer-to-peer networking, blockchain mechanism, and cryptography. Peer-to-peer networking offers a distributed censorship-resistant platform. Blockchain offers transparency, verifiable consistency, and consensus. Cryptography offers secure and tamper-proof transaction.

Ethereum is a decentralized platform that runs smart contracts which are applications that run exactly as programmed without any possibility of downtime, censorship, fraud or third party interference. Every node has a virtual machine that runs contracts on which users can call functions and execute a transaction. This is shown in Fig. 4.4. Contracts reside on the ethereum blockchain. Every contract has its own ethereum address and a balance. It can send and receive transactions wherein it gets activated upon the receipt of transaction and can be deactivated. An ethereum virtual machine (EVM) runs a Turing complete language. Turing completeness is a property of a programming language that allows a computer to simulate anything under the universe. It enables a computer to loop and process its own output in iteratively complex terms. This property is absent in almost all the public blockchains except ethereum. Smart contracts can be written using Solidity (a Javascript-like language), Serpent (a Python-like language), Mutan (C-like language), and LLL (Lisp-like language) and are compiled into bytecode before they are deployed in the blockchain. Smart contracts have a fee for every CPU step and an extra fee for storage. Users can run the application on the local blockchain. Ethereum gives priority to increasing the number of smart contracts, which in turn will raise ethereum's popularity and profitability.

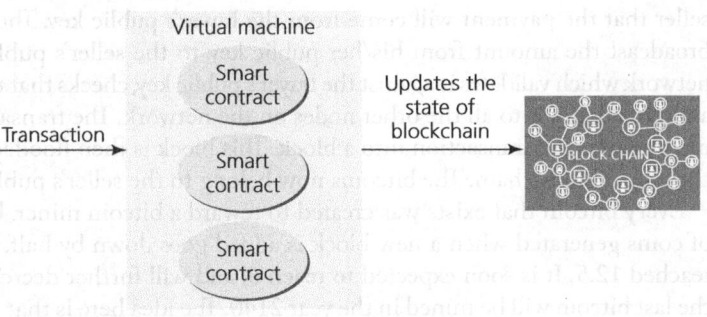

Fig. 4.4 Ethereum transaction

Ethereum operates on the basis of proof-of-stake, which replaces miners in bitcoin management by *validators*, and the reward system is changed. Every validator owns ether (currency) and puts his/her owned ether on the line to certify that a block is valid. Hence, it provides a barrier against malicious behaviour. There is a transaction fee for the validators, for the smart contract they validate. So it is energy efficient and also promotes collaboration rather than competition.

In addition to the use of EVM for the execution of the contract logic, ethereum provides two protocols for peer-to-peer support for the exchange of messages and static files. Whisper is the peer-to-peer protocol for exchanging messages. It provides a powerful, distributed, and private messaging capability with support for single cast, multicast and broadcast messages. Swarm is another peer-to-peer protocol that provides an incentive-based approach for the distribution of static content among peers and to exchange them efficiently.

Ethereum is based on trust. Trust, on the Internet, can be established either by mutual agreement by both the parties or by a third party. But both are susceptible to all sorts of abuse, which blockchain-based technology can mitigate or remove entirely. Blockchain facilitates two communicating parties to collaborate without the need for a central authority. It itself acts as a machine building trust.

The following are the advantages with Ethereum: (a) uptime is less, (b) offers security, (c) is almost free, and (d) ensures transparency. Some of its limitations are as follows: (a) EVMs are slow and they are not suitable for large computations, (b) storage on the blockchain is expensive, and (c) problems are introduced with scalability.

4.5 COMPARISON BETWEEN BITCOIN AND ETHEREUM

The ranking of cryptocurrencies changes continuously. New cryptocurrencies with new advantages will appear, whereas the old ones will vanish over time. But the one cryptocurrency that will remain for long is the bitcoin because it is considered a safe haven in the cryptocurrency exchange system. Next to bitcoin, ethereum is the most popular cryptocurrency. Table 4.1 compares and contrasts the two most popular cryptocurrencies, namely bitcoin and ethereum.

Table 4.1 Comparison between bitcoin and ethereum

Attribute	Bitcoin	Ethereum
First release	3 January 2009	30 July 2015
Exchange procedure	Blockchain technology	Blockchain technology
Design	Secure peer-to-peer decentralized payment system	Decentralized platform that runs smart contracts
Priority	Security, speed	Increasing the number of smart contracts which in turn will raise ethereum's popularity and profitability
Operation basis	Proof of work which involves solving complex mathematical problems (cryptographic challenges) by miners	Proof of stake replaces miners with validators and the reward system is changed; focuses on collaboration and provides consequence for malicious behaviour
Miners	Used to add transaction records to blockchain; processing power is used to complete transactions and bitcoins are rewarded as incentive.	Not applicable
Validators	Not applicable	Own ether and put their own ether on the line to certify that a block is valid.
Reward	25 bitcoins per block	5 ethers per block
Method of reward	Miners are rewarded for every block	Transaction fee is there for validators for the smart contract they validate, hence energy efficient; also it promotes collaboration rather than competition
Block time	10 minutes	14 seconds
Programming language used for implementation	C++ and has less than 70 commands	Turing complete language (7 different programming languages)
Vulnerability to attack	Secure as of 2017	Witnessed DAO attack
Possibility of regulation	Remote	Remote
Legal status	Few countries have accepted but countries such as India, China, and Russia have heavily restricted how bitcoins can be used	Few countries have accepted but countries such as India, China, and Russia have heavily restricted how Ethereum can be used

4.6 BLOCKCHAIN

Blockchain is a digital transaction log or distributed ledger which digitally logs all electronic purchases that are carried out using virtual currencies such as bitcoins and ether. The blockchain resides on a user's device and is referred to as node. Every node must authenticate every transaction based on advanced cryptography because a blockchain is unique. Every node has a public key, which is similar to an email address, and a private key, which is similar to a password. Authentication is based on the digital signature accompanying the transaction message. After the digital signature of a transaction is authenticated, it is pooled with other authenticated transactions in the form of a block, thereby creating a chain of block known as blockchain. To prevent fraud and double spending, the order of transactions is also authenticated. This order cannot be manipulated in the blockchain unless most of the computing nodes that maintain the blockchain in the network are compromised. Transparency and integrity are ensured through mathematics during transactions. Figure 4.5 illustrates a blockchain transaction.

There are three types of blockchain, namely public, private and hybrid. A *public blockchain* is fully decentralized and is a platform where anyone on the platform can read or write to it, provided he/she is able to show proof-of-work for the same. It is widely used. In a *private blockchain*, the owner is the centralized authority and can read or write to the platform and has the rights to make any changes (change the rules, revert transactions, etc.) based on the need. It is used for building proprietary systems to reduce cost and increase efficiency. *Hybrid blockchain* is a mix of both public and private blockchains wherein the ability to read and write to the platform is extended to a certain number of people/nodes. It is used by groups of organizations/firms, who get together to work on developing models by collaborating with each other. As a result, they gain a blockchain with restricted access.

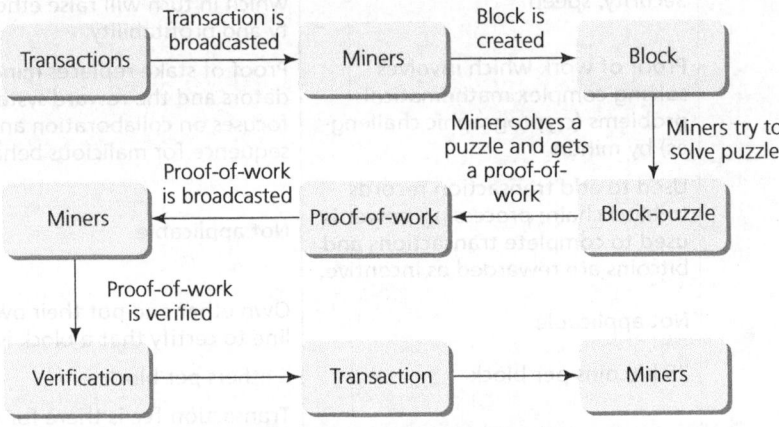

Fig. 4.5 Blockchain transaction

The issue with blockchain is that it uses decentralized, distributed technology for transactions. On an average, it takes nearly half an hour for a transaction because the size of the block is very small. The blockchain data is not checked during client start-up time. Thus, blockchain takes a longer time for first-time synchronization with the bitcoin network.

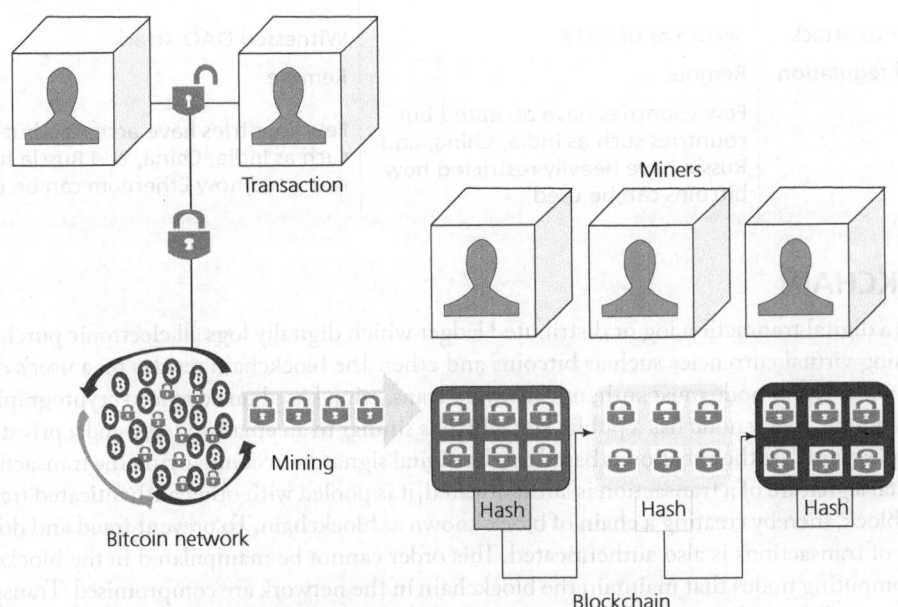

Fig. 4.6 Association between bitcoin and blockchain

4.6.1 Association between Bitcoin and Blockchain

A bitcoin client software is required to create a virtual wallet, a private key, and a public key to authenticate and secure every transaction. A user generates a request to perform a bitcoin transaction using a mobile phone or a computer. The bitcoin transaction request floats on the bitcoin network until miners pick it up for processing. During mining, bitcoin transactions are packed into data blocks and then randomly assigned with a header. Miners compete to match the header of a data block with a *nonce*, which is an arbitrary number used only once to get an alphanumeric code called *hash*. The miner who successfully generates the hash is accepted by the bitcoin network and he/she is rewarded with bitcoins. The hash value is then added to the next data block, which results in the creation of a blockchain. The association between bitcoin and blockchain is illustrated in Fig. 4.6.

4.7 RANSOMWARE

Ransomware is a scary, malicious computer program that restricts the user's access to the system in the form of locking the system's screen, locking the user's files, interrupting the normal boot process of a system, encrypting web servers, infecting mobile devices via drive-by downloads, and demanding ransom to remove the restrictions. Ransomware is also known as *scareware* or *rogueware*. Examples of ransomware include Reveton, CryptoLocker, TorrentLocker, CryptoWall, Fusob, and WannaCry.

4.7.1 Evolution of Ransomware

The ransomware that have evolved over the recent years are presented in a timeline in Fig. 4.7.

1986 **AIDS** is the first known ransomware.
2005 **Extortion** ransomware evolved.
2006 **TROJ.RANSOM.A**, a worm started spreading using RSA encryption schemes and increasing key size.
2011 A ransomware that imitated the appearance of the Windows Activation Notice.
2012 **Reveton**, a type of ransomware that impersonated law enforcement agencies. It was also referred to as Police Ransomware or Police Trojans. Reveton showed a notification page stating that the targeted individuals are performing an online illegal activity from the victim's local police agency.

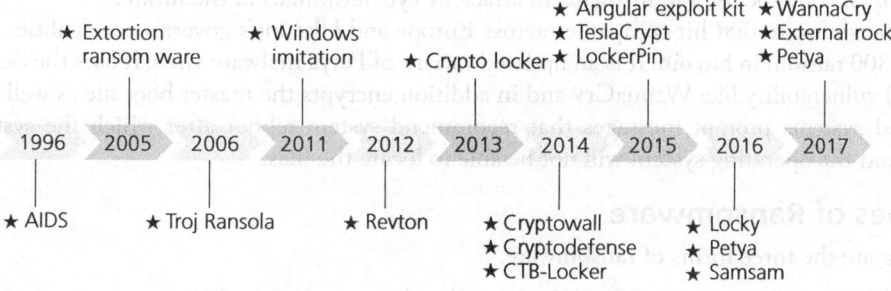

Fig. 4.7 Evolution of ransomware

2013 **CryptoLocker** acted as a source of inspiration for other ransomware variants. It was based on public key algorithm. The private key was known only to the cybercriminal who created the CryptoLocker software. It was impossible for the victim to decrypt any file without the private key.
2014 **CryptoWall** was the first ransomware variant that accepted ransom payments in bitcoin. It was spread via spam email and infected victims through drive-by-downloads and malvertising.

Cryptodefense used an inbuilt Windows encryption method. It stored the private key in plain text on the infected computer. It used Tor network and bitcoin for payment.

CTB-Locker used Tor exclusively for communication, making it more efficient and harder to detect. It spreads through drive-by downloads and spam emails.

2015 Multiple variants of ransomware on multiple platforms started to cause major damage.

The **Angler Exploit Kit** was the most popular ransomware which was used in a series of malvertisement attacks using media.

TeslaCrypt targeted the video game community by encrypting gaming files. Once encrypted it proceeded by deleting shadow volume copies (a technology included in Microsoft Windows that allows taking manual, automatic backup copies, or snapshots of computer files or volumes, even when they are in use) and system restore points to prevent file recovery. It was distributed through Angler, Sweet Orange, and Nuclear exploit kits.

LockerPin was an android ransomware which used a built-in android locking screen mechanism. It sets a PIN in the android device or changed the original PIN if it was set already.

2016   **Locky** ransomware was initially distributed via Microsoft Word Document. Later it started to spread via Adobe Flash Player and Windows Kernel Exploits.

Petya overwrote the Master Boot Record of the targeted system which was delivered through cloud storage device such as Dropbox.

Samsam exploited the vulnerabilities on Java-based web servers. It used open source tools to identify and compile a list of hosts reporting to the user's active directory. Then an executable file was used to distribute the malware and encrypt the files on the system. Payment was demanded in the form of bitcoin.

2017   **WannaCry** was the largest cyber attack ever. It is a new strain of ransomware. It has hit hundreds of thousands of computer systems since its emergence on 12 May 2017. The period May 1 to May 19 is referred to as WannaCry week. WannaCry is also referred to as WannaCrypt, WanaCryptor, WCrypt, WCry. WannaCry disabled hospitals, business enterprises, and transportations globally. It has an ability to spread through the network of an organization, and can exploit vulnerabilities in the Windows Operating System which is patched by Microsoft. The exploit is referred to as Eternal Blue.

External Rock exploits the same vulnerability in Windows which WannaCry used to spread. It uses the National Security Agency (NSA) tool Eternal Blue to spread from one system to the other. It is more powerful than WannaCry as it uses six other NSA tools such as EternalChampion, EternalRomance, and DoublePulsar. It does not inject malicious content, or lock or corrupt files but leaves an entry point that is vulnerable to remote commands which could be used to instigate an attack by cybercriminals in the future.

Petya is a ransomware that hit companies across Europe and Ukraine's government facilities in June 2017 demanding $300 ransom in bitcoin. It is an updated variant of Petya malware virus. It uses the Server Message Block (MSB) vulnerability like WannaCry and in addition encrypts the master boot file as well as other files. Petya infected-systems prompt messages that recommend system reboot after which the system becomes inaccessible and the operating system will not be able to locate the files.

4.7.2 Types of Ransomware

The following are the three forms of ransomware:

Scareware   This includes rogue security software and tech support scams. The system may pop up a message claiming that a malware has been discovered and payment is required to get rid of it. If the user does not respond, he/she gets bombarded with pop-up messages but the data remains safe.

Screen lockers   This class of ransomware prompts the user with a message that an illegal activity has been detected on the computer and the user has to pay a fine upon starting up the computer. To regain control over the PC, a system restore may help. In the worst case, running a scan from a bootable CD or USB drive will help.

Encrypting ransomware   This class of ransomware encrypts files once the system is compromised and demands payment to decrypt it.

4.7.3 Entities Affected by Ransomware Attack

Home users, businesses, and public institutions get affected by ransomware because of the following reasons:

Home users
1. No data backup is taken.
2. People lack awareness on online safety.
3. There is no knowledge on cyber security education and cyber protection.
4. There is reluctance to invest in cyber security solutions.
5. There is a notion that antiviruses can eliminate all kinds of attacks. This is not true as antivirus can fail to spot or stop a ransomware attack.
6. The count of Internet home users is huge and so any attack would result in more money.

Businesses
1. Its impact on business can affect many people and so there is more likelihood of getting ransom.
2. The amount of ransom demanded can be high as any such attack imposes a threat to reputation.
3. The gateway to an attack is more, as more human factors are involved which can be compromised via social engineering.

Public institutions
1. There is more scope for sensitive and confidential information in large databases which will definitely yield a huge ransom for cybercriminals.
2. Any such attack to a huge database can almost collapse everyday activities.
3. The hardware and software used will mostly be outdated and have security lapses.
4. Most of the employees will not have much knowledge of cyberattacks and usually do not take any countermeasures.
5. There is reluctance to adopt cyber security measures, as a result of budget constraints.

4.7.4 Mode of Infection of Users with Ransomware

Cybercriminals look for ways to infect a system or network and use that backdoor entry to spread malicious content. The following are some of the ways through which computers and mobile devices get infected:

Emails, links in emails, or messages on social networks In this type of attack, the victim clicks a malicious link in an email attachment or a message on a social networking site.

By sending emails disguised as legitimate messages, cyber criminals trick users into either opening an infected attachment or clicking a link that takes the user to an infected website. Ransomware is usually hidden as email attachments. The attachments usually take the form of an MS Office document, a PDF file, or a JavaScript. The success of the ransomware attack by a cybercriminal, however, lies in making the email appear legitimate to the target user.

The emails need not be spam mails. Cyber criminals may even target a phishing attack (which requires a user to click on a poisoned browser or email link) called spear phishing. For instance, an attack towards an academician may take the form of a LinkedIn name of a colleague to appear more convincing.

The **Locky** ransomware of early 2016 relied on macros of Word documents. If macros were not enabled, the user would not be able to properly read the document, and they would be asked to enable them. Once macros are enabled, it allows malicious code in the document to download and execute the actual ransomware payload. In the same way, JavaScript file attachments are also used to deliver ransomware.

Security exploits in vulnerable software Cyber criminals look for vulnerable entry points in the target systems to either attack the system and the data or to launch attacks to some other system on the network.

Pay per install This popular method attacks computers that are already a part of a botnet (a group of infected computers under the control of criminals called *botmasters*), further infecting them with additional malware. The criminals who look for security vulnerabilities are called *bot herders* and they are paid to find these opportunities.

Drive-by downloads When a victim clicks on a compromised website this class of ransomware gets installed. There is an increase in drive-by downloads as observed by McAfee Labs researchers. It has hit the users of some streaming video portals. Cyber criminals launch a ransomware attack by inserting malicious code in subtitle files that are played in a video. This is done by bypassing the security software and it can infect the device and its data. The vulnerability lies in the way the media players (VLC, Stremio, Popcorn Time) process subtitle files. Once the media player parses the malicious subtitle files, cyber criminals gain access to the system and the data.

Redirection of Internet traffic to malicious websites This refers to links available, either explicitly or even hidden, to redirect traffic to malicious websites.

Compromised websites or legitimate websites that have malicious code injected on their websites Exploit kits allow criminals to upload malicious code to any web page they have access to. That code is designed to exploit specific vulnerabilities in browsers or other software the visitor may be running. If the vulnerability is present, the exploit kit can leverage it to download ransomware.

Malvertising Cyber criminals compromise ad networks, so that even visits to legitimate, mainstream websites with a malicious ad therein can result in a ransomware attack.

Malicious ads containing the Angler exploit kit appeared on The New York Times, the BBC, AOL, and MSN homepage and made several tens of thousands of visitors vulnerable to the ransomware attack.

Through printers and scanners Printers are often not treated as being as much of a priority to patch and update as regular PCs. Hence they are vulnerable to attack as it allows cybercriminals to compromise the device and replace a genuine printer driver for one containing a malicious payload. Further, a cybercriminal could advertise a fake printer on the network, and wait for an unsuspecting user to connect.

Cloud infected by ransomware Enterprises, small companies, and everyone in between adopt cloud-based tools and environments for their business and personal needs. RANSOM_CERBER.CAD is a ransomware that targets home and business users of Microsoft's cloud-based productivity platform and is capable of encrypting 442 file types. It uses AES-265 and RSA to Zone settings of IE, deletes shadow copies, disables Windows Start-up Repair, and terminates a range of widely used software products like Outlook and Word. Another variant RANSOM_CERBER.A plays an audio message that the files in the system are encrypted and a ransom payment is essential to withdraw the attack.

Ransomware-as-a-service This is on the rise and works by generating ransomware with just a few clicks from an anonymous website.

4.7.5 Events in Ransomware Attack

The following is the sequence of events in any ransomware attack:
1. Initially, the victim (user) receives an email which includes a malicious link or a malware-laden attachment. Alternatively, any security exploit that originates from a malicious website can create a backdoor by using vulnerable software on the victim's PC.
2. If the victim clicks on the link or downloads and opens the attachment, a downloader (payload) will be placed on the affected PC.
3. The downloader uses a list of servers controlled by cybercriminals to download the ransomware program on the system, which responds with the requested data.
4. The entire contents on the hard disk are now encrypted by the malware including the data stored in cloud accounts (Google Drive, Dropbox) if synced to the infected PC. Even the files on other computers connected to the local network are also encrypted.
5. A warning pops up on the screen with instructions on how to pay for the decryption key.

4.7.6 Post-delivery of Ransomware

The following actions take place after a ransomware payload is available in the victim computer:
1. Once the ransomware payload is delivered, the attack is executed in a fraction of time, scanning local and connected drives for files to encrypt. Some ransomware such as Locky and DMA Locker can encrypt unmapped network shares, thereby spreading the infection on a large scale.
2. Once the encryption process is complete, all the files become inaccessible. The user is then notified in the form of Ransom notes which are typically .TXT files, a web page, or even Windows wallpaper demanding a ransom amount, how the user has to pay it (typically with Bitcoin), or simply directs them to a web page for further instructions.

4.7.7 Preventing Ransomware from Full Execution

To reduce the impact of a ransomware attack, early identification is necessary, which in turn requires awareness about it. However, one of the following steps can prevent the ransomware from being fully executed and reduce its impact:
1. Identify the actions a ransomware needs to accomplish, to successfully complete the infection process, and then stop those actions. For instance, an attack by Locky ransomware can be stopped by disabling MS Office macros.
2. Identify the location of specific executables; these can be blocked by software restriction policies.
3. Identify and install a tool that actively blocks ransomware behaviour.

4.7.8 Steps to Carry Out in Event of Infection with Ransomware

The Federal Bureau of Investigation's (FBI's) Internet Crime Complaint Centre suggests those attacked by ransomware to do the following after an attack:
1. Isolate the affected computer immediately and power it off if it is not completely corrupted.
2. Secure the backup data or systems by taking them offline.
3. Implement the security incident response and business continuity plan.
4. Contact law enforcement by filing a complaint at the FBI's Internet Crime Complaint Centre (IC3).
5. A user can contact a reputable computer expert to assist with removing the malware, if they are unable to do so on their own.
6. Keep operating systems and legitimate antivirus and antispyware software updated.
7. Users need not pay the ransom as there is no guarantee that the files will be restored. Further, refusal to pay the ransom would quickly reduce its occurrence in the future.
8. Collect and secure a partial portion of the ransomed data for investigation.
9. Delete registry files to stop the program from loading.

4.7.9 Best Practices to Adopt

The following are some of the best practices one can adopt to be safe from ransomware attacks:
1. Invest in cyber security measures such as (a) using an automated patching tool to keep the software updated and a traffic scanning tool to keep the user from accessing infected web locations, (b) using an application to block advanced forms of malware, which the antivirus cannot detect or block, and (c) using an antivirus if infected.
2. Regularly back up the contents of the computer on an external hard drive or CD-ROM. Disconnect backups from PCs and desynchronize backup accounts maintained on the cloud when not in use.
3. Keep the operating system, the installed software, and the antivirus patched and up to date.
4. Enable pop-up blockers. Criminals regularly use pop-ups to spread malicious software. Preventing pop-ups is easier than accidentally clicking on or within them.
5. Avoid suspicious mails and websites especially the ones that are alerted by security companies as source of malicious content.

6. Avoid using potentially vulnerable applications and software, and uninstall unused software.
7. Avoid free online offers for screen savers and games unless the download is from trusted websites.
8. Install Adblocker as it can protect users from malicious ads (malvertising) that can infect even mainstream, legitimate websites.

4.7.10 Role of Antivirus

Antivirus software maintains a massive database of digital signatures of known viruses and conducts pattern matching on the background for the presence of any such pattern on the system where it is installed. This software, upon finding a file with a known signature, while scanning the hard disk, will either quarantine or delete it. It is efficient only against known viruses and its behaviour is uncertain with new viruses. The worst part is that antivirus software cannot detect a virus when it is encrypted as the signatures do not match.

Some antivirus software have sandboxing functionality and dedicated apps to create virtual machines to test untrusted files which pose unnecessary overheads.

Antivirus software that uses heuristic analysis relies on the behaviour of the software running on a user's machine to detect previously unknown viruses. If any suspicious activity, like the encrypting of the user's documents, is observed, such software can be stopped and removed. Some of the antivirus software vendors such as Trend Micro, Cisco, and Kaspersky Lab fall under this category.

The antivirus software may handle ransomware in one of the following ways: (a) To defend against ransomware, the first step is to disinfect, which is possible by mere reinstallation of the operating system. This is of course time-consuming. (b) It is necessary to retrieve the data, and not just the system, after neutralizing the infection. This necessitates decrypting the data, for which recovering the key database is essential. In some cases, the decryption key may even be located in the computer if the mechanism adopted by the cyber criminals is weak. (c) Behavioural analysis is important to look for events that indicate the presence of ransomware and eliminate it before there is potential damage.

Kaspersky Internet Security handles ransomware by creating fresh copies of files when it detects some strange modification and prevents the files from being encrypted. Thereafter, it examines the software that attempted the encryption and blocks it if it seems suspicious.

Though the most common form of ransomware infection is either via exploit kits or phishing emails, other approaches like the use of programming languages such as C, C++ C#, Java, JavaScript and Windows .bat files have also been observed. AVG Internet Security offers multiple layers of protection taking into account the ransomware variants.

Apart from installing antivirus and antimalware, following the best practices in cyber security will help in preventing ransomware attacks.

4.7.11 Prevention and Response Team

US Government The US Government has given the following best practices and mitigation strategies, focusing on prevention and response to ransomware incidents, to the Chief Information Security Officers (CISO) of small, medium, and large organizations. The following are some of the preventive measures suggested:

1. Implement awareness and training program on ransomware to all the individuals.
2. Enable strong spam filters to prevent phishing emails.
3. Scan all incoming and outgoing emails. This is just to detect any threat and to prevent executable files from reaching end users.
4. Configure firewalls to block malicious sources.
5. Patch operating systems, software, and firmware on devices.
6. Set antivirus and antimalware programs to conduct regular scans, periodically and automatically.
7. Manage and use privileged accounts based on the principle of least privilege.
8. Configure access controls.

9. Disable macro scripts from office files transmitted via email.
10. Implement software restriction policies (SRP) to prevent programs from executing from common ransomware locations.
11. Use applications that are known and permitted by security policy to allow systems to execute programs.
12. Back up data regularly and secure backups. Ensure that they are not connected permanently to computers and networks.
13. Conduct an annual penetration test and vulnerability assessment.

Government of India Similarly the Indian Computer Emergency Response Team (CERT), Ministry of Electronics and IT, and Government of India have framed the following guidelines for organizations involving ICT [Ministry of Electronics and Information Technology, Government of India, 2017]

1. A central information security officer (CISO) has to be appointed in each ministry, department, and organization to eliminate cyber breach.
2. Assessment of risks, threats to organization's information assets, vulnerabilities, and the likelihood of occurrence have to be estimated and evaluated.
3. Information security management system (ISMS) encompassing cyber security as well as physical and logical security controls have to be implemented.
4. The CISO must be given the mandate and resources to establish an information security program and coordinate security policy compliance efforts across the organization.

Some of the roles and responsibilities of CISO, as dictated by them, are as follows:

1. Maintaining and updating the threat landscape for the organization on a regular basis.
2. Staying up-to-date about the latest security threat environment and related technology developments.
3. Establishing a cybersecurity programme and business continuity programme for drafting various security policies.
4. Ensuring review of the information security policy by internal and/or external subject matter experts.
5. Reviewing and updating the cyber security policy documents.
6. Establishing and reviewing the risk assessment methodology and selection of appropriate controls for risk mitigation by leveraging technology.
7. Log review, analysis, and exception reporting.
8. Vulnerability assessment and penetration testing (VAPT) of all websites, portals, and IT systems, on a quarterly basis at a minimum.
9. Web application security assessment (WASA) and white-listing of all web applications in use by the organization, annually, at a minimum.
10. Ensure no unsupported operating systems are in use in the department (Most of the systems and software in government organizations are apt only to be kept in a museum, but are still in use!).
11. CISO prescribes hardening guidelines, patch management guidelines, antivirus/malware guidelines, no privilege access on endpoints, regular review of access privileges, acceptable configuration guidelines, and proper implementation of procedures.
12. Periodic assessment/audits of third party service providers to assess risks to the organization.
13. Issuing and periodic review of device hardening guidelines, patch management guidelines, antivirus/malware guidelines, user access management guidelines, privilege access management guidelines, end-point management guidelines, connectivity guidelines for trading partners and external agencies, controls on mobile devices, and wireless technology.
14. Authorizing an acceptable use policy for software packages and freeware in compliance with the organization's risk/threat landscape, business objectives, and security policy and procedures.
15. Ensuring that the IT infrastructure deployed for online operations is kept up-to-date as per policy and is always under maintenance and technical support.
16. To coordinate response to security incidents.

17. To prepare evidence for legal action following an incident.
18. To analyse incidents so as to prevent their recurrence.

4.8 DEEP WEB AND DARK WEB

The deep web is a subset of the Internet in which the content of databases and web services cannot be indexed by conventional search engines, but provides full access to the Internet. Mostly, the deep web uses dynamic databases that lack hyperlinks. The dark web is a part of the deep web which is not indexed and provides restricted access to the Internet. The dark web hides the identity, action, and location of the Internet user. The dark web is also used for performing illegal activities by linking it to terrorist plots, drug details, and child pornography.

In other words 'dark web' is the encrypted network that exists between Tor servers and their clients, whereas the 'deep web' refers to the content of databases and other web services which are not indexed by search engines. The Tor network is an important portion of the dark web. The *Tor network* is a group of volunteer-operated servers that allows users to improve their privacy and security on the Internet. Traffic analysis is possible because any packet on the Internet can be either a header or the actual payload and this payload alone is encrypted, whereas the header is left unencrypted leaving a lot of information like the source to which the user is connected with, the duration, and so on, to hackers and attackers. This header can be a valid source of information for cybercriminals in ascertaining the pattern and interests of the user. Tor offers protection from traffic analysis on the Internet, which would otherwise enable anyone to monitor the traffic and analyse the behaviour and interests of an individual.

It is possible that more cybercrimes and attacks are centred around the deep web and dark web. Hence, research in this area is the need of the hour to combat cybercrimes and attacks in the future.

4.9 DEEP WEB AND ITS CHALLENGES

The modus operandi of crimes has evolved with time. If it was burglary then, it is hacking now. Of late, there have been increasing incidents of ransomware attacks, in addition to other cybercrimes. Crimes are now committed over the deep web of cyberspace.

4.9.1 The Internet

The Internet is comprised of network protocols, machines, switches, routers, fibre optic cables, and junction boxes over which data is transferred. It is a collection of stand-alone computers loosely linked together, mostly using the telephone network. Machines that store information and pass it on when requested for are called *servers*. Those that hold ordinary documents are called *file servers*, the ones that hold people's mail are *called mail servers*, and the ones that hold web pages are *web servers*. A *client* is a computer that gets information from a server. Most data move over the Internet by way of *packet switching*. Browsers help the user's computer to connect to the Internet via an *Internet service provider (ISP)*. Search engines list the web pages stored on different servers.

The web consists of HTML and HTTP protocols which are used by browsers to talk to servers. File and content sharing over the Internet is made possible with HTTP protocols and web page codes.

The *surface web* is easily searchable with search engines, whereas the *dark web* has hidden content that cannot be accessed easily. The term deep web is a metaphor for fishing in the deep end of an ocean of information, where traditional fishing nets cannot reach. This is shown in Fig. 4.8.

Surface Web

The surface web, also called *visible web*, is made of static and fixed web pages whose contents do not change. They reside on a server and can be retrieved any time. A change can only be made to static and fixed web pages by directly editing the HTML code, and a new version of the page is uploaded onto the server. It includes websites whose domains end with .com, .org, .net, etc. The content of those websites do not require any special configuration to access. Thus the surface web consists of all web contents that can be found using search engines such as Google, Bing, or Yahoo.

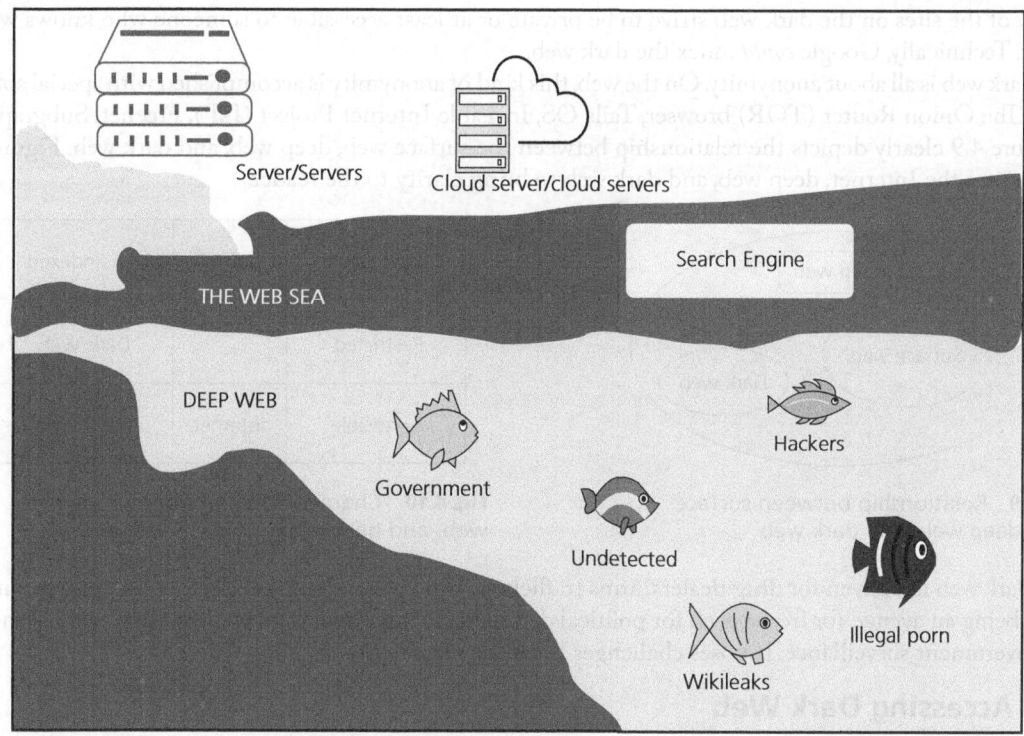

Fig. 4.8 Surface web and deep web

Deep Web

The deep web refers to online databases and other dynamic pages that are accessible through standard Internet browsers and methods of connection, but not indexed by major search engines (e.g., Google). Often the deep web is not indexed by search engines because of website or service misconfiguration, search listing opt-out requests, pay walls, registration requirements, or other content access limitations. To access such content, a user has to perform a form submission with valid input values. The deep web is also believed to be the biggest source of structured data on the web. Multiple technologies such as ubiquitous computing, distributed/cloud computing, mobile computing, and sensor networks have all contributed to the expansion of the deep web.

Dark Web

The dark web comprises websites that are outdated, broken, abandoned, or inaccessible using standard browsing techniques. The dark web corresponds to a relatively small portion of the deep web that includes web services and pages that are intentionally hidden. These services and pages cannot be directly accessed through standard browsers alone; they rely on the use of an overlay network requiring specific access rights, proxy configurations, or dedicated software. Dark web is a framework where access is restricted at the network level, for example TOR or I2P. Private virtual private networks (VPNs) and mesh networks also fall into this category. Network traffic over these frameworks is masked in such a way that snooping shows only to which darknet the node is connected to and how much data is moved, without necessarily revealing what sites are visited or the content of said data. This is in contrast to unencrypted surface and deep web services, wherein ISP and network operators can openly see the content of the traffic generated.

A darknet is a routed allocation of IP address space that is not discoverable by any usual means. Technically, a darknet is a variation on a VPN with additional measures in place.

Many of the sites on the dark web strive to be private or at least accessible to someone who knows what to look for. Technically, Google *could* index the dark web.

The dark web is all about anonymity. On the web, this kind of anonymity is accomplished with special software such as The Onion Router (TOR) browser, Tails OS, Invisible Internet Project (I2P), Freenet, Subgraph OS, etc. Figure 4.9 clearly depicts the relationship between the surface web, deep web, and dark web. Figure 4.10 differentiates the Internet, deep web, and dark web to bring clarity to the reader.

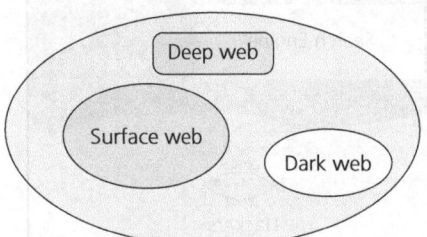

Characteristic	Indexed	Non-Indexed
Restricted	–	Dark web
Accessible	Internet	Deep web

Fig. 4.9 Relationship between surface web, deep web, and dark web

Fig. 4.10 Characteristics of Internet, deep web, and dark web

The dark web is a haven for drug dealers, arms traffickers, child pornography collectors, and other criminals, besides being an avenue for free speech for political dissidents living under oppressive regimes and a sanctuary from government surveillance. It poses challenges to law enforcement.

4.9.2 Accessing Dark Web

Accessing the dark web is a simple two-step process which is as follows:
1. The first step is to download 'TOR Browser Bundle' from TorProject.org (Box 4.4). Firefox browser is preferable with additional precautions to ensure that the system software is up-to-date, the firewall is running, and JavaScript and Cookies in the TOR browser settings are turned off, besides turning off all other Internet applications.
2. The second step is to look for onion website addresses or domain names as there is no well-indexed search engine on the darknet. However, the surface net has links to nearly forty marketplaces in DeepDotWeb.com, and an up-to-date list of .onion URLs in www.ahmia.fi/address. Some popular market places include Silk Road 2.0, Evolution and Agora with 22,576 users, 12,444 users and 11,904 users respectively as of 2014.

4.9.3 Size and Scale of Deep Web

Google being the largest search engine has only indexed one-fifth of the content of the surface web. It is technically very difficult to estimate accurately the size of the deep web. As reported by Bergman [1], the size and scale of the deep web is estimated as follows:

- The deep web contains 7500 terabytes of information as against only 19 terabytes of information on the surface web.
- The deep web contains 550 billion individual documents as against 1 billion on the surface web.

> **Box 4.4 Tor Browser**
>
> The Tor browser has been downloaded 120 million times in a period of one year, with 2 million users from 110 countries accessing the darknet every day. The darknet is growing very very fast and the closure of Silk Road (It was once known as 'the eBay of illegal goods', the popular darknet marketplace. It was taken down by the Federal Bureau of Investigation in October 2013.) seems to have had no impact on illicit activities there.

- The deep web is the largest growing category of new information on the Internet.
- More than half of the deep web content resides in topic-specific databases.
- 95 per cent of the deep web is publicly accessible information which is not subject to fees or subscriptions.
- The deep web is approximately 400 to 600 times more massive than the Internet.

4.9.4 Onion Router—TOR

The most widely used network to access the dark web is the onion router (TOR). It was initially developed by the U.S. Naval Research Laboratory as a tool for keeping private government communications. TOR transmits a user's web traffic through a series of computers located around the globe, called nodes, making a user's online movements difficult to track.

The TOR network is a peer-to-peer network but with some extra goodies to better ensure anonymity. When a TOR site is set up it picks several random peers to be interaction points, which advertise the service across the network without revealing its original location. To connect to the service, the TOR network establishes random rendezvous points, or computers essentially, where the source files get handed off. Because the transmissions are encrypted and decrypted at every juncture, the content cannot be traced back to its origin. The browser can only ascertain the first computer's IP address that it is connected to. It has to know this in order to talk, but there are many more steps after that which are hidden.

Other Darknets

Some of the other existing darknets are listed here:
- Invisible Internet Project (I2P), like TOR, is a network that sits on top of the Internet and provides some masking of its user's identities. The open-source I2P protocol is implemented using Java and supports web surfing, chatting, blogging, and file sharing.
- Freenet is probably the most well-known darknet after TOR and I2P. It is a P2P platform with the goal to be censorship and surveillance resistant.
- Zeronet is another such framework, but it is based on BTC cryptography and the BitTorrent network. Similar to Freenet, Zeronet aims to be a decentralized P2P network that resists orders to be taken down and cannot be knocked offline.
- Netsukuku is another P2P mesh network under development which, unlike TOR, is not another overlay network. It is a separate physical network and dynamic routing system.
- Riffle is an MIT project that is reported to be faster than TOR. It provides stronger security guarantees, bandwidth, and computation-efficient communication systems with strong anonymity.

4.9.5 Search Engines vs Deep Web

Search engines such as Google, AltaVista, Yahoo!, and Lycos have created technologies to *crawl* through websites and index them as a way for users to identify pages of interest and rank them to yield the best results for the user. They are now capable of indexing millions of pages in a short span of time.

In the deep web, *harvesting* is the process of extracting all the text-based content from each of the results pages and making the content suitable for some type of analysis, depending on the needs of end-users. A *harvester* acts on behalf of search engines. Specialized software, for example, BrightPlanet's Deep Web Harvester, is a tool for harvesting.

4.9.6 Deep Web Source Repository

The source repository is a library of deep web sources/websites. Once new sources are entered into the repository, they are indexed and saved for future harvests. Harvesters collect and group the sources according to their types.

4.9.7 Challenges
Some of the challenges associated with the deep web are presented here:
1. Dark web is an increasingly popular corner of the Internet where thousands of electronic communication device (ECD) users from around the globe interact anonymously and in many cases, illegally. For instance, 10,000-plus valid credit cards can be bought for an average price of $2–$10 each.
2. TOR network employs a special browser for encrypted Internet traffic. This poses difficulties for evidence collection by law enforcement agencies.
3. Deep web corresponds to the *invisible web* where the sites are not listed explicitly, as done by search engines on the surface web. Harvesters are required for listing.
4. The dark web is a privileged place for cybercriminals which, under specific conditions, could operate on anonymity.
5. Law enforcement agencies in many countries are still not in a position to deal effectively with the illegal activities that leverage infrastructures in the dark web.
6. Law enforcement agencies across the globe should share their strategy for legislation, technical abilities, and capacity building so as to deal with illegal activities on the dark web (Box 4.5).
7. The dark web poses new and formidable challenges for law enforcement agencies across the world who have been dealing, for decades, with more conventional international drug trafficking and arms sale. The reach and anonymity of these Internet operations are difficult to penetrate.
8. TOR networks are not easily visible through popular Internet search sites. The buyers and sellers don't exchange cash, dealing instead in often untraceable cryptocurrencies, usually bitcoins. So there are no banking records for investigators to subpoena.
9. Those who created and support the TOR network claim it as a means to protect online users' privacy and anonymity in the digital age. They do not condone its use for illegal activities as it was originally created for the U.S. Navy for end-to-end encrypted transaction. The same browser has unfortunately been misused by terrorists.
10. Strategies to monitor and counter nefarious activities on the dark web were intensely debated at the 'Cyber 360' conference organized by Synergia Foundation where it was concluded that agencies had no alternative but to disrupt such networks, rather than tracking them down and prosecuting criminals.
11. The issue of pharmaceutical crime on darknet marketplaces poses a great challenge to law enforcement and a severe risk to public health on a global scale.
12. Lack of training also poses challenges to handle the deep web.

Box 4.5 Silk Road
Silk Road is an online black market and the first modern darknet market. It is the best known platform for selling illegal drugs. As part of the dark web, it is operated as a Tor hidden service, such that online users are able to access it anonymously and securely without potential traffic monitoring. The website was launched in February 2011. There were only a limited number of new seller accounts available in the beginning and such an account had to be purchased at an auction. Later, a fixed fee was charged for each new seller account. In October 2013, the Federal Bureau of Investigation (FBI) shut down the website. Silk Road 2.0 came online on 6 November 2013 and was run by former administrators of Silk Road. It too was shut down on 6 November 2014 as part of the so-called 'Operation Onymous'. Silk Road 3.0 took it from there. However in January 2017, the site went down but has been fully operational since 7 May 2017. The revamped Silk Road 3.0 has some fancy new design features and the promise of increased security.

4.9.8 Counter Measures to Overcome Challenges with Deep Web

The following are some of the measures to overcome the challenges associated with deep web:

1. Periodical training of law enforcement agencies is absolutely necessary.
2. INTERPOL global complex for innovation (IGCI) offers capacity development support for member countries that want to create or further develop cybercrime units with specialized training on analysing darknet infrastructure.
3. Global legal solutions in the form of mutual legal assistance treaty (MLATs) for first incident response (IR) teams for exclusively dealing with the deep web should be framed.
4. Intelligence agencies should analyse the software available in the market, and utilize their R&D teams to combat deep web challenges by developing in-house software solutions.
5. Good practices to combat this menace must be shared globally, through conferences, webinars, etc.
6. Public education is another weapon that could help trademark counsel to mitigate the threat to the brands they protect. There is evidence, for example, that buying counterfeits on the darknet directly funds organized crime.

Issues such as the Y2K problem and encryption are steadily getting resolved. This clearly shows that challenges associated with the deep web can also be managed effectively.

POINTS TO REMEMBER

- *Cyber war* is a global web-based battle that is intended to attack computer systems and networks in the name of patriotism, political motivation, or revenge.
- *Cyber war tool* is a logical tool (cyber war weapon) that is used to conduct cyber war.
- Cyber war tools are categorized as reconnaissance tools, scanning tools, access and escalation tools, exfiltration tools, sustainment tools, assault tools, and obfuscation tools.
- *Cryptocurrency* is an encrypted digital or virtual currency which is not issued by a central authority.
- The characteristics of cryptocurrencies are virtual, vulnerable, cryptographic, decentralized, digital, open source, and using proof-of-work.
- Some of the existing cryptocurrencies include Bitcoin, Ether, Monero, Lite Coin, Ripple, Golem, Dash, Maid Safe Coin, Lisk, and Ethereum Classic.
- Bitcoin is the world's biggest digital currency. Its features as a *public* (anybody can use it), *permissionless* (anybody can build on top of it), *highly censor-resistant* (nobody can block your transactions), and *un-seizable* form of money all stem from its decentralized architecture.
- Bitcoin cash (BCH) is the method adopted by the bitcoin community (BTC) to scale this currency to more users. It is the child of a user-activated fork of the bitcoin blockchain ledger that split the bitcoin blockchain on 1 August 2017 with upgraded consensus rules that allow it to grow and scale.
- Blockchain is a digital transaction log or distributed ledger which digitally logs all electronic purchases that are carried out using virtual currencies like bitcoins/ether.
- The term smart contract was conceived by Nick Szabo, a computer scientist, law scholar and cryptographer, in 1997 long before the arrival of the bitcoin. He wanted to use a distributed ledger to store contracts.
- Turing completeness is a property of a programming language that allows a computer to simulate anything under the universe. It enables a computer to loop and process its own output in iteratively complex terms. This property is absent in almost all the public blockchains except ethereum.
- Ransomware is a scary malicious computer program that restricts user's access to the system in the form of locking the system's screen, locking the user's files, interrupting the normal boot process of a system, by encrypting web servers, or by infecting mobile devices via drive-by downloads and demands ransom to remove the restriction.
- Ransomware may be Scareware, Screen Locker, or Encrypting Ransomware.
- Home users, businesses, and public institutions are affected by ransomware.

- The sources of a ransomware attack are emails, links in them or message on social networks, security exploits in vulnerable software, pay per install, and drive-by downloads. Internet traffic is redirected to malicious websites, compromised websites, malvertising, printers and scanners, cloud-based tools, and ransomware-as-a-service.
- Ransomware can be stopped from being fully executed.
- Adopting the best practices can keep organizations and businesses safe from ransomware.
- Besides antivirus and antimalware, following the best practices will help in preventing a ransomware attack.
- The Government of India has framed guidelines for organizations involving ICT which includes appointment of a central information security officer (CISO), assessment of risks and threats, implementation of information security management system (ISMS), and establishment of an information security program. Besides this, the roles and responsibilities of CISO have been framed.
- The deep web is a subset of the Internet in which the content of databases and web services cannot be indexed by conventional search engines but provides full access to the Internet.
- The dark web is a part of deep web which is not indexed and provides restricted access to the Internet.
- The most widely used network to access the dark web is the onion router (TOR). It was initially developed by the U.S. Naval Research Laboratory as a tool for keeping private government communications.
- In the deep web, *harvesting* is the process of extracting all the text-based content from each results page and making the content to be suitable for some type of analysis depending on the needs of end-users.

KEY TERMS

Bitcoin It refers to a virtual currency which facilitates instant payment using peer-to-peer (P2P) technology.

Blockchain It is a digital transaction log or distributed ledger which digitally logs all electronic purchases that are carried out using virtual currencies such as bitcoins and ether.

Cryptocurrency This refers to encrypted digital or virtual currency which is not issued by a central authority. It is transferred between peers and confirmed in a public ledger via mining, which is a process of confirming transactions and adding them to the ledger.

Cyber war This refers to a global web-based battle that is intended to attack computer systems and networks in the name of patriotism, political motivation, or revenge.

Dark web It is a part of the deep web which is not indexed and provides restricted access to the Internet. The dark web hides the identity, action, and location of the Internet user.

Deep web It is a subset of the Internet in which the content of databases and web services cannot be indexed by conventional search engines, but provides full access to the Internet.

Ethereum It is a decentralized platform that runs smart contracts which are applications that run exactly as programmed without any possibility of downtime, censorship, fraud, or third party interference.

Ransomware It is a scary malicious computer program that restricts the user's access to the system in the form of locking the system's screen, locking the user's files, interrupting the normal boot process of a system, encrypting web servers, infecting mobile devices via drive-by downloads, and demanding ransom in order to remove the restriction.

Smart contracts These are self-executing contracts with the terms of agreement between the buyer and the seller directly written into lines of code. These are small computer programs that are stored across the distributed, decentralized blockchain network.

MULTIPLE-CHOICE QUESTIONS

1. Cyber war tool is a _____ that is used to conduct cyber war.
   (a) physical tool
   (b) logical tool
   (c) logical tool or cyber war weapon
   (d) none of these
2. Cryptocurrency transactions are facilitated through _____ for security.
   (a) public key
   (b) private key
   (c) public and private keys
   (d) Internet
3. _____ is a security-centric system which provides space on an individual's computer during coin exchange.
   (a) Lite Coin
   (b) Ripple
   (c) Dash
   (d) Maid Safe Coin
4. Bitcoin is a _____ form of money.
   (a) permissionless
   (b) highly censor-resistant
   (c) un-sizable
   (d) all of these
5. Which protocol(s) is/are supported by ethereum for the exchange of messages and static files?
   (a) Whisper protocol
   (b) Swarm protocol
   (c) Whisper and Swarm protocols
   (d) Client–Sever protocol
6. CryptoLocker is based on _____ algorithm.
   (a) public key
   (b) private key
   (c) public and private keys
   (d) none of these
7. _____ includes rogue security software and tech support scams.
   (a) Scareware
   (b) Screen Lockers
   (c) Encrypting Ransomware
   (d) None of these
8. Ransomware can be prevented from being fully executed by _____.
   (a) identifying the actions to be accomplished to successfully complete the infection process
   (b) identifying the location of specific executables
   (c) identifying and installing a tool that actively blocks behaviour
   (d) all of these
9. Some antivirus software have _____ to create virtual machines to test untrusted files.
   (a) dedicated apps
   (b) sandboxing functionality
   (c) all of these
   (d) none of these
10. _____ is a group of volunteer-operated servers that allow users to improve their privacy and security on the Internet.
    (a) Tor network
    (b) Peer-to-peer network
    (c) Client–Server network
    (d) Point-to-point network
11. Blockchain offers _____
    (a) transparency
    (b) verifiable consistency
    (c) consensus
    (d) all of these
12. _____ is the largest cyberattack ever.
    (a) WannaCry
    (b) Samsam
    (c) Petya
    (d) External Rock
13. _____ is one of the reasons for public institutions to be affected by ransomware.
    (a) Reluctance to adopt cyber security measures
    (b) Lack of data backup
    (c) No knowledge in cyber security education
    (d) None of these
14. _____ is a ransomware that targets home and business users of Microsoft's cloud-based productivity platform.
    (a) RANSOM_CYBER_CAM
    (b) RANSOM_CYBER_CAD
    (c) Both (a) and (b)
    (d) RANSOM_CYBER_AUTO
15. _____ the affected computer immediately and _____ if the computer is not fully affected.
    (a) Isolate, restart
    (b) Isolate, hibernate
    (c) Isolate, power off
    (d) Integrate, restart

REVIEW QUESTIONS

1. What is cryptocurrency? Explain the types of currencies.
2. Explain bitcoin and ethereum. Compare and contrast them.
3. Explain blockchain and its relationship with bitcoin.
4. Discuss in detail the (a) safety measures to be followed against a ransomware attack (b) steps to be carried out after a ransomware attack.
5. Explain the role of antivirus in ransomware.
6. Explain the role of national and international prevention, and that of the response team for ransomware.
7. What is deep web? What is dark web? How does deep web differ from dark web?
8. Write notes on the following: (a) victims of ransomware (b) the ways in which victims are infected by ransomware (c) consequences of a ransomware attack (d) post-delivery of ransomware.
9. Give an account of the deep web and its challenges.

APPLICATION EXERCISES

1. Imagine a situation when the moment you turn on your computer, a message prompt displays that the files are encrypted and you need to pay a ransom. What will be your next course of action?
2. Analyse cryptocurrencies such as bitcoin, bitcoin cash, and ethereum and the reasons for their price fluctuation in the market over a period of time.
3. Having learnt the purpose of TOR browser, explain a scenario of how it can be used productively or beneficially. Further, explain the cons of the TOR browser with an example, and suggest the actual challenges and the measures to be adopted to handle this. (*Hint*: Pharmaceutical crimes on the darknet marketplace employ TOR as a tool.)

BIBLIOGRAPHY

1. U.S. Department of Justice, *How to Protect Your Networks from Ransomware*, available at: https://www.justice.gov/criminal-ccips/file/872771/download (Accessed 8 January 2018)
2. Ministry of Electronics and Information Technology, Government of India (2017), *Key Roles and Responsibilities of Chief Information Security Officers (CISOs) in Ministries/Departments and Organisations Managing ICT Operations*, available at: http://meity.gov.in/content/key-roles-and-responsibilities-chief-information-security-officers-cisos (Accessed 8 January 2018)
3. Graham Cluely (2016), *How Boobytrapped Printers Have Been Able to Infect Windows PCs for over 20 years*, available at: https://www.tripwire.com/state-of-security/featured/boobytrapped-printers-windows-malware (Accessed 8 January 2018)
4. Limor Kessem (2017), *WannaCry Ransomware Spreads Across the Globe, Makes Organizations Wanna Cry About Microsoft Vulnerability*, available at: https://securityintelligence.com/wanna-cry-Ransomware-spreads-across-the-globe-makes-organizations-wanna-cry-about-microsoft-vulnerability/ (Accessed 8 January 2018)
5. McAfee—Antivirus, Endpoint Security, Encryption, Firewall, Email Security, Web Security, Network Security (1992), available at: https://www.mcafee.com/in/security-awareness/articles/Ransomware.aspx (Accessed 8 January 2018)
6. McAfee Labs (2016), *Taking Steps to Fight Back Against Ransomware*, available at: https://securingtomorrow.mcafee.com/mcafee-labs/taking-steps-to-fight-back-against-Ransomware/ (Accessed 8 January 2018)
7. Dan Turkel (2016), *Even the Best Antivirus Likely Can't Save your Files from a Ransomware Infection*, available at: http://www.businessinsider.in/Even-the-best-antivirus-likely-cant-save-your-files-from-a-Ransomware-infection/articleshow/50709985.cms (Accessed 8 January 2018)
8. Andra Zaharia (2017), *What is Ransomware - 15 Easy Steps To Protect Your System [Updated]*,

available at: https://heimdalsecurity.com/blog/what-is-Ransomware-protection/ (Accessed 8 January 2018)
9. Techsupportall (2016), *Top 5 Best Anti-Ransomware & Anti-Exploit Software 2017 to Avoid Ransomware Attacks (Free & Paid)*, available at: https://www.techsupportall.com/best-anti-Ransomware-software/ (Accessed 8 January 2018)
10. John E. Dunn and Christina Mercer (2017), *Best Anti-ransomware tools: How can I Remove Ransomware from My Computer?*, available at: http://www.techworld.com/security/best-Ransomware-removal-tools-how-clean-up-cryptolocker-cryptowall-extortion-malware-3626974/ (Accessed 8 January 2018)
11. Locky Ransomware, available at: https://www.escanav.com/en/articles/escan-locky-Ransomware.asp (Accessed 8 January 2018)
12. *Investopedia—Sharper Insight. Smarter Investing, Cryptocurrency*, available at: http://www.investopedia.com/terms/c/cryptocurrency.asp (Accessed 8 January 2018)
13. Introduction to Cryptocurrency—CryptoCurrency Facts, *What is a Cryptocurrency Wallet?*, available at: http://cryptocurrencyfacts.com/what-is-a-cryptocurrency-wallet/ (Accessed 8 January 2018)
14. *Ethereum for Beginners*, available at: http://www.coinscrum.com/wp-content/uploads/2016/06/EthereumForBeginners.pdf (Accessed 8 January 2018)
15. Julianne Harm, Josh Obregon, and Josh Stubbendick, (2015), *Ethereum vs. Bitcoin*, available at: http://www.economist.com/sites/default/files/creighton_ university_kraken_case_study.pdf (Accessed 8 January 2018)
16. Blockchain Lab (2016), *The DAO Attack Grid View – Deloitte*, available at: https://www2.deloitte.com/content/dam/Deloitte/ie/Documents/Technology/ie_the_dao_attack.pdf (Accessed 8 January 2018)
17. Let's Talk Payments (2015), *Know More About Blockchain: Overview, Technology, Application Areas and Use Cases*, available at: https://letstalkpayments.com/an-overview-of-blockchain-technology/ (Accessed 8 January 2018)
18. Steve (2016), *Surface Web, Deep Web, Dark Web - What's the Difference?*, available at: https://www.cambiaresearch.com/articles/85/surface-web-deep-web-dark-web----whats-the-difference (Accessed 8 January 2018)
19. INTERPOL (2015), *Pharmaceutical Crime on the Darknet - A study of Illicit Online Marketplaces*, available at: https://www.gwern.net/docs/sr/2015-interpol-pharmaceuticals.pdf (Accessed 8 January 2018)
20. Benjamin Brown (2016), *State of the Dark Web*, available at: https://www.akamai.com/cn/zh/multimedia/documents/state-of-the-internet/akamai-2016-state-of-the-dark-web.pdf (Accessed 8 January 2018)
21. Peter Biddle, Paul England, Marcus Peinado, and Bryan Willman, *The Darknet and the Future of Content Distribution*, available at: http://msl1.mit.edu/ESD10/docs/darknet5.pdf (Accessed 8 January 2018)
22. Daniel Sui, James Caverlee, and Dakota Rudesill (2015), *The Deep Web and the Darknet: A Look Inside the Internet's Massive Black Box*, available at: https://www.wilsoncenter.org/publication/the-deep-web-and-the-darknet (Accessed 8 January 2018)
23. BrightPlanet Deep Web Intelligence (2013), *Whitepaper on Understanding the Deep Web in 10 Minutes*, available at: https://brightplanet.com/2013/03/whitepaper-understanding-the-deep-web-in-10-minutes/ (Accessed 8 January 2018)
24. World Trademark Review (2014), *Shining a Light on the Darknet*, available at: http://www.worldtrademarkreview.com/Magazine/Issue/51/Features/Shining-a-light-on-the-Darknet (Accessed 8 January 2018)
25. Chris Woodford (2016), *The Internet*, available at: http://www.explainthatstuff.com/internet.html (Accessed 8 January 2018)
26. Eser Hangul, *How a Search Engine Works? An Explanation in 3 steps*, available at: https://tinobusiness.com/how-a-search-engine-works-an-explanation-in-3-steps/ (Accessed 8 January 2018)
27. Karl Hindle (2012), *How Search Engines Work Using a 3 Step Process*, available at: https://www.business2community.com/seo/how-search-engines-work-using-a-3-step-process-0322639 (Accessed 8 January 2018)
28. *Dark Side of the Web—Subcultures and Deviancy on the Dark Web*, available at: https://davidenew-

media.wordpress.com/workingterms/deep-web/ (Accessed 8 January 2018)
29. Mohit Kumar (2012), *What is the Deep Web? A first Trip into the Abyss*, available at: https://thehackernews.com/2012/05/what-is-deep-web-first-trip-into-abyss.html (Accessed 8 January 2018)
30. Bitcoin Economics (2016), *Silkroad 3.0 Is Back... Will It Last?*, available at: https://www.cryptocoinsnews.com/silk-road-3-0-back-will-last/ (Accessed 8 January 2018)
31. Jake Frankenfield (2017), *Bitcoin vs Bitcoin Cash: What's the Difference?*, available at: https://www.investopedia.com/tech/bitcoin-vs-bitcoin-cash-whats-difference/ (Accessed 8 January 2018)
32. Spencer Bogart (2017), *Bitcoin vs Bitcoin Cash: A Story Of Prioritization & Healthy Competition in Money*, available at: https://www.forbes.com/sites/spencerbogart/2017/11/13/bitcoin-vs-bitcoin-cash-a-story-of-prioritization-a-healthy-competition-in-money/#577aad0b4bcc (Accessed 8 January 2018)
33. Bitcoin.com (2017), *Difference Between Bitcoin Cash and Bitcoin*, available at: https://www.bitcoin.com/info/differences-between-bitcoin-cash-bcc-and-bitcoin-btc (Accessed 8 January 2018)
34. Sebastian Kettley (2017), *Bitcoin: What is the Difference between Bitcoin and Bitcoin cash?*, available at: https://www.express.co.uk/finance/city/838166/Bitcoin-price-bitcoin-cash-difference-explained-news-cryptocurrency-fork-value (Accessed 8 January 2018)
35. Investopedia—Sharper Insight, Smarter Investing, *Smart Contracts*, available at: https://www.investopedia.com/terms/s/smart-contracts.asp (Accessed 4 January 2018)
36. Mike Bursell (2017), *What Is a Blockchain Smart Contract?*, available at: https://opensource.com/article/17/12/whats-blockchain-smart-contract (Accessed 4 January 2018)
37. Shivdeep Dhaliwal (2017), *Ethereum Classic News*, available at: https://cointelegraph.com/tags/ethereum-classic (Accessed 5 January 2018)
38. Chris DeRose (2016), *Who Will Pay for Turing-Complete Smart Contracts?*, available at: https://www.coindesk.com/turing-complete-smart-contracts/ (Accessed 5 January 2018)
39. Mark Hedge (2018), *Crypto Creator Who is Satoshi Nakamoto? Bitcoin Creator Whose Identity is Unknown but Could Be One of the World's Richest People*, available at: https://www.thesun.co.uk/news/5037060/satoshi-nakamoto-bitcoin-inventor-worlds-richest/ / (Accessed 10 January 2018)
40. Michael K. Bergman (2001), *The Deep Web: Surfacing Hidden Value*, available at: https://quod.lib.umich.edu/j/jep/3336451.0007.104?view=text;rgn=main (Accessed 25 January 2018)
41. S. Murugan (2018), *Deep Web: Challenges to Law Enforcement Agencies*, available at: Souvenir – 61st All India Police Duty Meet, pp. 45–53.

Answers to Multiple-choice Questions

1. (c)	2. (c)	3. (d)	4. (c)	5. (c)
6. (a)	7. (a)	8. (d)	9. (b)	10. (a)
11. (d)	12. (a)	13. (a)	14. (b)	15. (c)

Introduction to Cyber Forensics

Learning Objectives

This chapter provides an introduction to cyber forensics. It presents the components of security, and the definitions of cyber forensics. It highlights forensic investigation and forensic examination processes besides describing their benefits. The chapter explains the types of forensics, namely disk forensics, network forensics, wireless forensics, database forensics, malware forensics, mobile forensics, GPS forensics, and memory forensics. Email forensics is explained in detail. Incident and incident handling approaches, and the role of CSIRT are also discussed. The reader will be familiar with the following after studying the chapter:

- Relationship between cybercrime, cyber forensics, and cyber security
- Computer forensics
- Disk forensics
- Network forensics
- Database forensics
- Email forensics
- Malware forensics
- Memory forensics
- Building a forensics computing lab
- Incident
- Incident handling
- CISRT

5.1 INTERRELATION AMONG CYBERCRIME, CYBER FORENSICS, AND CYBER SECURITY

Cybercrime refers to any criminal offence that involves a computer/network or an electronic communication device (ECD), where the computer is used to either commit the crime or is the target of the crime.

Computer forensics, alternatively referred to as cyber forensics, focuses on the investigation of cybercrimes. It deals with the acquisition, analysis, and admissibility of digital evidence from a computer or any ECD after the occurrence of a cybercrime. The evidence gathered will be used in criminal proceedings. Thus, cyber forensics attempts to determine what has happened to the digital media, as a result of the incident, and to determine the parties associated with the incident and the commission of cybercrime. Computer forensics could be used to gather evidence not only in cybercrimes but also in other crimes.

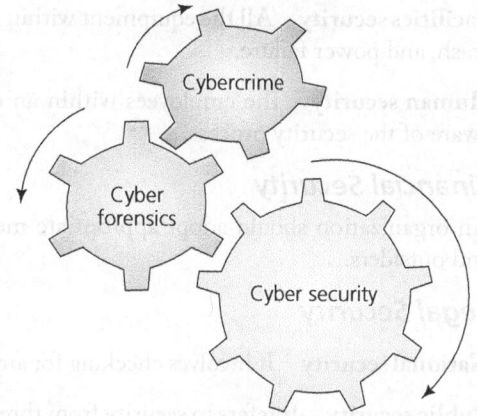

Fig. 5.1 Interrelated techniques and technologies

Cyber security refers to the technologies, processes, and practices designed to protect networks, computers, programs, and data from attack, damage, or unauthorized access. It deals with the determination of the vulnerabilities that exist in networks, computers, programs, and data and patches the loopholes. Any breach in cyber security makes commission of cybercrime successful.

Cyber forensics is a response to cybercrime to investigate if anything adverse has happened and to determine the source of the incident, the associated damage, and to recover the data. For example, with hacking, cyber forensics attempts to determine the hacker, the data lost, and to recover the data. On the other hand, cyber security prevents incidents of cybercrime with the implementation of security measures. Thus cybercrime, cyber forensics, and cyber security are interrelated, as shown in Fig. 5.1, where every cybercrime drives the cyber forensics team and the cyber security team to work in tandem, to respond to and to prevent cybercrimes respectively.

In a computing context, security can be viewed from different dimensions.

5.1.1 Security

The security of an organization has four dimensions: IT security, physical security, financial security, and legal security.

IT Security

Application security Applications should be developed to overcome vulnerabilities and threats. Any loophole in the application and any security weakness will be exploited by the offender.

Computing security Computing systems in an organization are exposed to threats such as viruses, Trojans, and other intentional attacks. An efficient security policy should be in force so as to avoid threats to business operations.

Data security Data pertaining to the organization should be secured from unauthorized manipulation, theft, loss, and secrecy. Any deviation could result in huge revenue loss to the business and reputation.

Information security Information security attempts to protect information from illegal use, access, and modification and ensures confidentiality, integrity, and availability of information.

Network security An organization's network should be secure enough to facilitate safe data transfer. Any lapse in network security exposes the organization, leading to system collapse and disruption of business.

Physical Security

Facilities security All the equipment within an organization should be secured from physical damage, system crash, and power failure.

Human security The employees within an organization should be provided with security training and be aware of the security process.

Financial Security

An organization should adopt appropriate measures so as to be financially immune to threats from insiders and outsiders.

Legal Security

National security It involves checking for any lapse in security from threats that arise out of nationwide issues.

Public security It refers to security from threats associated with societal issues such as riots, strikes, or clashes.

This chapter focuses on cyber forensics, but incident handling is also discussed later in the chapter.

5.2 CYBER FORENSICS

Cyber forensics is an electronic discovery technique that is used to determine and reveal technical criminal evidence. It often involves the extraction of electronic data for legal purposes. Cyber forensics is also known as computer forensics. The term encompasses activities such as acquiring, retrieving, preserving, and preparing data for presentation, usually at a forensic lab, on the data that is processed electronically and stored on media present in an ECD.

5.2.1 Definition

According to the McGraw-Hill Dictionary of Scientific and Technical Terms,

Computer forensics is the study of evidence from attacks on computer systems in order to learn what has occurred, how to prevent it from recurring and the extent of damage.

Judd Robbins, a computer forensics investigator (Lunn 2001) has defined computer forensics as,

Computer Forensics is the application of computer investigation and analysis techniques in the interests of determining potential legal evidence.

Rodney McKemmish (McKemmish 1999) has defined it as,

Forensic Computing is the process of identifying, preserving, analyzing and presenting digital evidence in a manner that is legally acceptable.

5.2.2 Need

The need for cyber forensics is summarized here:

1. Many approaches such as fingerprinting and DNA extraction have already been proposed to establish liability and guilt in the event of occurrence of an incident or a crime. Such traditional approaches are either insufficient to prove an incident or end in deadlock during investigation. This necessitates investigation procedures like cyber forensics to match the current situation and prove an incidence of crime.
2. With the advent of the Internet, offences and crimes now span a diverse range from hacking, pornography, spamming, identity theft, cyber terrorism, etc. Hence, cyber forensics is essential to curb and control cybercrime in an effective manner.
3. Compared to traditional crimes, perpetrators of cybercrimes have changed the modus operandi of the crime, but there still remains some similarities. Hence, an investigation must be performed by a cyber forensics body after framing statutory and non-statutory measures.
4. Cybercrime can spread across boundaries in no time. Since every service is built on top of an IT infrastructure, cyber forensics is necessary to address incidents that conflict with the legal provisions. Cybercriminals and terrorists from different parts of the globe can be tracked using the IP addresses they operate from.
5. The integrity and existence of an organization's computer system and network infrastructure have to be ensured. Cyber forensics is essential to prosecute a criminal if a compromise of some sort is observed. It is even more needed to interpret the actual evidence so as to prove the attacker's action and the organization's innocence.

5.2.3 Objectives

The objectives of cyber forensics are as follows:

1. To identify the evidence associated with a malicious activity in a short span of time
2. To recover and analyse the evidence and related materials from computers and other ECDs
3. To present the collected evidence in a court of law
4. To estimate the potential impact of the malicious activity
5. To assess the intention and identity of the offender

5.2.4 Computer Forensics Investigations

Cybercrime investigation works in phases. In the first phase, the forensic investigator does a preliminary analysis and gathers information from the crime scene. In the second phase, he/she works on forensic copy acquisition and recovery, and in the third phase he/she performs a detailed analysis and prepares a comprehensive report. This is shown in Fig. 5.2. For doing these, the forensic investigator should possess extensive knowledge in this area and highly specialized skills. The evidence in the case of a cybercrime has to be gathered from ECDs.

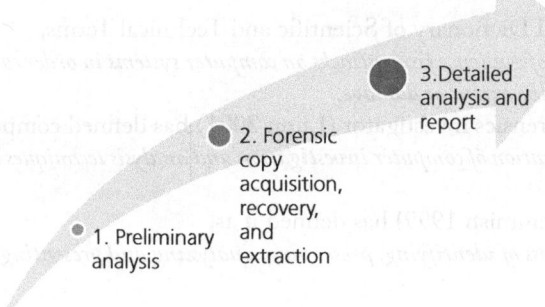

Fig. 5.2 Phases of cybercrime investigations

5.2.5 Steps in Forensic Investigation

Cyber forensic investigation should ensure the integrity of the evidence while handling and analysing so that the evidence is admissible in court. The steps in forensic investigation are explained here, and is shown as a flowchart in Fig. 5.3.

1. The investigation starts when a crime is reported or a complaint is received.
2. In response to the complaint, the following are made:
   (a) If evidence has to be gathered from a third party, a notice is served to preserve it.
   (b) If it is a criminal offence, a first information report (FIR) is filed.
   (c) A search warrant (if required) is obtained from the court.
3. First responder or computer emergency response team (CERT) procedures are performed (Boxes 5.1 and 5.2).
4. Evidence is seized from the crime scene. This includes photographing the scene and marking the evidence. Necessary documentation is also done. Witnesses present during the seizure of evidence and the suspect himself/herself can be interviewed. If there are any complications in evidence collection, or if the investigating officer (IO) does not possess evidence collection expertise, a third-party expertise may be called in. Evidence involving a third party must also be collected. The chain of custody has to be documented.

> **Box 5.1 First Responder**
>
> The first responder is the person who first reports to the crime scene and assesses the ECD present at the crime scene. He/She can be a network administrator, a law enforcement officer, an investigating officer, or a person from the forensic lab. He/She is responsible for protecting, integrating, and preserving the evidence from a crime scene. The first responder should be competent enough to handle the forensic investigation procedures. He/She has to collect evidence in a forensically secure manner and ensure the admissibility of the evidence.

Introduction to Cyber Forensics 153

> **Box 5.2 Computer Emergency Response Team**
>
> The CERT is an expert group that handles computer security incidents. It is also called computer security incident response team (CSIRT). It was first formed in 1988, when worms and viruses hit the Internet at Carnegie Mellon University, under the US government contract so as to handle computer security incidents. Now CERT/CSIRT is an integral and essential component of any organization taking care of information security operations.

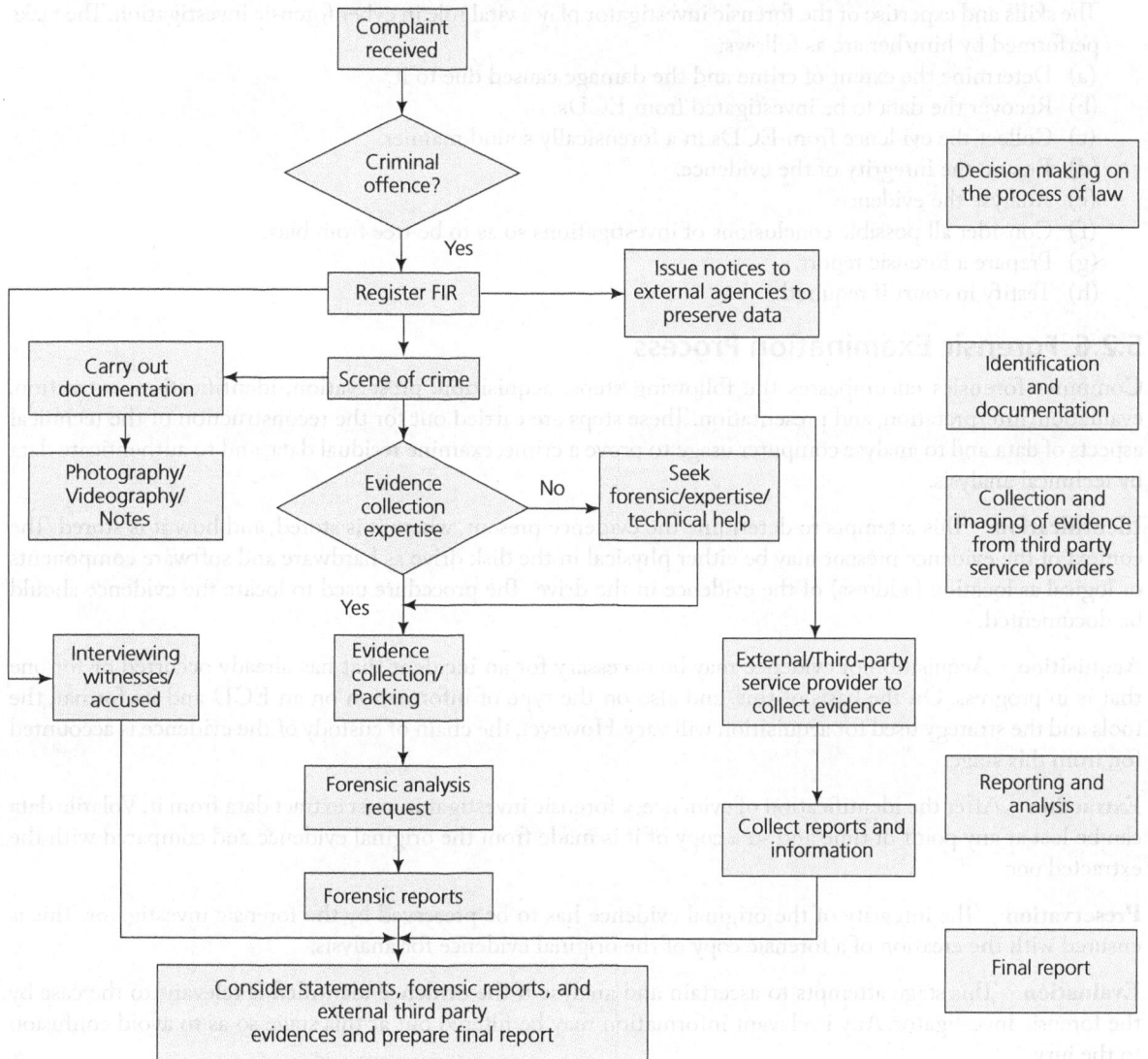

Fig. 5.3 Flowchart of cyber forensic investigation

5. The collected evidence is numbered and securely transported to the forensic laboratory for analysis.
6. The following are done at the forensic lab:

(a) Two bit stream copies of the evidence are created. The hash values of the original and forensic copies are verified.
   (b) Chain of custody is maintained.
   (c) The original evidence is stored in a secured location.
   (d) The forensic copy is analysed for evidence.
   (e) A forensic report is prepared stating the methods and the recovery tools used, the potential evidence, and the findings.
   (f) The report is presented to the client.
7. In some situations, the forensic investigator may even be called to testify in court as an expert witness. The skills and expertise of the forensic investigator play a vital role in cyber forensic investigation. The tasks performed by him/her are as follows:
   (a) Determine the extent of crime and the damage caused due to it.
   (b) Recover the data to be investigated from ECDs.
   (c) Collect the evidence from ECDs in a forensically sound manner.
   (d) Ensure the integrity of the evidence.
   (e) Analyse the evidence.
   (f) Consider all possible conclusions of investigations so as to be free from bias.
   (g) Prepare a forensic report.
   (h) Testify in court if required.

5.2.6 Forensic Examination Process

Computer forensics encompasses the following steps: acquisition, preservation, identification, extraction, evaluation, interpretation, and presentation. These steps are carried out for the reconstruction of the technical aspects of data and to analyse computer usage to prove a crime, examine residual data, and to authenticate data by technical analysis.

Identification This attempts to determine the evidence present, where it is stored, and how it is stored. The context of the evidence present may be either physical in the disk drive as hardware and software components or logical as location (address) of the evidence in the drive. The procedure used to locate the evidence should be documented.

Acquisition Acquisition of evidence may be necessary for an incident that has already occurred or for one that is in progress. On the basis of this, and also on the type of information on an ECD and its format, the tools and the strategy used for acquisition will vary. However, the chain of custody of the evidence is accounted for, from this stage.

Extraction After the identification of evidence, a forensic investigator must extract data from it. Volatile data can be lost at any point of time and so a copy of it is made from the original evidence and compared with the extracted one.

Preservation The integrity of the original evidence has to be preserved by the forensic investigator. This is ensured with the creation of a forensic copy of the original evidence for analysis.

Evaluation This stage attempts to ascertain and analyse if the evidence identified is relevant to the case by the forensic investigator. Any irrelevant information may be filtered out at this stage so as to avoid confusion to the jury.

Interpretation The forensic examiner should interpret what is found during analysis in an easily understandable manner.

Presentation The FE presents the suitability of the evidence with respect to the case before the court. Documentation related to evidence should be prepared—chain of custody and evidence analysis. He/She has to defend the claims raised with respect to evidence handling and processing. Such evidence will be subjected to scrutiny by the opponent. The chain of custody helps to substantiate claims when the authenticity of the evidence is challenged.

5.2.7 Methods Employed in Forensic Analysis

The following are the methods involved in forensic analysis:

Data recovery   This includes recovering and analysing deleted files that have not been overwritten as yet, as well as carving out portions of text from the unallocated and slack space.

String and keyword searching   This attempts either to identify readable text within a binary file or a specific string within a file. Search is made with known and unknown files as well as unallocated and slack space.

Volatile evidence analysis   This provides the analyst details about the state of the system by looking into connections, processes, and cache tables gathered from the RAM.

Timeline analysis   This attempts to create a timeline of events and makes an analysis on the basis of modified, accessed, and changed times associated with files that are imaged.

System file analysis   This reveals to the analyst, any unauthorized changes that are made to system binaries.

5.2.8 Classification of Cyber Forensics

The field of cyber forensics can be broadly classified as follows:
1. Disk forensics
2. Network forensics
3. Wireless forensics
4. Database forensics
5. Malware forensics
6. Mobile device forensics
7. GPS forensics
8. Email forensics
9. Memory forensics

The following sections explain each one of these briefly.

5.2.9 Benefits of Cyber Forensics

The benefits offered by cyber forensics are as follows:
1. The ECD subjected to forensic examination is protected from alteration, damage, data corruption, and viruses.
2. Files, hidden files, and password-protected files are discovered, and deleted data is recovered from the ECD.
3. The contents of hidden files and swap files used by the application programs as well as the operating system are revealed.
4. The contents of password-protected and encrypted files are accessed using tools .
5. All possible and relevant data present in special areas of the disk are analysed.
6. All the possible relevant files are discovered.
7. It offers expert consultation and testimony, when required.

5.3 DISK FORENSICS

Disk forensics is the process of extracting forensic information from storage media such as hard disk, USB drive, CD, DVD, flash drive, and floppy disk, among others. The steps in disk forensics are as follows:

Identification of evidence   This step locates the source of evidence (storage media/disk) at the crime scene.

Seizure and acquisition of evidence   At the crime scene, the hash value of the original evidence in the storage media, after seizure, is computed using a forensic tool. The hash value is stored and the evidence is packed and sealed. Acquisition is the process of taking a bit-by-bit copy of the original evidence which itself is write-protected. This is done in a forensic laboratory.

Authentication and analysis of evidence This step is conducted at the cyber forensics lab where the hash values of both the original media and forensic copy created are compared to make sure that they are the same.

Preservation of evidence After acquisition and authentication of evidence, it is kept in a place that is secured from magnetic and other radiation sources.

Analysis of evidence Analysis is the process of collecting the evidence from the storage media, as the case may be.

Report on findings The people associated with the investigation should prepare a case analysis report that includes all the details: examination, analysis, and authentication. It should also include the observations of the examiner and the conclusion. The report should be simple and well written in a way that is easy for even a non-technical person to comprehend.

Documentation Every activity at every step is documented to make the case admissible in court.

5.3.1 Challenges

Disk forensics poses the following challenges:

1. Text search utility is usually used by examiners of media to find keywords which would serve as evidence. However, if the keyword is misspelt, files in the media are encrypted, or stored as graphic, it becomes impossible to gather evidence. Certain graphic files will open only if it is extracted from the image file and opened with the respective software. In such circumstances, it is the responsibility of the examiner to look for other alternatives to gather the evidence rather than concluding that the evidence does not exist.
2. Hidden files, encrypted files, files with disguised names, and files whose extensions are altered make it difficult to gather evidence. Similarly, hidden areas in the storage media provide room for hiding evidential data unless specialized tools are used to analyse them.

5.4 NETWORK FORENSICS

Network forensics refers to the capture, recording, and analysis of network events so as to discover any malicious activity, security attack, or any violation. It finds applications in cases relating to hacking, fraud, data espionage, data theft, defamation, narcotics trafficking, credit card cloning, software piracy, sexual harassment, etc.

5.4.1 Tools for Analysis

Some of the common network forensic analysis tools are as follows:

1. Intrusion detection system (IDS) is a device or a software application that monitors networks and the systems under it for malicious activity or policy violations and maintains a record of the activity. Any such detected activity or violation will be reported to the administrator of the network. The records of IDS can be used to reconstruct a security incident.
2. Logging gathers and records the activity on a network with the help of IDS which can help in tracking an offender (hacker).
3. Packet capturing tools can gather and record every bit exchanged between any two designated hosts. Since a large amount of data is generated by these tools in a short span of time, it cannot be used to capture data for a longer period.
4. NetFlow data collector gathers and records data about every network connection—for example, the source, the destination, and the volume of data. Since it preserves only a summary of every network connection, it can be used to capture data over a long period.

5.4.2 Challenges

Some of the challenges associated with network forensics are listed here:

1. A large volume of data, in the order of gigabytes, is generated by the network every day and so it becomes tedious to search for evidence in case an incident is reported or discovered long after its occurrence.
2. The inherent anonymity of Internet protocols with MAC address at the data link layer, IP address at the network layer, and an email address at the application layer and the possibility that all these three addresses can be spoofed poses the biggest challenge in identifying the source of the incident. However, network forensic analysis tools make it possible to figure out the network activity.
3. Single-purpose tools perform network forensic tasks such as collection, filtering, and stream reassembly from applications, routers, firewalls, and other authentication systems. However, it is insufficient to figure out the network activity. Raw network packets should be captured to gather the highest level of traffic detail. This is possible with sniffing.
4. Sessioning is the act of assembling raw data packets between specified points as a complete stream which helps in gathering information about a specific communication (who was part of the communication, when it took place, and what was communicated). Protocol analysis tools can be used to produce a tree-oriented view of sessions and such a visual presentation gives a clear picture of what happened on the network.

5.5 WIRELESS FORENSICS

Wireless forensics is a discipline associated with network forensics. The wireless forensic process involves capturing the data moving over the network and analysing network events so as to uncover network anomalies, discover the source of security attacks, and investigate breaches on computers and wireless networks to determine whether they have been used for illegal or unauthorized activities. Its main goal is to provide the methodology and tools required to collect and analyse (wireless) network traffic that can be presented as valid digital evidence in a court of law. The evidence collected can correspond to plain data or, with the broad usage of voice-over-IP (VoIP) technologies, especially over wireless, and can include voice conversations. The stages in wireless forensics are of course the same with disk forensics and network forensics.

Traffic analysis in wireless networks involves the following stages:

1. Data normalization and mining to search through the data
2. Traffic pattern recognition for identifying suspect patterns
3. Protocol dissection for analysing the header fields
4. Reconstruction of application sessions for visualization

However, before analysis, data from multiple sensors have to be merged.

5.5.1 Forensic Tools

A network forensic analysis tool (NFAT) is available for network forensics, but there is no such alternative for wireless forensics. However, the following tools can help investigators in collecting useful evidence:

1. The graphical Wireshark protocol dissector is used to inspect every field of the frames captured.
2. ngrep (network grep) is used to search for specific strings in the contents of the frames.
3. Text-based tcpdump or tshark sniffers are used to automate and script the analysis of certain tasks, for example, filtering traffic based on specific conditions.

5.5.2 Challenges

The challenges with wireless forensics are as follows:

1. The first challenge is with respect to radio frequency communication and the complexity of the medium. To gather evidence, the forensic tool should support 802.11. Any standard wireless equipment has a single radio component that can handle one channel at a specific point of time. If there are multiple access points, wherein a suspect is located, the tool should be capable of listening to all channels.
2. Wireless clients roam from one access point to the other when its network interface card (NIC) determines the signal strength of the former to be weaker than the latter. A wireless forensic tool should keep track of

data during roaming only if it has the capability to monitor all the channels with multiple traffic capture sensors. Otherwise, some portion of the session will be lost affecting the evidence collected.
3. With wireless traffic, control and management frames are exchanged in addition to data frames for the purpose of synchronization. This leads to processing overheads and considerable amount of storage is needed to save the data collected.

5.6 DATABASE FORENSICS

A database is a collection of records. Usually critical and sensitive information are stored in databases (e.g., bank account of customers, health data of patients, etc.). Threats to database security may be (a) from an insider who tampers with the database with fraudulent transactions and performs malicious actions, or (b) from an outsider who performs an intentional unauthorized attempt to access the database to destroy data. Such security incidents cause loss to databases. The goal of database forensics is to determine the security breach to a database, when it occurred, and to revert any unauthorized data manipulation operations.

Thus, the objective of database forensics is five-fold:

1. To prove or disprove the occurrence of a data security breach
2. To determine the scope of database intrusion
3. To retrace the user DML and DDL operations
4. To identify the data before and after transactions
5. To recover previously deleted database data

The first step in database forensics is to determine whether the database system is actually breached. Error logs and failed logins usually raise suspicion of breach. The next step is to determine which data records were retrieved. This could be ascertained from (a) the data cache, which holds recently accessed data pages, (b) the plan cache, which holds the cached database statements, and (c) the server state, which gives the most recently executed statement by session.

Thus, the primary information sources for any database breach are as follows:

1. Files where the metadata resides, for example, MAC gives the time the file was last modified, last accessed, and when it was changed or created. A timeline analysis can be used to determine any breach.
2. Internal structures refer to the SQL server artifacts in the form of cached data. This can be traced with forensic tools (e.g., Windows Forensic Toolchest and other automated scripts).
3. Logical structures or the index files also provide information in the form of trees for different node entry sequences.

5.6.1 Forensic Approaches

The two approaches to database forensics are explained here:

Reactive Approach

In this, forensics begins after a data breach has occurred and works with the goal of reconstructing or recovering the original state of the database. A reactive approach relies on traditional forensic analysis such as imaging and data carving which are not fully compatible with the database. Hence, it does not ensure evidence integrity which in turn has an impact on admissibility in legal proceedings. However, reactive approaches are not effective in responding to an incident because of limited time and resources, and the need to rely on ad hoc practices depending on the DBMS. Thus reactive approaches, though more developed, are not fully admissible for forensic purposes.

Proactive Approach

This is a formal approach to the forensic analysis of databases, and has two characteristics, resilience and provenance. Resilience ensures accountability (auditing and forensics), and deploys security configurations and controls to detect and prevent security incidents caused by insiders. Provenance is a property of accountability to trace any activity back to its source (time and locations). Chain of custody is ensured by provenance of evidence during its recording and storage.

Proactive approaches are an emerging research trend but are more flexible when it comes to responding to an incident and has higher likelihood of admissibility in legal proceedings. Proactive approaches ensure the integrity of the evidence, both during and after a security event, thus making it admissible. They meet both audit requirements and the forensic requirements, where the former includes generating reliable evidence by logging and monitoring user action on the database and the latter includes investigating incidents by identifying, preserving, acquiring, and analysing the evidence to finally report on and reflect upon the events.

5.6.2 Forensic Methodology

Database forensics is achieved in four stages:

Investigation Preparedness

In this stage, a forensic workstation is first configured. An SQL server forensics incident response team (IRT) is created to develop SQL server incident response scripts. These scripts can be integrated with automated live forensic suites if necessary.

Incident Verification

This stage attempts to determine whether an incident has occurred by looking for signs of penetration. Identification of signs of penetration is made from any one or more of the following:
1. SQL server penetration
2. Active unauthorized SQL server connections
3. Past unauthorized SQL server access from SQL server error logs, plan cache, and other session details.

Collection of Artifacts

Artifacts portray the occurrence of the incident and are collected from various sources. Artifacts may be either resident SQL server artifacts or non-resident SQL server artifacts.
1. *Resident SQL server artifacts* may be either volatile or non-volatile. Volatile artifacts include data cache, plan cache, cache clock hands, active virtual log files (VLFs), server state, and ring buffers. Non-volatile artifacts include reusable VLFs, table statistics, SQL server logins, authentication settings, authorization catalogs, schemas, native encryption, databases, database users, database objects, jobs, triggers, SQL server error logs, trace files, data files, endpoints, CLR libraries, AutoEXEC procedures, time configuration, server versioning, server hardening, server configuration, collation settings, and data types and data page allocations.
2. *Non-resident SQL server artifacts* include system event logs, external security controls, and web server logs.

Analysis of Artifacts

1. The first step in this stage is to carry out a pre-analysis by creating an image employing write blockers to create a repository, the database.

> **Box 5.3 Detecting Database Breaches with Honeypots**
>
> SQL injection is the most popular hacking method, and occurs using an application. Usually such a breach is detected after a long period and honeypot provides an opportunity to detect a breach happening currently. The database honeypot is much like a table populated with data that is not necessary for auditing but makes the hacker less suspicious, and should appear as if it is part of the application schema. The table can even have an index to a different honeypot table. The table is audited with vendor specific standard auditing. Any action detected on the table raises a notification to the administrator. The purpose of the table is to make the hacker act on the wrong data.

2. Then a security audit can be performed using honeypots. Besides this, SQL logs and system event viewer logs can be analysed. In addition, forensic tools and profiler-trace, or monitoring software like Idera can be used to carry out a thorough analysis (Box 5.3).

5.7 MALWARE FORENSICS

Malware is a tool commonly used by cybercriminals. Malware forensics is the process of examining the computer system so as to find the malicious code, determine how it got there, and what changes it has caused to the system. End users play a major role in malware forensics and they should be trained to identify malware threats so that they don't go unidentified and spread across the network. Hence, when users shut down or restart their machines, information needed for forensics will be lost. However, such information actually helps FEs to locate the source of malware, or the user's most recently visited websites.

The more a system is susceptible to malware, the higher the risk of data leakage, corruption, and loss. If malware is detected early, malware forensics can recover the system memory and end user system usage. The potential indicators of malware attack are (a) abnormal performance exhibited by the machine, (b) random rebooting of the machine, (c) opening of files without user input, (d) drives becoming suspiciously full, and (e) message pop-ups by the malware detection software.

The malware forensic process begins with the examination of the following: master boot record, volatile data after obtaining information about the operating system and its configuration, files on the system that were identified in volatile data, hash of the files on the system, programs that run on the system, auto-start locations, host-based logs, file system artifacts, web browsing history, and suspected malicious files. Besides these, a timeline analysis and keyword search can also be performed.

Prior to forensic examination, it is advisable to disconnect the system from the network so as to control the malware infecting other systems.

5.7.1 Malware Analysis

Malware analysis is performed with the following action plan:

Removal of malware Once the presence of malware is ascertained, after a thorough forensic examination, the hard disk is cleared and the machine is re-imaged so as to prevent other systems in the network from getting affected.

Scanning of machine for malware This is done to ascertain whether there is infection by malware since the last scan. Scanning the machine should be performed with the updated version of the malware detection software.

Gathering of data/evidence Data is collected from the machine and the user, and includes logs from anti-malware, browser cookies, history, and email history. This information can be used to figure out the origin of malware and to alert other users in the network and protect machines in the network.

5.8 MOBILE FORENSICS

The term mobile forensics refers to the recovery of digital evidence from mobile devices. It is a branch of digital forensics. Some of the challenges with mobile forensics are technical, legal, and administrative and are as follows:
1. Data on mobile devices can be stored, accessed, and synchronized with multiple devices. Since the data is volatile and can be transformed or erased in a fraction of time, specialized skills are expected from an examiner to acquire and analyse the evidence on it.
2. There are many models of mobile devices in the market, each one varying in size, hardware, features, and the operating system. Therefore, the examiner is expected to remain updated on the models and their respective forensic techniques.

3. Most mobile platforms have built-in security features, for example, default encryption mechanisms that necessitate the examiner to break through the encryption mechanisms to extract data from the mobile device.
4. The presence of anti-forensic techniques such as data hiding and remote wiping may pose restrictions to the examiner in the acquisition of evidence from mobile devices.
5. The data on a mobile device may be altered intentionally or unintentionally. An accidental reset of the mobile device by the examiner may result in loss of data. Further, a well-versed suspect can attempt to alter the data by moving a file, renaming a file, etc. Any malicious program from within a mobile device or over a wired or wireless interface may attempt to destroy data on it.
6. Legal issues refer to the issues that confuse a forensic examiner (FE), as regards the findings. For example, if the examiner ascertains that an upload/download of a file or installation of virus has been done on a suspect's mobile device, a competent opposing lawyer may claim that to be a result of a Trojan. A powerful defence by the FE, with an in-depth analysis and a detailed report of the evidence, is required.
7. The administrative issues are because there are no accepted standards tied to legislations and so the competence and integrity of FEs are questionable.

5.8.1 Stages

Mobile forensics is done in five stages: (a) seizure, (b) preparation, (c) acquisition, (d) examination or analysis, and (e) presentation and reporting.

Seizure The examiner observes the state of the mobile device at the crime scene. If the device is in 'on' state, its network connection is disabled and the flight mode is enabled to avoid any exchange of data using Bluetooth or Wi-Fi access points. Besides this, it will prevent the data in a mobile device from being erased with a remote wipe command. The mobile device should be packaged in a Faraday bag (which blocks radio signals to or from the mobile device), and disabling its network connectivity prevents draining of battery and also protects from leaks within a Faraday bag. If the mobile device is password-protected, the examiner may have to either determine the PIN or bypass the lock. If the mobile device is in 'off' state, it is packaged in a Faraday bag.

Preparation The following information has to be identified by the FE for any mobile device:

Legal authority Before acquisition, the FE should ascertain what legal options exist to acquire data from the mobile device and also analyse potential limitations.

Goals of examination The FE should identify what data is essential and the depth to which examination is to be carried out. Depending on this, the tools are selected for extraction of evidence, which in turn increases the efficiency of the examination stage.

Make, model, and identifying information of device This helps in the selection and usage of appropriate tools to gather evidence from the mobile device.

Removable and external data storage If any such storage, for example, a memory card or SD card is present, it is detached from the mobile device and analysed separately using appropriate forensic techniques.

Acquisition This stage may be either physical where the raw memory data is extracted while the mobile device is powered off, or logical where the file system of the mobile device is acquired. Physical extraction directly accesses the device's flash memory and acquires a bit-by-bit copy of the file system including deleted data and unallocated space. It is done using forensic tools. On the other hand, logical acquisition is done with the device manufacturer's API to synchronize the mobile device with the forensic workstation. It is easy to perform and can also be done using forensic tools but acquires only the files on the mobile device and not the data contained in the unallocated space.

Table 5.1 Potential evidence in mobile devices

Forms of evidence	Description
Address book	Contact names, numbers, and email addresses
Call history	Dialled, received calls with call duration, and missed calls
SMS	Sent and received text messages
MMS	Media files such as sent and received photos and videos
Email	Sent, received, and drafted email messages
Web browser history	History of websites visited
Photos/Videos	Photos/Videos captured using the camera in the mobile device, the ones downloaded from the Internet, and those received from other devices
Music	Music files downloaded from the Internet and those received from other devices
Documents	Documents created using the application in the mobile device, those downloaded from the Internet, and the ones received from other devices.
Calendar	Calendar entries and appointments
Network communication	GPS locations
Maps	Searched and downloaded maps
Deleted data	Information deleted from the mobile device

As explained, during preparation, appropriate tools are used for the extraction of data from the mobile device. Mobile operating systems play a significant role in determining the way the FE could access the mobile device. For example, Android (a Linux-based OS and Google's open source platform for mobile phones) operating systems facilitate terminal-level access while this is not supported by iOS or the iPhone operating system (the OS by Apple Inc. present in mobile devices such as iPad, iPod touch, and iPhone). More details on mobile operating systems are available in Chapter 8. The extracted data is compared with the device to ascertain the accuracy of extraction. Instead, multiple tools can be used for extraction and their results compared. In case of file systems, hash values of the extracted and the original data may be checked for a match to ascertain the integrity.

Evidence examination The potential evidence from mobile devices are listed in Table 5.1. The evidence extraction step requires the FE to document the extraction process and to ensure that the results are repeatable and defendable. The documentation should include the following: the physical condition of the mobile device, photograph of the mobile device with its components, the state of the mobile device, the make and model of the mobile device, tools used for acquisition, tools used for examination, and the data and observations made during forensic examination. The data extracted from the device has to be preserved for presentation before the court.

Presentation and reporting In this stage, the FE prepares a structured report on the findings after examination of the digital evidence present in the mobile device. The report should be well written with appropriate terminologies and in such a way that the end reader, even if not technically sound, is able to interpret it. The FE should be prepared to participate in meetings to elaborate on the report.

5.8.2 Analysis Tools

The tools used for acquisition and analysis may be either manual or automated and are listed here:

Manual extraction During this process, information is photographically documented by simply scrolling through the device using its keypad or touch screen. Lack of familiarity with the device and its interface may leave room for omission of data as well as to miss data. Any deleted information, however, cannot be extracted with this approach.

> **Box 5.4 Case Studies on Mobile Forensics**
>
> *Case 1: Mobile forensics to locate kidnapped minors*
> In Minnesota, 2014, two thirteen-year-old girls went missing. Detectives looked for digital clues in their ipods and smartphones as they serve as the log of a person's movement and activities. It helped in locating the girls in this case. The criminal was then charged with felony criminal sexual conduct, kidnapping, and solicitation of children. This was possible because digital evidence from electronic devices is a treasure trove of information and in this case, left footprints of whom the girls were talking to or tweeting and the events in the past few hours and days. Mobile forensics drastically reduced the time to locate the girls since they were reported missing.
>
> *Case 2: Mobile forensics in texting and driving case*
> In Texas, a driver was charged with vehicular homicide for killing another driver in a car accident. He swerved into an oncoming lane of traffic hitting the other car head-on, killing the driver on the spot. It was suspected that the driver had been texting on his mobile while driving though he claimed that he did not use the phone. However, he had sent and received texts prior to the accident. Mobile forensics was used to determine if the device was in use at the time of accident. In-depth forensic examination of text messages, emails, call history, and web browsing around the time of accident were made after imaging the device, during which process even deleted data was extracted. A snapshot of data at the specific point of time was obtained. Further, investigation confirmed that the driver was on the phone and surfing the web at the time of the accident. Thus, by gathering evidence using mobile forensics, the cause of the accident was proved.

Logical extraction In this case, the mobile device is connected to a forensic workstation that sends commands to the device which are then interpreted by the device and the requested data is sent back from the device to the workstation. Most forensic tools adopt this process as extraction is fast and it is easy for the FE to handle the data with only minimum training. However, it poses the risk of data getting written onto the mobile device, leading to questions about its integrity. Moreover, any data deleted as a result cannot be recovered.

Physical extraction It is also called hex dump where the mobile device is connected to a forensic workstation and a bootloader is pushed into the mobile device that dumps the memory of the mobile device into the workstation. This raw image is in binary format and requires technical expertise to process it further. However, it provides the FE with considerable information (e.g., information about deleted files and data residing in unallocated spaces).

Chip-off This refers to acquiring data directly from the mobile device's memory chip by physically removing the chip and using a chip reader to extract data from it. This requires expertise in hardware and any mishandling may damage the chip. The information, if extracted successfully, is in raw format, and hence has to be parsed, decoded, and interpreted.

Micro read This refers to the usage of an electron microscope to analyse physical gates on the chip and translate them into ASCII characters, thus interpreting the data as present on a chip. However, this requires expertise at the chip level and the file system and the process is time-consuming. This is usually used if all other approaches to data extraction have failed.

Two case studies on mobile forensics are presented in Box 5.4.

5.9 GPS FORENSICS

GPS is now present in smartphones, PDAs, and tablets, besides being fitted to vehicles. They serve as evidence in many criminal cases—for example, drug trafficking, death due to negligent driving, burglary, etc. The information stored in such devices includes home location, as well as stored and entered locations. In case of GPS fitted to vehicles, it provides information such as details of the journey with data and time, paired mobile phone details, deleted data, etc.

As part of forensic examination, live and deleted data can be recovered from different navigation devices. Forensic tools can be used to extract information from trip logs which hold information about a device's recorded location.

5.10 EMAIL FORENSICS

Email is the most popular form of communication. It is a way of transmitting electronic messages in the form of text, graphic files, sound, and even animated images from one computer to the other. It works on store-and-forward model. Common crimes associated with emails include narcotic trafficking, extortion, child pornography or abductions, fraud, terrorism, and sexual harassment. Besides this, email scams and frauds involving phishing or spoofing are on the rise. Phishing emails are in HTML format, permitting the creation of a link to text on a web page. The most popular email scam is 419 or the Nigerian scam. Crimes associated with emails depend on their jurisdiction. For example, a spam mail is illegal in Washington State, but not in other parts of the world. Hence, consultation with legal authorities provides clarity. Examiners and investigators should be aware of the process of examination and interpretation of the unique content of email messages.

The goal of email forensics is to identify the person behind the crime, collect evidence, and present the findings to build a strong case. Email tracing and email tracking can be achieved with email forensics, where the former aims to determine the source of the email, whereas the latter monitors email delivery to the recipient. Tracing is done when an email header is available, whereas tracking is done even when no information is available about an email ID (Box 5.5).

5.10.1 Client and Server in Email

Email is used in either of these environments, Internet and controlled, which includes LAN and WAN. Both the environments use the client/server architecture where the central server distributes the email to clients that are distributed. The client's email software can be installed either separately from the OS and have its own directories and data files or may use existing elements, for example, browsers. Email clients run programs such as Outlook Express, Eudora, or Pine. The servers run specialized software, for example, Windows Server 2003 or Novell Netware and run programs such as Exchange, GroupWise, or Sendmail. Email accounts are protected, requiring a username and password, and is meant to protect personal email from unauthorized access. Public email services such as AOL and Hotmail are Internet-based and are open to everyone. They do not follow standard naming conventions and so are not informative because of non-standard names. On the contrary, LAN and private email services are part of the local network and are specific to an organization, and have naming conventions of the form name@company.com where 'name' can be a standard name assigned by a local administrator which makes tracing of emails easier.

5.10.2 Structure of Email

Every email has a header, message body, and provisions for attachments.
1. The header is the potential information source of the email. The fields of the header include From, To, CC, BCC, Subject, Date, Message ID, Received, X-Originating-IP and, X-Mailer. The IP address of the sender can be obtained from X-Originating-IP or the last received MTA.

> **Box 5.5 Email Tracing versus Email Tracking**
>
> *Email tracing* is done when there is a need to analyse a mail for multiple purposes, such as source IP, date/time, time zone, and intermediate email servers.
>
> *Email tracking* is done to monitor email delivery to a recipient (email ID of the recipient is required), especially to get information such as IP address, date/time, and location at the receiving end where the email is opened. This helps to ascertain whether the email is actually delivered to the intended recipient and if it has been opened, and if it has, the IP address of the machine in which it was opened along with the exact date/time.

2. The message body is compiled by the user and is stored in the email client and server as binary data. No data is added by email servers. A string search using an analysis tool can be made on it to check for the presence of any relevant information during investigation.
3. Attachments contribute to 80% of email data. These are difficult to handle during analysis. However, email standard allows text attachments to be encoded with MIME/base64 formats. Attachments are the most frequent 'vectors' for viruses and so they are scanned before reviewing.

5.10.3 Working of Email

An email is composed using a mail client, for example, Gmail, Yahoo mail, etc. The client sends the message to a mail transfer agent (MTA) which is a server that runs the simple mail transfer protocol (SMTP). As each message passes through the MTA, it puts header information on the top (received header). Every MTA the message passes through adds a timestamp to the message. These timestamps are very useful in investigation. The MTA locates the advertised mail server for the recipient of the mail and passes the message along. Finally the recipient accesses his/her mail server using either POP3 or IMAP and downloads the message to their email client.

5.10.4 Email Protocols

An email client interacts with the email server using the following protocols:

1. Post office protocol (POP)
2. Internet message access protocol (IMAP)
3. Microsoft's mail API (MSMAPI)

The mail server stores incoming mail and distributes it to the appropriate mail box. However, its behaviour depends on the protocol being used. This influences the way in which investigation has to be done at the server. The POP stores only incoming messages and so it is sufficient to investigate the workstation. IMAP and MSMAPI store all the messages; copies are stored on the workstation, server, or both. HTTP is used for web-based sending and receiving of messages that are stored on the server and may be archived on the workstation. However, it is easy to spoof identity.

Table 5.2 Examining email header

Fields	Description
Originator	From, Sender, Reply to. The From field should contain the name of the host as presented in the EHLO command and an address literal containing the IP address of the source as determined from the TCP connection.
Destination address	To, cc, bcc
Identification	Message ID is an optional field but is essential while tracing email using server logs
Informational	Subject, Comments, and Keywords. This field may contain @.
Resent	Resent-date, Resent-from, Resent-sender, Resent-to, Resent-cc, Resent-bcc and Resent-msg-id. This field is an optional one but it is added by most servers.
Tracing	When an SMTP server receives a message either for delivery or forwarding, trace information is added to the beginning of the header.

Messages are relayed between servers using simple mail transfer protocol (SMTP) which can also be spoofed easily. Every mail server on the path through which the email traverses adds information to the top of the email header. Hence, during examination the header should be read from the bottom. The SMTP header consists of fields listed in Table 5.2. The server that makes the final delivery inserts the return-path line to the header. The content of the SMTP header provides clues to spoofing during examination.

5.10.5 Examining Email Messages

Examination of email messages can begin with accessing the victim's computer and retrieving the evidence as the first step. Investigation of the victim's email using an email client is for one or more of the following: (a) look for, open, and copy the evidence in the email along with the header, and (b) look for protected and encrypted material. GUI makes it possible to copy email, either by dragging it to a storage medium or saving it in a different location. The copied mail can be sent as an attachment to another email address. The evidence email can even be printed, including the header. Standard procedures may be used for different types of email clients. For example, for command line client like Pine, the email message has to be opened and the copy option should be used. So as to include text in reports, cut and paste will help. In the case of a suspect's computer, deleted emails should also be considered during examination.

5.10.6 Viewing Email Headers

Email headers contain information such as unique identifying numbers, IP address of the sending server, and the time the mail was sent. Headers can be viewed using GUI clients, command-line clients, and web-based clients. The procedure for viewing email headers with different client applications are listed here:

1. Hotmail Classic, Windows Live Hotmail
    (a) Log into Hotmail.
    (b) Click on the 'Options' tab on the top of the navigation bar.
    (c) To view the full email message header, right-click the suspect email message displayed in the list of messages. A menu will pop-up.
    (d) Click on the 'View source' option in this menu, and a new window will open. This window will display the full email header.
2. Yahoo Mail Classic
    (a) Log into the Yahoo! Mail account.
    (b) Click on the email and open it.
    (c) On the bottom right corner is a link called 'Full Header'.
    (d) Click on 'Full Header' for the header to show up at the top of the email message.
3. New Yahoo Mail
    (a) Click on the Inbox to see the list of messages.
    (b) Click on a message and open the email.
    (c) On the top right corner of the email message is the 'Standard header' and an arrow next to it. Click on this arrow and then click 'Full headers'.
    (d) A new window will open with the header information.
4. AOL or AIM
    If the mail is sent from anywhere other than AOL, do the following:
    (a) Open the email to be traced.
    (b) Look at the link details of the header (just below the To: email in the email message).
    (c) If the mail is sent from an AOL user, search the Reverse AOL screen name to deduce the source location.
5. Gmail or Googlemail
    (a) Log into Gmail or Googlemail account.
    (b) Open the email whose headers have to be viewed.
    (c) Reply can be seen at the top right of the message pane.
    (d) A little arrow pointing down next to Reply can be seen. Click on this down arrow next to Reply.
    (e) A drop down menu will open up. Select 'Show original' in this menu.
    (f) The full header will now appear in a new window.
6. Rediffmail
    (a) Login to the Rediffmail account.

(b) Open the inbox and right-click on the mail.
(c) Right-click the message to view headers and select 'Properties'.
(d) The full header appears in a new window.
7. Xtramail
    (a) Log into XtraMail.
    (b) Click 'Options' in the left-hand navigation bar.
    (c) Click on the 'Display' button.
    (d) Change the 'Message headers' option to 'Full'.
    (e) Click on the 'OK' button.
8. Thunderbird (Mozilla Firefox)
    (a) Go to 'View'.
    (b) Go to Headers.
    (c) Select 'All' to view email headers
9. Outlook
    (a) Open the 'Message options' dialog box.
    (b) Copy the headers.
    (c) Paste them onto any text editor.
10. Outlook Express
    (a) Open the message properties dialog box.
    (b) Select 'Message source'.
    (c) Copy and paste the headers onto any text editor.
11. Eudora
    (a) Click the BLAH BLAH BLAH button.
    (b) Copy and paste the email header.
12. Pine and ELM
    (a) Check enable-full-headers.
13. AOL headers
    (a) Open the 'Email details' dialog window.
    (b) Copy and paste the headers.
14. Juno
    (a) Click 'Options' and select 'Show headers'.
    (b) Copy and paste headers.
15. Hotmail
    (a) Click 'Options' and then click the 'Mail display settings'.
    (b) Click the 'Advanced option' button under 'Message headers'.
    (c) Copy and paste the headers.
16. Apple Mail
    (a) Click 'View' from the menu, point to 'Message', and then click 'Long header'.
    (b) Copy and paste headers.
17. Yahoo
    (a) Click the mail options.
    (b) Click 'General preferences' and 'Show all headers on incoming messages'.
18. WebTV
    (a) Send the message to yourself.
    (b) Open it with your regular email client.
    (c) The message will contain the headers.

5.10.7 Examining Email headers

The email headers are examined to gather supporting evidence which actually helps to track the suspect. The following information can be gathered by examining the headers:

1. Return path (can be easily spoofed)
2. Recipient's email address
3. Type of sending email service
4. IP address of the server from where the mail has been sent
5. Name of the email server
6. Unique message number
7. Date and time at which the mail has been sent
8. Information related to attached files

The email headers can be viewed differently with different clients and some of these are listed here:

1. Hotmail
   (a) Login and select 'Options'.
   (b) Select 'Preferences' and scroll down the list to 'Message headers'.
   (c) Select 'Advanced'.
   (d) Scroll up or down and select 'OK'.
2. Yahoo
   (a) Login and select 'Options'.
   (b) Select 'Mail preferences'.
   (c) Scroll down and select 'All' at the 'Message headers' option.
   (d) Scroll up or down and select 'Save'.
3. Outlook 2000
   (a) Open the email and click 'View' in the menu bar and select 'Options' in the drop-down list.
   (b) The header is displayed at the bottom of the window that opens up.
4. Outlook Express 5.5
   (a) Open or select the email.
   (b) Select 'File' and choose 'Properties'.
   (c) Select the 'Details' tab to view the header.
5. Netscape Communicator 4.77
   (a) Open the email.
   (b) Select 'View' in the menu bar, choose 'Headers' in the drop-down list, and select 'All'.

5.10.8 Examining Additional Email Files

Email messages usually get saved on the client side at the server. For example, Microsoft Outlook saves them in .pst and .ost files where the former includes sent, received, deleted, and draft messages and the latter contains offline files. Besides this, the personal address book also provides valuable information for investigation.

In case of UNIX, email groups can be created by an administrator where all the members can read the same messages. In such a situation, the investigator can be added as a member in the suspect's group so as to have the same access as the suspect. This will help in the investigation of crime.

With web-based client applications such as AOL and Hotmail, history, cookies, cache, and temp files provide evidence as well. A string search may be employed to gather evidence from such files. Besides this, specialized software, for example, a cookie reader can be employed to gather evidence from the cookies file.

5.10.9 Tracing Email Messages

The following are the steps in tracing an email:
1. Given a suspicious email, it has to be read to determine whether any crime/violation has been committed. It has to be checked for any opened attachments. The header is examined and the IP address of the sender is recorded.
2. The server through which the email is sent is obtained from the email header and has to be contacted. The domain names point of contact can be used to gather information about the server, for example, the

American Registry of Internet Numbers (ARIN) available at www.arin.net. Different countries and regions have separate IP address databases for WHOIS look-up and some are listed in Table 5.3. This helps to determine a specific IP address. Figure 5.4 shows the results of ARIN for the IP address 210.212.253.247. Since it does not have the details, it instead tells the database from where information about that IP address is available. Figure 5.5 shows the results for the said IP from the APNIC database.

3. The suspect's contact information can be obtained from the header. However, it need not be true always. Hence it has to be verified with the network log especially the router/firewall log. Router logs record all incoming and outgoing traffic to a network. Similarly, email traffic can be filtered from a firewall log to verify whether the email has passed through. Any text editor or specialized tools can be employed for this purpose. In certain situations, it may even be necessary to verify the route through which the email has traversed.

Table 5.3 Links to database of IP addresses

Database name/Country/Region	Link to resource
Asia Pacific Network Information Centre (APNIC)	http://www.apnic.net/
American Registry of Internet Numbers (ARIN)	http://www.arin.net/whois/arinwhois.html
Dragon Star IP Index	http://ipindex.dragonstar.net/index.html
Germany	http://www.denic.de/servlet/Whois
HIS	http://www.his.com/utilities/
Hong Kong	http://www.hknic.net.hk/hknic/whois.html
IP Index	http://www.flumps.org/ip/
Korea	http://www.nic.or.kr./www/english/
Netherlands	http://www.domain-registry.nl/bestaat.Ip
European Network Coordination Centre (RIPE)	http://www.ripe.net/cgi-bin/whois
Russian Institute for Public Networks (RIPN)	http://www.ripn.ru:8080/nic/whois/en/index.html
Brazilian Registry (RNP)	http://registro.fapesp.br/

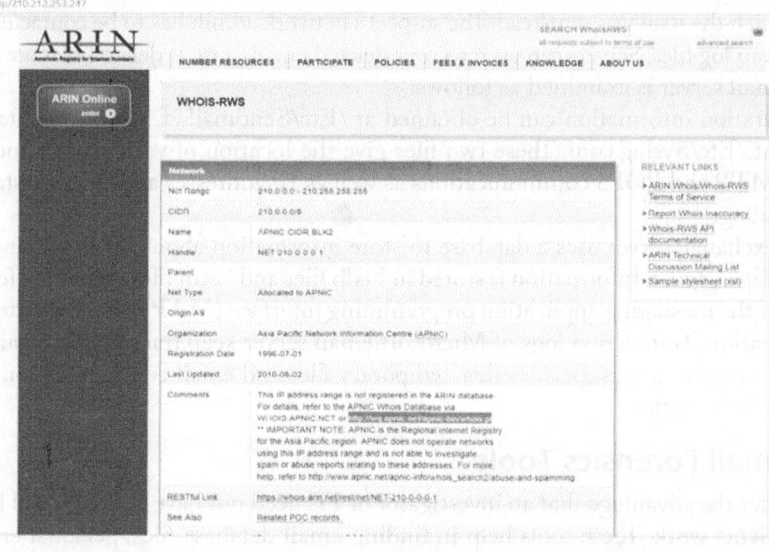

Fig. 5.4 Search result of ARIN

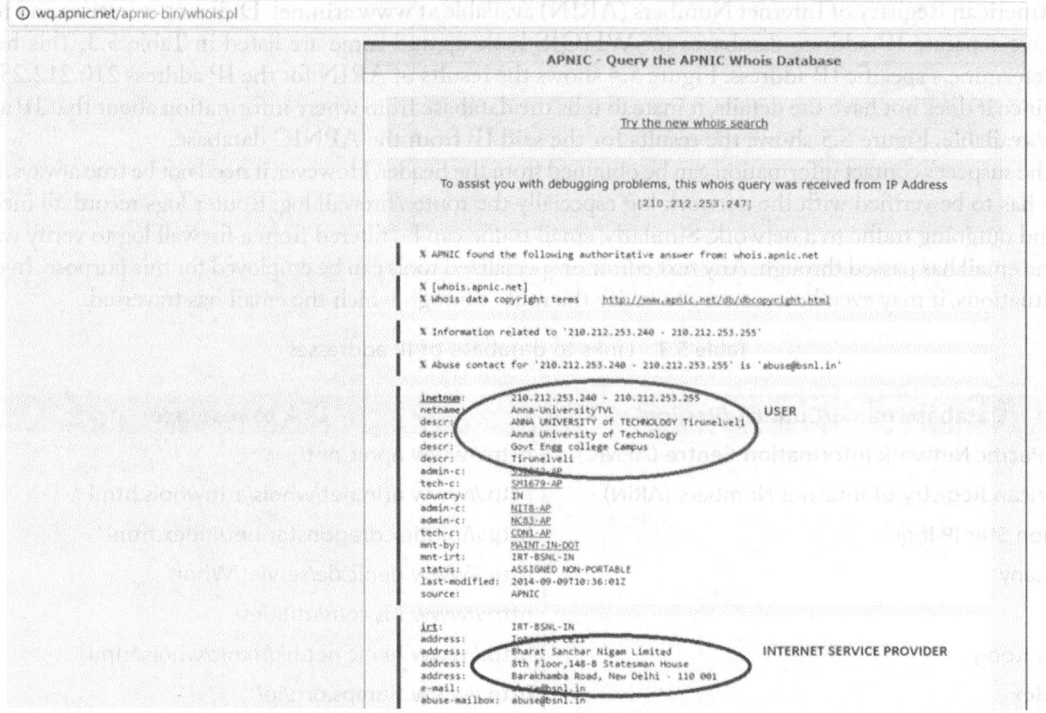

Fig. 5.5 Search result of APNIC

Source: www.arin.com, last accessed on 28 November 2017

5.10.10 Email Servers and their Examination

Email servers are computers which run a server OS and an email package. Emails are stored on them either as a database or as a flat file. Logs in the server can be either default or configurable and can be continuous or circular (overwritten). A log file maintains information such as the content of the email, IP address of the sender, and the date and time at which the mail was sent/read. The suspect's network admin has to be contacted at the earliest to gather information from log files. Servers can even recover deleted emails, just as deleted files are recovered on a hard disk.

A UNIX email server is examined as follows:

The configuration information can be obtained at /Etc/Sendmail.cf. All the events are available at the log file available at /Etc/Syslog.conf. These two files give the location of various logs and their rules. Particulars and logs of SMTP and POP3 communications as well as IP addresses and the timestamps can be obtained at /Var/Log/Maillog.

Microsoft exchange server uses a database to store information about the email and is based on Microsoft extensible storage engine. Information is stored in *.edb files and *.stm files where the former holds information responsible for the messaging application programming interface (MAPI) and the latter is responsible for non-MAPI information. Transaction logs of Microsoft email server keep track of the email database. Checkpoints keep track of transaction logs. Besides this, temporary files and email communication logs (RES#.log) can be used to gather information.

5.10.11 Email Forensics Tools

These tools offer the advantage that an investigator or FE need not have the technical know-how of how email servers and clients work. These tools help in finding email database files, personal email files, offline storage files, and log files. Some of the tools are listed here:

(a) FINALeMAIL is a tool that scans email database files and can recover deleted emails.
(b) FTK is an all-purpose tool that filters and finds files specific to email clients and servers.

5.10.12 Tracking Emails

Tracking email can be done by services such as Readnotify, DidTheyReadIt, and getnotify.

For example, tracking can be done with Readnotify as follows:

1. The investigator has to register an email ID for tracking with Readnotify at www.readnotify.com.
2. An email to be sent to the recipient is created and .readnotify.com is added at the end of the recipient's email address (this cannot be seen by the recipient).
3. When the recipient opens the sent mail, a tracking report is sent to the registered mail ID of the investigator. The report is also available at the server which can be accessed when the investigator signs in to his/her readnotify account.

Case studies on email forensics are presented in Box 5.6.

Box 5.6 Case Studies on Email Forensics

Case 1: Email investigation for extortion

Pranab Mitra, a former executive of Gujarat Ambuja Cement, posed as a woman, Rit Basu, and created a fake email ID through which he contacted one V.R. Ninawe, an Abu Dhabi-based businessman. According to the FIR, Mitra trapped Ninawe online and sent emotional messages, thereby indulging in online sex. Later, Mitra sent an email stating that 'she' would commit suicide if Ninawe ended the relationship. He also gave him another friend Ruchira Sengupta's email ID which was in fact his second bogus address. When Ninawe sent a mail to the other ID, he was shocked to learn that Mitra had died. Then Mitra began emotionally blackmailing him by calling

Abu Dhabi to say that the police here were searching for Ninawe. Ninawe panicked on hearing the news and asked Mitra to arrange for a good advocate for his defence. Ninawe even deposited a few lakhs in the bank as advocate fees. Mitra even sent emails as high court and police officials to extort more money, thereby managing to extort Rs96 lakh from him. Ninawe finally came down to Mumbai to lodge a police case. Email investigation was used to prove the case and Mitra was booked for cheating, impersonation, blackmail, and extortion under Sections 420, 465, 467, 471, and 474 of the IPC, read with the newly formed Information Technology Act.

Case 2: Email investigation to prove defamation

X, the sales manager of Company A, gave a four-week notice, and left soon after. Soon after, Company A received notification from a number of clients stating that they had received emails from an unknown Hotmail account containing defamatory information about Company A. Investigation of email received by the clients necessitated a search for evidence on X's PC, as the emails originated from this location.

Examination of the PC also provided evidence of confidential data having been copied to removable external media during the preceding four weeks. Acquisition of data from X's hard disk and an analysis of data that was deleted, as well as the system files that were recovered, showed that email data was created at the date and time that X was known to be operating the PC.

Detailed analysis also showed that confidential data of Company A was copied to a USB drive. The files and detailed report enabled company A to take legal action against X.

5.11 MEMORY FORENSICS

Memory forensics is the examination of volatile data in a computer's memory dump and is also called memory analysis. Volatile data includes browsing history, clipboard contents, and chat messages present in the short-term memory storage. A memory dump is the data captured from the random access memory or RAM, usually before system crash. Memory dump is usually used by experts as it provides diagnostic information at the time of the crash and contains a code that caused the crash.

Memory forensics is used to figure out criminal activity such as hackers or insider threats. Each program or data which is created, examined, or deleted is stored in the RAM. This includes images, all web-browsing activity, encryption keys, network connections, or injected code fragments. Besides this, certain artifacts are present only in RAM, for example, open network connections present during a crash. Some malwares reside only in the memory, rather than the disk. All these have increased the demand for memory forensics tools.

5.11.1 RAM Artifacts

The speciality of RAM is that any actively used information or data by the computer program or the hardware device will run through the system's RAM at the time it is being used. So RAM becomes an integral part of computer forensics. RAM artifacts include any piece of data that is used by a software application or the hardware device. Some of the artifacts are given here:

Network connections This information is useful to gather the remote IP address and the port used in network connections which in turn will help an investigator or the FE to identify any computer intrusion, any remote destination the malware is communicating with, if the system is infected with malware, the source of child pornography, etc. The port is used to determine the type of traffic—HTTP, FTP, and so on.

Running processes A list of active programs perceived through the task manager will provide the FE with an idea of how the system is being used. However, the presence of a rootkit or a hidden Trojan cannot be ascertained with visual inspection.

Usernames and passwords All information related to usernames and passwords used by a user for authentication with different applications are stored in the RAM.

Dynamic link libraries RAM includes all the dynamic link libraries (DLLs) associated with running processes as well as malware. Identification of malicious DLL that has been injected can be tracked with memory forensics.

Contents of open window All keystrokes corresponding to different applications, for example, social networking sites, email, etc., are available in RAM.

Open registry keys of process Registry keys associated with every process is available in RAM. It is useful to figure out malware and expel it from a system by looking for registry keys corresponding to malware and then restarting the system and locating the registry key that survives a restart.

Open files for process All open files will be preserved in the RAM. This will help in locating any resident file pertinent to the malware that records keystrokes, usernames, and passwords.

Memory resident malware Certain malware resides only in the memory and its presence is not registered even on the hard drive. Hence, any data collected from RAM could help the FE in ascertaining the presence of malware.

5.11.2 RAM Analysis

Memory analysis tools should be employed to collect networking connections and process information (registry and file information). Two popular memory forensic tools used to collect such information are Volatility, which is free and open-source, and HBGary, which is proprietary. For example, Network connections can be analysed from the Internet history which gives the frequency of visit to a particular domain or IP address and from network logs which give the amount of inbound and outbound traffic. The mapping of the IP address and the owner can be made from WHOIS queries. From the network connections, the process ID can be fetched, which in turn helps to gather the name of the process. Once the network connection, process ID, and process name are obtained, any suspicious malicious activity can be figured out. All the three have timestamps and so a timeline analysis can also be performed by the FE. Similarly, forensic tools could be employed to gather data about process information, the registry, and file information. Some of the forensic tools can even carve out an executable file from memory.

5.11.3 Forensic Tools

Some of the popular forensic tools used for RAM analysis include [Raj Chandel] the following:
1. Magnet RAM Capture is a free imaging tool that can capture the physical memory of a suspect's computer, facilitating an FE to recover and analyse artifacts that are found in the memory.
2. Belkasoft Live RAM Capturer is a free forensic tool that is used to extract the entire content of the computer's volatile memory, even if protected by an active anti-debugging or anti-dumping system.
3. MoonSols Dump is a forensic tool where a double-click on the executable can generate a copy of the physical memory in the current directory.
4. FTK Imager can acquire live memory and paging file on 32-bit and 64-bit systems.

5.12 BUILDING FORENSIC COMPUTING LAB

The functionalities expected in a lab vary with the suspect's device, operating system, software applications used, the hardware platform underlying it, etc. However, there are some basic requirements that govern a forensic lab. These are listed here:
1. A log register should be maintained at the entrance of the lab as a layer of monitoring for protection.
2. The lab area should be secured by cipher combination locks to ensure that the chain of custody is maintained.
3. The lab should be equipped with fire safety measures.
4. The work area should be equipped with necessary infrastructure such as work tables, chairs, and storage capability.
5. The evidence storage area should have a strongly constructed metal shelf, be non-destructive, and fire-proof.
6. A forensic toolkit should contain disassembly and removal tools, packaging and transport supplies, etc., facilitating the examiner to collect evidence from the crime scene.
7. A forensic lab should have the following: workstations, UPS, book racks with necessary reference materials, necessary software and tools, safe locker, LAN, and Internet connectivity.
8. The hardware equipment necessary for a forensic lab includes computer systems or a forensic workstation which could be either a desktop or a laptop with optical drive, network interface, uninterrupted power supply, write blocker, scanner, printer, and provision for evidence back-up. The forensic workstation should be shielded from electromagnetic signals. A minimum of two drives with different operating systems, for example, Windows and Linux, is preferable. Other peripherals that are required include cables, detachable storage media, CD/DVD readers and writers, IDE hard drives, RAM, SCSI cards, power cords that are compatible with different ECDs, laptop hard drive connectors, etc.
9. The software requirements include the following:
   (a) Boot disks to pull up a computer in a non-Windows environment.
   (b) Imaging tools to capture data bit-by-bit, from the suspect's device. This serves the purpose of data preservation and duplication besides verifying the integrity of data after duplication. (e.g., FTK imager)
   (c) A data recovery or an extraction utility to recover digital information even revealing obscure information from verified images. Even hidden and manipulated data have to be extracted using such a utility. (e.g., Diskedit). Data extraction can be either physical or logical. In physical extraction, the data across the entire drive is identified without any regard to the file system. The methods used may include a search

for keywords that may not be accounted for by the operating system, file carving to extract useable files not accounted for by the operating system, and extraction of the partition table and unused space on a drive to estimate the physical size of the disk. On the other hand, logical extraction will extract data from active files, deleted files, encrypted files, compressed files, file slack, and any unallocated space.

(d) Data analysis software may be either manual or automated, and comes with specific functionality such as text search, file viewer, time frame analysis (where every file is provided with a date and time stamp which actually provides information to the investigator about when the suspect last accessed the respective file), application analysis and indexing, hidden data analysis, and ownership analysis.

(e) Network analysis software to trace connections, identify ISPs, ping specific IP addresses, and the like.

(f) Reporting and presentation software for the interpretation of the analysis, so as to prepare proper documentation and a report of the findings, including a presentation.

(g) Other software like registry viewer to generate reports after viewing the Windows registry, password recovery toolkit for the recovery of data from encrypted files, wiping software to reuse the storage equipment in lab, and antivirus software to prevent evidence and the equipment from destruction.

10. In addition to the aforementioned means, Unix platforms are handled in the following ways: data dumper (dd) utility for taking a forensic copy of the disk, grep program for text searching, and the Coroner's toolkit for the forensic analysis of Unix machines.

11. Internet connectivity with sufficient bandwidth for the workstations is necessary.

12. The various forensic tools used by forensic investigators, both at the lab and on the field, are as follows:

(a) Storage bags for storing the evidence collected and transporting it to a lab

(b) Remote chargers to power-up devices carrying evidence if they run out of power

(c) Write-blocker which prevents alteration and erasure of data during investigation

(d) Data acquisition tools such as cables of different kinds, rapid action imaging devices (RAID) to quickly copy data from a suspect's drive to a forensic secure drive, SIM card readers, and video capture devices.

(e) Forensic archival devices to copy forensic data from many CDs/DVDs

(f) Mobile forensic laptops with all specialized software and tools which a forensic investigator may require on field

(g) Forensic workstations with all necessary software, write blocking, and hot swapping (there is no need to turn off the workstation while attaching a suspect's drive) facility

(h) A separate workstation designated as an imaging workstation to exclusively image different storage devices with provisions to image data even from password-protected devices

(i) Software for extracting and analysing data from storage devices

5.13 INCIDENT AND INCIDENT HANDLING

This section explains incident, incident handling, and incident response. Incident is any unacceptable act on computers, networks, and other ECDs or violation of policy or law associated with information assets. While incident handling refers to the logistics, communications, coordination, and planning functions needed to resolve an incident, incident response refers to the technical components required to analyse and contain an incident.

5.13.1 Incident

An incident is an event or set of events that threaten the security of computing systems and the network. An incident, for example, can be repudiation (impersonation of another person), reconnaissance (where lapses in security are used to bring about an attack), harassment, extortion (threatening to reveal personal information either for financial benefit or with the intention to defame), pornography trafficking, organized crimes such as illegal passport and visa creation. The outcome of any incident can be a system crash, packet flooding in the network, or unauthorized use of another person's account over the Internet.

Some incidents related to security include tampering of data, unauthorized access to an organization's resources, threats and attacks, malicious code that damages an organization's computing resources, denial of service attack, defaced web pages, and any other incident that weakens the trust of financial systems.

The incidents can be categorized as low-, mid-, or high-level according to its intensity and impact. Low-level incidents include, for example, password compromise, presence of a virus or a worm, and misuse of a peripheral and should be addressed within a day. Mid-level incidents are more serious, for example, violation of access privileges, illegal access to a network, unauthorized processing of data, etc., that should be handled within hours of it being reported. High-level incidents should be handled as soon as possible and include, for example, pornography, denial-of-service attack, etc.

Certain signs give an alert on the occurrence of an incident, for example, a log file with suspicious entries, modified files, unusual system behaviour, and services on ports, more packets to a network than usual, inaccessible drives, etc. Besides this, an intrusion detection system (IDS) alerts administrators if any security breaches are noticed.

The following can help administrators to prevent and minimize incidents. They are mentioned here:

1. Use of scanning tools to look for vulnerabilities periodically
2. Auditing to ensure whether proper measures have been adopted for the vulnerabilities identified
3. Periodical review of server logs to detect intrusions, if any, and to ensure remote access is made only by authenticated users
4. Employing defence-in-depth, that is, a layered security approach so that one layer can compensate for the drawbacks or failure of the other layer while offering protection

5.13.2 Incident Handling

Incident handling refers to the set of procedures used to handle an incident. It comprises three functions: incident reporting, incident analysis, and incident response. The procedures check for the occurrence of events and analyse the cause, impact, and the damage caused by the events and define the security measures to be adopted so that the incident does not occur in the future.

Incident handling is achieved in stages as shown in Fig. 5.6, which are as follows:

Preparation It refers to making an organization aware of security threats and ways to prevent and handle them. It includes installation of specialized tools to avoid damage due to worms and viruses, making employees aware of and training them on the latest threats and trends in security management.

Identification It includes the analysis of an incident to ascertain its nature, intensity, and impact on systems and the network. Training is given to the employees of the organization on how to identify an incident.

Containment It refers to adopting measures that reduce the intensity and impact of the incident. In other words, it is an attempt to control and curb the incident, for example, in case of data loss, change of passwords to the systems holding data and taking backup, and in case of a viral attack, detecting and removing it using software tools.

Eradication This stage attempts to put an end to the incident after understanding the salient points of the containment stage. There is also further training to prevent the incident from recurring.

Recovery This stage attempts to restore the affected systems to their normal state after completely eliminating the vulnerability. This is done by determining the appropriate course of action to restore the system, ensure that the system is performing well, and is free from vulnerability (according to the log files). These actions are performed by the IRT.

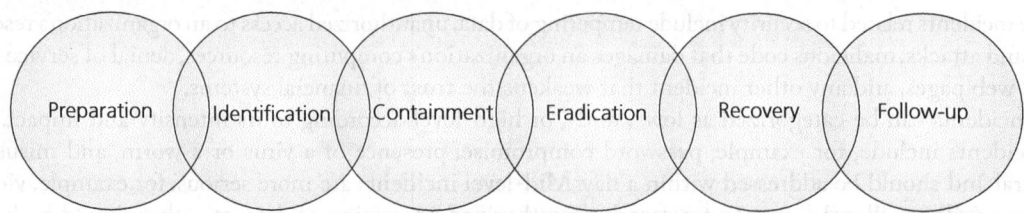

Fig. 5.6 Stages in incident handling

Follow up At this stage a cost analysis on the incident that occurred is made for ascertaining data and hardware loss. Then, appropriate revisions to the security policy and procedures in force are made. This aims to avoid incidents in the future.

The IRT learns from every incident it faces. As a post-incident activity, the organization should conduct a meeting to discuss what, when, why, and how the incident occurred, how combating the incident has been successful, and how to prevent another incident in the future. All these are documented and can be used as training material for new members.

The collected incident data is used for calculating the number of incidents handled and the time per incident. The period for evidence retention should be decided by the organization depending on its objective. Further, the employees should be educated and trained with awareness programmes to handle computer-related incidents.

5.13.3 Incident Reporting

An organization or individual who witnessed a security breach can report the incident to the CERT Coordination Center, a law enforcement agency, or to the CSIRT if the organization has one. While reporting, the following details have to be furnished, namely the intensity of the breach, the incident which revealed the vulnerability, what shortcomings have been exploited, log files which reveal the footprints of the attacker, and the time of breach. In case of a security incident, the date, time, and location of the incident, when and how it occurred, who was behind the incident, the type of data that was affected as a result, the damage incurred because of the incident and the remedial measure adopted, if any, should be reported.

In general, organizations hesitate to report an incident for various reasons, some of which are (a) assumption that the incident is unique, (b) fear that reporting of the incident will tarnish its reputation, (c) lack of awareness about the incident, (d) fear of losing customers, and (e) the need to handle issues internally.

5.13.4 Incident Response

Once a security incident is reported, a response to it is based on the uncorrupted data and documentation. The response is achieved in stages, as listed here:

1. In this stage, the resources that are affected as a result of the incident are ascertained by the incident investigator and coordinator (IIC). This is an iterative stage where the result is achieved after several iterations.
2. After ascertaining the impact of the incident, the incident is assessed according to its severity level as low, mid, or high by the IIC.
3. The event is now assigned a unique identity. The severity level of the incident determines the procedures and resources to be used to respond to and recover from the incident.
4. The IIC coordinates with the task force including managers of resources and other personnel to resolve the incident.
5. The IIC, after determining the risks, contains the resources that are suspected of containing threats.
6. Information related to the incident is the evidence and it is collected from the systems and digitally signed.
7. Forensic analysis of the collected evidence is performed following due procedures and the evidence is preserved carefully.

Incident Response Policy

An incident response policy should include the following:

1. The management's support for the policy is essential. The financial resources, training to implement the policy, and the countermeasures have to be endorsed by the management.
2. The organization's approach to the incident may be any one of the two depending on the objective. If there is a need for compensation, the incident should be documented properly so as to be produced in court as evidence. If the objective is to restore the services quickly, the portion corresponding to the damage is contained and countermeasures are taken so as to deny further access to the attacker.
3. The policy should include remote network connections and define why it is necessary to disconnect the users during a security incident.
4. The policy should define partner agreements where partners may be service providers and the customers, and the right to monitor the network and disconnect when needed should be discussed.
5. The members of the incident team, and their roles and responsibilities are clearly defined. Not all the members need to be from within an organization. If necessary, some people may be hired or outsourced.
6. A communication plan must be defined—whom to contact and how to contact him/her. This will avoid confusion in times of emergency.
7. Provision for reporting the incident and archiving it so as to ensure that the countermeasures developed for the incident are sufficient to prevent the incident from recurring are a must.

5.14 COMPUTER SECURITY INCIDENT RESPONSE TEAM

Computer security incident response team (CSIRT) is a service organization that receives reports, reviews, and responds to computer security incidents. It could be a formalized team or an ad hoc one where the former performs the response activity as its major function, whereas in the latter the team is formed when there is a need to respond to a security incident. CSIRT services are performed for a defined constituency, where the constituency could be an organization, a country, or a network.

CSIRT comprises the following members with responsibilities as follows:

Incident investigator and coordinator (IIC)

1. Decides how to recover from the incident
2. Determines the severity of the incident
3. Performs investigation and technical analysis
4. Supervises evidence collection
5. Serves as expert witness for testimony

Incident liaison (IL)

1. Coordinates with IIC
2. Coordinates resources such as hardware, personnel, and fund allocations
3. Serves as secondary expert witness

Senior system manager

1. Controls access to accounting modules
2. Responsible for security of the systems
3. Informs system security manager
4. Responsible for the establishment and maintenance of security for risk management
5. Ensures information resources for audit requirements
6. Ensures participation of all levels of employees to implement policies and procedures

Information system security officer

1. Responsible for the identification of threats and vulnerabilities
2. Responsible for the identification of restricted, unrestricted, and sensitive resources
3. Develops and maintains a risk management process, disaster recovery, and contingency plan as well as updated security procedures.

5.14.1 Forensic Readiness

It involves an organization having incident response procedures in place along with trained personnel to handle any investigation. This facilitates the organization to collect and preserve evidence efficiently at minimum cost. Forensic readiness ensures the following: (a) the evidence collected is forensically sound; (b) demonstrates a possible crime or an incident that may have an impact on the organization; (c) ensures that the evidence collected will influence the outcome of the legal proceeding.

POINTS TO REMEMBER

- Cybercrime refers to any criminal offence that involves a computer/network, where the computer is used to either commit the crime or is the target of the crime.
- Computer forensics, alternatively referred to as cyber forensics, focuses on finding digital evidence in a computer or any electronic communication device (ECD) after the occurrence of a crime.
- Cyber security refers to technologies, processes, and practices designed to protect networks, computers, programs, and data from attack, damage, or unauthorized access.
- The phases of cybercrime investigations are preliminary analysis, image acquisition, recovery, extraction, and finally preparing a detailed report after analysis.
- First responder is the person who first reports to the crime scene and assesses the ECD where the crime has been committed.
- CERT is an expert group that handles computer security incidents.
- Forensic examination includes the following steps: acquisition, preservation, identification, evaluation, extraction, interpretation, and presentation.
- Cyber forensics can be classified as disk forensics, network forensics, wireless forensics, database forensics, malware forensics, mobile device forensics, GPS forensics, email forensics, and memory forensics.
- Disk forensics is the process of extracting forensic information from storage media (e.g., hard disk, USB drive, CD, DVD, flash drive, floppy disk, etc.).
- Network forensics refers to the capture, recording, and analysis of network events so as to discover any malicious activity, security attack, or violation.
- Network forensic analysis tools include IDS, log files, packet capturing tools, etc.
- The wireless forensic process involves capturing all data moving over the network and analysing network events to uncover network anomalies, discover the source of security attacks, and investigate breaches on computers and wireless networks to determine whether they are or have been used for illegal or unauthorized activities.
- Traffic analysis in wireless forensics involves data normalization and mining, traffic pattern recognition, protocol dissection, and reconstruction of application sessions.
- The goal of database forensics is to determine what kind of security breach occurred at a database, when the incident happened, and to revert any unauthorized data manipulation operations.
- The primary information sources for the database breach are files, internal structures, and logical structures.
- The forensic approaches to database forensics are reactive and proactive.
- Malware forensics is the process of examining the computer system so as to find the malicious code, determine how it got there, and what changes it has caused to the system.
- The potential indicators of malware attack are (a) abnormal performance exhibited by the machine, (b) machine rebooting randomly, (c) files opening

up on their own without user input, (d) drives becoming suspiciously full, and (e) message pop-ups by the malware detection software.
- Mobile forensics refers to the recovery of digital evidence from mobile devices.
- GPS forensics attempts to gather information stored in mobile devices which include home location, and stored and entered locations.
- The goal of email forensics is to find out who is behind the crime, collect the evidence, and present the findings so as to build a case.
- Email tracing and email tracking can be achieved with email forensics, where the former aims at finding out the source of the email, whereas the latter monitors email delivery to the recipient.
- Email client interacts with the email server using protocols—post office protocol (POP), Internet message access protocol (IMAP), and Microsoft's mail API (MSMAPI).
- Memory forensics is the examination of volatile data in a computer's memory dump and is also called memory analysis.
- The hardware equipment necessary for a forensic lab include a computer system or a forensic workstation which could be either a desktop or a laptop with optical drive, network interface, uninterruptible power supply, write-blocker, scanner, printer, and provision for evidence backup.
- The software requirements for a forensic lab include boot disks, imaging tools, data recovery and extraction utility, data analysis software, network analysis software, reporting and presentation software, and other essential software.
- An incident is an event or set of events that threaten the security of computing systems and the network.
- Incident handling refers to the set of procedures used to handle an incident. It comprises three functions: incident reporting, incident analysis, and incident response.
- Computer security incident response team is a service organization that receives reports, reviews, and responds to computer security incidents.
- Forensic readiness ensures the following: (a) the evidence collected is forensically sound; (b) demonstrates a possible crime or an incident that may have an impact on the organization; (c) ensures that the evidence collected will influence the outcome of legal proceedings.

KEY TERMS

Chip-off This refers to acquiring data directly from the mobile device's memory chip by physically removing the chip and using a chip reader to extract data from it.

Containment This refers to adopting measures that reduce the intensity and impact of the incident.

Cyber forensics This refers to an electronic discovery technique used to determine and reveal technical criminal evidence.

Cybercrime This refers to any criminal offence that involves a computer/network or ECD, where the computer is used to either commit the crime or is the target of the crime.

Disk forensics This refers to the process of extracting forensic information from storage media such as hard disk, USB drive, CD, DVD, flash drive, and floppy disk.

Malware forensics This refers to the process of examining the computer system so as to find the malicious code, determine how it got there, and what changes it has caused to the system.

Memory forensics This refers to the examination of volatile data in a computer's memory dump and is also called memory analysis. Volatile data includes browsing history, clipboard contents, and chat messages present in the short-term memory storage.

Micro read This refers to the usage of an electron microscope to analyse physical gates on the chip and to translate them into ASCII characters, thus interpreting the data as present on a chip.

Network forensics This refers to the capture, recording, and analysis of network events so as to discover any malicious activity, security attack, or any violation.

Sessioning This is defined as the act of assembling raw data packets between specified points as a complete stream which helps to gather information about a specific communication (who was part of the communication, when the communication took place, and what was communicated).

MULTIPLE-CHOICE QUESTIONS

1. The objective of cyber forensics is to _____ .
   (a) estimate the potential impact of the malicious activity
   (b) assess the intention and identity of the offender
   (c) identify the evidence associated with a malicious activity in a short span of time
   (d) all of these

2. A person who first reports to the crime scene and assesses the electronic communication device where the crime has been reported is known as _____.
   (a) first reporter
   (b) first responder
   (c) first investigator
   (d) first evidence collector

3. Identification, acquisition, extraction, _____, evaluation, interpretation, and presentation are the steps in forensic examination.
   (a) precaution   (b) prevention
   (c) preservation   (d) production

4. _____ provides analysts with the state of the system by looking into connections, processes, and cache tables.
   (a) Timeline analysis
   (b) Volatile evidence analysis
   (c) Data recovery analysis
   (d) System file analysis

5. A forensic tool that helps to collect useful evidence is _____.
   (a) ngrep   (c) sshark
   (b) mgrep   (d) nshark

6. The security of an organization has three dimensions, namely _____.
   (a) computer security, virtual security, economic security
   (b) computer security, physical security, financial security
   (c) IT security, virtual security, economic security
   (d) IT security, physical security, financial security

7. _____ is an electronic discovery technique used to determine and reveal technical criminal evidence.
   (a) Cyber analysis   (c) Digital evidence
   (b) Cyber forensics   (d) None of these

8. Cyber forensic investigation should ensure the _____ of the evidence while handling and analysing so that the evidence is admissible in court.
   (a) uniqueness   (c) confidentiality
   (b) integrity   (d) completeness

9. _____ reveals to the analyst any unauthorized changes made to system binaries.
   (a) System file analysis
   (b) Timeline analysis
   (c) Volatile evidence analysis
   (d) Data recovery analysis

10. The contents of _____ used by the application programs as well as the operating systems are revealed as a result of forensics.
    (a) hidden files
    (b) swap files
    (c) hidden and swap files
    (d) none of these

11. _____ can gather and record every bit exchanged between any two designated hosts.
    (a) Intrusion detection system
    (b) Logging
    (c) Packet capturing tools
    (d) NetFlow data collector

12. _____ process involves capturing all data moving over the network and analysing network events to uncover network anomalies, discover the source of security attacks, and investigate breaches on computers and wireless networks to determine whether they have been used for illegal or unauthorized activities.
    (a) Disk forensics
    (b) Network forensics
    (c) Wireless forensics
    (d) Database forensics

13. The more a system is exposed to malware, the higher the risk of _____.
    (a) data leakage and data corruption
    (b) data leakage and data loss
    (c) data corruption and data loss
    (d) data leakage, data corruption, and data loss

14. _____ is also called hex dump.
    (a) Physical extraction
    (b) Logical extraction
    (c) Manual extraction
    (d) Automatic extraction

15. _____ is a free forensic tool that is used to extract the entire content of the computer's volatile

memory even if protected by an active anti-debugging or anti-dumping system.
 (a) Belkasoft Live RAM Capturer
 (b) Magnet RAM Capture
 (c) MoonSols
 (d) FTK Imager
16. _____ is a service organization that receives reports, reviews, and responds to computer security incidents.
 (a) Component security incident response team (CSIRT)
 (b) Computer security incident response team (CSIRT)
 (c) Computer security internet response team (CSIRT)
 (d) Computer security incident receive team (CSIRT)

REVIEW QUESTIONS

1. Define computer forensics.
2. What are the four different dimensions of an organization's security?
3. Explain the phases of cybercrime investigation with a diagram.
4. Briefly discuss the steps involved in forensic investigation.
5. Explain in detail network forensics and wireless forensics, and their challenges.
6. Explain (a) forensic approach, and (b) forensic methodology.
7. What is mobile forensics? List and explain its challenges and stages.
8. Explain email forensics. Distinguish email tracing from email tracking.

APPLICATION EXERCISES

1. A company named X has noticed many suspicious entries and IP addresses while observing the log record. Further, some clients have reported that they have been receiving a message prompt, redirecting them to a payment gateway that does not look genuine. You are now the forensic investigator. How will you ascertain if malware activity has taken place or malware is present in the system. How will you disinfect malware-infected machines? How will you ensure that the malware activity has not spread to all other systems in the network? Perform forensic investigation and trace the causes for the incident.
2. The sales manager, A, of Company X leaves his job and soon after, clients of that company start receiving defamatory messages from A against Company X. How can you prove that the mail originated from the sales manager in your capacity as the cyber FE?
3. The manager of a company notices that the work output of employee X has dropped over the past few weeks. The reason for this is that X has been spending many hours surfing the Internet. However, Internet surfing is prohibited under X's terms of employment. Even after warning, X continues with this unauthorized activity. The manager wants to expel X from the company with valid proof and seeks the help of an FE. Discuss how such a situation must be handled by the FE.
4. The trade secrets and sensitive information from a company were sold to a competing company. Employee X who is working in that company is suspected to have sold the information. Employee X claims that his system does not have any USB ports. Explain how such a suspicion was raised. How can it be proved that employee X was associated with the cybercrime?
5. The manager of a company suspects that the pay bills maintained in the system have been tampered with and that some of the associated files have been deleted. How can the files be recovered and how do you prove that the pay bill has been tampered with?

BIBLIOGRAPHY

1. Marjie Britz, T. (2013), *Computer Forensics and Cyber Crime – An Introduction*, 3rd edn: Pearson Education India
2. Robert Newman, C. (2007), *Computer Forensics – Evidence Collection and Management*, Auerbach Publications, Taylor & Francis Group
3. Heather Mahalik (2014), *Introduction to Mobile Forensics*, available at: https://www.packtpub.com/books/content/introduction-mobile-forensics (Accessed 01 December 2017)
4. eForensics, *Introduction to Mobile Forensics*, available at: https://eforensicsmag.com/introduction-to-mobile-forensics/ (Accessed 01 December 2017)
5. Andrew Martin (2008), *Mobile Device Forensics*, available at: https://www.sans.org/reading-room/whitepapers/forensics/mobile-device-forensics-32888 (Accessed 01 December 2017)
6. Mile Chapple, *How to Perfrom a Network Forensic Analysis and Investigation*, available at: http://searchsecurity.techtarget.com/answer/How-to-perform-a-network-forensic-analysis-and-investigation (Accessed 01 December 2017)
7. Cyber Forensics, *Network Forensics*, available at: http://www.cyberforensics.in/(A(NjkCEh4LzwEkAAAAOTYwM2QzZjct-NjYzZS00ODUwLWI3NDItY2I5Nzhm-MGYzMWM0s6BsKUFQYS8R3DfJFfg-7fipNxUE1))/Research/NetworkForensics.aspx?AspxAutoDetectCookieSupport=1 (Accessed 02 December 2017)
8. Raul Siles (07 January 2007), *Wireless Forensics: Tapping the Air – Part Two*, available at: https://www.symantec.com/connect/articles/wireless-forensics-tapping-air-part-two (Accessed 02 December 2017)
9. Edgar Weippl, *Database Forensics*, available at: https://www.nii.ac.jp/issi/pdf/2/4Johannes_Heurix.pdf (Accessed 02 December 2017)
10. Marcel Jans (2011), *Detecting Breaches in an Oracle Database with a Honeypot*, available at: https://mjsoracleblog.wordpress.com/2011/05/29/detecting-breaches-in-an-oracle-database-with-a-honeypot/ (Accessed 02 December 2017)
11. Paresh Motiwala (2015), *Db forensics for Sql Rally*, available at: https://www.slideshare.net/PareshMotiwala2200co/db-forensics-for-sql-rally (Accessed 02 December 2017)
12. Denys A. Flores (20 January 2016), *Research on Database Forensics*, available at: https://www.slideshare.net/DenysAFlores/database-forensics (Accessed 03 December 2017)
13. Charles Coker, Nicole Michaells, and Andrew Akker, *The Barebones Approach to Malware Forensics*, available at: https://eforensicsmag.com/the-barebones-approach-to-malware-forensics/ (Accessed 03 December 2017)
14. Benjamin Franklin (2014), *Improving your Malware Forensics Skills*, available at: http://journeyintoir.blogspot.in/2014/06/improving-your-malware-forensics-skills.html (Accessed 03 December 2017)
15. Raj Chandel, *Four Ways Capture Memory for Analysis (Memory Forensics)*, available at: http://www.hackingarticles.in/4-ways-capture-memory-for-analysis-memory-forensics/ (Accessed 03 December 2017)
16. LIFARS (14 June 2017), *The importance of Memory Forensics*, available at: https://lifars.com/2017/06/memory-forensics-tools/ (Accessed 03 December 2017)
17. Mark Wade (06 December 2011), *Memory Forensics: Where to Start*, available at: https://www.forensicmag.com/article/2011/06/memory-forensics-where-start (Accessed 03 December 2017)
18. Computer Forensics Evidence Collection and Preservation, Course Technology, Cengage Learning, EC-Council Press. available at: https://zodml.org/sites/default/files/Computer_Forensics.pdf (Accessed on 03 December 2017)
19. Sonya Krakoff (2018), *What's the Difference between Cybersecurity and Computer Forensics?*, available at: https://www.champlain.edu/champlain-college-online/blog-topics/cybersecurity/whats-the-difference-between-cybersecurity-and-computer-forensics (Accessed 12 February 2018)
20. Dorothy A. Lunn (2001), *Computer Forensics – An Overview*, available at: https://www.giac.

org/paper/gsec/559/computer-forensics-overview/101340 (Accessed 12 February 2018)
21. Rodney Mckemmish (1999), *What is Forensic Computing?*, available at: https://isis.poly.edu/kulesh/forensics/ti118.pdf (Accessed 12 February 2018)
22. Shannon Prather (2014), *Minnesota Detectives Crack the Case with Digital Forensics*, available at: http://www.startribune.com/when-teens-went-missing-digital-forensics-cracked-case/278132541/ (Accessed 12 February 2018)
23. McCann Investigations, *Case Study: Mobile Device Forensics in Texting and Driving Cases*, available at: http://www.mccanninvestigations.com/media/6310/case_study_texting_and_driving.pdf (Accessed12 February 2018)

Answers to Multiple-choice Questions

1. (d) 2. (b) 3. (c) 4. (b) 5. (a)
6. (d) 7. (b) 8. (b) 9. (a) 10. (c)
11. (c) 12. (c) 13. (d) 14. (a) 15. (a)
16. (b)

Digital Evidence

Learning Objectives

This chapter provides an overview of digital evidence, the collection procedure, and the obstacles to the collection process. The objective of this chapter is to provide an introduction to digital evidence, a deeper insight into the sources of evidences, namely various operating systems and their artifacts, the Windows registry, and various file systems. The chapter briefly explains the disk structure. Besides this, the chapter also talks about the sources of digital evidence in mobile devices and the Internet. Finally it elaborates on the challenges associated with digital evidence. The reader will be familiar with the following after studying the chapter:

- Evidence collection procedure
- The sources of evidence in computer systems
- The sources of evidence in mobile devices and on the Internet
- The challenges and obstacles in the digital evidence collection process

6.1 INTRODUCTION TO DIGITAL EVIDENCE AND EVIDENCE COLLECTION PROCEDURE

Digital evidence is defined as information and data that is stored, received, or transmitted by an electronic device and is of value to an investigation. It can be found on a computer hard drive, a thumb drive, a mobile phone, a personal digital assistant (PDA), a CD, floppy disk, DVD, flash card in a digital camera, memory stick, memory/SIM cards, fax machines, answering machines, cordless phones, pagers, caller-ID, scanners, printers, copiers, and CCTV equipment, among other places. Thus evidence can be found on the Internet, on standalone computers, or electronic communication devices (ECDs) and mobile devices. It is acquired when electronic devices are seized and secured for examination. Digital evidence is in binary form. Even though digital evidence stored in ECDs is in the binary form (machine language) and not understandable by humans, it should be presented in human readable form so as to be relied upon in court. It is very different from physical evidence. Figure 6.1 compares and contrasts digital and physical evidence.

6.1.1 Types of Digital Evidence

Digital evidence may exist in two forms:

Volatile evidence It refers to the frequently changing information (e.g., information about running processes or network, contents on the clipboard and some data in memory which are usually lost when the power for the ECD is turned off).

Non-volatile evidence It refers to the contents that can be recovered from an ECD even if it is not powered on. Some examples of both the forms are presented in Fig. 6.2.

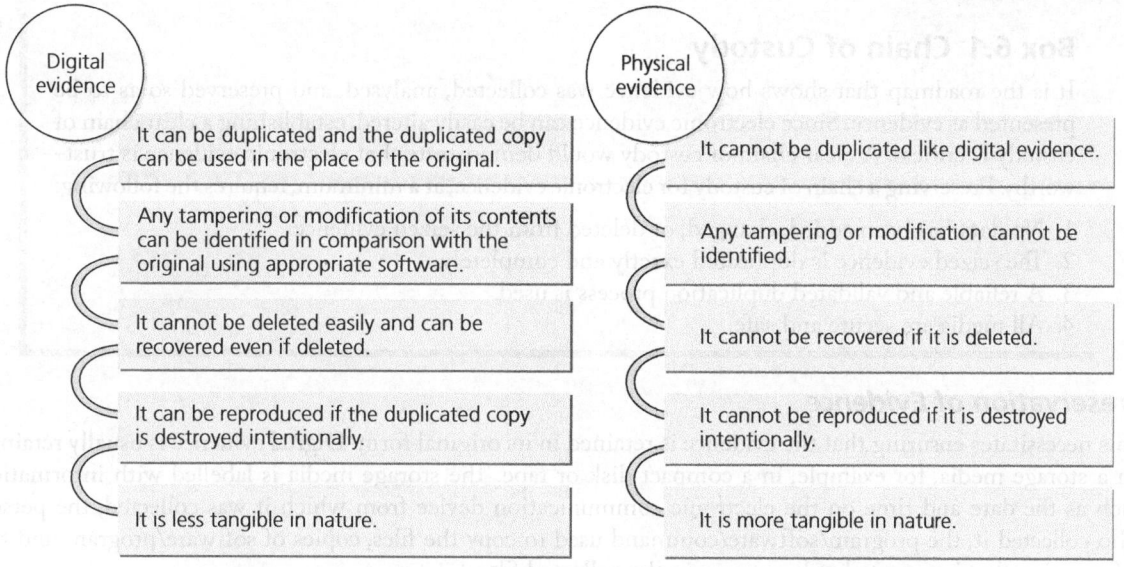

Fig. 6.1 Comparison between digital evidence and physical evidence

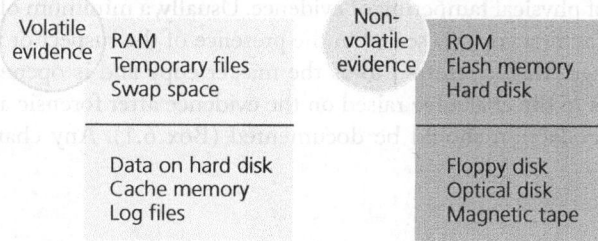

Fig. 6.2 Types of digital evidence

6.1.2 Evidence Collection Procedure

Evidence collection involves five phases which are explained here. This is shown in Fig. 6.3.

Identification of Evidence

Evidence has to be found by determining exactly where it is stored in the ECD. This necessitates distinguishing between the actual evidence and junk data. Identification implies what the information available in the data is, where it is located, and how it is stored. Consequent to this, it has to be ascertained if the evidence is relevant to the offence committed or the case. The order of volatility of evidence is important to determine the order of gathering the evidence and to minimize any loss of data relevant to the case or corruption of the same. Evidence should be collected using the right tools and by a person who has expertise in it and is legally entitled to.

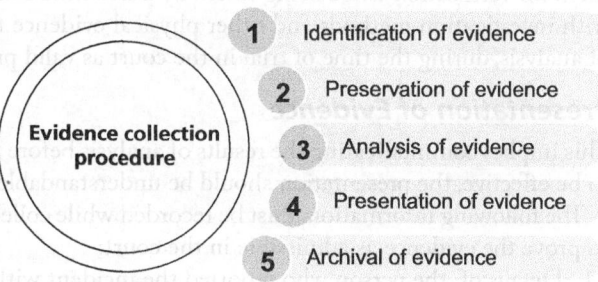

Fig. 6.3 Evidence collection procedure

> **Box 6.1 Chain of Custody**
>
> It is the roadmap that shows how evidence was collected, analysed, and preserved so as to be presented as evidence. Since electronic evidence can be easily altered, establishing a clear chain of custody is critical. A clear chain of custody would demonstrate that electronic evidence is trustworthy. Preserving a chain of custody for electronic evidence, at a minimum, requires the following:
>
> 1. No data has been added, changed, or deleted from the seized evidence.
> 2. The seized evidence is duplicated exactly and completely.
> 3. A reliable and validated duplication process is used.
> 4. All media are secure and safe.

Preservation of Evidence

This necessitates ensuring that the evidence is retained in its original form. Digital evidence is usually retained on a storage media, for example, in a compact disk or tape. The storage media is labelled with information such as the date and time on the electronic communication device from which it was collected, the person who collected it, the program/software/command used to copy the files, copies of software/program and the information that is expected to be present in the collected files.

Evidence collected should be free from alteration and from any external avenues of change. Tampering of evidence should be avoided. It is always safe to have a forensic image or forensic copy of the evidence collected which could help at times of physical tampering of evidence. Usually, a minimum of two copies of the evidence are created, one of which (the first copy) is sealed in the presence of the suspect or the owner of the ECD and is placed in a secured storage. This is referred to as the master copy and is opened for examination only on the direction of the court as to the challenge raised on the evidence after forensic analysis on the second copy.

Every step of evidence collection should be documented (Box 6.1). Any change made on it should be documented and justified.

Analysis of Evidence

The stored evidence has to be analysed to extract the relevant information and recreate the chain of events. Analysis should be performed in a dedicated host that is clean, secure and isolated from the network. During analysis, every action performed on the evidence is documented to ensure that the procedure is repeatable and capable of yielding the same results at any later point in time.

Analysis of evidence includes assessment of the means and motivation of the committed crime while evaluating the collected digital evidence, experimentation of all possibilities of methods and techniques and documenting them for verification while being tested by a scientific community and the court, correlation of analysis results with investigation methods and other physical evidence to map the crime scene and validation of the results of analysis, during the time of trial in the court as valid proof.

Presentation of Evidence

This implies communicating the results of analysis before the court to either prove or disprove a crime. In order to be effective, the presentation should be understandable even to a layman.

The following information must be recorded while collecting digital evidence through seizure memo in order to prove the evidence as admissible in the court:

1. Details of the person who reported the incident with date and time
2. Name of the investigating officer (IO) and his/her team involved in the investigation
3. Case number and Section of the Law
4. Reasons for search and seizure

5. List of ECDs seized from the crime scene deemed to be associated with the crime incident with its detailed specification
6. Network diagram from the crime scene if the ECD is connected to a network
7. Applications running on the ECD if it is in ON state.
8. Access control policies in force on the ECD at the time of collecting evidence from it
9. List of steps involved in the collection of evidence
10. Chain of custody of collected evidence
11. Date and time of evidence collection
12. Place of seizure
13. Names of witnesses

Archival of Evidence

Storage of collected evidence in some storage medium is essential. Logs in the system can be used to keep the evidence secure as every activity is time-stamped but it should be digitally signed and encrypted to prevent contamination of evidence. If logs are retained locally, their integrity lies susceptible. Network traffic should also be monitored for any malicious activity apart from monitoring logs.

6.1.3 MECHANISMS ASSOCIATED WITH DIGITAL EVIDENCE COLLECTION

The mechanisms associated with the collection of digital evidence are as follows:

Seizure It refers to taking into custody the ECD associated with a crime by the law enforcement agency for the purpose of investigation.

Acquisition It refers to gathering the data from the seized ECD. This is done as a bit stream backup copy of the original, possibly on a forensically sterile storage media. Acquisition is termed successful only if the message digest (hash value) of the original matches with the message digest of the bit stream backup copy made.

Acquisition can be done either by imaging or by cloning. In imaging, the contents of the original (from the seized ECD) are copied bit-by-bit or sector-by-sector as an image file format to the forensically sterile storage media. On the other hand, cloning is done either for a partition or a whole disk usually on a disk of the same geometry and is identical to imaging except that the forensically secure storage media can be used to replace the original for booting.

More details on these two mechanisms are available in Chapter 8.

6.2 SOURCES OF EVIDENCE

The sources of evidence can be broadly categorized into three:
1. Digital evidence from standalone computers/ECDs
2. Digital evidence from mobile devices
3. Digital evidence from the Internet

6.3 DIGITAL EVIDENCE FROM STANDALONE COMPUTERS/ELECTRONIC COMMUNICATION DEVICES

Digital evidence can be collected from computers/ECDs from the following:
1. Address books and contact lists
2. Audio files and voice recordings
3. Backup of various programs, including backup of mobile devices
4. Bookmarks and Favorites
5. Browser history

6. Calendars
7. Compressed archives (ZIP, RAR, etc.) including encrypted archives
8. Configuration and .ini files (may contain account information, last access dates, etc.)
9. Cookies
10. Databases
11. Documents
12. Email messages, attachments, and email databases
13. Events
14. Hidden and system files
15. Log files
16. Organizer items
17. Page files, hibernation files, and printer spooler files
18. Pictures, images, digital photos
19. Videos
20. Virtual machines
21. System files
22. Temporary files

However the main sources of evidence for investigators in the standalone computer systems/ECDs are the operating systems and the system files, its artifacts, the storage media, and the file system. They are discussed elaborately in the following subsections.

6.4 OPERATING SYSTEMS AND THEIR BOOT PROCESSES

In computer systems, the hard disk forms the bootable media (contains software to start up a computer) and so it should be examined non-invasively (to avoid accidental writes to the media). This is done by altering the normal sequence of execution during the booting of the operating system.

The predominantly found operating systems include the Microsoft family of operating systems, Linux and Mac. Their boot processes are briefly explained as follows:

Microsoft Family

Power on Self Test (POST) is the first action performed by a computer while booting to ensure the presence and functionality of the core components.

Upon successful completion of POST, basic input output system (BIOS, which is a small segment of code stored in read only memory, ROM) reads the master boot record (MBR). The following actions vary with different versions of Windows:

MS - DOS

1. After BIOS, control is transferred to the boot code contained in the MBR and is executed.
2. Master boot code examines the master partition table for any extended DOS partition. If present, it looks for a bootable partition specified in the partition table.
3. Master boot code locates the extended DOS partition and loads the extended partition table to gather information about the first logical volume. This is done for all the other extended partitions (if present) in order for them to be loaded and recognized by the system.
4. Master boot code now loads the primary partition (boot partition). This master or volume boot sector now controls the remaining of the boot process.
5. The boot code also examines the disk structures for correctness.
6. The boot code then searches the root directory for operating system files such as IO.SYS, MSDOS.SYS, and COMMAND.COM. These files are loaded and control is transferred to them. Execution happens as

a nested chain, that is IO.SYS is executed which in turn executes MSDOS.SYS. Finally COMMAND. COM is executed followed by CONFIG.SYS and AUTOEXEC.BAT files. Now the computer is under the control of the operating system.

Windows XP The following are performed after POST:
1. MBR transfers the control to XPSetup to enable Windows to control the start-up process.
2. MBR reads the boot sector to find the code to start NTLDR (the boot loader) available in the root folder of the bootable partition. NTLDR starts file systems, runs Ntdetect.com to capture information on installed hardware, reads registry files, and loads device drivers.
3. Finally Ntoskrnl.exe starts Winlogon.exe to facilitate the log on process.

Windows Vista and Windows 7 These two operating systems support both BIOS-based and extensible firmware interface (EFI)-based booting. In EFI-based booting, the inbuilt boot loader loads the boot manager (bootmgr). In case of BIOS-based booting, BIOS transfers the control to the code in MBR; MBR reads the partition tables and enables the system to read the file system. It then locates and starts the boot manager in the root directory (Box 6.2).

After loading the boot manager, the following are common for both the forms of booting:
1. Boot loader loads and launches the kernel, services from system registry key, hardware application layer (HAL), and device drivers, and also enables paging.
2. Session manager takes control and the system switches from text mode to graphics mode.
3. Logon manager facilitates the user to log on.

Windows 8 and Windows 10 Windows 8 and 10 offer support for both BIOS and unified extensible firmware interface (UEFI). The boot process is the same as that of Windows 7. However, Windows 10 is more powerful in resuming from sleep and hibernate than Windows 7 and Windows 8 (Box 6.3).

Linux This is an open source operating system and the boot process is as follows:
1. When the computer is turned ON, BIOS scans the MBR for boot loader which may be either LInuxLOader (LILO) or GRand Unified Bootloader (GRUB) and loads and executes it.
2. Bootloader loads the kernel which in turn loads the primary process, init.
3. Init runs a series of shell scripts referred to as 'rc files' which spawn all other processes to end up in a running Linux system.

Box 6.2 Extensible Firmware Interface

Extensible Firmware Interface (EFI) provides communications between the hardware and the operating system as that of BIOS but has its own boot loader and supports the new GPT (**G**lobally **U**nique **I**dentifier (GUID) **P**artition **T**able) partitioning scheme in addition to the master boot record (MBR) partitioning scheme.

Box 6.3 Unified Extensible Firmware Interface

UEFI is the alternate to BIOS. Unlike BIOS which stores boot code in ROM, the boot code is stored in the EFI directory or in non-volatile memory. This provides a lot more space for language localization, boot-time diagnostics, and utilities. Besides this, UEFI offers more features such as supporting larger hard drives, faster boot times, more security, and convenience by way of graphics and mouse cursors.

Macintosh—Mac The boot process in Macintosh from Apple, popularly called Mac operating system, has four stages which are as follows:

1. Firmware initiates POST and selects the OS booter which may be either BootX for PPC or boot.efi for Intel.
2. Booter loads the OS kernel.
3. Kernel launches the core UNIX system and starts the process launchd. The launchd process starts all other processes required to run the operating system and the user environment.
4. Finally the parent launchd runs as root and initiates all other processes especially the login window application which facilitates user log in.

6.5 STORAGE MEDIUM

Storage medium refers to the technology used to hold, preserve, and retrieve data, which may be application or user data.

6.5.1 Disk Drive

The primary storage medium in computer systems is the disk drive which may be either a hard disk drive or a solid state drive.

Hard Disk Drive or HDD

Fig. 6.4 shows the logical view of a hard disk. Disk platters are the round, flat, magnetic metal or ceramic disks in a hard disk that hold the actual data. As the number of platters increases, the storage capacity rises. Tracks are the concentric circles on platters which are in turn divided into sectors where the information is stored. The read/write heads on both surfaces of a platter are tightly packed and locked together on an assembly of head arms. Hard disks with a large number of platters are more sensitive to vibrations, flaws in the surface of a platter, and head misalignment, and are hence prone to failure.

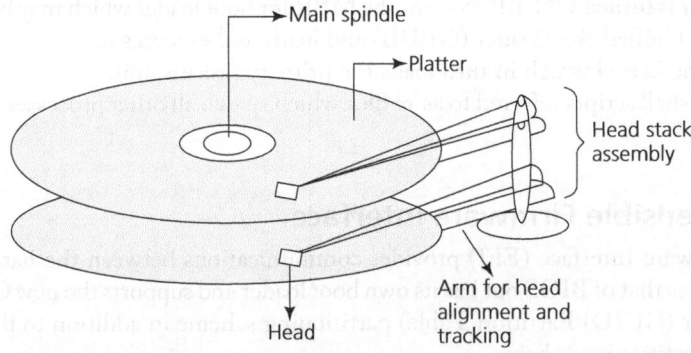

Fig. 6.4 View of HDD

Solid State Drive or SSD

In contrast to HDD, SSDs have no moving parts but non-volatile memory chips which can retain information even in the absence of power like the USB flash drives. Hence SSDs are noted for their less power usage, faster data access, high reliability, and lighter weight.

Sectors

The data storage area in a physical disk is divided into equal-sized blocks called sectors while manufacturing. Thus a sector is the basic physical unit, the smallest writeable, the smallest addressable, and the smallest allocable unit of a physical disk. Each sector is of 520 bytes; in older drives 512 bytes hold data and the remaining 8 bytes are used for internal drive control, drive management, error detection and correction, and sector identification. The last two bytes (bytes 510 and 511) of the sector are always hexadecimal 55 AA (0x55AA). The advancement in storage technology has raised the sector size to 4096 bytes in advanced format (AF) drives. Fig. 6.5(a) shows a sector which includes the following: sector header (Head) which includes ID information and synchronization bytes.

ID information This contains the sector number and the location that identify the sector on the disk. It also contains status information about the sector.

Synchronization fields This helps the drive controller to guide the read process.

Data This is the actual data in the sector.

ECC This is an error-correcting code that ensures data integrity.

Gaps Spaces are provided to give the drive controller time to continue the read process.

Fig. 6.5(b) shows a linear array of 8 sectors which correspond to 4K bytes of user data. Fig. 6.5(c) shows an AF sector of 4096 bytes of data with format efficiency improvement in storage size.

Logical Block Addressing

Logical block addressing (LBA) linearly numbers the sectors in sequential order without any relation to its physical location within a physical disk. LBA translation is however performed by the physical disk controller and BIOS.

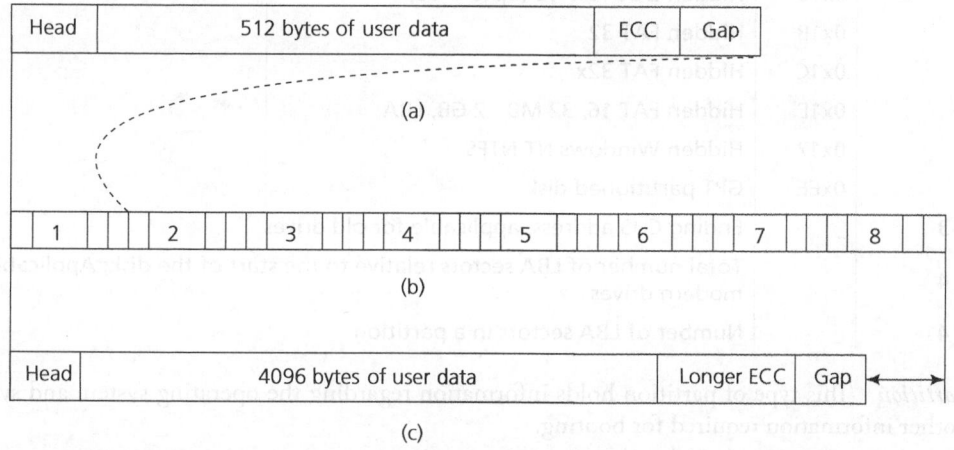

Fig. 6.5 Linear block addressing (a) Sector of 512 bytes (b) Linear array of 8 sectors making 4K bytes of user data (c) Sector of 4096 bytes (advanced format)

Hidden Sector Data

This refers to bytes in the physical disk that are used by 'itself' and are invisible to the user. It may also refer to 'spare' sectors within an SSD which are meant to replace damaged or bad sectors. Bad sectors are used to refer to areas of a disk that have become unusable.

Partitioning

Partitioning refers to logically dividing the physical drive into multiple pieces (logical volumes) called partitions. This makes the physical drive readable and writeable. Partitioning can be done with standard tools, for example, with Diskpart in Windows family of operating systems or with third party tools, for example, PowerQuestPartitionMagic. A partition can be one of the following two types:

Table 6.1 Data structure and interpretation of partition table

Byte			Description
Offset	Length	Value	
0x00	1	–	Boot indicator where LSBs 00 indicate that the partition is non-bootable and 80 indicates that it is bootable.
0x01	3	–	Starting Cluster, Head, Sector (CHS) address; Applicable for old drives
0x04	1	0x04	DOS FAT 16
		0x05	DOS Extended Partition, CHS
		0x06	FAT 16
		0x07	NTFS
		0x0B	FAT 32
		0x0C	FAT 32x, LBA
		0x0F	Extended partition, LBA
		0x83	Linux — System identifier of partition type
		0x14	Hidden DOS FAT 16 (upto 32 MB)
		0x15	Hidden DOS extended partition
		0x16	Hidden DOS FAT 16 (upto 2 GB)
		0x1B	Hidden FAT 32
		0x1C	Hidden FAT 32x
		0x1E	Hidden FAT 16, 32 MB - 2 GB, LBA
		0x17	Hidden Windows NT NTFS
		0xEE	GPT partitioned disk
0x05	3	–	Ending CHS address; applicable for old drives
0x08	4	–	Total number of LBA sectors relative to the start of the disk; Applicable for modern drives
0x0C	4	–	Number of LBA sectors in a partition

Primary partition This type of partition holds information regarding the operating system and system area, as well as other information required for booting.

Extended partition This partition holds the data and files that are stored on the disk.

A unique number referred to as disk signature (four bytes long) is written to the disk during initialization for the purpose of identifying it. In an unpartitioned physical drive, the entire storage space is considered as a single volume and by default, the operating system assigns the drive label 'C'. Except drive labels 'A' and 'B' reserved for floppy disks, other alphabets can be used as labels. Partitioning a drive offers the flexibility of having multiple logical volumes in a single physical drive each having either the same or different file system. Thus, there can be a maximum of twenty four partitions in a physical disk drive.

Master Boot Record

The first physical sector of the physical drive contains the MBR which is loaded into the memory by BIOS during the booting process. MBR includes a table (partition table) that contains information about each partition that the hard disk has been formatted into. The MBR also includes a program that reads the boot sector record of the partition containing the operating system to be loaded into RAM. The partition table within an MBR is called master partition table (MPT).

Master Partition Table

The master partition table (MPT) is 64 bytes in length and can accommodate four 16-byte entries. This means that in an MPT, there must be atleast a minimum of one 'primary partition' and a maximum of up to four 'primary partitions'. Primary partition implies that it is bootable and is used to boot the operating system in a computer. The partition table entry has a specific data structure which includes the type of partition, its location and its size as listed in Table 6.1.

Extended Primary Partition

Extended primary partition (EPP) enables greater number of partitions and logical drives by further division of a partition into multiple partitions called extended logical partitions (ELP). A physical drive can have only one EPP whose entry is within the MPT. There is no volume boot record in an EPP. EPP has no partition table to denote where each logical drive begins and ends. Rather, every ELP has a pointer to itself and the next ELP, which is referred to as daisy chaining. Every ELP has a partition table with two entries and is similar in data structure to that of the partition table in MBR. They are (a) an entry to itself to denote where the logical drive begins and there will be a volume boot record; and (b) pointer to the next ELP, which provides the size of ELP.

Hidden Partitions

A partition can be hidden by one of the following ways:
1. Removal of assigned drive label
2. Use of encryption
3. Altering the system identifier bytes in the partition table, for example, changing the system identifier from 0x0B to 0x1B makes FAT 32 as Hidden FAT 32.
4. Breaking the daisy chain by manipulating the values pointing to the next logical partitions in an ELP partition table.

GPT Partitions

UEFI uses a partition table called Globally Unique IDentifier (GUID). Each entry is 128 bytes long. It supports upto 128 partitions and uses 64-bit LBA. The characteristic of a GPT partitioned disk is shown in Table 6.2.

Table 6.2 Features in GPT partitioned disk

Characteristic	Description
Protective MBR	This is present in physical sector 0 but there is no boot code.
Primary GUID partition table (GPT) header	It follows MBR in physical sector 1. It begins with an 8-byte EFI signature. Its size is 92 bytes and it depicts the layout of the disk like the location of the partition table, partition data areas, and backup copies of GPT header/partition table.
GUID partition entries	Partition tables are located in Sector 2.
Partition type GUID	It holds both system and file system information.
Partition attribute flags	This has provisions to prevent a partition from being hidden.
Partition area	The partitions can accommodate only NTFS. GPT has a Microsoft reserved partition (MRP) used by an operating system, to retain certain temporary files.

Host Protected Areas

It refers to the area reserved on the hard disk to store information that cannot be modified and that is inaccessible to the user, BIOS/UEFI, and the operating system. This may even be a potential place to hide evidence.

6.5.2 Other Storage Media

Some other non-volatile storage media are as follows:

Floppy Disks

Floppy disks were popular with personal computers in the 1980s to distribute software, transfer data, and create backup. It was an affordable storage when compared to hard disks. Later it was replaced by the optical storage technology.

Optical Storage Media

This includes compact discs (CDs) and digital versatile discs (DVDs) where the following are present:

1. A physical sector in a CD is of 2064 bytes with the following fields, namely sync, sector head, error checking, and detection, and a user data area which is 2352 bytes in CD-DA, 2048 bytes in CD-ROM, and 2336 bytes in video CD. Each sector has its unique logical sector number (LSN).
2. In case of DVD, a sector is of 2064 bytes. The header is of 12 bytes and includes 4 bytes for sector ID, two bytes for ID error detection code, and six bytes for copyright protection information. This is followed by user data of 2048 bytes. The end of a sector is marked by 4 bytes of error detection code.

Flash Memory

It is a non-volatile storage chip that can be electrically erased and reprogrammed. It is widely used in memory cards, USB flash drives, MP3 players for storage and transfer of data. Most flash drives are preformatted with the FAT32 or exFAT file systems which are discussed in the later subsections.

6.6 FILE SYSTEM

A file system is used to most effectively store, organize, and access data in a computer. Data storage devices such as hard disks, CD-ROMs, flash memory devices, and floppy disks use file systems to store data. A file system provides the following: storage, hierarchical categorization, management, navigation, access, and data recovery features. File systems are organized in the form of tree-structured directories. An operating system's ability to access files on a volume depends on the file system with which the volume was formatted.

Some of the file systems are as follows:
1. File allocation table 16 (FAT16)
   (a) This 16-bit file system was developed for DOS and further supported by other operating systems. It consumes little memory and is simple and reliable.
   (b) The filename can be up to a maximum of 8 characters and file extension can be a maximum of 3 characters.
   (c) Its main shortcomings are that it supports a maximum of 64 KB allocation units and that it becomes less efficient on partitions larger than 32 MB.
2. FAT32
   (a) This is a 32-bit version of the FAT file system. The use of smaller clusters results in a more efficient storage capacity. It supports drive sizes up to 2 TB.
   (b) It can relocate the root directory and use the backup copy instead of the default copy. One of the main features is that it can dynamically resize a partition.
3. New technology file system (NTFS)
   (a) NTFS is very different from FAT and provides enhanced security, file-by-file compression, quotas, and encryption. It is designed to quickly perform standard file operations such as read, write, search, and even advanced operations such as file system recovery on very large hard disks.
   (b) When a volume is formatted as an NTFS volume, the master file table (MFT) and several system files are created. The MFT is the first file in an NTFS volume and contains information about all the files

and folders in the volume. The first information is about the partition boot sector which can be up to 16 sectors long starting from sector zero. NTFS has several versions:
  (i) V1.2: Found in Windows NT 3.51 and Windows NT 4
  (ii) V3.0 (sometimes called Version 5.0): Found in Windows 2000
  (iii) V3.1 (sometimes called Version 5.1): Found in Windows XP and Windows Server 2003
  (iv) Transactional NTFS (TxF): Found in Windows Vista

Table 6.3 lists some of the popular operating systems and the file system in force.

Table 6.3 Operating systems and their file system formats

Operating system	File system format of volume
Windows 10	NTFS
Windows 8	NTFS
Windows 7/Vista	NTFS
Windows XP	NTFS or FAT32
Windows 2000	NTFS
Linux	ext2
Debian GNU/Linux 7.0	ext4

6.6.1 FAT File System and its Components

The FAT file system can be viewed as two groups, namely the system area comprising reserved regions (holds boot record) and FAT region and data area comprising the root directory region and the file and subdirectories region. This is shown in Fig. 6.6. The operating system fetches information about the files and directories stored in the data area from the system area. The presence of root directory in the data area offers it the flexibility to expand and occupy more than one cluster in the storage media (Box 6.4).

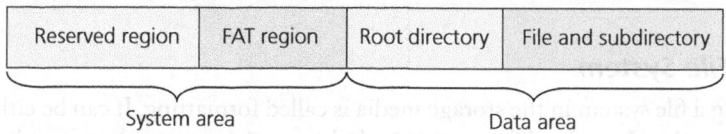

Fig. 6.6 Overview of FAT

Boot Record

1. A boot sector is always the first sector on the first track of memory that contains code for bootstrapping systems. Each valid boot sector has two bytes (0x55, 0xAA), called a boot sector signature.
2. There are two major kinds of boot sectors:
   (a) *Volume boot record—VBR*: It is the first sector of a data storage device that has not been partitioned, or the first sector of an individual partition on a data storage device that has been partitioned. It contains a code to load and invoke the operating system or other standalone programs installed on that device or within that partition.

> **Box 6.4 Cluster**
> It is the smallest logical storage unit in a disk. The file system divides the storage in a disk volume into discrete chunks of data for efficient disk usage and performance. These chunks are called clusters. A file is allocated a certain number of clusters.

(b) *Master boot record*: It is the first sector of a data storage device that has been partitioned. It contains a code to invoke the volume boot record and locate the active partition. An MBR contains the following structures:
  (i) *Master partition table*: It is a small bit of code that contains a complete description of the partitions that are contained in the storage device.
  (ii) *Master boot code*: The master boot code is a small bit of computer code loaded and executed by the BIOS to start the boot process.

Table 6.4 Data structure of FAT 32 directory entry

Offset (Hex)	Size (in Bytes)	Description
0x00	8	Filename (status byte)
0x08	3	File extension
0x0B	1	File attribute byte
0x0C	1	Reserved for Windows NT
0x0D	1	Time stamp in milliseconds at the time of file creation
0x0E	2	File creation time
0x10	2	File creation date
0x12	2	Last accessed date
0x14	2	High word of entries of starting cluster (0x00)
0x16	2	Time when the last write to file was made or created
0x18	2	Date when the last file was modified or created
0x1A	2	Low word of the entries of the starting cluster
0x1C	4	File size (this is zero if file is a directory)

Formatting FAT File System

The process of creating a file system in the storage media is called formatting. It can be either low-level formatting or high-level formatting. Low-level formatting includes creating a sector layout and sequencing of storage media, and is done at the factory. High-level formatting is used to refer to formatting done with a Windows GUI environment or specific tools.

Directory Structure

The FAT 32 directory is composed of 32 byte records, one for every file and subdirectory within a directory. The data structure of FAT 32 directory is illustrated in Table 6.4.

When a file is created, a directory entry is created and the actual data is written in the first cluster identified by FAT. FAT is updated on the clusters used by the file. When a file is deleted, the respective status byte in the directory entry is changed to 0xE5. The chain of clusters associated with the deleted file is zeroed in FAT for possible reuse of those clusters for writing. The process of deletion of a file however does not overwrite or change its contents.

The date and time values in hexadecimal little endian arrangement have to be converted to binary to interpret the data. Consider that the file created time at offset 0x0E and 0x0F are B6 61. This in little endian arrangement is 0x61 and 0xB6. In binary representation it will be 0110000110110110 and is interpreted (from MSB to LSB) as 01100 hours, 001101 minutes and 10110 seconds which means 12 hours 13 minutes and 44 seconds (multiplied by 2). Similarly date is interpreted with 7 bits for year, 4 bits for month and 5 bits for day.

Table 6.5 Status byte of FAT32 directory entry

Value of status byte	Interpretation
Legal character of file name	First character of filename; Normal directory entry
0xE5	Deleted directory entry
0x00	Empty directory entry and is ignored by operating system.

The data structure of status byte is interpreted as shown in Table 6.5. The attribute byte and its possible values and interpretations are shown in Table 6.6.

Table 6.6 Attribute byte of FAT32 directory entry

Attribute byte value (in hex)	Interpretation
01h	Read only
02h	Hidden
04h	System
08h	Volume label
10h	Directory
20h	Archive ON
40h	Reserved
80h	Reserved

Data Area

The data area comprises sectors grouped together to form clusters which hold files and directories. The size of a cluster is determined when the disk volume is partitioned. Larger volumes use larger cluster sizes. For hard disk volumes, each cluster ranges in size from 4 sectors (2,048 bytes) to 64 sectors (32,768 bytes). The sectors in a cluster are contiguous and so each cluster is a continuous block of space on a disk. A file may span many clusters which may either be contiguous or non-contiguous. Directories refer to the root directory (only one on a FAT volume) and subdirectories.

File Allocation Table

FAT is a table containing information about the location and status of file data in the partitions. It provides mapping between the file and its location (cluster) and helps in tracking all the clusters in a volume. FAT may include four types of entries as shown in Table 6.7 which correspond to unallocated clusters, allocated clusters, end of file (EOF) tag, and bad sectors.

Table 6.7 Entries in FAT32 table, its value, and interpretation

FAT 32 entry (Hex Value)	Inference	Description
0x00000000	Unallocated cluster	Cluster which remains free for storage.
0x00000002–0xFFFFFFEF	Next cluster	Current cluster is in use and indicates the next cluster that is part of the file.
0xFFFFFFF8–0xFFFFFF0F	End of file	It is either the last cluster of the file or the one and only cluster of the file.
0xFFFFFFF7	Bad cluster	Not meant for storing data as the 512 bytes of data cannot be held by the cluster.

File Allocation Table (2)
This is actually the secondary copy of FAT. Any changes made to the primary FAT is reflected in this secondary FAT too.

Slack
Slack space is the area of a disk between the end of the file and the end of the cluster allocated to the file. It may be one of the following:

File slack The area between the end of the actual file and the end of the cluster assigned to it.

RAM slack It is the area between the end of the file to the end of the sector. In older versions of the computer, excess data from RAM, such as directory structures and passwords, were pulled to this area and hence the name. However, modern operating systems pad these unused locations with 0x00.

Residual slack It is the area between RAM slack and the end of the cluster the file occupies. It holds contents from files that reside in the same cluster prior to the time the file has overwritten it. Residual slack may carry evidence in the form of documents, chat history, photographs, etc.

6.6.2 Extended File Allocation Table File System
Extended file allocation table (exFAT) file system is a successor of FAT 32 and has the capability to overcome file and volume size limits and supports future proofing. It is also designed to be suitable for portable media with large sizes besides being flexible. This file system can be used with Windows XP, Windows Vista, Windows 7, Windows 8, and Windows 10.

exFAT file system comprises two main regions: file system area and data region.

File System Area
File system area includes the main boot region, backup boot region, and FAT region.

The back boot region is an identical copy of the main boot region and follows it.

1. *Boot region (main boot region and backup boot region)*: The main boot region comprises of the following:
    (a) *Boot sector:* This is identical to volume boot record in FAT32. The boot signature for exFAT is 0xAA55. exFAT can be formatted to support sector size ranging from 512 to 4096 bytes, cluster size ranging from 512 bytes–32 mega bytes and a maximum volume size of 128 petabytes. The hardware requirement is that the hard drive should be greater than 32 GB.
    (b) *Extended boot sectors:* This extends to eight sectors and holds the bootstrapping code.
    (c) *OEM parameters:* This holds manufacturer-specific information. It is designed to accommodate 10 parameters, each 48 bytes long. The remainder of the sector remains unused. The first 16 bytes of each parameter contains Parameter GUID which defines how the rest of that field is structured.
    (d) *Reserved:* This sector is reserved for future use.
    (e) *Boot checksum:* It contains four byte checksum of four other regions of the main boot region.

The sectors reserved for main boot region and backup boot region in a volume are shown in Table 6.8.

Table 6.8 Sector reservation for main boot region and backup boot region

Main boot region	Sector	Backup boot region	Sector
Main boot sector	0	Backup boot sector	12
Main extended boot sectors	1	Backup extended boot sectors	13
Main OEM parameters	9	Backup OEM parameters	21
Main reserved	10	Backup reserved	22
Main boot checksum	11	Backup boot checksum	23

2. *FAT region*: This region begins at Sector 24. The primary difference between exFAT and FAT32 is that the file system holds entries for files that are allocated non-contiguous clusters. Whether a file is allocated contiguous clusters or is fragmented is marked by flags in the directory entry and is recorded in a Bitmap file. There is only one FAT region as opposed to two in FAT32 and it is not necessary to be present in Sector 24. The entries in an exFAT table are shown in Table 6.9.

Data Region

This region holds the file system structures, files, and directories in the form of heap of clusters. The difference from FAT32 is that exFAT has different directory entries and two additional system files, namely allocation bitmap and up-case table.

Table 6.9 Entries in exFAT FAT Table and its value

exFAT entry (Hex Value)	Inference	Description
0xFFFFFFF8	Media type	Pre-determined value.
0x00000002– 0xFFFFFFF6	Next cluster	Current cluster is in use and indicates the next cluster that is part of the file. (Fragmented file)
0xFFFFFFFF	End of file	It is either the last cluster of the file or the one and the only cluster of the file.
0xFFFFFFF7	Bad cluster	Not meant for storing data as the 512 bytes of data cannot be held by the cluster.

Allocation bitmap This system file records the allocation of clusters within a cluster heap starting from logical cluster number (LCN) 2. It uses a one-bit flag to mark the status of every cluster within a cluster heap where a zero indicates that the cluster is unallocated while a one indicates that it is allocated and in use.

Up-case table This may be present in LCN 3 or in the next subsequent cluster after allocation bitmap if it extends for more than one cluster. exFAT uses only Unicode characters for filenames in directory entries and is case insensitive. All search operations for files are performed after converting the filenames to uppercase. Formatting files with a very large cluster size, of course, results in a large slack space which provides room for hiding evidence.

Root directory and directory entries The starting cluster of the root directory is present in the master boot sector and is usually the next cluster after the up-case table, which may even span more than one cluster. The first cluster of the root directory is overwritten with 0x00 while formatting. Every file may have 19 directory entries to reflect its status and is referred to as directory entry set.

The first byte of a directory entry called *entry type* defines its type where the first five bits mark its type code (the role of the directory entry and the kind of information it will contain), the sixth bit marks its importance as either critical (denoted by 0) or benign (denoted by 1), the seventh bit marks type category as either primary (denoted by 0) or secondary (denoted by 1) and the eight bit indicates the usage as either not in use (denoted by 0) or in use (denoted by 1). The sixth bit (importance) is important in a forensic sense as a benign directory entry set is ignored by exFAT and provides room for hiding evidence.

The directory entries are as follows:

Volume label directory entry This is the first directory entry in the root of an exFAT volume. Volume label is a Unicode string of upto eleven characters. The data structure of volume label directory entry is shown in Table 6.10.

For example, an entry type of 0x83 (1000 0011 in binary) is interpreted (from LSB to MSB) as type code - 00011 holding volume label, importance - 0 being critical, category - 0 being primary and in use - 1 indicating that it is in use.

Table 6.10 Data structure of volume label directory entry

Offset (Hex)	Length (Bytes)	Field name	Description
0x00	0x01	Entry type	0x83 denotes label is assigned, 0x03 denotes no volume label
0x01	0x01	Character count	Number of Unicode characters in label
0x02	0x0B	Volume label	Volume label in Unicode
0x18	0x08	Reserved	Reserved for future use

Allocation bitmap directory entry This is the second directory entry in the root of an exFAT volume. It is a critical primary directory entry. It is located as a system file within the cluster heap and so it has a directory entry. The data structure of allocation bitmap directory entry is shown in Table 6.11.

Table 6.11 Data structure of allocation bitmap directory entry

Offset (Hex)	Length (Bytes)	Field name	Description
0x00	0x01	Entry type	0x81
0x01	0x01	Bitmap flags	0 denotes bitmap 1, 1 denotes bitmap 2
0x02	0x12	Reserved	Reserved for future use
0x14	0x04	First cluster	The starting logical cluster number (LCN) of allocation bitmap
0x18	0x08	Data length	Size of allocation bitmap file in bytes

Up-case table directory entry This is the third directory entry in the root of an exFAT volume. It is also a critical primary directory entry. It has a directory entry as it is located as a system file within the cluster heap. The data structure of up-case table directory entry is shown in Table 6.12.

Volume GUID directory entry This directory entry holds a GUID which is used by the operating system to distinguish volumes. It is a benign primary directory entry.

TexFAT padding directory entry It is used as part of transaction-safe exFAT file system. It is also a benign primary directory entry.

Table 6.12 Data structure of up-case table directory entry

Offset (Hex)	Length (Bytes)	Field name	Description
0x00	0x01	Entry type	0x82
0x01	0x03	Reserved	Reserved for future use
0x04	0x04	Table checksum	32-bit checksum of the contents of the table
0x08	0x1C	Reserved	Reserved for future use
0x14	0x04	First cluster	The starting LCN of up-case table
0x18	0x08	Data length	Size of up-case table file in bytes

Windows CE access control table directory entry This is used to provide support for Windows CE applications. It is also a benign primary directory entry in the root directory.

File directory entry This is the first entry of a directory set. It describes the files and directories which exist within a volume. It is a critical primary directory entry. Its data structure is illustrated in Table 6.13.

Table 6.13 Data structure of file directory entry

Offset (Hex)	Length (Bytes)	Field name	Description
0x00	0x01	Entry type	0x85 - in use; 0x05 - not in use
0x01	0x01	Secondary count	Number of secondary directory entries
0x02	0x02	Set checksum	Checksum of directory entry set
0x04	0x02	File attributes	Reserved for future use
0x06	0x02	Reserved	The starting LCN of up-case table
0x08	0x04	Create time stamp	
0x1C	0x04	Modified time stamp	32-bit DOS time format
0x10	0x04	Last accessed time stamp	
0x14	0x01	Create 10ms increment	
0x15	0x01	Modified 10ms increment	It is meant for working in the order of millisecond
0x16	0x01	Create UTC differential	
0x17	0x01	Modified UTC differential	7-bit signed integer, representing 15 minute intervals
0x18	0x01	Last accessed UTC differential	
0x19	0x07	Reserved	Reserved for future use

When a file is deleted, its file directory entry is not deleted but only the entry type field is set as not in use.

Stream extension directory entry This entry gives information such as where a file is located in a cluster heap and how long the filename is. This entry is a critical secondary directory entry and it immediately follows file directory entry. Its data structure is illustrated in Table 6.14.

Table 6.14 Data structure of stream extension directory entry

Offset (Hex)	Length (Bytes)	Field name	Description
0x00	0x01	Entry type	0xC0 - in use; 0x40 - not in use
0x01	0x01	General secondary flags	Reserved for future use
0x02	0x01	Reserved	32-bit checksum of the contents of the table
0x03	0x01	Name length	Unicode characters in filename
0x04	0x02	Name hash	16-bit hash of uppercase version of filename which is used while searching
0x06	0x02	Reserved	Reserved for future use
0x08	0x08	Valid data length	It defines how much of allocated file space is currently used.
0x10	0x04	Reserved	Reserved for future use
0x14	0x04	First cluster	The starting logical cluster number of file
0x18	0x08	Data length	Length of file/directory data in bytes

6.6.3 New Technology File System

Windows family of Operating Systems uses new technology file system (NTFS) as the default operating system for fixed disks. NTFS is an extensible operating system which makes it suitable for increasing size in drive volumes. NTFS does not have a reserved area or file area and all the system files (metadata) are stored along with data in the data area. Metadata files make up the NTFS file system. This means that NTFS uses metadata files to track information about every file in a volume, to track storage allocation, security issues, accessibility permissions, journaling, and encryption.

Versions of NTFS The different versions of NTFS differ in length and content of a file record header and therefore a forensic investigator should be aware of the version changes in order to locate evidence. Minimum information on versions is listed in Table 6.15.

Table 6.15 Versions of NTFS

Versions	Description
NTFS v1.0 - v1.1	It was released with Windows NT 3.1 operating system.
NTFS v1.2	It was released with Windows NT 3.51 operating system. It supported compressed files, alternate data streams, and user access control for security.
NTFS v3.0	It was released with Windows 2000 operating system. It featured administrative controls with disk quotas, encryption, and parse points.
NTFS v 3.1	It was released with Windows XP. It included added features in file system to recover damaged data. The NTFS driver version was NTFS.sys 5.0.
NTFS.sys 6.0	It was the NTFS drivers version released with Windows Vista to include features without any change in the NTFS structure.
NTFS.sys 6.1	It was the upgraded NTFS driver released with Windows 7 that included changes on disk data survival after a full format and partition resizing and without any change in the structure of NTFS.
NTFS v3.1-80	It was released with Windows 8 to include metadata files to facilitate faster data recovery process and reduce volume downtime. It includes provisions to format volume with GUI and no backward compatibility. The NTFS driver version was NTFS.sys 6.2.
ReFS	ReFS is resilient file system, which is a new file system built upon NTFS (however required for booting) to include additional features such as protection against data corruption with scrubbing, support for larger volumes, and performing functions faster. It is included as part of Windows 10 besides its availability in Windows Server 2012.

Volume boot record The first 16 sectors in a volume make up the VBR and it is the system file $Boot. $Boot includes volume and file system parameters, pointers to file system components, and bootstrap code. The data structure of $Boot file with only the details required for forensic examination is shown in Table 6.16.

Table 6.16 Data structure of $Boot file of NTFS with only required fields

Offset (Hex)	Length (Bytes)	Field name	Description
0x03	8	OEM ID	ASCII - NTFS
0x0B	2	Bytes per sector	Provides cluster size
0x0D	1	Sectors per cluster	
0x1C	4	Hidden sectors	Sectors before the start of the volume

(Contd)

Table 6.16 (Contd)

Offset (Hex)	Length (Bytes)	Field name	Description
0x28	8	Total sectors on the volume	Volume size
0x30	8	$MFT starting extent	LCN of master file table and its mirror copy
0x38	8	$MFTMirr starting extent	
0x40	1	Clusters per $MFT record	Indicates clusters per record if the value is positive; a negative value indicates record size in bytes raised to the power of the absolute value.
0x44	1	Clusters per index buffer	Size of index buffer
0x48	8	Volume serial number	Dictates the serial number of the volume
0x50	4	Checksum	Checksum of the boot Sector
0x54	426	Bootstrap code	Bootstrapping code
0x1FE	2	Boot signature	0xAA55

Table 6.17 Metadata files of NTFS

Metadata File	Filename	Record	Description
Master file table	$MFT	0	It contains a record for every file on the volume.
MFT mirror	$MFTMirr	1	It contains the backup of the first four records of MFT.
Log file	$LogFile	2	It helps in the recovery of corrupt files.
Volume	$Volume	3	It includes volume information such as volume label, NTFS version, and flags
Attribute definitions	$AttrDef	4	It includes attribute names, numbers, and description.
Root filename index		5	It is the root directory.
Cluster bitmap	$Bitmap	6	It highlights the allocated and unallocated clusters.
Boot sector	$Boot	7	It is the volume boot record.
Bad cluster file	$BadClus	8	It is the sparse file holding information about the bad clusters in a volume.
Security file	$Secure	9	It is the index of security settings applied to files within a volume.
Upper case table	$Upcase	10	It converts lower to matching Unicode uppercase characters.
Extended attributes	$Extend	11	It is the directory that contains optional extensions such as $Object Id, $Quota, $Reparse etc.
–	–	12-15	Reserved for future use.
–	–	16-23	Not in use.
Quota	$Quota	24	$Extend\$Quota provides users right on disk space usage.
Object identifier	$ObjId	25	$Extend\$ObjId is the index of all unique IDs of files and directories in a volume.
Reparse	$Reparse	26	$Extend\$Reparse is the index of all reparse points in a volume.

Table 6.18 Data structure of file record header

Offset (Hex)	Length (Bytes)	Field name	Description
0x00	4	Signature	FILE if error free and BAAD if fix-up errors are spotted.
0x04	2	Offset to fix up array	The last two bytes of a file record have a check value and the same is written to fix-up array as an error checking feature to detect single sector failure due to corruption of data.
0x06	2	Entries in fix up array	
0x08	8	$LogFile sequence number	It is used to identify the latest record written to $LogFile as a part of NTFS transaction and is useful in case of recovery of file system data and restoring to a healthy state.
0x10	2	Sequence count	It records the number of times file record entry has been used, being 0x01 initially and incremented by 1 after every reference till the maximum limit of 0xFFFF is reached after which it is reinitialized to 0x01.
0x12	2	Hard link count	It gives the count of filename attributes
0x14	2	Offset to first attribute	It gives the length of the file record header.
0x16	2	Allocation status flag 0 - Allocated 1 - Directory	This field is 0x00 for deleted file, 0x01 for allocated file, 0x02 for deleted directory, and 0x03 for allocated directory.
0x18	4	Logical size of $MFT Record	Length of the actually used bytes in the file record.
0x1C	4	Physical size of $MFT record	Length of the file record in bytes.
0x20	8	File reference to base record	Every file has an entry in $MFT called base record and if the file is not fragmented this field is zero. When a file is fragmented, it will have extended records which store the extended information and carry in its header the reference to the base record.
0x28	2	Next attribute Identification	It gives the identification number given to the next attribute that is written to the file record.
0x2A	-	Fix-up Array and Attributes	This is used with NTFS v3.0.
0x2C	4	$MFT File Record Number	These are used with NTFS Version 3.1 and above.
0x30	-	Fix-up Array and Attributes	

Master file table This is used by NTFS to track files within its volume and so every file has an entry called file record in the master file table (MFT). In fact MFT has an entry to itself in MFT. The first 26 file entries in MFT are metadata files of which the first four are crucial and so a backup copy of it is maintained in $MFT-Mirr. The metadata files of NTFS are listed in Table 6.17.

File record Every entry in MFT has a unique ID for the file it refers to. Addition of a file to a volume inserts a file record in MFT. When a file is deleted, its file record is reused by MFT by overwriting before creating new records. Every file record is 1024 bytes in length.

A file record has two parts: (a) information about the file record itself in the file record header, and (b) information about the file or directory it points to in attributes. The signature of the beginning of file record is "FILE" and the end is "0xFF FFFF FF".

File record header This provides information about the file record and it varies in length with different versions of NTFS. For example, it is 48 bytes in length with NTFS v1.x and 56 bytes in length with NTFS v3.x. The data structure of a file record header is given in Table 6.18.

Attributes An attribute comprises a header and a varying length content, which give information about the file and its contents. In some cases, the content may not be present within the attribute due to the limited size of the MFT file record. It follows the file record header and is of different types. The attributes header is 16 bytes in length and it contains information about the attribute. Its data structure is illustrated in Table 6.19.

Table 6.19 Data structure of attribute header and content of NTFS

Offset (Hex)	Length (Bytes)	Field name	Description	
0x00	4	Attribute type identifier	This defines the type of the attribute.	Attribute header
0x04	4	Length of the attribute	It defines the length in bytes.	
0x08	1	Content non-resident Flag	This flag dictates whether the attribute is entirely contained in the $MFT record (resident attribute) or moved to another area of volume because of limited space in $MFT (non-resident attribute).	
0x09	1	Length of the stream name	It specifies the number of Unicode characters.	
0x0A	2	Offset to the stream name	It specifies from the beginning of the attribute.	
0x0C	2	Flags	This field will be 0x0001 for compressed file, 0x4000 if encrypted, and 0x8000 if sparse.	
0x0E	2	Attribute identifier	It specifies the sequential order in which the attribute was added.	
0x10	4	Size of the content	It denotes the attribute content size in bytes.	Attribute content
0x14	2	Offset to the content	It denotes the offset from the start of the attribute in bytes.	

Attributes are of different types based on their contents, which reflect the different aspects of the file. Attribute types are given in Table 6.20.

Table 6.20 Attribute types of NTFS

Attribute identifier	Attribute name	Description
10 00 00 00	$Standard_Information	It is a resident attribute containing file permissions, security, and administrative information.
20 00 00 00	$Attribute_List	It dictates the location of all attributes that do not fit on a single file record entry (non-resident attributes).

(Contd)

Table 6.20 (Contd)

Attribute identifier	Attribute name	Description
30 00 00 00	$File_Name	It is a resident attribute representing the name of the file.
40 00 00 00	$Volume_Version	It specifies volume version in NTFS v1.x.
40 00 00 00	$Object_ID	It contains the GUID for the file in NTFS v3.x.
50 00 00 00	$Security_Descriptor	It contains the access control list and security properties of an individual file in NTFS v3.x.
60 00 00 00	$Volume_Name	It holds the volume name as Unicode with a maximum length of 127 characters. This attribute is present only in $MFT record for $Volume metadata file.
70 00 00 00	$Volume_Information	It stores the NTFS version of the volume and its state. This attribute is present only in the $MFT record for $Volume metadata file.
80 00 00 00	$Data	It contains the actual data or pointers to the actual data. Any file greater than 600 bytes will become non-resident.
90 00 00 00	$Index_Root	It is a resident attribute that lists a directory's child files as a sorted tree. It points to B-tree head node.
A0 00 00 00	$Index_Allocation	It points to the location of the index buffers of a large directory.
B0 00 00 00	$Bitmap	It represents the allocation status of either a location or an entity of file record entry.
C0 00 00 00	$Symbolic_Link	Soft link information in NTFS v1.2. This points to a volume where GUID is used to identify the mount point.
C0 00 00 00	$Reparse_Point	Soft link information in NTFS v3.x. A reparse point is a pointer to another location or file on the same volume.
D0 00 00 00	$EA_Information	Allows compatibility with HPFS.
E0 00 00 00	$EA	
00 01 00 00	$Logged_Utility_Stream	Contains information and keys for encrypted attributes in NTFS v3.x.

$Standard_Information attribute has the data structure defined in Table 6.21 which is very useful during a forensic examination in collecting evidence.

Table 6.21 Data structure of $Standard_Information attribute

Offset (Hex)	Length (Bytes)	Field name	Description
0x00	8	Create time	It denotes when a file/directory was created on a volume.
0x08	8	File modified time	It denotes when the file content was changed.
0x10	8	$MFT Modified Time	It denotes when the content of $MFT file record was changed.
0x18	8	Last accessed time	It denotes when a file was accessed by the user, application, or system activity. However a registry setting by the user can disable it.

(*Contd*)

Table 6.21 (Contd)

Offset (Hex)	Length (Bytes)	Field name	Description
0x20	4	File type flags	Value may be 0x0001—read only; 0x0002—hidden file; 0x0004—System file; 0x0020—archive; 0x0040—device; 0x0080—normal; 0x0100—temporary; 0x0200—sparse file; 0x0400—reparse point; 0x0800—compressed file; 0x1000—offline; 0x2000—content not indexed; 0x4000—encrypted
0x24	4	Maximum number of versions	It holds 0x00 if disabled and the maximum versions otherwise.
0x28	4	Version number	It holds the file's version.
0x2C	4	Class ID	It denotes the class ID.
0x30	4	Owner ID	It denotes the owner ID for quota.
0x34	4	Security ID	Reference to $Secure which is an index to permission settings in NTFS v3.x.
0x38	8	Quota charged	Bytes from user's quota in NTFS v3.x.
0x40	8	Update sequence number (USN)	Index to $USN journal in NTFS v3.x. $USN Journal records the filenames being changed, time to time, and the type of change.

$File_Name has the data structure defined in Table 6.22. This reveals that certain information about a file can be retrieved from one or more sources.

NTFS stores date and time in "FILETIME" structure which is a 64-bit unsigned value and so it should be added/subtracted from UTC in order to extract the local time zone set by the user.

Table 6.22 Data structure of $File_Name attribute

Offset (Hex)	Length (Bytes)	Field name	Description
0x00	6	$MFT record number of parent directory	It denotes the directory where the file resides.
0x06	2	Sequence number of parent directory	It denotes how many times a parent has been used.
0x08	8	File name creation time	It denotes when a file/directory was created in a volume.
0x10	8	File name modification time	It denotes when the file content was changed.
0x18	8	$MFT modification time	It denotes when the content of $MFT file record was changed.
0x20	8	Last access time	It denotes when a file was accessed by the user, application, or system activity. However, a registry setting by the user can disable it.
0x28	8	Allocated size of the index	These are used only if there is an index record.
0x30	84	Actual size of the index	

(Contd)

Table 6.22 (Contd)

Offset (Hex)	Length (Bytes)	Field name	Description
0x38	4	File type flags	Value may be 0x0001—read only; 0x0002—hidden file; 0x0004—system file; 0x0020—archive; 0x0040—device; 0x0080—normal; 0x0100—temporary; 0x0200—sparse file; 0x0400—reparse point; 0x0800—compressed file; 0x1000—offline; 0x2000—content not indexed; 0x4000—encrypted
0x3C	4	Reparse value	This field is used by extended attributes (EA).
0x40	1	Filename length	It specifies the number of Unicode characters.
0x41	1	Filename type (Namespace)	In NTFS, the first six characters of the LFN are used and then a tilde (~) and number are appended to the name, and the same extension is used as namespace.
0x42	varies	Filename	It denotes the filename.

Virtual cluster numbering and run list   Every cluster is given an address relative to its location within a volume which is called logical cluster number (LCN). Every cluster allocated to a file is given a virtual cluster number (VCN) relative to its location within a file. When a file is fragmented and is non-resident VCN helps to put the clusters back in order. Run list is a collection of pointers to data where each pointer (run) points to a contiguous content (cluster). End of run list is marked by 0x00. Only the first run list entry has a value corresponding to LCN of the volume. LCNs of the remaining runs are estimated relative to its previous starting extent.

File creation, deletion, and recovery   When a file directory is created, a file record is assigned if available in $MFT, bitmap is changed to show that this record is allocated, the allocation status flag in the record header is marked allocated, attributes of the file are written to $MFT file record, and content non-resident attributes are updated to reflect the clusters allocated to store the file contents.

When a file/directory is deleted, the record header sequence count is incremented by one, the allocation status flag in the record header is marked unallocated, bitmap is changed to show that this record is unallocated, and content non-resident attributes are updated to reflect that the clusters allocated earlier are unallocated now. The contents will therefore remain until it is overwritten by some other file.

It can be seen from Table 6.22 that $File_Name attribute contains the $MFT record number and sequence number of the parent directory of file and so the parent–child relationship between the directory and the file is said to exist. When a file is deleted, there are three possibilities:

1. The sequence number of the parent directory is incremented by one and the allocation status of the parent, that is, allocation status flag in file record header (refer to Table 6.18) is set as 0x02 which means that the parent directory has been deleted and not overwritten. Hence the parent–child relationship between the two $MFT records is said to exist. Hence the file/directory can be recovered.
2. The sequence number of the parent directory is incremented by one and the allocation status of the parent, that is, allocation status flag in file record header (refer to Table 6.18) is set to 0x01 and 0x03 which means that the parent directory has been deleted and overwritten by file or directory. Hence the parent–child relationship between the two $MFT records does not hold.
3. The sequence number of the parent directory is incremented by more than one which means that the parent directory has been deleted and overwritten many times. Hence the parent–child relationship between the two $MFT records does not hold.

Orphaned files are used to refer to files whose parent–child relationships are broken [possibility (2) and (3)] and so recovery becomes impossible.

6.6.4 ext family of File Systems

This file system is used by Linux/Unix operating system. Four versions of ext family of operating systems are described briefly here:

Extended File System

The extended file system (ext) is the file system primarily used in Unix and Linux systems. In order to handle physical devices, it uses virtual directories where fixed length blocks are used to store data. It uses a system called *inodes* to keep track of the files stored in its virtual directory. Thus, every physical device will have an inode table to store its file information. For every file, it maintains the following information: filename, file size and the number of allocated blocks, file owner with user identifier (UID), the group it belongs to with group identifier (GID), the access permissions for the file and the pointers to the disk block that hold the file contents. Ext supports fragmentation which reduces its performance as all the blocks of a file have to be accessed while a search for a specific file is made on the physical device. Every entry in the inode table is assigned a unique number called inode number assigned by the file system and is used by the file system to access and identify a file rather than filename and path. ext limits the maximum file size to 2 GB.

Second Extended File System

Second extended file system (ext2) brings change to the inode table to include added information about a file such as the time when the file was created, modified, and last accessed which actually helps administrators to track file access on a system. The maximum individual file size permitted by ext2 ranges from 16 GB to 2 TB. Besides this, ext2 allows the maximum file system size to 32 TB to help accommodate larger files in database servers. ext2 handles the drawback with fragmentation in ext by grouping together data blocks pertaining to files. The drawback with ext2 is the consistent updates to inode table upon every file access which is impossible during lack of synchronization as a result of power outage.

ext3

This file system is similar to ext2 but offers ordered mode of journaling by default. However, it offers the flexibility to change to other modes of journaling with commands. ext3 does not support data compression, file recovery in case of accidental deletion, and encryption. The maximum individual file size with ext3 varies from 16 GB to 2 TB and the overall size of ext3 file system can be from 2 TB to 32 TB (Box 6.5).

Box 6.5 Journaling File Systems

With this feature, instead of directly writing data to a storage device and updating the inode table, data is written to a temporary file called journal. Upon successful writing to a physical storage and inode table, the journal entry is deleted. Three types of journaling are offered: (a) data mode, (b) ordered mode, and (c) writeback mode. In data mode, both the inode and file data are journaled and so the chances of losing data are less but it results in poor performance as data is written twice, once in the journal and once in storage. In ordered mode, only inode data is written to the journal and the journal entry is deleted only after the file data is successfully written. Thus, it ensures a compromise between performance and safety. In write back mode, only inode data is written to the journal and there is nothing to keep track of when the file data is written. Hence there is a high risk of losing data but no loss in performance as no journaling of data is done.

ext4

Besides offering the features of ext3, it supports compression and encryption. ext4 offers a new feature called *extents* where storage is allocated in blocks in the storage device and only the starting location is recorded in the inode table (replaces the idea of storing blocks pertaining to a file as a list). Moreover, the unused space within the data blocks reserved for a file are filled with zeros and are not allocated for any other file. ext4 supports huge individual file size where the maximum size can range from 16 GB to 16 TB and also huge overall file system size of 1 EB (Exa Byte where 1 EB = 1024 PB (peta byte) and 1 PB = 1024 TB (tera byte)). The other new features introduced with ext4 to improve its performance and reliability when compared to ext3 are multiblock allocation, delayed allocation, journal checksum, fast fsck (file system checker runs faster than ext3), etc. Further, with ext4 the journaling feature can be turned on and off.

Components of ext File System

The components of ext file system are as follows:

Superblock The first 1024 bytes from the start of the file system is referred to as the superblock and contains the following information:

1. Layout of the file system
2. Block and inode allocation information (block size, number of blocks, and number of inodes)
3. Metadata which indicates when the file system was last mounted
4. Enabled file system features

Backup copies of superblock are also maintained. The data structure of ext superblock is given in Table 6.23.

Table 6.23 Data structure of ext superblock

Offset	Length (Bytes)	Field name	Description
0	4	Number of inodes	Total number of inodes, both used and free.
4	4	Number of blocks	Total number of blocks in the system.
8	4	Number of blocks reserved	Number of blocks reserved for usage by a superuser.
12	4	Number of unallocated blocks	Total number of free blocks.
16	4	Number of unallocated inodes	Total number of free inodes.
20	4	Block where block group 0 starts	ID of the block containing the superblock structure.
24	4	Block size	Field used in the computation of block size.
28	4	Fragment size	Field used in the computation of fragment size.
32	4	Blocks in each block group	Total number of blocks per group.
36	4	Fragments in each block group	Total number of fragments per group.
40	4	inodes in each block group	Total number of inodes per group.
44	4	Last mount time	Last time the file system was mounted.
48	4	Last written time	Last write access to the file system.
52	2	Current mount count	Denotes how many times the file system was mounted.
54	2	Maximum mount count	Maximum number of times the file system may be mounted.
56	2	Signature (0xEF53)	Bit value identifying the file system where 0xEF53 is EXT2_SUPER_MAGIC.

(Contd)

Table 6.23 (Contd)

Offset	Length (Bytes)	Field name	Description
58	2	File system state	Denotes whether file system was clearly mounted.
60	2	Error handling	Denotes error where 1 = continue, 2=remount read only and 3=cause a kernel panic.
62	2	Minor version	Indicates minor revision level.
64	4	Last consistency check time	Last file system check.
68	4	Interval between forced consistency checks	Maximum Unix time interval.
72	4	Creator OS	Identifier of the OS that created the file system.
76	4	Major version	Revision value.
80	2	UID that can use reserved blocks	Default user ID for reserved blocks.
82	2	GID that can use reserved blocks	Default group ID for reserved blocks.
84	4	First non-reserved inode in file system	Index to the first inode usable to standard files.
88	2	Size of each inode structure	Size of inode structure.
90	2	Superblock (backup copy)	Block number hosting superblock structure.
92	4	Compatible feature flags	Bit mask of compatible features.
96	4	Incompatible feature flags	Bit mask of incompatible features.
100	4	Read-only feature flags	Bit mask of read-only features.
104	16	File system ID	Volume ID.
120	16	Volume name	Volume name mostly unused.
136	64	Path where last mounted on	Directory path where file system is mounted.
200	4	Algorithm usage bitmap	Method of compression employed.
204	1	Number of blocks to pre-allocate for files	Number of blocks to be allocated for a new file
205	1	Number of blocks to pre-allocate for directories	Number of blocks to be allocated for a new directory
206	2	Unused	Not used
208	16	Journal ID	UUID of journal Superblock
224	4	Journal inode	Inode number of the journal file
228	4	Journal device	Device number of the journal file
232	4	Head of orphan inode list	Link to the first inode in the list of inodes to be deleted
236	4×4	Hash seed	Seed used for hash algorithm for directory indexing
252	1	Hash version	Default hash version for directory indexing
253	3	Unused	Unused
256	4	Default mount option	Default mount option for file system
260	4	GID of meta block group	GID of meta block group
264	760	Unused	Unused

Group descriptor tables Group descriptor tables (GDT) follow the superblock, holding information about the blocks that are grouped together as a block group. Every group descriptor contains the allocation status of the blocks. A backup copy of the GDT is stored in every block group. The data structure of GDT is given in Table 6.24.

Filenames For every file/directory, there is an associated entry in the directory containing the filename associated with a file, address of the inode associated with the file, and a flag to indicate whether it is a file or directory. ext3 allows different filenames (links) to point to the same file called hard links, whereas soft links or symbolic links have indirect link to the actual file by storing path to the file. Therefore, soft links have no actual data but have their own inode.

Table 6.24 Data structure of group descriptor table

Offset	Length (Bytes)	Field name	Description
0	4	Start address of block bitmap	Block ID of first block of the block bitmap
4	4	Start address of inode bitmap	Block ID of first block of the inode bitmap
8	4	Start address of inode table	Block ID of first block of the inode table
12	2	Unallocated blocks in group	Total number of free blocks for the represented group
14	2	Unallocated inodes in group	Total number of free inodes for the represented group
16	2	Number of directories	Total number of inodes allocated to directories of the represented group
18	2	Padding	Padding the structure on a 32-bit boundary
20	12	Unused	Reserved for future use

Metadata (inode) In ext3, there is one inode per directory or file and a set of inodes is assigned to each block group. Every inode has the following information apart from that contained in ext: timestamps and link count indicating the number of filenames pointing to this node. If all directory entries pointing to a given inode are removed, the inode has a link count of zero which means that the file or directory is deleted. Timestamp is updated as modified (M) when the file or directory is written, accessed (A) when the file or directory is read, changed (C) when the metadata of the file is modified, and deleted (D) when a file is deleted. Besides these, ext3 supports extended file attributes which are stored in a separate attribute block referenced by the inode, for example, attribute j denotes that journaling is enabled. The data structure of inode is given in Table 6.25.

Data unit or block Every block is identified by an address and is an entity of block allocation group as defined by the group descriptor table.

Journal ext3 records block-level changes when the metadata changes as a result of transactions. Every transaction has a sequence number and begins with a descriptor block, and commit blocks with old inode information to extract old timestamps and ownership information.

Some of the Linux commands which help in the collection of evidence are as follows:
1. *fdisk* command is used to obtain the geometry of the disk, like number of partitions and how each one is formatted.
2. *mount* command can be used to mount the designated partition.
3. The *stat* command can be used to obtain a variety of information related to timestamp of the file.

Table 6.25 Data structure of inode

Offset	Length (Bytes)	Field name	Description
0	2	File mode	File and access rights
2	2	Lower 16 bits of UID	UID associated with the file
4	4	Size	Size of the file in bytes
8	4	Access time	Number of seconds since the last time the inode was accessed
12	4	Change time	Number of seconds since the inode was created
16	4	Modification time	Number of seconds since the last time the inode was modified
20	4	Deletion time	Number of seconds since the inode was deleted
24	2	Lower 16 bits of GID	Value of POSIX group having access to this file
26	2	Link count	Denotes the number of times an inode is referred to
28	4	Sector count	Total number of blocks reserved to contain the data of this inode
32	4	Flags	Dictates how implementation behaves when accessing data for this inode
36	4	Unused	OS dependant value
40	15×4	Block pointers	Block numbers pointing to blocks containing data for this inode
100	4	Generation number	File version
104	4	Extended attribute block	Block number containing extended attributes
108	4	Size	Directory ACL
112	4	Fragment block address	Location of file fragment
116	1	Fragment index in block	
117	1	Fragment size	
118	2	Unused	OS dependent structure
120	2	Upper 16 bits of UID	
122	2	Upper 16 bits of GID	
124	4	Unused	

6.6.5 Hierarchical File System

The Mac operating system is based on hierarchical file system (HFS/HFS+) where HFS makes use of 16 bits and HFS+ makes use of 32 bits to address clusters on a disk. HFS/HFS+ consists of directories and subdirectories in which data is stored. The five data structures which make up the HFS+ file system are volume header, catalog file, extents overflow file, attributes file, and allocation bitmap file. This is shown in Fig. 6.7. The catalog file is analogous to MFT and contains records for every file and directory with attributes such as date and timestamps. Records in this file are stored as B-Tree which facilitates efficient searching where every record is uniquely identified by catalog node ID (CNID). A record may be any one of the four types, namely folder, file, folder thread, and file thread. The data structure of folder and file are given in Tables 6.26 and 6.27 respectively.

Mac file structure contains two important elements called data fork and resource fork which help in identifying and recovering deleted files. Data fork contains the actual data of the file and resource fork contains the resource map with special data structure information such as icons and menu items.

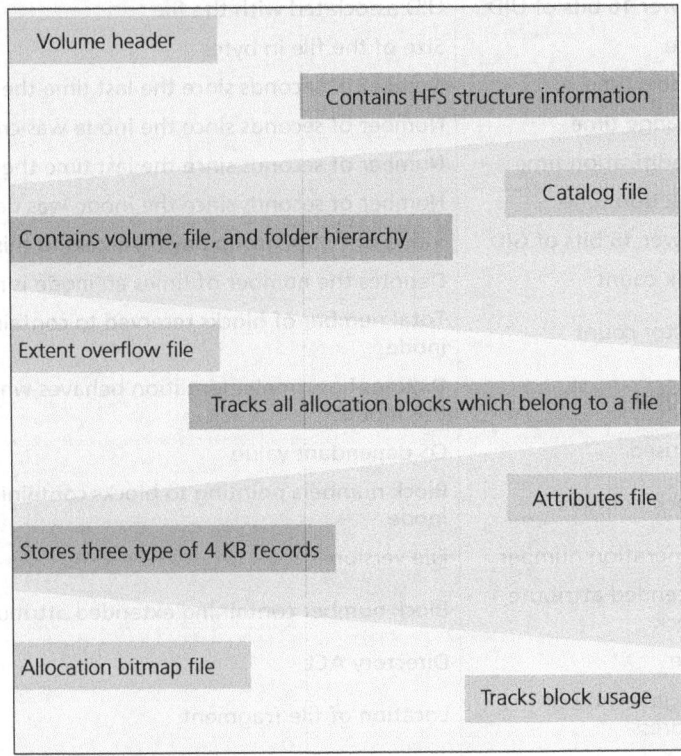

Fig. 6.7 Data structures of HFS+

Table 6.26 Data structure of folder record in HFS

Field name	Description
Record type	0x0100 for directory/folder
Name	Directory name
Valence	Number of files and folders within this directory
CNID	Catalog node ID (CNID) which is unique
Creation date	Date on which the directory was created
Modification date	Date on which a file or folder was created/deleted or modified within this folder
Access date	Date when a file or folder was last accessed by the system
Backup date	Date on which the folder was last backed up

Table 6.27 Data structure of file record in HFS

Field name	Description
Record Type	0x0200 for file
Name	Filename
CNID	Catalog Node ID (CNID) which is unique
Creation date	Date on which the file was created
Modification date	Date on which a file was extended, truncated, or modified
Access date	Not maintained by HFS
Backup date	Date on which the folder was last backed up
Data fork	Information about the location and size of data fork
Resource fork	Information about the location and size of resource fork

6.7 WINDOWS REGISTRY

The registry contains information about the profiles for each user, the applications installed on the computer, and the types of documents that each can create, property sheet settings for folders and application icons, what hardware exists on the system, and which ports are being used on which Windows OS. The registry contains three types of information which are of evidentiary value: (a) user-specific information such as desktop preferences, typed URLs, messenger contacts, (b) system-specific information such as network settings, time zone information, last shutdown date/time and hardware information, and (c) application specific information such as file associations, application registration information, etc. The core registry files reside in the Windows directory. The registry primarily contains keys (directories) and values and the keys may have subkeys. The five root-level keys are as follows:

1. HKEY_CURRENT_USER: It is abbreviated as HKCU and is scanned for information about the configuration of the user currently logged in.
2. HKEY_USERS: HKEY_CURRENT_USER is a subkey of HKEY_USERS. It can be checked for all the user profiles loaded on the computer.
3. HKEY_LOCAL_MACHINE: It is abbreviated as HKLM and can be searched for the configuration information of a particular computer.
4. HKEY_CLASSES_ROOT: It is a subkey of HKEY_LOCAL_MACHINE\Software. The information stored in this key ensures that the correct program opens when a file is opened in Windows Explorer.
5. HKEY_CURRENT_CONFIG: This key contains data about the hardware profile used by the local computer at start-up.

The possible registry type value is summarized in Table 6.28.

The registry has a group of supporting files that contain backup of its data. The extensions and the filenames of these files vary from operating system to operating system and are as follows:

1. HKEY_LOCAL_MACHINE\SAM Sam, Sam.log, Sam.sav
2. HKEY_LOCAL_MACHINE\Security Security, Security.log, Security.sav
3. HKEY_LOCAL_MACHINE\Software Software, Software.log, Software.sav
4. HKEY_LOCAL_MACHINE\System System, System.alt, System.log, System.sav
5. HKEY_CURRENT_CONFIG System, System.alt, System.log, System.sav, Ntuser.dat, Ntuser.dat.log
6. HKEY_USERS\DEFAULT Default, Default.log, Default.sav

Regedit (regedit.exe or regedt32.exe) is the tool which comes default with Windows operating system and allows the user to interface directly with the system registry. Regedit displays the keys in the left-hand pane in a hierarchical tree-view and the values in the right-hand pane for the currently selected key. For each value, the value name, value type, and value data can be viewed. Thus it helps in the live examination of the system.

Table 6.28 Registry type values

Field name	Description
REG_NONE	Data with no type written by system/application which is displayed as binary value in hexadecimal format
REG_SZ	A fixed length text string
REG_EXPAND_SZ	A variable length data string resolved when a program or service uses the data
REG_BINARY	Raw binary data corresponding to a hardware component which is displayed as binary value in hexadecimal format
REG_DWORD (or) REG_DWORD_LITTLE_ENDIAN	Data represented as a 32-bit integer for device drivers and services which are displayed in binary, hexadecimal, or decimal format (Little Endian)
REG_DWORD_BIG_ENDIAN	Data represented as a 32-bit integer for device drivers and services which are displayed in binary, hexadecimal, or decimal format (Big Endian)
REG_LINK	A Unicode string naming a symbolic link
REG_MULTI_SZ	A multiple string of values separated by spaces or commas
REG_RESOURCE_LIST	A series of nested arrays designed to store a resource list used by a hardware device driver or one of the physical devices it controls. This data is detected and written into the \ResourceMap tree by the system and is displayed in hexadecimal format.
REG_FULL_RESOURCE_DESCRIPTOR	A series of nested arrays designed to store a resource list used by a physical hardware device. This data is detected and written into the \HardwareDescription tree by the system and is displayed in hexadecimal format.
REG_RESOURCE_REQUIREMENTS	A series of nested arrays designed to store a device driver's list of possible hardware resources or one of the physical devices it controls can use, from which the system writes a subset into the \ResourceMap tree. This data is detected by the system and is displayed in hexadecimal format.
REG_QWORD	Data represented by 64-bit integer and displayed as a binary value

The evidential data widely procured from the registry includes the following:
1. Time Zone information can be obtained at HKEY_LOCAL_MACHINE\System\CurrentControlSet\Control\TimeZoneInformation
2. Information about the system's hardware is located at HKLM\System\CurrentControlSet\Enum
3. The protected storage system provider (PStore or PSSP) is an encrypted area of the Windows Registry devoted to store passwords. For every user there is a PStore area of registry which is located at Microsoft\Windows\Protected Storage System Provider.
4. SIDs are an alphanumeric character string which is assigned by a Windows domain controller during the log-on process. The SID uniquely identifies the user or group from all other users and groups on the network and is available at HKEY_LOCAL_MACHINE\Sam\SAM\Domains\Account\Users\Names.
5. The URLs for Internet Explorer typed by the user along with the time information is available within each user's NTUSER.DAT file which can be located with SID as SID of User \ Software \ Microsoft\ InternetExplorer\TypedURLs and SID of User\Software\Microsoft\Internet Explorer\TypedURLsTime.

6.8 WINDOWS ARTIFACTS

This refers to evidential data that is saved by the operating system as a result of human interaction or usage on the system and not by default during installation.

The following are considered as the artifacts of Windows OS:

Link Files (.LNK)

Link files or shortcut files are pointers to other files, folders, programs, or data. These files contain one or more useful information about the target file such as file MAC time, file size, volume details, original path to the file, the version of Windows, who created the application, and the file system in use. This information can be obtained by employing a file parser. In an evidential perspective, the volume details help to locate the exact storage media and MAC address to locate a specific computer.

The recent folder in Windows holds the .lnk file that is created automatically when a file is opened. This folder is located in Windows XP at C:\Documents and Settings\%Username%\Recent and inWindows Vista, 7,8 and 10 at C:\Users\%Username%\Appdata\Roaming\Microsoft\Windows\Recent.

Jumplists

This refers to a list of files that are opened by a particular application. Jump lists are automatically populated by the system as well as by the respective application. However, this can be examined only with forensic tools. Jumplists can be disabled by the user.

Recycle Bin

The Recycle Bin serves as a repository of files deleted by the user with provision for recycling the data, thereby preventing accidental deletion. On a FAT volume, the Recycle Bin is available under the root folder, whereas in an NTFS volume, for every user a folder named after user's security identifier (SID) exists in the Recycle Bin to hold the deleted files. INFO2 file located in \Recycled in FAT or \Recycler\%SID% in NTFS provides information such as file size, date and time of deletion, path, and the original filename of the deleted file.

Event Logs

These are log records which record information about errors, failure/success, information, and warnings pertaining to application, system, and user. These logs are available in C:\Windows\System32\winevt\logs. Examining the event logs facilitates collection of evidential data, such as whether a user performed a particular action at a particular time, whether any installation of software was performed, whether logging on to a network was performed and so on.

Thumbnail Cache

Thumbnail is a miniature view of the picture file. Windows supports this feature with a hidden system file named thumbs.db in the folder which facilitates the user to view the file or folder in thumbnail mode. However, the user can disable this feature. The thumbs.db file is updated whenever an image is added to a folder but not removed from it when a file is deleted. Thus, this could serve as evidential data as the remnant of the file remains to prove that the file has been deleted.

Virtualization

Windows 7 and higher versions have built-in support for virtualization. Windows 7 supports native virtual hard drive (VHD) boot where the VHD file acts as a hard disk and booting is done from the VHD file rather than the physical hard disk in the system. In such situations, the challenge lies in locating the VHD file. Its presence can be felt from the presence of a pointer to the VHD file in the boot control disk (BCD) registry hive file.

RAM Files

Windows support virtual memory with the use of pagefile to move data back and forth from the memory as and when required with the constraint that the entire content of the file does not fit into the memory at the same time. Windows 7 has pagefile.sys in the root folder, whereas Windows 8 has Pagefile.sys and Swapfile.sys to assist RAM and to swap in and out the applications. This provides room for evidence to be collected from Swapfile.sys. Besides this, C:\Hiberfil.sys which is responsible for providing hibernation support provides information about the files the user has been working on.

Backup and Restore

Backup feature facilitates taking an alias copy of all files (except system files, program files, and temporary files) on the storage media (usually an external disk) without user intervention which helps at the time of system restore. Thus backup and restore feature provide a way for analysis of previous versions of files and also the files that were deleted since the last backup.

Volume Snapshot Service

Windows Vista and higher versions provide support for volume snapshot service where differential backup is performed to create restore points called *volume snapshot* or *shadow copy*. This makes restore possible not from a single backup, but rather from many snapshots. The advantage is that disk space utilization is efficient with VSS as only incremental backup is performed every time except for the first snapshot. Hence, restore from single snapshot would reconstruct only a fragment and not files on entirety. The date and time of the snapshot may reveal evidential data when and where the snapshot was created and keyword search may also be performed to look for specific evidence.

Access Control Lists

This defines the way the user has control over files, namely to read, to write, and to modify. This provides evidential information in the sense that it gives information on the level of access every user account may be provided with.

Cloud Storage Facility

Skydrive is a cloud storage facility that helps users to store information in the cloud. However, the contents of the cloud storage are available offline in the systemat%UserProfile%%\AppData\Local\Packages\Microsoft.microsoftskydrive_8wekyb3d8bbwe \AC\INetCache.

Similar is the case with Onedrive where logs are available at %SystemDrive%\Users\{username}\AppData\Local\Microsoft\ OneDrive \Logs.

For Google Drive, it is available at %SystemDrive%\Users\{username}\AppData\Local\Google\Drive\Sync_Config.db.

Internet History Artifacts

Artifacts generated during browsing by the user are recorded as multiple files with .dat extension, whereas Windows 8 uses a single ESE database and is located at %UserProfile%\AppData\Local\Microsoft\Windows\WebCache\WebCacheV01.dat.

Email

In Windows 8 the sent and the received mails of the user are stored locally in *.eml file.

Sticky Notes

These are reminders for users and reside on desktop Windows Vista and higher versions. They are stored in
   %user_profile$\AppData\Roaming\Microsoft\Sticky Notes\ StickyNotes.snt. Examining the sticky note may provide evidential data at times.

Home Groups

This functionality supported by Windows 7 and higher operating systems facilitates easy sharing of files rather than configuring the computers individually with user accounts and shared resources. This provides evidential information of what data is shared and who actually created the data.

Wireless Network History

When a computer running on Windows is connected to a Wi-Fi network, a record is written to C:\ProgramData\Microsoft\Wlansvc\Profiles\Interfaces besides being available in the registry. Previous network connections of the user can serve as evidence.

6.9 BROWSER ARTIFACTS

The Internet is widely used by everyone and so it plays a major role in both the execution of crime and as evidence. The data that gets stored in the system as a result of browsing the Internet is termed as browser artifacts and it may serve as evidence.

In any browser, the History folder provides information about the sites visited by the user with date and time. The Cookie History folder also maintains information such as the sites visited by the user with date and time along with login names. The Cache or the Temporary Internet History folder also provides this information. If multiple user logins are available in a system, then the operating system maintains a separate history profile for every user.

The browser artifacts that can be collected from some of the popular browsers are listed here.

Internet Explorer (IE)

History By default, it records the history for 20 days, which is hidden from the user. In earlier versions of IE (up to IE9), history was stored in the index.dat binary file in a subfolder MSHIST followed by a series of numbers and located in C:\Users\User\AppData\Local\Microsoft\Windows\History\History.IE5C:\Users\User\AppData\Local\Microsoft\Windows\History\Low\History.IE5

Four record types are present in this file, namely HASH, URL, REDR, and LEAK where URL and LEAK provide information about date and timestamps and LEAK provides information about server-side redirections which can be used to confirm whether a user has visited specific sites. Instead of index.dat, IE 10 and above store the cached information in a database WebCacheV01.dat which is an extensible storage engine (ESE) database format and is located in C:\Users\User\AppData\Local\Microsoft\Windows\WebCache directory. Analysis of this file with necessary tools may reveal potential evidence. IE 11 run on Windows 7 has WebCacheV01.dat located in C:Users\User\AppData\Local\Microsoft\Windows\TemporaryInternetFiles\Content.IE5\

and if run onWindows 8 has WebCacheV01.dat located in C:\Users\User\AppData\Local\Microsoft\Windows\INetCache\Low\IE\.

Cache files The cache files are available in the following location in Windows 7 and above:

C:\Users\User\AppData\Local\Microsoft\Windows\TemporaryInternetFiles\Content.IE5
C:\Users\User\AppData\LocalLow\Microsoft\Windows\AppCache in subdirectories.

Cookies files Cookies are small files generated by the web browser and are stored in a user's computer holding data specific to the user and the website, for access either by the web server or the developer of the web page or the user himself. Most cookies contain the URL of the website that was visited along with the date and time stamp. IE stores cookies in individual files in the following location:

C:\Users\User\AppData\Roaming\Microsoft\Windows\CookiesC:\Users\User\AppData\Local\Microsoft\Windows\Cookies\Low

However, cookies can be disabled by the user.

Firefox

1. The version of this browser can be obtained from install.log file at the following locations:
   C:\Program Files (x86)\Mozilla\Firefox\install.log
2. Firefox creates a unique user profile for every user where every profile is assigned a profile folder, bookmarks history, stored passwords, and history. This profile is available in the location C:\Users\\AppData\Roaming\Mozilla\Firefox\profiles.ini.

3. The History folder stores all the browsed websites and downloaded files in a database places.sqlite which can be located at:
C:\Users\User\AppData\Roaming\Mozilla\Firefox\Profiles\\places.sqlite.
4. The cache files of specific users can be obtained from their profile names known at the following locations:
C:\Users\User\AppData\Local\Mozilla\Firefox\Profiles\\Cache (or)
\AppData\Local\Mozilla\Firefox\Profiles\.default\Cache
5. The cookies are available in the location:
C:\Users\User\AppData\Roaming\Mozilla\Firefox\Profiles\\cookies.sqlite

Opera

1. Opera saves the history information in the folder as SQLite database at C:\Users\User\AppData\Roaming\Opera Software\Opera Stable\History which can be viewed with a third party tool named SqLiteBrowser.
2. Cookies are stored, within the Opera browser, at C:\Users\User\AppData\Roaming\Opera Software\Opera Stable\Cookies as SQLite database similar to the History file.
3. Opera records passwords which the user might have used during web logins for websites or mail. The same is available at C:\Users\User\AppData\Roaming\Opera Software\Opera Stable\Login Data.
4. The browsing activity during every session is captured and stored in C:\Users\User\AppData\Roaming\Opera Software\Opera Stable\session.db.
5. The cache information is stored at the following two locations:
C:\Users\User\AppData\Local\Opera Software\Opera Stable\Cache
andC:\Users\User\AppData\Local\Opera Software\Opera Stable\Media Cache
6. Finally the web data pertaining to Opera is available at:
C:\Users\User\AppData\Roaming\Opera Software\Opera Stable\Web Data

Google Chrome

1. With Chrome, History files and cookies are stored as SqLite databases in the folders at C:\Users\User\AppData\Local\Google\Chrome\User Data\Default
2. The cache data is available at the following location and can be analysed further using specialized tools:

C:\Users\User\AppData\Local\Gooege\UserData\Default\Cache (or) C:\Users\User\AppData\Local\Google\User Data\Default\Media Cache.

Apple Safari

1. This is the default browser with Mac OS and the user directory is located at /Users/$USER/Library/Safari/ in a machine that runs on Mac OS and at C:\Users\{user}\AppData\Roaming\Apple Computer\Safari\ on a Windows System.
2. The Cache directory is located at /Users/$USER/Library/Caches/com.apple.Safari in a machine that runs on Mac OS and at C:\Users\{user}\AppData\Local\Apple Computer\Safari\Cache.db on a Windows System as database files with .db extension and have to be analysed with specialized tools.
3. Besides this, the history is available in the user directory as a property list (plist) file (much like XML) History.plist and the downloads as Downloads.plist in the user directory.

6.10 MACINTOSH ARTIFACTS

In this operating system, information is not available in the log, property list, or registry as in Windows systems. Rather these data can be collected only while the system is running. All the running applications can be captured from the system area of the toolbar. Live acquisition requires the SOP to be followed for the collection of evidence.

1. Recently accessed files are available in the home directory at ~/Library/Preferences/com.apple.recent.items.
2. Web activity can be traced from the respective files of the browsers being used, for example, from ~/Library/Caches/Metadata/Safari for Safari and ~/Library/Caches/com.apple.Safari for graphical preview of web pages.
3. Further browsing history can be tracked from "~/Library/Safari" folder in History.plist file and the downloaded file can be traced from Downloads.plist file.
4. The cache data can be obtained from Library/Caches/MS Internet Cache. Tracing of email can be done with ~/Library/Preferences/com.apple.mail.plist file.

The following items can be obtained from the finder sidebar: shared computers/devices, external/cloud storage, filevault, non-bootable hard drives, or partitions.

Filevault is a system that encrypts files on Macintosh. Filevault uses an encrypted file system that encrypts the user's home folder using AES encryption (varies with different OS versions) which enables file access only with valid user password/master password and decrypted on the fly. This necessitates collection of evidences such as size of the home folder, login/master password, and logical copy of home folder to external media (for analysis) while filevault is disabled. In Versions 10.7 and above Filevault2 is in use where a master password is created to handle a situation where a user loses his/her password, thereby losing access to the system.

About This Mac is a native application that lists the processor, OS version, installed RAM, and startup disk. Besides this, upon clicking the More Info information tab, information about hardware, software, and network, storage, memory, services, and display can be obtained.

System profiler provides information about the hardware (hard drive) and network settings (MAC and IP address) of the system and the user and system logs which can be either printed or exported to other file formats. Thus these sources of evidence provide information for performing analysis so as to prove a crime.

6.11 LINUX ARTIFACTS

1. In Linux, the artifacts pertaining to user accounts can be obtained from /etc/ where passwd file contains the username, hashed password field, user ID, primary group ID, path of user's home directory, etc.
2. A user's activity during a session is recorded as shell history (.bash_history) in the user's home directory. Whenever a user has attempted to connect to the Internet, the host name and IP address are recorded in the known_hosts file located in the home directory.
3. The log files are located in /var/log.
4. Browser artifacts, for example, Firefox, retains the history data at /home/$USERNAME/ .mozilla/firefox/ as SQLite database. For example, Cookies.sqlite records data related to cookies, Downloads.sqlite stores information about data downloaded using Firefox as a browser, key3.db is a record of stored passwords, signons.sqlite stores the passwords saved by the user, and Places.sqlite holds data about visited pages.

6.12 WHOLE DISK ENCRYPTION OR FULL DISK ENCRYPTION

In contrast to file encryption which encrypts specific files, a whole disk encryption (WDE) encrypts the entire hard drive. It is like taking a disk drive and coating it with a layer of impenetrable paint so that everything on the drive is soaked with encryption. Because everything is encrypted, including the operating system, during booting of the computer the encrypted drive has to be unlocked with a password or a key set beforehand. This pre-boot authentication before system start-up stops unauthorized users from accessing the data. This is true even if an encrypted disk drive is removed from one computer and installed in another. It is an efficient data protection scheme which is primarily designed to protect laptops and hard drive from being stolen; however, it poses difficulties with the collection of evidence.

Windows EFS is a feature available with NTFS by default wherein a user is given the choice to encrypt the files with a file encryption key (FEK) and keep the data secure. He/She can decrypt the files as the key is available on demand.

Windows Bitlocker is a full volume encryption tool available by default in Windows Vista, Windows 7 and 8 operating systems but is disabled. It is based on pre-boot authentication if enabled.

Truecrypt is a free encryption software with WDE capability that works by mounting a large file on the system as volume and assigns a drive letter wherein the user can place his/her files to be encrypted. Thus, when the drive is dismounted, files and other contents in that volume are encrypted and protected. When the volume is mounted again, entry into it is forbidden till the password is provided to gain access. This sort of encrypted volume can be primarily identified by looking for volumes with .tc extension. In general, it is good to suspect any file of larger size to be an encrypted volume and to look for passwords (from registry or saved passwords list) to perform a dictionary attack against the file. Apple's Filevault supports WDE with operating system versions of 10.7 and above. To access a Filevault encrypted disk, a key is necessary. Apple requires the user to remember the login password which otherwise will make the drive inaccessible, and data will be irrecoverably lost. To help prevent users from losing their data, certain versions offer an option to store the encryption key with Apple which can be obtained by following an appropriate legal process.

6.13 EVIDENCE FROM MOBILE DEVICES

Mobile devices have become an integral part of our life and inevitable not only for making calls but also for sending and receiving text messages. Besides that, these devices run on operating systems that support diverse applications (apps). They are also equipped with Wi-fi, Bluetooth, and GPS technologies to connect with the Internet, other devices, and to benefit from location-based services. They are also equipped with camera and media players. The presence of these features poses a downside as well.

Before ensuring the presence of evidence, it is necessary to gather information about the device which includes identification of the vendor and service provider, the type of network connection (GSM/CDMA), and the model number. The physical components of a cell phone include a handset containing RAM, cellular radio, input and output components; battery, SIM card, and memory card. Embedded applications store their information in SQLite format.

Mobile devices, when synced, will back up the data in some storage, for example, at Users\(username)\AppData\Roaming\Apple Computer\MobileSync\Backup when synced to system that runs Windows operating system and at X: ~/Library/Application Support/MobileSync/Backup when run on Mac operating system. Specialized tools are required to collect evidence from mobile devices, especially flash memory chips.

Evidence from mobile devices such as mobile phones and smartphones include the following: from the carrier and provider, for example, call detail record (CDR) analysis and site ID or location analysis, from the phone's memory, from memory card and from SIM card. Table 6.29 presents the digital evidences that can be collected from mobile devices and its locality within the device. More detailed information on the mobile operating system and the location of evidence are presented in Chapter 8.

Table 6.29 Digital evidence on mobile devices

Device	Locality of evidence	Digital evidence
Baseline phone	Hardware	Handset date and time; international mobile equipment identity (IMEI)
	User-created information	Address book; SMS; calendar, memos; to-do lists
	Phone-created information	Call register (received, sent, missed)
Smartphone	User-created information	Photographs (including EXIF data); video/audio; maps, MMS; GPS waypoints; stored voicemail; files stored on system; connected computers

(Contd)

Table 6.29 (Contd)

Device	Locality of evidence	Digital evidence
Local workstation	Internet-related information	Online accounts; purchased media (often discoverable in embedded metadata); email; Internet usage; social networking information
	Installed third-party applications	Alternate messaging and communication systems; additional capabilities; malware applications; penetration testing; other applications—anything that can help provide an alibi to an individual or tie him/her to a crime
	Transferred information	Tethered mobile devices; backed up phone data; backed up third-party applications; store accounts; purchased media
	Tracking information	Connected cell towers over time; location at different times; current location (inaccurate)
Carrier SIM card	Usage information	Billing information; call register over time; Internet/data usage; messages not delivered (after radio isolation); be warned—SIM cloning does occur and information is not to be taken at face value
	Identifiers	Subscriber identifier (IMSI); SIM card identifier (ICC-ID)
SIM card	Usage information	SMS; abbreviated dial names/numbers; last dialed numbers; location areas

6.14 DIGITAL EVIDENCE ON THE INTERNET

The primary sources of evidence from the Internet are listed here:

Evidence from Email

Email is an important communication tool these days and the source of crime for harassment, espionage, threat etc. It leaves room for evidence collection at different locations, for example, the ECD used, a server on a LAN, and also with the Internet service provider.

Email communication makes use of protocols, namely simple mail transfer protocol (SMTP), post office protocol (POP)/Internet message access protocol (IMAP). Once the sender has composed the mail and clicks on send, SMTP (utilizes Port 25) transfers the mail from the sender's side (either from the sender's email client like Outlook or a web browser) utilizing the Internet, to the receiver's email server where it remains in the mailbox of the receiver. The receiver can access, view/download the mail to his/her local machine using either POP (utilizes Port 100) or IMAP (utilizes Port 143). IMAP supports offline mode and enables the receiver to manipulate the messages.

The email header contains fields such as Content-Type, Content-Description, and Content-Transfer-Encoding. SMTP sends text data as ASCII and the presence of base64-encoded sections implies the presence of non-ASCII characters in the email which can be analysed from the email header (Content-Transfer-Encoding). In reality, message transfer agents (MTA) which are the equivalent of post offices for electronic mail add the current time and name of the MTA along with some technical information (the sender's machine name, IP address, and the date and time at which it was sent). Therefore the sender can be tracked by simply retracing the route that the email has travelled by reading through the email's headers.

Further, along the path the email traverses, every machine embeds its location in the mail header. Except when the mail has been deleted, the mail can be retrieved from the server upon the issue of a search warrant for the stored mail from the service provider. Besides this, mail artifacts can be extracted from the history, cache, and temporary Internet files. Thus, employing suitable analysis tools provide room for collecting evidence.

The most popular email clients are Microsoft Outlook and Windows Live Mail. Others include AOL, ThunderBird, Opera etc. The email clients store information about the mails sent and received in their folders which serve as evidence. For example, Microsoft Outlook stores messages, contacts in messaging application programming interface (MAPI) folders which are available in the personal folders file (.pst) in hard drive, in mailbox at the server as .mdbfile, and in offline folders as .ost file in the hard drive. Specific email analysis tools can be employed to extract useful evidence from them.

Peer-to-peer Networking (P2P)

In a peer-to-peer network, there is no notion of client and server which means any node can serve either as client or server at different instances of time. The primary objective is to facilitate file sharing without the need for the owner of the network to own servers. Instead, the user's computer acts as super nodes which maintain the list of files available for sharing where the list dictating the availability of file is populated continuously as the sharing rights change and the nodes enter and leave the network. P2P networks do not have a central database but can make file sharing possible employing ultrapeers. Some popular P2P networks include BitTorrent, eDonkey, Gnutella, and so on.

For example, BitTorrent relies on the use of torrent metafile which is smaller in size (upto 100 KB) holding information such as filename, size, hash of each block, number of pieces, and the client instructions necessary to download files. These torrents are uploaded in torrent tracker sites which act like a centralized server. These trackers usually suggest to the peers holding the file from where it can be downloaded. Trackers update themselves with additional client available to share the file. Thus gathering information will help in identifying the person and the node involved in the distribution of any illegal content.

Traces of P2P file sharing may be noticed in the application files and logs, operating system, and from the unallocated space. For example, the shared folder, downloads folder, and temp folder may provide traces of evidence for P2P file sharing. Besides, this registry and the unallocated space as a result of file deletion may also serve as valid evidence.

Evidence at Application Layer

Besides this, the contact information for a given host can be obtained by searching the Whois databases (http://whois.arin.net). A program called *traceroute* provides a list of routers that information passes through to reach a specific host. This program helps to determine the computers that were involved in the transfer of information on the Internet. Intermediate routers sometimes have relevant digital evidence in log files.

Evidence at Network and Transport Layer

Two protocols are used for assigning IP addresses and to combat IP spoofing: bootstrap protocol (BOOTP) and dynamic host configuration protocol (DHCP). BOOTP and DHCP are quite similar. Both the protocols require hosts to identify themselves using a MAC address before obtaining IP addresses. When a computer boots up, it sends its MAC address to the BOOTP or DHCP server which recognizes the MAC address and sends back an IP address. The server assigns a specific IP address to a specific MAC address thus giving the effect of static IP addresses. A criminal could reconfigure his/her computer with someone else's IP address to hide his/her identity. Investigators can still identify the computer using the MAC address. This is because the criminal's computer must use the address resolution protocol to receive data from a router which requires the actual MAC address of the computer. The router would have an entry in its ARP cache showing a particular computer is using someone else's IP address. This serves as proof and helps in locating the criminal's computer and the criminal himself/herself.

6.15 DIGITAL EVIDENCE AS ALIBI

An alibi is merely evidence that demonstrates that a defendant in a criminal case was somewhere other than the scene of the crime at the time of crime. The most common forms of alibi with digital evidence are time and location. When an individual does anything involving a computer or network, the time and location are often noted, generating digital evidence that can be used to support or refute an alibi.

The date and time in a computer system can be altered either with a valid reason or in order to disguise the presence of evidence. The latter case will have an impact on the log files created. Similarly, in the case of networks, in order to falsify evidence, a user may attempt to modify the IP address and the date and time timestamps but however could not compromise all the machines. This would otherwise pose hardships with email tracking when it is used to commit a crime. Besides this, information can be obtained from the ISPs at both the ends to traverse the route the mail has undergone along with date and timestamp from the email header.

6.16 IMPEDIMENTS TO COLLECTION OF DIGITAL EVIDENCE

The following are some of the obstacles to the collection of digital evidence:

Changing default location of files or renaming files or folders Desired files will always not be available in the default location and so a search needs to be made on the entire hard disk. Most applications store their file's location in the Windows registry or in their respective configuration files. Expertise and skill in working with different applications are required for forensic analysts to handle this situation. As an alternative, automated tools may also be relied upon.

Hiding and/or protecting files with file system attributes and File system attributes and permissions are primarily used to protect the information in files. However, a forensic analysis tool can figure this out and by-pass security attributes, thereby providing a way to access hidden files which may host any valid evidence.

Deleting files Deleted files are usually available in the recycle bin in a Windows PC. If it is not present there, they may be recovered using data recovery tools. Data recovery is possible because the contents of the file are not wiped. Scanning the entire or analysing the file system not only aids in recovering deleted files but also to locate any temporary files that hold any valid piece of evidence (Box 6.6).

Formatting entire hard drive in attempt to destroy evidence Formatting a hard drive can be done in either one of the two ways: full format or quick format. *Quick format* initializes the disk partition by creating a new, empty file system, whereas a full format wipes the disk clean by writing zeros on the disk and in addition checks the disk and marks any unreliable sectors as bad sectors. Thus, quick format is non-destructive and recovering data from a quick formatted hard drive is possible by file carving. In some situations, specialized applications can wipe the data in a quick formatted drive, thus making carving impossible. Live RAM analysis may be performed at such a juncture in an attempt to recover at least some of the recent activities (Box 6.7).

Formatting solid state drive in attempt to destroy evidence Solid state drive (SSD) is a storage technology much different from hard drives as it adopts to store information differently from hard drives. This makes the information in SSD easily erasable but more difficult to recover. The reason behind is the use of TRIM command which actually zeros information once a file is marked deleted by the operating system. Any write blocker also cannot prevent the impact of the TRIM command. Loss of a file or information can be prevented if TRIM is not supported by the operating system, communication interface, or the file system (Box 6.8).

> ### Box 6.6 Deleting a file versus Wiping a file
> **Deleting a file** When a file is simply deleted or erased, pointers to the file are 'zeroed' (i.e., alterations are made to the FAT or MFT) so that at the logical level the file does not appear to the user, but at the physical level the file data is still intact on the media and may be recovered.
> **Wiping a file** When a file is wiped, the entirety of the file is overwritten by a known or random hex character or pattern rendering it irrecoverable.

Box 6.7 File Carving

File carving is a term used in computer forensics to denote the identification and extraction of file types from unallocated clusters using file signatures. A file signature (also referred to as magic number) is a constant numerical or text value used to identify a file format. Detecting such constants in files is a simple way of distinguishing between file formats. Basically every file has a header and a footer in order to get correctly recognized. For example, a pdf file starts with "%PDF" and ends with "%EOF" whereas a jpeg image file begins with "0xFFD8" and ends with "0xFFD9". These constants are called magic numbers. The objective of file carving is to identify and extract or carve a file based on its signature information alone.

Box 6.8 TRIM versus Popular Operating Systems

Windows supports TRIM on NTFS volumes but not on FAT formatted disks.

Linux supports TRIM on all types of volumes including those formatted with FAT.

Disabling all logging (if supported by an application) Disabling logging functionality if supported by an application can prevent history files from being retained in the system which otherwise would have provided information about the users who logged on to the application.

6.17 CHALLENGES WITH DIGITAL EVIDENCE

Some of the challenges experienced with digital evidence are listed here:

1. Digital evidence obtained from a storage media and other components of an ECD is huge. Of course not all the information will be useful and relevant. Extracting the right piece of evidence is the biggest challenge.
2. Not all portions of evidence can be collected manually. Certain evidence in the form of logs in the systems can reveal very little information about the crime. Further, more information has to be collected by employing appropriate forensic tools. Therefore, a forensic investigator who is professionally sound should be involved in the collection of evidence.
3. The digital evidence collected should be able to completely map a crime scene which otherwise would make it worthless and difficult in proving the crime.
4. Digital evidence is more susceptible to alteration and hence appropriate tools should be employed to ascertain whether it has been modified or a valid evidence had been deleted.

POINTS TO REMEMBER

- Digital evidence is defined as information and data that is stored, received, or transmitted by an electronic device and is of value to an investigation.
- Evidence collection involves five phases: identification of evidence, preservation of evidence, analysis of evidence, presentation of evidence, and the archival of evidence.
- The sources of evidence may be standalone computers, mobile devices, and the Internet.
- Predominant operating systems include Windows, Linux/Unix, and Macintosh.
- The primary storage medium can be a hard disk Drive or a solid state drive.
- The data storage area in a physical disk while manufacturing is divided into equal sized blocks called sectors.
- The advancement in storage technology has raised the sector size from 512 bytes to 4096 bytes in advanced format (AF) drives.
- Hidden sector data refers to bytes in the physical disk that are used by themselves and are invisible to the user.

- Partitioning refers to logically dividing the physical drive into multiple pieces (logical volumes) called partitions.
- The first physical sector of the physical drive contains the master boot record (MBR) which is loaded into the memory by BIOS during the booting process.
- An operating system's ability to access files on a volume depends on the file system with which the volume was formatted.
- The FAT file system can be viewed as two groups, namely the system area comprising reserved regions (holds boot record) and FAT region and data area comprising the root directory region and the file and subdirectories region.
- Slack space is the area of a disk between the end of the file and the end of the cluster allocated to the file.
- exFAT file system comprises two main regions: file system area and data region.
- NTFS is an extensible operating system suitable for the increasing size in drive volumes.
- Orphaned files are used to refer to files whose parent–child relationships are broken.
- Linux/Unix operating systems use ext family of file systems.
- Mac operating system is based on hierarchical file system (HFS/HFS+) where HFS makes use of 16 bits and HFS+ makes use of 32 bits to address clusters on a disk.
- The registry contains information about profiles for each user, the applications installed on the computer, and the types of documents that each can create, property sheet settings for folders and application icons, what hardware exists on the system, and which ports are being used on the Windows that it operates on.
- Windows artifact refers to evidential data that is saved by the operating system as a result of human interaction or usage on the system and not by default during installation.
- A whole disk encryption encrypts the entire hard drive.
- Evidence can be collected from mobile devices, from the phone itself, from a local workstation, or from its SIM card.
- Evidence can be collected from the Internet, from email, from peers, and from the application and network layers.
- An alibi is merely evidence that demonstrates whether a defendant in a criminal case was present somewhere other than the scene of a crime when the crime was committed. Most common forms of alibi with digital evidence are time and location.
- Digital evidence collection has obstacles to face and challenges to address.

KEY TERMS

Access control lists This defines the way the user has control over files, namely to read, write, and to modify.

Acquisition This refers to the gathering of data from the seized ECD.

Alibi This refers to evidence that demonstrates that a defendant in a criminal case was somewhere other than the scene of a crime when the crime was committed.

Digital evidence This is defined as the information and data that is stored, received, or transmitted by an electronic device and is of value to an investigation.

Event logs This refers to log records which record information about errors, failure/success, information and warnings pertaining to application, system and user.

Extended partition This refers to a partition that holds the data and files that are stored on the disk.

File allocation table (FAT) It is a table containing information about the location and status of file data in the partitions.

Hidden sector data This refers to bytes in the physical disk that are used by 'itself' and are invisible to the user. It may also refer to 'spare' sectors within an SSD which are meant to replace damaged or bad sectors.

Host protected areas It refers to the area reserved on the hard disk to store information that cannot be modified and that is inaccessible to the user, BIOS/UEFI, and the operating system.

Jumplists This refers to a list of files that are opened by a particular application which are automatically populated by the system as well as by the respective application.

Non-volatile evidence This refers to the contents that can be recovered from an ECD even if it is not powered on.

Partitioning This refers to logically dividing the physical drive into multiple pieces (logical volumes) called partitions.

Primary partition This refers to a type of partition regarding the operating system and system area, as well as other information required for booting.

Recycle Bin It refers to a repository of files deleted by the user with provision for recycling the data thereby preventing accidental deletion.

Seizure This refers to taking into custody the ECD associated with a crime by the law enforcement agency for the purpose of investigation.

Volatile evidence This refers to the frequently changing information, for example, information about running processes or the network, contents in the clipboard, and some data in the memory which are usually lost when power for the ECD is turned off.

Windows artifacts This refers to evidential data that is saved by the operating system as a result of human interaction or usage on the system and not by default during installation.

MULTIPLE-CHOICE QUESTIONS

1. _____ refers to the contents that can be recovered from an electronic communication device.
   (a) Physical evidence
   (b) Volatile evidence
   (c) Non-volatile evidence
   (d) None of these

2. The minimum requirement(s) to preserve a chain of custody is/are _____.
   (a) No data has been added, changed, deleted from the seized information evidence.
   (b) The seized/information evidence is duplicated exactly and completely.
   (c) A reliable and validated duplication process is used.
   (d) All of these

3. Hidden partitioning can be done by the _____.
   (a) use of encryption   (c) (a) and (b)
   (b) use of decryption   (d) none of these

4. _____ is the secondary copy of FAT.
   (a) Data area   (b) File allocation table
   (c) FAT region   (d) Data region

5. Which portion of the ext file system is referred to as super block?
   (a) The first 128 bytes from the start of the file system
   (b) The first 1024 bytes from the start of the file system
   (c) The first 256 bytes from the start of the file system
   (d) The first 512 bytes from the start of the file system

6. Which one of the following procedures is correct for evidence collection?
   (a) Identification, analysis, preservation, presentation, and archival of evidence
   (b) Identification, preservation, analysis, presentation, and archival of evidence
   (c) Identification, analysis, presentation, preservation, and archival of evidence
   (d) Identification, archival, presentation, preservation, and analysis of evidence

7. Acquisition can be done by _____.
   (a) imaging   (c) imaging or cloning
   (b) cloning   (d) imaging and cloning

8. The source(s) of evidence is/are _____.
   (a) standalone computers/electronic communication device
   (b) mobile devices
   (c) Internet
   (d) (a), (b), and (c)

9. _____ is the first action performed by a computer while booting to ensure the presence and functionality of the core components.
   (a) Power on Self Test (POST)
   (b) Plug on Self Test (POST)
   (c) Power on System Test (POST)
   (d) Plug on System Test (POST)

10. Boot code searches the root directory for operating system files like _____.
    (a) DOS.SYS   (c) CMD.SYS
    (b) COM.SYS   (d) IO.SYS

11. Windows 8 and 10 offer support for _____.

(a) BIOS (c) BIOS and UEFI
(b) UEFI (d) none of these
12. _____ is noted for less power range, faster data access, high reliability, and lighter weight.
(a) Hard disk drive (c) Track
(b) Solid state drive (d) Disk platters
13. _____ is found in Windows XP and Windows Server 2003.
(a) V1.2 (c) V3.1
(b) V3.0 (d) Transactional NTFS
14. Timestamp is updated as _____ when the file or directory is written in inode.
(a) Updated (U) (c) Modified (M)
(b) Wrote (W) (d) Changed (C)
15. _____ make up HFS+ file system.
(a) Volume header, catalog file, overflow file, attributes file, and volume footer
(b) Volume header, catalog file, overflow file, attributes file, and volume footer
(c) Volume header, catalog file, overflow file, and attributes file
(d) Volume header, catalog file, extents overflow file, attributes file, and allocation bitmap file

REVIEW QUESTIONS

1. Compare and contrast digital and physical evidence.
2. Explain evidence collection procedure.
3. List and discuss in detail the classification of sources of evidence.
4. Discuss briefly about components of FAT file system.
5. Write in detail about new technology file system (NTFS).
6. Explain how the browser artifacts of different operating systems can be collected as evidence.
7. What are the challenges with digital evidence?
8. Define digital evidence as alibi.
9. Mention in detail the obstacles in digital evidence collection.
10. Define the types of digital evidence.

APPLICATION EXERCISES

1. From Fig. 6.8, manually parse the file creation date and time.

```
Offset     0  1  2  3  4  5  6  7  8  9  A  B  C  D  E  F
01C271E0   E5 59 52 45 4C 4C 7E 31 4A 50 47 22 00 B7 13 6F
01C271F0   EA 42 EA 42 00 00 9B A3 E2 42 FB 1B C7 7F 00 00
```

Fig. 6.8

2. In Fig. 6.9, FAT 32 directory entry is given. Determine the filename, file extension, file attributes, last written/modified date, last written/modified time, starting cluster, and file size.

```
0003E260   41 53 00 30 00 79 00 31  00 65 00 0F 00 DF 6E 00
0003E270   74 00 2E 00 70 00 64 00  66 00 00 00 00 00 FF FF
0003E280   53 30 59 31 45 4E 54 20  50 44 46 20 00 C5 69 6F
0003E290   EB 42 EB 42 00 00 EC 61  EB 42 4A 1C 00 5A 00 00
```

Fig. 6.9

3. Assume that a machine has NTFS file system. How can you determine whether there is track fragmentation? Justify your answer.
4. From the master boot record shown in Fig. 6.10, determine OEM ID, bytes per sector, sectors per cluster, total sectors, starting Cluster $MFT, starting cluster $MFTMirr, $MFT file record size, volume serial number, and signature.

Fig. 6.10

5. From Fig. 6.10 in Question 4, determine the number of primary partitions and extended partitions, the drive letters assigned to the drive and the file system in each partition.
6. As an examiner you are asked to obtain the history records of Internet Explorer. Explain how you will do it. Specify the path to locate it.

BIBLIOGRAPHY

1. Forensicsciencesimplified, *A Simplified Guide to Digital Evidence*, available at: http://www.forensicsciencesimplified.org/digital/ (Accessed 18 December 2017)
2. Yuri Gubanov (05 August 2012), *Retrieving Digital Evidence: Methods, Techniques and Issues: Part 1*, available at: https://www.forensicmag.com/article/2012/05/retrieving-digital-evidence-methods-techniques-and-issues-part-1 (Accessed 18 December 2017)
3. Yuri Gubanov (30 May 2012), *Retrieving Digital Evidence: Methods, Techniques and Issues*, available at: https://www.forensicmag.com/article/2012/05/retrieving-digital-evidence-methods-techniques-and-issues (Accessed 18 December 2017)
4. Forensicexplorer, *Forensic Explorer Data Carving*, available at: http://www.forensicexplorer.com/data-carve.php (Accessed 18 December 2017)
5. Antonio Merola (10 November 2008), *Data Carv-*

ing Concepts, available at: https://www.sans.org/reading-room/whitepapers/forensics/data-carving-concepts-32969 (Accessed 18 December 2017)
6. Kb, *Whole Disk Encryption*, available at: http://kb.mit.edu/confluence/pages/viewpage.action?pageId=6391171 (Accessed 18 December 2017)
7. Nextstep4it (10 October 2014), *Best Linux File System Tutorial – ext2, ext3, ext4, JFS & XFS*, available at: https://www.nextstep4it.com/basic-linux-file-system-tutorial/ (Accessed 18 December 2017)
8. Shabbathster (15 July 2014), *Linux File Systems: Ext2 vs Ext3 vs Ext4 vs Xfs*, available at: http://shabbathster.blogspot.in/2014/07/linux-file-systems-ext2-vs-ext3-vs-ext4.html (Accessed 18 December 2017)
9. Ramesh Natarajan (16 May 2011), *Linux File System: Ext2 vs Ext3 vs Ext4*, available at http://www.thegeekstuff.com/2011/05/ext2-ext3-ext4/ (Accessed 18 December 2017)
10. Dave Poirier, *The Second Extended File System*, available at http://www.nongnu.org/ext2-doc/ext2.html#SUPERBLOCK (Accessed 19 December 2017)
11. Basic Computer Forensic Examiner's Guide by the International Association of Computer Investigative Specialists.
12. Brian Carrier (2005), *File System Forensic Analysis*, Addison Wesley Professional.
13. Cybercrimes - Training Material for Investigators, Data Security Council of India (DSCI), India.
14. Powerdatarecovery, The Best Data Recovery Software, available at: https://www.powerdatarecovery.com [Accessed on 19 December 2017]

Answers to Multiple-choice Questions

1. (c)	2. (d)	3. (a)	4. (b)	5. (b)
6. (b)	7. (c)	8. (d)	9. (a)	10. (d)
11. (c)	12. (b)	13. (c)	14. (c)	15. (d)

7
CYBER FORENSICS — THE PRESENT AND THE FUTURE

Learning Objectives

This chapter provides an insight into the cyber forensic tools, its types, and categories. It presents the free and open-source as well as proprietary forensic suites, imaging and validation tools, integrity verification tools, tools for data recovery and RAM analysis, tools for analysis of registry, encryption and decryption, password recovery, and network analysis. It highlights other miscellaneous tools used for forensic investigation besides explaining the tools used for UNIX system analysis. The chapter also explains the forensic analysis tools for mobile devices and email. The current requirement for forensic investigators is explained with the career prospects in this field; the available certifications and training are elucidated. The reader will be familiar with the following after studying the chapter:

- Forensic tools, its types, and categories
- Cyber forensic suite
- Imaging and validation tools
- Forensic tools for integrity verification and data recovery
- Forensic tools for RAM and registry analysis
- Forensic tools for network analysis and password recovery
- Forensic tools for UNIX system analysis
- Miscellaneous tools
- Forensic tools for mobile devices and email analysis
- Need for forensic investigators
- Career prospects for forensic investigators
- Forensic training and certifications

7.1 FORENSIC TOOLS

Forensic tools are used by forensic investigators to acquire digital evidence from electronic communication devices (ECDs), index them, and perform a detailed analysis. The uses of forensic tools are two-fold. They are as follows:

Criminal investigation If the damage caused by an incident is severe, say, web defacing or an intrusion, then criminal charges have to be filed against the offender. Forensic tools help to locate and identify the offenders so that the crime may be reported and, consequent to that, to collect evidence and present them in the court for prosecution.

Cleaning up and rebuilding After an incident, its impact has to be ascertained to figure out the damage and to rebuild the system. Forensic tools help to determine the root cause of the incident, remove the gap responsible for the incident, and rebuild the system. For example, if an offender has deleted the files in the computer system, then appropriate forensic tools may be used to recover the same.

Forensic tools are either proprietary or open-source and freely available. The choice of forensic tools depends on the following factors: operating system(s) supported, the preferences of user interface, budget, and the functionality/capabilities of the tool itself.

7.1.1 Types

Cyber forensic tools can be classified into two: (a) hardware forensic tools and (b) software forensic tools. Hardware forensic tools include specific components of the server used for investigation. Software forensic tools include command line applications and technical applications meant for forensic investigation, which either perform a single task or a collection of tasks.

7.1.2 Categories

Computer forensic tools can be categorized according to their working (Fig. 7.1) and are as follows:

Acquisition

The first step after identification of evidence in the forensic investigation process is acquisition. It refers to creating an exact bit stream copy of the original storage media that exists on the subject's computer. It aims at making a copy from the primary source of data or the original evidence. Mere copying of files does not recover all the areas of data on the device for examination. Creating a forensic image and using it for forensic examination offers the following benefits:

1. Preserves the original evidence
2. Prevents inadvertent alteration of the original evidence during examination
3. Allows recreation of another image, if necessary

Acquisition may be physical or logical. In physical acquisition, the entire drive is copied, including the deleted files and file fragments, whereas in logical acquisition only the active data, that is, the data excluding deleted space, deleted files, and fragments are copied.

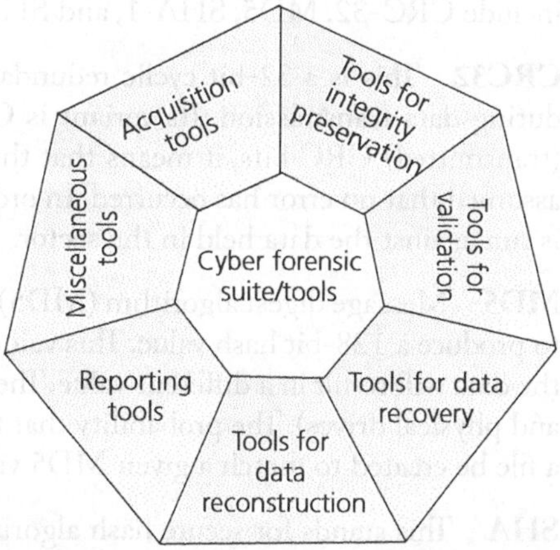

There are two ways of copying files from the original storage media: (a) logical backup and (b) bit stream imaging.

Logical backup copies the directories and files of a logical volume. It does not capture other data that may be present on the media, such as deleted files or residual data stored in the slack space. Further, it also modifies the timestamps of the data, thus contaminating the timeline.

Bit stream imaging, disk imaging, or cloning refers to the generation of a bit-for-bit copy of the original media including hidden and residual data, for example, free space, slack space, swap, residue, unused space, and deleted files. It requires more storage space and takes a longer time than logical backups.

Bit stream imaging performs either a disk-to-disk or a disk-to-file copy. A disk-to-disk copy copies the contents of the media directly into another media and is mainly used to test booting, whereas a disk-to-file copy copies the contents of the media into a single logical file which is compressible and results in faster searches. However, image file-to-disk is used to restore an image.

Drive imaging tools help in the acquisition stage.

Fig. 7.1 Categories of cyber forensic tools

Integrity Preservation

It is necessary to ensure the integrity of the original media during backup and imaging. Integrity of the data preserved during acquisition is ensured by comparing the secondary with the primary or the original copy of the data.

A write-blocker is used to ensure that the backup or imaging process does not alter the data on the original media. A write-blocker is a hardware- or software-based tool that prevents a computer from writing to the computer storage media connected to it. Hardware write-blockers are physically connected to the computer and the storage media being processed to prevent any write to that media.

In case of a hardware write-blocker, the suspected storage media (original evidence) should be directly connected to the write-blocker and the write-blocker should be directly connected to the computer or the device used to perform backup or imaging.

In the case of a software write-blocker, the software should be first loaded on the computer and then the device used to perform backup or imaging has to be connected to the computer.

Forensic image files are created as a result of the imaging process with extensions varying with the drive imaging tool being used. For example, '.p01' with Cyber Check Suite, '.e01' with Encase, and '.001/.SFB' with SafeBack.

Validation and Discrimination

Validation includes processes for comparison, such as hashing and analysing the file header after filtering. After backup or imaging, it is necessary to verify whether the data copied is an exact duplicate of the original data. The message digest is computed so as to verify its integrity. This message digest is a hash that uniquely identifies data and has the following property: changing a single bit of the data will result in a completely different message digest.

Hash algorithm is used as a utility to create a binary or hexadecimal number that represents the uniqueness of a data set such as a file or a disk drive. This unique number is also referred to as a digital fingerprint. If two files have the same hash values, they are 100% identical. Utility algorithms that are used to produce hash values include CRC-32, MD5, SHA-1, and SHA-512.

CRC32 This is a 32-bit cyclic redundancy check (CRC) code, mainly used as an error detection method during data transmission. Its variant is CRC16. If the computed CRC bits are different from the original (transmitted) CRC bits, it means that there has been an error in the transmission. If they are identical, it is assumed that no error has occurred. In order to ensure that the data present in a sector is without errors, CRC is run against the data held in the sector.

MD5 Message digest algorithm (MD5) is a mathematical algorithm that uses a cryptographic hash function to produce a 128-bit hash value. This value can be used to demonstrate the integrity of data. Changes made to the data will result in a different value. The function can be performed on various types of data (files, partitions, and physical drives). The probability that two files have the same message digest is in the order of 2^{64} and that a file be created to match a given MD5 value is in the order of 2^{128}.

SHA This stands for secure hash algorithm. It is an algorithm for computing the condensed representation of a message or a data file. The SHA hash functions are a set of cryptographic hash functions designed by the National Security Agency (NSA) and published by the National Institute of Standards Technology NIST as a US Federal Information Processing Standard. SHA-1 produces a message digest that is 160 bits long, whereas in SHA-256 and SHA-512 the number denotes the bit length of the digest they produce.

Upon verification of the integrity, forensic image files are written to brand new freshly formatted media or forensically sterile media. Any number of copies can be made for forensic examination.

Discrimination is essential to separate the suspicious data. Searching and comparing file headers help in discrimination. Filtering tools and keyword search utility assist in this phase.

Image validation and verification tools are available as a part of the forensic suite and as standalone tools, for example, hash generation and hash verification tools.

Extraction

The required data is extracted during investigation using recovery task, which could involve viewing data, searching for keywords, decompressing data, carving, decrypting, etc. Data recovery tools are available to handle this stage.

Reconstruction

This attempts to reconstruct the suspect's drive to analyse what happened during the crime. This requires acquisition and analysis of the evidence by forensic investigators. It involves activities such as disk-to-disk copy, image-to-disk copy, partition-to-partition copy, and image-to-partition copy. Tools for copying are available.

Reporting

The last step in forensics is reporting, where a step-by-step process carried out during analysis is presented in detail. Reporting tools, either standalone or those present in the forensic suite can be used for this purpose.

Besides these, password recovery tools, mobile forensic tools, network analysis tools, and other miscellaneous tools are presented in the following section.

7.2 CYBER FORENSIC SUITE

A forensic suite has specialized tools for RAM analysis, hard disk imaging and analysis, verification and validation, registry analysis, data recovery and reconstruction, password cracking, and a lot more, as a collection apart from tools for reporting.

7.2.1 Free and Open-source Forensic Suite

The following are some of the free and open-source forensic suites:

OSForensics

This tool is used for discovering the evidence, and identifying and reporting it. It discovers relevant data faster using file searching and indexing. It can recover deleted files, extract passwords, and decrypt files irrespective of the operating system and the file system in force.

OSForensics can identify evidence and any suspicious activity using features such as hash matching and drive signature analysis. It can create a timeline of user activity. OSForensics has come out with new reporting features such as building custom reports, adding narratives, and attaching another tool's report to the OSF report.

A collection of tools are provided as free tools for use with OSForensics, namely OSFMount for mounting dd image files in Windows, OSFClone which is a self-booting disk cloning tool, ImageUSB to write an image to multiple USB drives, and VolatilityWorkbench which is a Windows GUI for volatility and memory analysis.

Nirsoft

This is a website available at http://www.nirsoft.net/ and provides a unique collection of freeware utilities. It has password recovery utilities, network monitoring tools, Internet-related utilities, MS-Outlook tools, command line utilities, desktop utilities, and freeware system tools. A few representative tools from the Nirsoft website are presented in Table 7.1.

Table 7.1 Example tools from Nirsoft

Tool	Purpose
WebBrowserPassView	To store the passwords stored by the web browser
WirelessNetView	To view the details of a wireless network in a given area
BluetoothView	To monitor Bluetooth activity in an area
AdapterWatch	To display information about network adapters
IPNetInfo	To find all available information about an IP address
MyLastSearch	To view the latest searches with Google, Yahoo, etc.
OutlookAttachView	To view and extract attachments in the Outlook mailbox
SearchMyFiles	To search for files and folders
USBDeview	To view all installed and connected devices to the system

Windows Sysinternals Live

Sysinternals Live from Microsoft is a service that provides a collection of Sysinternals tools to be directly executed from the Web (https://live.sysinternals.com/) without manually downloading them. Some of the tools from the suite are listed here:

Sysmon v6.10 This is a background monitor that records an activity to the event log for use in security incident detection and forensics, monitors WMI filters and consumers (which is an autostart mechanism commonly used by malware, etc.).

Process Monitor v3.40 Process Monitor is a file system registry, process and network real-time monitor that includes a runtime switch for terminating monitoring after a specified amount of time. In hexadecimal mode, it shows the process tree with proceCraig Wrss IDs in hexadecimal. It can fix a bug in automated boot log conversion.

Autoruns v13.80 This is a utility for viewing and managing autostart execution points (ASEPs). It has asynchronous file saving, shows the display name for drivers and services, and fixes bugs in offline total virus scanning.

AccessChk v6.1 This is a command-line utility that shows effective and actual permissions for file, registry, service, process object manager, and event logs. It can report Windows 10 process trust access control entries and token security attributes.

Autopsy

The Autopsy Forensic Browser [Craig Wright, (2009)] is a graphical interface to the command line digital investigation tools in The Sleuth Kit. Together, they help in the investigation of file systems and volumes of a computer. It is used by law enforcement, military, and corporate examiners to investigate what happened on a computer. Autopsy supports web artifact analysis and registry analysis which other commercial tools do not provide (Box 7.1).

Dead analysis autopsy and The Sleuth Kit are run in a trusted environment, typically in a lab, to examine the data from a suspect system. In *live analysis*, the suspect system is analysed while it is run with Autopsy and The Sleuth Kit, from a CD in an untrusted environment. This is frequently used during incident response while the incident is being confirmed. Following confirmation, the system is acquired and a dead analysis is performed.

The Autopsy browser provides the following evidence search functionalities:

File listing This is used to analyse the files and directories including the names of deleted files and files with Unicode-based names.

File content The contents of files can be viewed in raw or hex, or the ASCII strings can be extracted. When data is interpreted, Autopsy disinfects so as to prevent the local analysis system from damage.

Hash databases Lookup of unknown files in a hash database helps to quickly identify it as good or bad. Autopsy makes use of the NIST National Software Reference Library (NSRL) and user created databases of known good and known bad files.

File type sorting The files are sorted based on their internal signatures to identify files of a known type. Autopsy can extract only graphic images (including thumbnails). The extension of the file will also be compared with the file type to identify files that may have had their extension changed to hide them.

> **Box 7.1 The Sleuth Kit—TSK**
>
> The Sleuth Kit is a collection of command line tools and C libraries, which allows investigators to analyse disk images and recover files for them. It is used behind the scenes in Autopsy and many other open-source and commercial forensic tools.

Timeline of file activity   A timeline of file activity helps to identify areas of a file system that may contain evidence. Timelines that contain entries for the modified, access, and change (MAC) times of both allocated and unallocated files can be created by Autopsy.

Keyword search   Keyword searches of the file system image can be performed using ASCII strings and grep regular expressions. Searches can be performed on either the full file system image or just the unallocated space. An index file can be created for faster searches. Strings that are frequently searched for can be easily configured into Autopsy for automated searching.

Metadata analysis   Metadata structures contain details about files and directories. Autopsy allows viewing of the details of any metadata structure in the file system. This is useful for recovering deleted content. Autopsy will search the directories to identify the full path of the file that has allocated the structure.

Data unit analysis   Data units are where the file content is stored. Autopsy allows viewing of the contents of any data unit in a variety of formats including ASCII, hexdump, and strings. Given the file type, Autopsy can search the metadata structures to identify the allocated data unit.

Image details   File system details can be viewed, including on-disk layout and times of activity. This mode provides information that is useful during data recovery.

Autopsy provides a number of functions that aid in case management. Some of these are as follows:

Event sequencer   Time-based events can be added from the file activity or IDS and firewall logs. Autopsy sorts the events so that the sequence of incidents associated with an event can be easily determined.

Notes   Notes can be saved on a per-host and per-investigator basis. This allows the investigator to make quick notes about files and structures. All notes are stored in an ASCII file.

Image integrity   Autopsy generates an MD5 value for all files that are imported or created by default so as to ensure that data is not modified during analysis. The integrity of any file that Autopsy uses can be validated at any time.

Reports   Autopsy can create ASCII reports for files and other file system structures. This enables the investigator to promptly make consistent data sheets during the course of investigation.

Logging   Audit logs are created on a case, host, and investigator level so that all actions can be easily retrieved. The entire Sleuth Kit commands are logged exactly as they are executed on the system.

Computer Aided Investigative Environment

Computer aided investigative environment (CAINE) [John Kehr] is an Italian GNU/Linux live distribution created as a digital forensics project. It offers a complete forensic environment integrating existing software tools as software modules besides providing a user friendly graphical interface. Some of the tools include Nirsoft Suite plus launcher, WinAudit, MWSnap, Arsenal Image Mounter, FTK Imager, Hex Editor, JpegView, Network tools, NTFS Journal Viewer, QuickHash, USB Write Protector, VLC, Windows File Analyzer, etc. For example, a tool with a GUI named BlockON/OFF present in CAINE can block all devices and assures that all disks are preserved from accidental writing operations by locking them in read-only mode. It has got Windows IR/Live forensics tools. The speciality with CAINE is that it can be controlled from remote, can boot fast, and can boot to RAM.

PALADIN

PALADIN [Sumuri] is a modified 'live' Linux distribution based on Ubuntu. It simplifies various forensics tasks in a forensically sound manner via the PALADIN toolbox. The PALADIN toolbox combines the power of several court-tested open-source forensic tools into a simple interface that can be used by anyone. It is used by forensic examiners from law enforcement, military, federal, state, and corporate agencies.

Autopsy is a full feature GUI-based forensic suite with all the features expected in a forensic tool, and even contains advanced features that are not found in other forensic suites. It is combined with PALADIN so as to allow the user to conduct a forensic exam from beginning to end, and perform reporting on different operating systems such as Mac, Windows, Linux, and Android file systems.

SUMURI remote service mode is included with every version of PALADIN and it can be activated at boot. In remote services mode, SUMURI experts can perform a variety of services remotely from anywhere in the world.

PALADIN has 33 categories of utilities, namely encryption tools, file differential tools, file system tools, forensic suite, hardware analysis, hashing tools, hex editor, imaging tools, Internet analysis, log analysis, mail analysis, malware analysis, memory analysis, messenger forensics, metadata analysis, mobile device analysis, network analysis, password discovery, photo analysis, PLIST analysis, Recycle Bin analysis, reporting tools, social media analysis, steganography tools, thumbnail analysis, timeline analysis, TSK, virtual machine, Windows registry, and more than hundred tools.

Hiren's Boot CD

Hiren's Boot CD is an all-in-one bootable CD, the latest version being Hiren's Boot CD 15.2. It has utilities that can be categorized as follows: antivirus tools, backup tools, BIOS/CMOS tools, browsers/file managers, cleaners, device driver tools, editors/viewers, FileSystems tools, hard disk tools, master boot record (MBR) tools, network tools, optimizers, other tools, partition tools, password/key tools, process tools, recovery tools, registry tools, remote control tools, startup tools, system information tools, testing tools, and tweakers.

Digital Evidence and Forensic Toolkit

Digital evidence and forensic toolkit (DEFT) is a customized distribution of the Ubuntu live Linux CD. It is an open-source distribution of Linux built around the digital advanced response toolkit (DART) software and based on the Ubuntu operating system. DEFT is easy to use besides including the best hardware detection feature and some open-source application recovery services. All types of hardware and software are supported, and its expert software engineers have developed powerful tools for maximum data recovery.

The mobile forensic workstation is designed to capture and process extremely high volumes of computer data quickly and efficiently with absolute assurance of data integrity. With its dedicated shipping container, it can be safely transported to remote locations that require a full computer forensic facility.

Besides this, the processing software and investigation software accomplish all the other tasks associated with the forensic investigation.

7.2.2 Proprietary Forensic Suites

The following are some of the forensic suites whose licences have to be purchased to work with:

Paraben Tools

The Paraben tools include a forensic replicator for acquiring data from a wide range of electronic media, chat examiner for the analysis of chat logs, network mail examiner for the examination of MS exchange and lotus notes files, email examiner for examining mailboxes, and device seizure which includes necessary hardware and software to aid in mobile forensics.

Computer Online Forensic Evidence Extractor

Computer online forensic evidence extractor is a USB 'thumb drive' that has been developed by Microsoft as a small plug-in device that investigators can use to quickly extract forensic data from computers associated with crimes.

The device contains 150 commands which dramatically reduce the time taken to gather digital evidence, thus making it suitable for analysing cybercrimes. It can decrypt passwords and analyse a computer's Internet activity as well as data stored in the computer.

Vogon

In the field of computer evidence, Vogon is the main supplier of systems and services to many government agencies in Europe and North America. Vogon helps companies to maintain business continuity in the face of data loss, by providing a world-leading data recovery service. All types of hardware and software are supported and its expert software engineers have developed powerful tools for maximum data recovery.

The imaging software has been developed by Vogon over the past decade to offer very high imaging performance together with comprehensive auditing and anti-repudiation techniques. This imaging process is independent of the file system(s) on the hard drive under investigation, and is recognized by courts around the world as the only valid means of capturing computer evidence.

7.3 DRIVE IMAGING AND VALIDATION TOOLS

Table 7.2 presents some of the drive imaging and validation tools with a brief description about each one of them.

Table 7.2 Drive imaging and validation tools

Tools	Description
Intelligent Computer Solutions Image Master Solo Forensics Hard Drive Duplicator	It is used to create an image of a hard drive including the MBR, the unallocated space, and all the files contained on the hard drive. It is very fast as there is no operating system or software. It is a modified hard disk controller with a simple menu screen from which the investigator can enter commands. This device is small and easy to use, and multiple suspect hard drives can be copied on a single target media. The target media must be large enough to hold the aggregate of suspect data acquisitions. A drawback to this hardware copy is that the data is written 'raw' to the target media and therefore no target image file is created. A subsequent image file acquirement will have to be performed on the target media in a lab environment at a later date.
Norton Ghost	This software package is used to create image files. These image files contain all the information necessary to recreate the hard drive or logical partition. Norton Ghost verifies possible data error by doing cyclic redundancy checking (CRC32). Some version of the Norton Ghost image contains only the active files on a hard drive and does not contain the unallocated space, which is potentially the most important part for forensic examination as a majority of the evidence found on the hard drive is in the unallocated space.
Symantec Ghost	This is used to take reliable backups of PC drives including applications and critical data. Ghost filters out what it considers unused space making it forensically unsound.
Safeback	This software is capable of making a bit stream backup, and is predominantly used by law enforcement agencies. The precision of the backup is guaranteed by mathematical CRC. The CRC process validates the comparison of the data on the hard disk which has been copied to the restored data. An advantage of the image files created by this particular software is that it can be written to many different media such as magnetic tape, hard drive, CD-ROM, and DVD. Safeback also copies all the data from the source hard drive including the file and disk slack.
Encase Forensic Imager	Encase Forensic Imager is based on trusted, industry standard Encase forensic technology. It facilitates forensic examiners with forensically sound acquisition of data from entire volumes or selected folders and investigation of the same. It makes it possible for forensic examiners to acquire data from a wide variety of devices such as tablets, hard drives, and removable media, unearth potential evidence with disk-level forensic analysis, view and browse for potential evidence files including folder structures, file metadata, and craft comprehensive reports on their findings, while maintaining the integrity of the evidence.

(Contd)

Table 7.2 (Contd)

Tools	Description
	Encase Forensic Imager is freely downloadable and requires no installation. It can be deployed via a USB stick so as to perform acquisition of a live device. This software creates an image file of the suspect media (hard drive, Zip™ and Jazz™ drive, floppy, CD, DVD, flash card). After the creation of the image file, it is possible to export any suspect files or documents. Encase produces a complete report as well. The report includes the disk geometry, hash set (to confirm the exactitude of the image file), display the file structure tree and list all the incriminating files. It supports different types of keyword search, thus saving the investigator's time. Encase gives the opportunity to preview the media before imaging. Previewing is done by booting the suspect computer with an Encase boot disk and previews the hard drive through a parallel port or network cable. Encase is able to read and image data from different operating systems such as MAC OS, Windows, Linux, and Solaris. Encase offers two possibilities: (a) accessing all the active files on the disk with which the investigator has the possibility to select the View All Files button and sort them by name, file extension, creation date, and last access date (This way, it is possible to have a quick overview of the file system.); and (b) offering a search tool that allows searching the entire disk including the file slack, the unallocated space, and the active files.
FTK Imager	This is a free tool for acquisition of data from AccessData. It creates a forensic image as a bit stream. Some other features of FTK Imager include the following: collection of volatile data from RAM, pre-analysis of data, information search, etc. This tool takes a snapshot of the entire disk drive and then copies every bit for analysis. It supports files systems such as FAT 12/16/32, NTFS, NTFS Compressed, and Linux ext2 and ext3. It is a full suite of forensic applications.
ProDiscover Incident response (IR)	This software has the ability to acquire a full image of the target. It provides both traditional and network forensic capabilities. It has a remote agent that runs in stealth mode and guarantees the integrity of the target in the presence of the software. The deployed agent facilitates the collection and analysis of a variety of data. Live analysis supports capturing RAM in Windows Vista and Windows Server 2008.
X-Ways Forensic (XWF)	This computer forensic environment supports the Windows family of operating systems. It has features for disk cloning and imaging, the ability to view and dump physical RAM and virtual memory of running process, and managing RAID and native support for a variety of file systems, etc.
DriveSpy	This DOS-based imaging tool developed by Digital Intelligence provides an MS-DOS command line along with additional commands. The program is 125 KB and fits in a floppy disk. It has facilities to create a disk-to-disk copy, create an MD5 hash, copy a range of sectors, search a drive, wipe a disk, etc.
Forensic Replicator	This disk imaging tool from Paraben Forensic tools runs on Windows operating systems and accommodates many types of electronic media and removable media. It has the ability to compress and split drive images for efficient storage.
SMART Acquisition Workshop (SAW)	This standalone utility from ASR Data Acquisition and Analysis creates images from storage devices and supports Windows, Linux, and Mac computers.
SMART	This suite from ASR Data Acquisition and Analysis includes two packages, namely SMART Acquisition for disk imaging and SMART Authentication for verification.
WinHex	This is a Windows-based universal hexadecimal editor and disk management utility from X-Ways Software Technology. It is used to recover lost or damaged files and edit disk contents.

7.4 FORENSIC TOOL FOR INTEGRITY VERIFICATION AND HASHING

The following are some of the free and open-source tools available to verify and validate the integrity of forensic images:

HashMyFiles

HashMyFiles is a small utility and freeware from NirSoft which allows the calculation of MD5 and SHA1 hashes of one or more files in the system. The MD5/SHA1 hashes list can be copied and saved in text/html/xml file. It can be launched from the context menu of Windows Explorer, and the MD5 and SHA1 of the selected folder can be viewed.

HashCalc

HashCalc is a calculator program and a free open-source utility developed by SlavaSoft, Inc. It is used for computing HMACs, messages digests, and checksums for files, text, and hex strings. It allows the calculation of hash (message digest), checksum, and HMAC values based on the most popular algorithms: MD2, MD4, MD5, SHA1, SHA2 (SHA256, SHA384, SHA512), RIPEMD160, PANAMA, TIGER, CRC32, ADLER32, and the hash used in eDonkey (eDonkey2000, ed2k) and eMule tools. It supports three input data formats: file, text string, and hexadecimal string. It is a very fast, easy-to-use application, and can work with large sized files and supports file drag-and-drop functionality. HashCalc generates hash, check sum, and HMAC for files of any type which makes it a valuable utility to test for corruption. This tool can compare music, audio, sound, video, film, game, image, icon, document, and other files, verify CD and hard drive files, perform checking of files of .mp3, .mpeg, .mpg, .avi, .vcd, .iso, .zip, .gif, .jpg, and .doc extensions and other downloads.

The following are some of the proprietary tools:

CRCMD5

CRCMD5 calculates the CRC-32 checksum for a DOS file or group of files and a 128-bit MD5 digest. Its syntax is given here:

crcmd5 <options> file 1, file 2..

where options may be /s which means that the files in the current directory and all the files in the subdirectory that match the stated file specification are included in the calculation. /h means that the output is header less text that consists of filename lines only.

DiskSig

DiskSig is used to compute CRC checksum and MD5 digest for an entire hard drive. The checksum and the digest include all the data on the hard drive, including erased and unused areas. By default, the boot sector of the hard drive is not included in this computation.

MD5summer

The MD5summer is a GUI application for generating and verifying MD5 checksums of files. MD5summer generates MD5 checksums for multiple files and stores the results in a text file. It can also take a test file and check the files in it. Entire directory structures can be summed recursively. Input and output files are compatible with those of the GNU MD5sum application.

7.5 FORENSIC TOOLS FOR DATA RECOVERY

The following are some of the tools available for data recovery:

Recuva

Recuva is a user friendly recovery tool. It can recover accidentally deleted files (pictures, music, documents, videos, and emails), images, and data in rewriteable media (memory card, external hard drives, and USB sticks). Recuva has an advanced deep scan mode that searches the disk thoroughly to find any traces of files that have been deleted. Those files that require permanent deletion are handled by Recuva with a secured overwriting feature that uses industry- and military-standard deletion techniques to ensure that the files are erased.

Byte Back

ByteBack is a data recovery and investigative tool developed by Tech Assist Inc. The features of this tool are quick cloning or imaging of physical sectors, automated recovery of files including deleted files, automatic repair of partitions and boot records, recovery of individual files in volumes, quick overwrite of every sector of a drive, and a media editor for viewing and searching through raw data. This tool also provides write blocking of the source drive and automatic CRC and MD5 hash calculation to verify copy operations.

IsoBuster

IsoBuster is capable of recovering data from CD, DVD, BD, HDD, flash drive, USB stick, media card, Compact CF, MMC, SD, diskette, SSD, and more via NTFS, UDF, FAT, HFS, ISO, IFO/VOB, and file signatures. It is a highly specialized and easy-to-use file data recovery software.

7.6 FORENSIC TOOLS FOR RAM ANALYSIS

The following are some of the free and open-source tools available for RAM analysis:

Live RAM Capturer

Belkasoft Live RAM Capturer is a free forensic tool. It is used to reliably extract the entire contents of a computer's volatile memory even if it is protected by an active anti-debugging or anti-dumping system. This is possible as it runs in the systems' most privileged kernel mode, whereas other acquisition tools fail as they operate in the systems' user mode.

Memory dumps are a valuable source of evidence and volatile information. Memory dumps may contain passwords to encrypted volumes (TrueCrypt, BitLocker, PGP Disk), account login credentials for webmail, and social network services such as Gmail, Yahoo Mail, Hotmail, Facebook, Twitter, Google Plus, and file sharing services such as Dropbox, Flickr, and SkyDrive.

Memory dumps captured with Belkasoft Live RAM Capturer can be analysed with Live RAM Analysis in Belkasoft Evidence Center. Belkasoft Live RAM Capturer does not require installation and can be launched in seconds from a USB flash drive. It is compatible with all versions and editions of Windows including XP, Vista, Windows 7, 8 and 10, and the 2003 and 2008 servers.

Volatility

Volatility from Volatility Framework is a command line-based memory forensic analysis platform. It supports memory dumps from all major 32- and 64-bit Windows versions and service packs including XP, 2003 Server, Vista, Server 2008, Server 2008 R2, and Seven. It can work with the memory dump in any one of the forms, namely raw format, Microsoft crash dump, hibernation file, or virtual machine snapshot. It supports different operating systems such as Windows, Linux, and Mac.

Volatility Workbench is a free, GUI-based open-source forensic tool for extracting artifacts from memory dumps. It runs on Windows. Some of its features include dumped information stored in a file, dropdown list of available commands with a short description for each, etc.

DumpIt

DumpIt is a tool from Comae Technologies [Lenny Zeltser, (2017)] that can take a memory snapshot of a host when the host is suspected to be compromised or infected. It is a fusion of two trusted tools, win32dd and win64dd. It is simple to use—a double click of the executable by the investigator runs the tool, takes a snapshot of the host's physical memory, and saves it in the folder where the executable is located. Later, memory forensics tools such as the Volatility Framework can be used to examine the memory file's content for any malicious artifacts.

The speciality of DumpIt is that it eliminates the need for the investigator to sit before the host for memory acquisition. Besides this, it is simple and easy to use, even for a naive user.

Magnet RAM Capture

Memory analysis [Jamie McQuaid, (2015)] can reveal information about a system and its users, malware, incidents of intrusion, and the evidence stored only in memory in pagefile.sys or hiberfil.sys but never written to the hard drive. Running processes and programs, active network connections, registry hives, passwords, keys, and decrypted files are just a few examples of the evidence that can be found in the memory. Many web apps, for example, Gmail, store data in the memory meaning that the evidence associated with it cannot be recovered from the hard disk.

Magnet RAM Capture supports both 32- and 64-bit Windows systems including XP, Vista, 7, 8, 10, 2003, 2008, and 2012. The standalone executable of Magnet RAM Capture can be run from either a USB stick or from the local machine.

Magnet RAM Capture creates a raw data dump with a .DMP extension. The data dump can be analysed with any memory analysis tool.

7.7 FORENSIC TOOLS FOR ANALYSIS OF REGISTRY

The following are some of the free and open-source tools available for registry analysis:

Regshot

Regshot is an open-source registry compare utility that helps to quickly take a snapshot of the registry and compare it with a second one. This is usually done after doing system changes or installing a new software product. The changes report can be produced in text or HTML format and contains a list of all modifications that have taken place between the two snapshots. In addition, the folders (with subfolders) that have to be scanned for changes can be specified.

RegRipper

RegRipper written in Perl is a Windows registry data extraction tool, and is an open-source forensic software application developed by Harlan Carvey. It is the fastest, the easiest, and the best tool for registry analysis in forensics examinations. It is not a registry viewer but is used to perform Windows registry hive file analysis. This tool is specifically intended for Windows 2000, XP, and 2003 hive files.

RegRipper can be customized to the examiner's needs through the use of available plugins or by users writing plugins to suit specific needs.

RegRipper bypasses Win32API and uses James McFarlane's Parse::Win32Registry module to access a Windows registry hive file in an object-oriented manner. This module is used to locate and access registry key nodes within the hive file as well as value nodes and their data. When accessing a key node, the LastWrite time is retrieved, parsed, and translated so that it is readable by an examiner. Data is retrieved in the same manner and if necessary, the plugin that retrieves the data will also perform translation of that data into something readable.

7.8 FORENSIC TOOLS FOR ENCRYPTION/DECRYPTION

The following are the free and open-source tools available for encrypting the contents in a media or to decrypt the files:

VeraCrypt

VeraCrypt from **IDRIX** is an open-source utility used for on-the-fly encryption. It is a free disk encryption software that is based on TrueCrypt 7.1a. It can create a virtual encrypted disk within a file or encrypt a partition or the entire storage device with pre-boot authentication.

VeraCrypt adds enhanced security and makes partitions encryption immune to brute-force attacks. This enhanced security adds some delay only to the opening of encrypted partitions but without any performance impact to the application use phase. This is acceptable to the legitimate owner but makes it much harder for an attacker to gain access to the encrypted data.

Encrypted Disk Detector

Encrypted Disk Detector (EDD) is a free command-line tool used to quickly and non-intrusively check for encrypted volumes on a computer system during incident response. This could help the forensic investigator to decide whether a live acquisition needs to be made in order to secure and preserve the evidence that would otherwise be lost if the plug was pulled.

EDD checks the local physical drives on a system for TrueCrypt, PGP, or Bitlocker encrypted volumes. If no disk encryption signatures are found in the MBR, EDD also displays the original equipment manufacturer ID (OEM ID) and, where applicable, the volume label, for partitions on that drive, and checking for Bitlocker volumes.

7.9 FORENSIC TOOLS FOR PASSWORD RECOVERY

The following are some of the tools that facilitate the recovery of passwords:

Passware Kit Forensic

Passware Kit Forensic is the complete electronic evidence discovery solution that reports all password-protected items on a computer and decrypts them. It reduces the time spent on recovering passwords, improves recovery rates, and gets more control over the password recovery process. Some of its features are as follows:

1. Recovers passwords for more than 200 file types and decrypts hard disks providing an all-in-one user interface
2. Scans computers and the network for password-protected files
3. Acquires memory images of the seized computers
4. Retrieves electronic evidence in a matter of minutes from a Windows desktop search database
5. Supports distributed password recovery
6. Runs from a USB thumb drive and recovers passwords without installation on a target PC

ElcomSoft

ElcomSoft offers GPU-accelerated password recovery and decryption tools and supplies a range of mobile extraction and analysis tools for iOS, Android, etc. It offers a range of products, which are listed here:

ElcomSoft Password Recovery Bundle (Forensic Edition) comes with all the password recovery tools in a single value pack. It helps to unlock documents, decrypt archives, and break into encrypted containers.

ElcomSoft Distributed Password Recovery breaks complex passwords, recovers encryption keys, and unlocks documents in a production environment.

Elcomsoft Mobile Forensic Bundle can perform physical, logical, and over-the-air acquisition of smartphones and tablets, break mobile backup passwords and decrypt encrypted backups, and view and analyse the information stored in mobile devices.

Elcomsoft Cloud eXplorer extracts everything from the Google Account, downloads users' location history, contacts, Hangouts messages, Google Keep, Chrome browsing history, search history and page transitions, Calendars, images, etc.

Ophcrack

Ophcrack is a free Windows password cracker based on a time-memory trade-off using rainbow tables. This is a new variant of Hellman's original trade-off with better performance. It recovers 99.9% of the alphanumeric passwords in seconds. It comes with a GUI and can run on multiple platforms.

7.10 FORENSIC TOOLS FOR ANALYSING NETWORK

The following are some of the forensic tools which either help in the detection and analysis of an incident, or to prevent it:

Wireshark

Wireshark is the most widely used network protocol analyser. It was formerly known as Ethereal. It captures packets in real time and displays them in human readable format. It has a power LAN analyser, can access very large pcap files, generate reports, and has provisions for triggers and alerts.

Packet Tracer

Cisco Packet Tracer is a software with which the complete network can be simulated by adding and connecting different network devices.

It is a powerful network simulation program that provides a platform to experiment with network behaviour and ask 'what if' questions. As an integral part of the networking academy comprehensive learning experience, Packet Tracer provides simulation, visualization, authoring, assessment, and collaboration capabilities and facilitates the teaching and learning of complex technology concepts.

OpenVPN

OpenVPN is a robust and highly flexible tunneling application that uses all of the encryption, authentication, and certification features of the OpenSSL library to securely tunnel IP networks over a single TCP/UDP port.

Network Mapper

Network Mapper (Nmap) is a free and open-source (licence) utility for network discovery and security auditing. Systems and network administrators employ Nmap for tasks such as network inventory, managing service upgrade schedules, and monitoring host or service uptime. Nmap uses raw IP packets to determine what hosts are available on the network, what services (application name and version) those hosts are offering, what operating systems (and OS versions) they are running, what type of packet filters/firewalls are in use, etc. It was designed to rapidly scan large networks but works fine against single hosts as well. Nmap runs on all major computer operating systems, and official binary packages are available for Linux, Windows, and Mac OS X. In addition to the classic command-line Nmap executable, the Nmap suite includes tools such as an advanced GUI and results viewer (Zenmap), a flexible data transfer, redirection, and debugging tool (Ncat), a utility for comparing scan results (Ndiff), and a packet generation and response analysis tool (Nping).

Firewall

It is a tool that employs traceroute-like techniques to analyse IP packet responses to determine gateway ACL filters and map networks. It sends packets to a system behind the firewall with incremental TTL values to determine what ports on a packet-filtering device are open. Firewalk does not have a Windows equivalent but has an easy-to-use GUI.

Tripwire

Tripwire is a host-based IDS that focuses on monitoring files for changes. It has been open-source since 2000. It generates a table of hashes of the contents of all the critical system files. Periodically, the program runs through the file system, generates hashes of all of the critical system files that have been configured, and compares them against the database. If there are changes, Tripwire generates an alert indicating that the file has been changed. It is predominantly used to correlate attacks rather than network traffic analysis. This is because when an attacker gains entry into the system through a backdoor for remote access, for example, there will be changes in configuration. Tripwire can raise alerts necessary to provide correlation.

Snort

Snort is an open-source network security and protection tool developed for UNIX and now applicable for Win32. It is an intrusion detection/prevention system that is capable of real-time traffic analysis and packet logging. It is a command line tool to watch network traffic, looks for rule-based intrusion signatures, alerts, and logs when a match is made, performs protocol analysis, troubleshoots the network, and controls unauthorized applications. This tool leaves only a small memory footprint and requires very little processing power. It can listen to all the traffic to one computer or put the network adaptor in promiscuous mode and listen to all the traffic on the wire.

NetAnalysis

NetAnalysis is a Windows-based application for the forensic recovery and examination of browser history and cache data. It is a state-of-the-art application for the extraction, analysis, and presentation of forensic evidence relating to the Internet browser and user activity on computer systems and mobile devices.

Alicence for HstExis was issued while buying NetAnalysis. HstEx is an advanced professional forensic data recovery solution designed to recover browser artifacts and Internet history from a number of different source evidence types. HstEx supports all the major forensic image formats (such as EnCase e01, ex01, or flat file dd).

NetAnalysis is designed for performing forensic analysis of web browsers and supports all the major desktop and mobile browsers. It supports the analysis of history, cache, cookies, and other artifacts with reporting capabilities to produce evidence related to user activity. The software also has analytical tools to decode and understand the data. Searching, filtering, and identifying items of interest/evidential value is easy with NetAnalysis because of searching and filtering features. An offline HTML5-compliant viewer is included which is capable of displaying cached web pages, videos, images, and other contents; it can also play audio files. The NetAnalysis v2 reporting suite offers reporting, data analysis, and visualization. The report manager provides the capability to save a report template to file and then re-use, as and when required.

7.11 FORENSIC UTILITY FOR METADATA PROCESSING

The following are some of the tools available for working with metadata:

PhotoMe

PhotoMe is an easy-to use utility that unlocks and organizes a photograph's metadata. Depending on the camera used for acquisition, metadata is automatically added to a photo when it is captured and includes details about how the picture was taken (camera model, f-stop, shutter speed, ISO value, colour model, colour temperature, lens used, and other settings), where (if the camera has a GPS module), and when. Metadata fields serve copyright status, copyright owner, caption, how the picture has been used and by whom, licensing issues, keywords, categories, and much more.

PhotoMe supports EXIF and IPTC/NAA (the major metadata standards). At default settings, PhotoME displays literally scores of fields containing valuable information about pictures.

Metadata Assistant

Metadata Assistant from PayneGroup is a security tool for removing electronic metadata. It offers a new user interface that can blend with Microsoft Office. It is compatible with the Microsoft family of operating systems, can be integrated with document management systems and email programs, and supports PDF cleaners.

Metadata Assistant provides quick access to the Worklist for document selection. Another key component is Outlook integration which provides the user the ability to remove metadata from all outgoing attachments sent using Outlook.

7.12 MISCELLANEOUS TOOLS

The following are the forensic tools used during various stages of forensic investigation to accomplish specific tasks:

CD/DVD Inspector

CD/DVD Inspector is an optical media analysis tool from InfinaDyne. It is a freely downloadable professional software used for intensive analysis and extraction of data from CD-R, CD-RW, and all types of DVD media, including HD DVD and Blu-Ray.

CD/DVD Inspector reads all major CD and DVD file system formats including ISO-9660, Joliet, UDF, HSG, HFS, etc. When the disc being examined contains more than a single file system, all file systems found are displayed.

CD/DVD Inspector includes a flexible report generator which can be tailored to specific requirements.

Mandiant Redline

Mandiant's Redline is a handy utility that allows the detection of newly released viruses and other types of malware that are likely to be missed by standard antivirus solutions. It provides a host of investigative capabilities to users to find signs of malicious activity through memory and file analysis and the development of a threat assessment profile. This utility is designed to target executable files that are not signed or verified and permits browsing of files, directories, processes, registry keys, semaphore, mutant, event, and sections associated with the process.

For advanced users, this can help differentiate between a false positive or an actual threat that made its way into the system. In addition to the said files, the tool also allows preview of the strings within each process space along with the network connections it has opened. For example, if a malware is present on the computer and the antivirus is updated, it is likely that it gets quarantined before any real damage can be done. On the other hand, if the system gets attacked by a brand new specimen of malware, there is high chance that it can be found using Mandiant Redline.

Redline facilitates the following:

1. It thoroughly audits and collects all running processes and drivers from memory, file-system metadata, registry data, event logs, network information, services, tasks, and web history
2. Redline's Timeline functionality, namely the TimeWrinkle and TimeCrunch features, can be used to analyse and view imported audit data. It also includes the ability to filter results around a given timeframe.
3. Streamline memory analysis with a proven workflow for analysing malware based on relative priority.
4. Identify the processes most worth investigating using the Redline Malware Risk Index score.

Filter_I

Filter_I is an intelligent fuzzy-logic filter used to filter out data that makes no sense during the analysis of ambient computer data. This saves days spent on evidence processing. It is a pattern recognition tool that relies on pre-programmed artificial intelligence to recognize patterns of text, especially to identify fragments from email, word documents, encryption, and network passwords, thereby identifying potential passwords and leads. It also reveals security leakages and violations in policy which are useful in computer security reviews.

GetFree

GetFree is a tool marketed by New Technologies Inc. (NTI), the maker of SafeBack to recover files in the unallocated space. It is specifically designed for law enforcement and forensic specialists to capture data in the unallocated space on computers running the Microsoft family of operating systems.

GetSlack

This tool from NTI is used to capture the file slack contained on the hard drives of the Microsoft family of operating systems which are further analysed with standard computer utilities or specialized NTI tools. Since memory dumps are stored in slack (encryption passwords are stored in file slack as memory dump) it raises security concern. File slack can also contain evidence in the form of fragments of word documents, email communications, and other browsing and social networking activity.

GetTime
This program is used to document the CMOS system date and time associated with allocated files and previously deleted files immediately after the seizure of evidence. However, except for low battery power, the reliability of date and time correspond to the seized system's settings for date and time.

NTI-DOC
NTI-Doc is a forensic documentation tool used to record electronic snapshots of files and subdirectories which includes file dates, times, and attributes that contribute as evidence. The documentation created by this program can be printed, viewed, and pasted into forensic reports.

Seized
This tool limits access to computers by locking and thereby securing the evidence seized which otherwise would result in overwriting of evidence stored in Swap files and in the erased file space in Windows. This program is designed to be installed on a DOS system and AUTOEXE.BAT is configured to call it. Once called, the program locks the computer and displays a warning message on the screen advising the user that the computer contains evidence and so the user should not operate without authorization.

TextSearch Plus
It is one of the tools of NTI Tool Suite and is also available as a standalone tool. TextSearch Plus is used to quickly search hard disk drives and other media for occurrences of keywords and specific patterns of text in the data stored in files, file slack, and unallocated file space. It operates either at the logical or at the physical level. It can quickly search huge hard disk drives. It works on the Microsoft family of operating systems. The keyword to be searched should be directly entered while starting the program. Depending on the configuration settings by the user, TextSearch Plus will either stop after finding a keyword or complete the search and finally list all occurrences of the keyword.

AnaDisk Diskette Analysis Tool
AnaDisk is a specialized floppy diskette analysis tool for use in security reviews to identify data storage pattern anomalies. It is an important tool for a forensic investigator to analyse a system for file anomalies and odd formats, tracks, and extra sectors. It also helps to identify data-hiding techniques which can otherwise go unnoticed.

CopyQM Plus
CopyQMPlus is a diskette duplication software wherein diskettes are formatted, verified, and copied in a single pass. It copies files, file slack, and unallocated storage space. It also creates a self-extracting executable program tied to specific diskette images which when run will restore the diskette image on to the diskette. Such disk images are password protected which are useful when disks are shared over the Internet where security is a concern. This is useful for forensic investigators when CDs have to be preconfigured for specific uses and duplicated.

DiskScrub Data Overwrite Utility
DiskScrub is a data overwrite utility that is used to eliminate all traces of data from a disk drive. The data overwritten cannot be recovered using any recovery or computer forensics software. The process involves writing data on the hard drive from the first sector to the last sector.

DiskSearch Pro
This is a command line text search engine from NTI. It can simultaneously search upto 250 keywords from active files, free, and unallocated space by employing fuzzy logic technology. However, the keyword search is restricted to MS-FAT and NTFS file systems.

Net Threat Analyzer

Net Threat Analyzer (NTA) is an Internet usage analysis and Internet threat identification tool from NTI. It is a command line search tool designed to detect text strings related to Internet usage including email, web browsing, and file downloads. It relies on artificial intelligence and fuzzy logic to automatically identify file downloads from the Internet, and Internet communications involving pornography, narcotics, bomb making, sex crimes, and hate crimes.

M-Sweep Data Scrubber

M-Sweep from NTI removes remnants of deleted files by overwriting disk space that is not used by current files. It can do so with hard drives lesser than 8GB. It works by cleaning out the slack space first, followed by cleaning unused, that is, unallocated or erased space that once held complete files on the current volume.

TeleDisk

This is a diskette imaging tool from Safeback.

FileCNVT

FileList Conversion Utility (FileCNVT) is a forensic tool that is used to quickly catalogue the contents of one or more computer hard disk drives.

DM

DM is a database analysis tool that is compatible with dBASE III file structure. It can be imported to other database file structures and spreadsheet applications.

Palm dd

Palm dd (pdd) is a free Windows-based forensic tool for Palm OS, memory imaging, and forensic data acquisition of Palm OS-based family of PDAs. It is a command line driven application and does not support graphics libraries, report generation, and search facility. The console mode acquires memory card information and a bit-to-bit image of a selected memory region. As a result, two files are generated—a file containing device information and the other, the bit image. Those files can be imported to other forensic tools such as Encase or Hexeditor for analysis. The data retrieved includes all user applications and databases. Such data are useful for forensic investigators, incident response teams, and criminal and civil prosecutors.

WinHex

WinHex is a forensic software from X-Ways that is used for data recovery. Besides this, it is a hex editor and disk editor.

SIFT

SANS Investigative Forensic Toolkit (SIFT) is a computer forensics VMware appliance that is pre-configured with all necessary tools to perform a thorough digital forensic examination. SIFT can securely examine raw disks, multiple file systems, and evidence formats.

X-Ways Forensics

X-Ways Forensics is a portable work environment that can run from a USB stick without installation. It offers the features of many forensic tools and can run under the Microsoft family of operating systems. It provides facilities for the examiners to share their data and collaborate with investigators.

dtsearch

dtSearch from dtSearch Corporation facilitates searching terabytes of text from the image of the suspect media.

QuickView Plus FileViewer

QuickView Plus is a file viewer that facilitates viewing parts of the file and searches for text strings within files. It supports viewing of files in over 300 formats.

Paraben Porn Stick

This tool is distributed on a thumb drive with preloaded software that searches a target computer for pornographic images.

Snagit

Snagit from TechSmith Corporation is used to capture and manage screenshots.

7.13 FORENSIC TOOLS FOR UNIX SYSTEM ANALYSIS

The following are some of the commands and tools which help in acquiring digital evidence from Unix systems:

Linux dd Command Utility

This tool is used for copying data and is also called data duplicator. It can be used for backing up and restoring an entire hard drive or a partition, creating a virtual file system to back up images on a CD and DVD as image files, and for copying regions of raw device files like backing up the MBR. It takes two inputs. The syntax for dd command is as follows:

```
dd if = <source file name> of = <target file name> [Options]
```

where 'source file name' represents the input file and 'target file name' represents the output file.

The functionalities of this tool are explained here:

Backing up and restoring entire hard drive or partition The data from one disk can be copied in its entirety to another disk without any regard to the file system and partitions on the disk. However, the disk in which the output data is written should either be of the same size or larger than the disk containing the input data. For example,

```
dd if = /dev/sda  of = /dev/sdb  bs=4096 conv=noerror, sync
```

copies data from disk /dev/sda to disk /dev/sdb. 'bs' refers to the block size to be used while reading/writing is specified explicitly and is 512 bytes by default. Block size should always be in multiples of 1024. Parameters 'conv' instructs the tool to copy data even if any errors are encountered and 'sync' allows the use of synchronized I/O.

A disk image or a file image facilitates faster copying of the exact data besides making restoration easier. For example,

```
dd if = /dev/sda  of = /tmp/sdadisk.img
```

where, sdadisk.img is the resultant image file with .img extension.

The disk image can be compressed if the exact content of the disk occupies much space. This is done as

```
# dd if = /dev/sda | gzip -c > /tmp/sdadisk.img.gz
```

where | command redirects the input file to be copied and -c writes the output on standard output without changing any files.

In order to restore a partition, the operation is performed in reverse.

```
# dd if = /tmp/sdadisk.img  of = /dev/sda
```

The compressed image can be restored as follows,

```
# gzip -dc /tmp/sdadisk.img.gz | dd of=/dev/sda
```

Creating virtual file system to back up images of CD and DVD as image files A virtual file system is a file system that exists as a file on the physical disk. For example,

```
# dd if = /dev/sda of = /file bs=1024 count = 500
```

creates a virtual file system on the disk named 'file' which is of the same size as the input file /dev/sda but with no data. 'count' specifies the number of blocks to be copied and the size of every block is specified explicitly with 'bs' parameter.

For example, backing up of the contents of a CD sector by sector to a directory can be done as follows:

```
# mount -o loop /cd.iso /mnt/cd
```

where the .iso can be mounted as follows:

```
# dd if = /dev/cdrom of = /cd.iso
```

where, the -o loop option enables mounting of the file like a normal device.

Backing up and restoring of MBR MBR corresponds to the first 512 bytes of data where the bootloader occupies almost 466 bytes of storage and the remaining is occupied by a partition table for that drive. So as to copy the contents of MBR, 'bs' = 512 bytes and count = 1, which means only one block is copied. This is shown here:

```
# dd if = /dev/sda of = /tmp/sdadev.imgbs = 512 and count =1.
```

Restoring the data is, however, similar to that of restoring the contents to disk.

The Coroners Toolkit

The Coroners Toolkit (TCT) is a collection of programs by Dan Farmer and Wietse Venema for digital forensic analysis of a UNIX system after break-in. Some of the important components of the tool include the grave-robber tool that captures information, the ils and mactime tools that display the access patterns of files dead or alive, the unrm and lazarus tools that recover deleted files, and the findkey tool that recovers cryptographic keys from a running process or from files. Parts of TCT can be used to aid analysis of and data recovery from other computer disasters. The suite runs under several Unix-related operating systems: FreeBSD, OpenBSD, BSD/OS, SunOS/Solaris, Linux, and HP-UX.

7.14 FORENSIC TOOLS FOR OTHER MEDIA

In general, any ECD that comes with a storage component necessitates the use of forensic tools to gather evidence, if the said device is used either as a tool or as a target of a crime. The following are the ways forensic tools are used to acquire evidence from some of the media:

USB Drives

It can be imaged using drive imaging tools. Prior to this, software- or hardware-based write-blockers were used to connect to the forensic machine.

Digital Cameras

The memory card and the internal memory in the camera can be acquired using the technique used for USB drives.

Non-detachable Hard Disk Drives

Laptops and some network attached printers and CD/DVD duplicators have hard drives that are not easily removable. In such a situation, the entire device becomes evidence. Imaging can then be done using network acquisition wherein the evidence computer is connected to the forensic computer using a special ethernet cable called 'cross cable' or 'network crossover cable'. On a crossover cable, on one end, only the positive and negative 'receive' pair are switched with the positive and negative 'transmit' pair respectively with regard to the positive and negative to maintain polarity. This facilitates machines to talk with each other over the network crossover cable. Once the computers are connected, the evidence computer is booted from a forensic distribution like Helix or Linen and data can be acquired like a regular hard drive acquisition from the evidence computer to the forensic computer using forensic tool like Encase.

7.15 FORENSIC HARDWARE

This section presents some of the forensic hardware from different vendors.

The following are the forensic hardware from *Digital Intelligence*:

Forensic Systems

The Forensic Recovery of Evidence Device (FRED) family of forensic workstations are highly integrated, flexible, and modular forensic platforms and now include DI's exclusive UltraBay 4d Write Protected Imaging Bay. This offers the ability to easily duplicate evidence directly from IDE/SAS/SATA hard drives, USB devices, Firewire devices, CDs, DVDs, LTO-4 tapes and PC Card/smartmedia/SD-MMC/memory stick/compact flash media in a forensically sound environment. It is available in mobile, stationary, and laboratory configurations. These systems are designed for both the acquisition and examination of computer evidence.

FRED Forensic Network

A forensic network is a series of processing and imaging computers directly connected and integrated with a high-speed, high-capacity server to share resources. The file server operates as the core of the forensic network. It can be used as a central storage facility for forensic images as well as applications software for use by the client processing and imaging stations. Workstation clients on the network perform the actual imaging and processing tasks, whereas the central file server stores the images and case works.

Forensic Write Blockers

Digital intelligence designs and offers parallel IDE, serial ATA, USB, Firewire, and SAS hardware write blockers as well as other custom solutions to maintain the integrity of data.

Besides these, standalone forensic devices, signal and power cables, accessories such as hard drive trays, cables, and other products account for forensic hardware.

Forensic Computers Inc.

Forensic Computers Inc. came up with products such as desktop workstations, mobile workstations, forensic duplicators, forensic bridges, cables, drive adapters, accessories, and ultimate kits.

7.16 FORENSIC ANALYSIS TOOLS FOR MOBILE DEVICES

This section presents the various tools available, both free and open source, as well as the proprietary tools used for acquiring evidence from mobile devices. The forensic hardware for mobile devices is also presented.

7.16.1 Free and Open-source Forensic Tools for Mobile Devices

This section presents some of the free and open source forensic tools available for gathering and analysing digital evidence from mobile devices.

BitPim

BitPim is distributed as open-source and free software under the GNU General Public Licence. It is a program that facilitates the viewing and manipulation of data on many CDMA phones from LG, Samsung, Sanyo, and other manufacturers. The data includes the phone book, calendar, wall papers, ring tones, and the file system for most Qualcomm CDMA chipset-based phones.

Mobile Phone Examiner Plus

Mobile Phone Examiner Plus (MPE+) is a standalone mobile device investigation solution that includes enhanced smart device acquisition and analysis capabilities. With a different approach to digital mobile forensics, MPE+ provides unique tools necessary to quickly collect, easily identify, and effectively obtain key data that other solutions miss, thereby facilitating mobile forensic examiners to take control of the investigation. Some of its features are as follows:

1. It supports 10,000 plus mobile devices.
2. It allows physical extractions of Android devices, with password bypass capabilities.
3. It permits the logical extraction of Android devices without the need to know the manufacturer or model.
4. It is possible to physically extract iOS devices and logically extract iOS devices without the need of iTunes.
5. Utilization of the iLogical and dLogical data collection capabilities to enable logical extraction of iOS and Android devices is 30% faster than that of leading competitors
6. SQLBuilder allows examiners to parse the data of all applications containing SQLite database.
7. pythonScripter provides users with the ability to build python scripts to parse anything from a mobile device with an easy-to-use interface. Unlike other solutions, MPE+ pythonScripter acts upon a copy of the evidence, not the original binary or evidence file.
8. Advanced analytics with graphical data visualization allows users to get a timeline view of device communication that can be customized to the year, month, day, or hour, quickly filtering relevant data.
9. Output all the data or selected data in several different report formats with customizable reports.
10. Advanced alert manager gives the examiner the ability to import or create predefined words, phrases, numbers, etc. This in turn, searches for on the entire data set, helps to uncover key evidence in a matter of minutes. Alerts can be run per case or for all collections, helping to triage large datasets for specific criteria.

MOBILedit Forensic Express

MOBILeditForensic Express is an all-in-one solution including a phone and cloud extractor, data analyser, and report generator. It uses multiple communication protocols and advanced techniques to get maximum data from each phone and operating system. Data from a phone that can be extracted with Forensic Express includes deleted data, call history, contacts, text messages, multimedia messages, photos, videos, recordings, calendar items, reminders, notes, data files, passwords, and data from apps such as Skype, Dropbox, Evernote, Facebook, WhatsApp, Viber, Signal, WeChat, and many others. It combines all the data found, removes any duplicates, and presents it all in a complete, easily readable report.

A powerful 64-bit application using both the physical and logical data acquisition methods, Forensic Express is excellent for its advanced application analyser, deleted data recovery, wide range of supported phones including most feature phones, fine-tuned reports, concurrent phone processing, and easy-to-use user interface. The password and PIN breaker offers the capability to gain access to locked ADB or iTunes backup with GPU acceleration and multi-threaded operations for maximum speed.

SIMCon

SIMCon is a program that allows the users to securely image all files on a GSM/3G SIM card to a computer file with the SIMCon forensic SIM card reader. The investigator can then analyse the contents of the card including stored numbers and text messages. It is used by investigators for securing evidence on SIM cards and presenting them in court. It is available only in Microsoft Windows. Some of the features of SIMCon include the following:

1. It can read all available files on a SIM card and store it in an archive file.
2. It analyses and interprets contents of files including text messages and stored numbers.
3. It recovers deleted text messages stored on cards which are not readable on phones.
4. It manages PIN and PUK codes (Box 7.2).
5. It facilitates printing of reports that can be used as evidence based on user selection of items.

6. It secures the file archive using MD5 and SHA1 hash values.
7. It supports international charsets.

AFLogical
AFLogical is an Android forensic tool developed by **viaForensics**. This tool performs logical acquisition of any Android device running either Android 1.5 or later versions. It allows the extracted data to be saved to the examiner's SD card in CSV format. There are two editions of this tool: AFLogical Open Source Edition (OSE) and AFLogical Law Enforcement (LE).

AFLogical Open Source Edition is a free and open-source software. It pulls out all available MMSes, SMSes, contacts, and call logs from an Android device.

7.16.2 Proprietary Forensic Tools for Mobile Devices
The following are some of the proprietary forensic tools available for the forensic analysis of mobile devices:

XAMN
XAMN is an analytical tool from MSAB designed specifically for visualizing mobile device forensic files from multiple handsets in one simple and easy to use interface. XAMN allows users to view the contents of upto 50 different XRY files in one place, to compare data from different devices simultaneously, and to look for connections because it is not just about the data, but about the meaning.

XAMN allows phone examiners with access to relatively low-powered computers to get high-powered analysis. With a unique ability to directly import native XRY files and import other files, XAMN allows the comparison and contrast of multiple phone files. It helps with a quicker understanding and assessment of all the information in a flexible, intuitive software package.

XAMN is optimized for normalization of times and dates from phones and quicker searching of multiple files for critical information. Some of its features are as follows:

1. Creation of a combined list view of all the data from multiple handsets
2. Creation of link analysis–connection view
3. Creation of three different types of timelines: overview, individual, and calendar
4. Creation of map views using geo-data
5. Visualization of conversations from smartphone apps
6. Addition of user data and tag phone data to aid understanding
7. Performing of language translations
8. Search for data across 50 XRY files, simultaneously

Oxygen Forensic Detective
Detective is a suite that provides more advanced forensic software to extract and analyse data from cell phones, PDAs, smartphones, and other mobile devices. It supports a wide variety of mobile devices. It uses proprietary low-level protocols to extract data from smartphones. It can be used to import backup or image files obtained using other forensic tools.

> ### Box 7.2 PIN and PUK Code
> A personal unblocking code (PUC), also known as personal unlocking key (PUK) is used in 3GPP mobile phones to reset a personal identification number (PIN) that has been lost or forgotten. The PIN and PUK codes are intended to protect SIM cards from unauthorized use. The blocking of the card with the PIN code can only be overridden by the use of a PUK code. A card blocked by the PUK code cannot be unblocked and cannot be used further but has to be replaced.

This tool extracts data, for example, phonebook with assigned photos, calendar events and notes, call logs, messages, camera snapshots, videos and music, voicemails, passwords, dictionaries, geo-positioning data, Wi-fi points with passwords and coordinates, IP connections, etc.

Some of the other features of this tool include the following:

1. It supports logical acquisition, file system, and physical acquisition.
2. It supports password recovery.
3. It aids in cloud data extraction and decryption.
4. It reads backup or images obtained using forensic tools.
5. It provides rooting and jail breaking evidence.
6. It can recover deleted data.
7. It recovers deleted data automatically.
8. It provides access to raw files for manual analysis.

Paraben's Device seizure

Paraben's Device Seizure is a software package that assists forensic investigators in the examination of mobile devices, allowing data acquisition, both logically and physically. Once data acquisition is completed, it is easy to export the report to a wide variety of formats to view the organized results.

Device Seizure is a forensic acquisition and analysis tool. Device Seizure from Paraben is used to perform logical and physical forensic extractions of mobile phones and device data such as iPhone, iPads, Androids, and BlackBerry. It does not change data on the device, and supports cell phone manufacturers, namely LG, Motorola, Nokia, Siemens, Samsung, and Sony Ericsson. It supports GSM SIM cards with the use of SIM card reader. The tool can acquire the following types of data: SMS messages, including deleted messages, phonebook information, call history, received calls, dialled numbers, missed calls, call dates and durations, datebook information, scheduler, calendar, to-do list information, file system, system files, multimedia files, Java files, deleted data, quicknotes, GPS, RAM/ROM, PDA databases, email, and registry data.

XRY

XRY from Micro Systemation is a software application designed to run on the Windows operating system and allows secure forensic extraction of data from a wide variety of mobile devices such as smartphones, GPS navigation units, 3G/4G modems, portable music players, and the latest tablet processors like the iPad.

7.16.3 Forensic Hardware for Mobile Devices

The following are some of the mobile forensic hardware:

CellDEK

CellDEK is a registered trademark of Forensic Science Service Ltd. It is built to perform on field (not just in the lab) so that investigators can immediately gain access to vital information, thereby eliminating the need to wait for a report from a crime lab. It is portable and compatible with over 950 of the most popular cell phones and PDAs. This advanced cell phone data extraction device is a self-contained system that features a touch-screen display allowing the user to quickly identify devices by brand, model number, dimensions, and/or photographs.

Once the device is identified, connectivity options are offered, suggesting the most suitable options. A 'smart adapter' feature then illuminates the correct USB adapter. Connectivity by infrared and Bluetooth is also built-in. The CellDEK then captures all stored data within five minutes. Upto 40 adapters may be stored in the system's built-in rack.

The CellDEK software automatically performs forensic extraction of the following data: handset time and date, serial numbers (IMEI, IMSI), dialled calls, received calls, phonebook (both handset and SIM), SMS (both handset and SIM), deleted SMS from SIM, calendar, memos, to do lists, pictures, videos, and audios. CellDEK displays the data on the screen and prompts for downloading to a portable USB device.

Cellebrite

Cellebrite offers the universal forensic extraction device (UFED) line of software and hardware to copy as much data as possible from seized smartphones. The Cellebrite UFED system is the only hand-held, cellular exploitation device worldwide that requires no PC or associated phone drivers. The UFED system will quickly extract phonebook, pictures, videos, SMS messages, call histories, ESN/IMEI information, and deleted SMS/call histories of the SIM/USIM for rapid analysis. Cellebrite supports all the major technologies (TDMA, CDMA, GSM, IDEN) including smartphone operating systems and PDAs (Apple Phone, Blackberry, Google Android, Microsoft Mobile, Palm, and Symbian) for over 95% of all handset models worldwide.

Cellebrite UFED offers several products that support data acquisition and analysis of Android devices. It is a popular commercial tool that provides the examiner with both logical and physical acquisition support as well as an analytical platform to examine data. Cellebrite Physical Analyzer, the analytical platform, allows the examiner to perform keyword search, bookmark, carve data, and create customized reports to support their investigation.

Cellebrite recently released a new product called UFED Cloud Analyzer which allows users to use authentication codes and passwords saved by mobile apps to automatically log in to Gmail, Google Drive, Facebook, Twitter, Dropbox, and Kik. Cloud Analyzer can also download emails, message history, files, and contact lists as available. Cellebrite claims that UFED Cloud Analyzer acts like these providers' apps by using their application programming interfaces (APIs) to access data.

7.17 FORENSIC TOOLS FOR EMAIL ANALYSIS

The tools listed in Table 7.3 help to identify the origin and destination of the message, trace the path traversed by the message, identify spam and phishing networks, etc. They provide information in an easy-to-use browser format, automated reports, and other features which serve as evidence.

Table 7.3 Tools for acquisition of evidence from emails

Tool	Use
Aid4Mail	This proprietary tool is used to analyse emails stored in hard disks. It supports email filtering based on text, time, date, keywords, logical operators, and regular expressions. Searches can be made by date, header content, and by message body content. Aid4Mail also offers the ability to process unpurged (deleted) email from mbox files and can restore unpurged email during exportation. It supports both offline and online modes of analysis.
Digital forensics framework (DFF)	This open-source framework is used to analyse emails stored in hard disks while it can also perform virtual machine disk reconstruction. For each message, the header information can be displayed and their filtering is possible with the use of regular expressions and based on the email content, tags, and timeline. It can support various formats allowing interoperability with other tools.
eMailTrackerPro	This can be used to analyse email files stored in the local disk and supports automatic email analysis for the identification of spamming incidents. It is capable of recovering the IP address from where the message is sent along with its associated geographical location (city) to determine the validity of an email message. It can find the ISP of the sender. A routing table is provided to identify the path between the sender and the receiver of an email. It can also check a suspected email against domain name server blacklists to safeguard against spam and malicious emails. It also displays whether any port is open in any of the HTTP or FTP servers in the tracked IP addresses.
Paraben Email Examiner	This tool can recover deleted emails from exchange (EDB), lotus notes (NSF), and groupwise email even if it is deleted from the deleted items folder.

(*Contd*)

Table 7.3 (Contd)

Tool	Use
EmailTracer	This tool developed by Resource Centre for Cyber Forensics (RCCF), India, traces the originating IP address and other details from the email header, generates a detailed HTML report of email header analysis, finds the city-level details of the sender, plots the route traced by the mail, and displays the originating geographic location of the email. Besides these, it has keyword searching facility on email content including attachment for its classification.
Adcomplain	This tool is used to analyse both the header of the message and the body. It automatically analyses the message, composes an abuse report, and mails the report to the offender's Internet service provider by performing a valid header analysis.
MailXaminer	This tool by SysTools allows the examination of messages from both web- and application-based email clients. It loads messages from the chosen email storage source and arranges them hierarchically for the purpose of evidence analysis and extraction. It could carve out deleted evidence or evidence from damaged sources in cases of evidence spoliation.
AbusePipe	This tool analyses abuse complaint emails and determines which of the ISP's customers is sending spam based on the information in the emailed complaints. AbusePipe can be configured to automatically respond to people reporting abuse. It can assist in meeting legal obligations such as reporting on the customers connected to a given IP address at a given date and time.
Internet Evidence Finder (IEF)	This tool by Magnet Forensics allows the definition of specific profiles for the recovery and detection of emails contained on physical drives. Once a search is completed, the results are shown and for each recovered email the investigator may observe details related to the artifact, header information, as well as the email body and attachments.
FINALeMAIL	This tool has the capability to restore lost emails to their original state. It can also recover full email database files. This is useful when such files are attacked by viruses or are damaged by accidental formatting.
Forensics Investigation Toolkit (FIT)	This is used to read and analyse the content of the Internet raw data in Packet CAPture (PCAP) format. It is used to perform content analysis and reconstruction on pre-captured Internet raw data from wired or wireless networks. The uniqueness of the FIT is that the imported raw data files can be immediately parsed and reconstructed. It supports case management functions, detailed information including date-time, source IP, destination IP, source MAC, WhoIS, and Google Map integration functions. Analysing and reconstruction of various Internet traffic types which include email (POP3, SMTP, IMAP), webmail (read and sent), IM or chat, file transfer (FTP, P2P), telnet, and HTTP (content, upload/download, video streaming, request).

7.18 NEED FOR COMPUTER FORENSIC INVESTIGATORS

Any organization that relies on an ECD may require a digital forensic investigator. Typically, government organizations, accounting firms, law firms, banks, and software companies require one. The following are some of the domains which demand computer forensic investigators:

Law enforcement officials seek the assistance of forensic investigators for pre-search warrant preparation and handling of the digital evidence post seizure.

Legal professionals require forensic investigators for cases such as divorce, fraud, and harassment involving digital evidence to be obtained from ECDs. Criminal prosecutors require forensic investigators to collect evidence in cases of financial fraud, pornography, etc.

Insurance companies look for forensic investigators to collect computer evidence in the case of frauds related to accidents and to arrive at a compensation.

Corporate human resources professionals demand forensic investigators to obtain evidence in the case of events such as theft and misappropriation of trade secrets.

System administrators and security consultants providing incident response services expect forensic investigators to determine the source of the incident and to respond to it.

Private investigators require forensic investigators to support claims of sexual harassment and age discrimination.

7.19 CAREER PROSPECTS FOR FORENSIC INVESTIGATORS

Computer forensics investigators play the role of gathering digital data from investigating ECDs and presenting information for legal cases to figure out and communicate how a crime has been committed using it. Besides this, he/she protects the computer system, recovers files (including those that were deleted or encrypted), analyses data, and provides reports, feedback, and even testimony when required. The career prospects for a forensic investigators include the following:

1. Computer forensics analyst
2. Computer forensics investigator
3. Computer forensics specialist
4. Computer forensics technician
5. Digital forensics specialist
6. Forensic computer examiner

The careers in computer forensics can even be in one of the following areas:

Local law enforcement The police department in every state has a crime division which requires a forensic examiner who should possess a degree with a major in law enforcement or a similar field.

Federal law enforcement A forensic position in this case would involve working on high-profile cases of national or international interest, and therefore he/she would require a high level of expertise besides possessing a degree and appropriate certification.

Armed forces Armed forces require computer forensic experts for special investigations due to the overlap of computer crime incidents with civilian matters.

Security department Almost in every department where security is a primary concern, new positions and departments are created, wherein full-time computer forensic staff are employed. Besides this, every company/organization that uses ECDs, constitute a forensic department and hire people to look after the security aspects.

However, the requirements for a forensic investigator include either a degree in computer forensics or any degree with experience in a related field. Additional certifications may help in finding suitable positions in reputed organizations. Besides this, being up to date with the latest technologies, analytical and problem solving skills, and the ability to communicate verbally and in written form are essential to be a successful forensic investigator. Any work experience in the computer field or in a law enforcement agency will be an added advantage.

7.20 FORENSIC TRAINING AND CERTIFICATIONS

Some of the popular certifications that forensic investigators should have are listed here and are also shown in Fig. 7.3.

AccessData Certified Examiner

AccessData certified examiner (ACE), a vendor-specific certification, is offered by AccessData. AccessData Digital Investigations Training is designed to educate forensic professionals and incident responders on FTK BootCamp, advanced FTK, password recovery, applied decryption, collaborative computing with AD Lab and network and incident response investigation with AD Enterprise, and to use AccessData's Forensic Toolkit, AD Enterprise, and AD lab collaborative technologies.

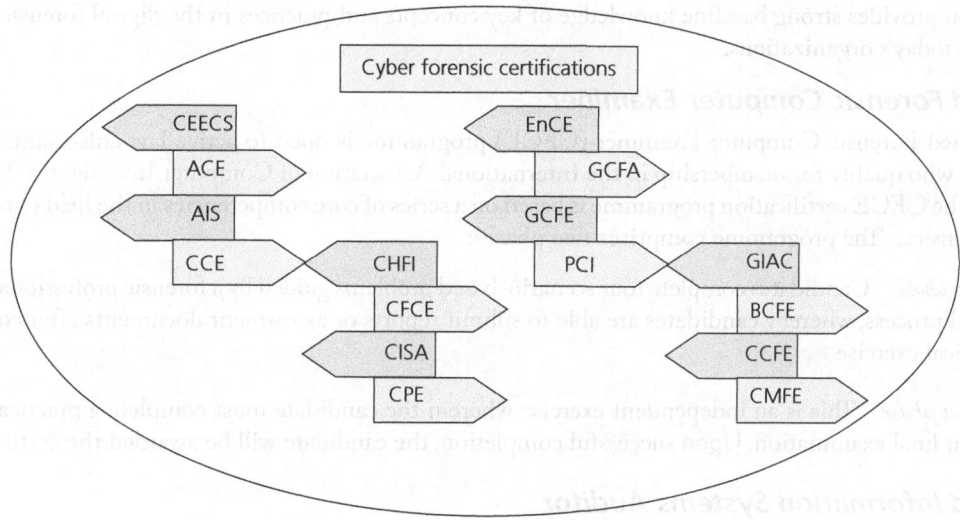

Fig. 7.2 Cyber forensic certifications

Advanced Information Security

The Security University Advanced Information Security (AIS) certification is a hands-on computer security certification created for network, IT, and security professionals. The AIS certification is designed to provide proficiency in information security methodologies to persons with an existing information technology background. Emphasis is on security risk assessment, security in Windows and Linux platforms, and information security tools.

Certified Computer Examiner

The International Society of Forensic Computer Examiners (ISFCE) is a private organization dedicated to providing an internationally recognized, computer forensics certification that is available to anyone who qualifies for it, at a reasonable cost. The CCE certification by ISFCE is widely considered to be the most prestigious non-vendor specific forensic certification available to all examiners working publicly or privately. The purpose of the Certified Computer Examiner (CCE) certification is for the following:

1. Professionalize and further the field of computer forensics.
2. Provide a fair, vendor neutral, uncompromised process for certifying the competency of forensic computer examiners.
3. Certify computer forensic examiners solely based on their knowledge and practical examination skills and abilities as they relate to the practice of digital forensics.
4. Set high forensic and ethical standards for forensic computer examiners.
5. Provide a universally recognized certification.

Certified Hacking Forensic Investigator

EC-Council's Certified Hacking Forensic Investigator (CHFI) certifies individuals in the specific security discipline of computer forensics from a vendor-neutral perspective. The CHFI certification will fortify the application knowledge of law enforcement personnel, system administrators, security officers, defence and military personnel, legal professionals, bankers, security professionals and anyone who is concerned about the integrity of the network infrastructure. The CHFI certification validates the candidate's skills to identify an intruder's footprints and to properly gather the necessary evidence to prosecute in a court of law. The CHFI certification provides strong baseline knowledge of key concepts and practices in the digital forensic domains relevant to today's organizations.

Certified Forensic Computer Examiner

The Certified Forensic Computer Examiner (CFCE) programme is open to active law enforcement officers and others who qualify for membership in the International Association of Computer Investigative Specialists (IACIS). The CFCE certification programme is based on a series of core competencies in the field of computer/digital forensics. The programme comprises two phases:

Peer review phase Candidates complete four scenario-based problems guided by a forensic professional through a mentored process, whereby candidates are able to submit reports or assessment documents after completing each practical exercise.

Certification phase This is an independent exercise wherein the candidate must complete a practical exercise and written final examination. Upon successful completion, the candidate will be awarded the certification.

Certified Information Systems Auditor

The Certified Information Systems Auditor (CISA) by the **Information Systems Audit and Control Association (ISACA)** is a globally recognized certification for IS audit control, assurance, and security professionals. The CISA certification showcases an individual's audit experience, skills, and knowledge and demonstrates the capability to assess vulnerabilities, and report on compliance and institute controls within the enterprise.

Certified ProDiscover Examiner

The Certified ProDiscover Examiner (CPE) is one of three vendor-specific computer forensic certifications such as the AccessData ACE and the EnCase Certified Examiner. This certification focuses on the ProDiscover incident response toolset. ProDiscover offers a Windows-oriented forensic application that covers collection, analysis, management, and reporting of computer disk evidence for Windows computers (FAT12, FAT16, FAT32, and all NTFS file systems including Dynamic Disk and Software RAID) as well as Sun Solaris UFS file system and Linux ext2/ext3 file systems.

EnCase Certified Examiner Programme

The EnCase Certified Examiner (EnCE) programme certifies both public and private sector professionals in the use of Guidance Software's EnCase computer forensic software. EnCE certification acknowledges that professionals have mastered computer investigation methodology as well as the use of EnCase software during complex computer examinations. Recognized by both the law enforcement and corporate communities as a symbol of in-depth computer forensics knowledge, EnCE certification demonstrates that an investigator is a skilled computer examiner.

GIAC Certified Forensic Analyst

The Global Information Assurance Certification Forensic Analyst (GCFA) certification is for professionals working in the information security, computer forensics, and incident response fields. The certification focuses on core skills required to collect and analyse data from Windows and Linux computer systems.

GCFA certifies that candidates have the knowledge, skills, and ability to conduct formal incident investigations and handle advanced incident handling scenarios, including internal and external data breach intrusions, advanced persistent threats, anti-forensic techniques used by attackers, and complex digital forensic cases.

GIAC Certified Forensics Examiner

The GIAC Certified Forensics Examiner (GCFE) certification is for professionals working or interested in the information security, legal, and law enforcement industries with a need to understand computer forensic analysis.

The certification focuses on core skills required to collect and analyse data from Windows computer systems.

GCFE certifies that candidates have the knowledge, skills, and ability to conduct typical incident investigations including e-Discovery, forensic analysis and reporting, evidence acquisition, browser forensics, and tracing user and application activities on Windows systems.

Professional Certified Investigator

The Professional Certified Investigator (PCI) credential provides demonstrable proof of an individual's knowledge and experience in case management, evidence collection, and preparation of reports and testimony to substantiate findings. Those who earn the PCI are ASIS board-certified in investigations.

SANS GIAC

GIAC certifications from SANS training ensure mastery in critical, specialized InfoSec domains. GIAC certifications provide the highest and most rigorous assurance of cyber security knowledge and skill available to industry, government, and military clients across the world. GIAC certifications build true hands-on experience, and test the pragmatics of cyber defence, penetration testing, digital forensics and incident response, developer, and information security management.

Certified Electronic Evidence Collection Specialist

The Certified Electronic Evidence Collection Specialist (CEECS) training and certification is conducted by the International Association of Computer Investigative Specialists (IACIS). The CEECS training has been incorporated into the IACIS Basic Computer Forensic Examiner (BCFE) Training programme.

Certified Computer Forensics Examiner

The Certified Computer Forensics Examiner (CCFE) by Information Assurance Certification Review Board (IACRB) tests a candidate's fundamental knowledge of the computer forensics evidence recovery and analysis process. Candidates are evaluated on their relevant knowledge of both hard and soft skills. Candidates will be tested on soft skills; they must prove that they have the requisite background knowledge of the complex legal issues that relate to the computer forensics field.

ASCLD/LAB

The ANSI-ASQ National Accreditation Board (ANAB) has signed an affiliation agreement with the American Society of Crime Laboratory Directors/Laboratory Accreditation Board (ASCLD/LAB), merging ASCLD/LAB into ANAB. Like ANAB, ASCLD/LAB provides accreditation based on international standards for public and private sector crime laboratories. Both ANAB and ASCLD/LAB are grounded in conducting scientific and technical assessments and committed to assuring competent and credible tests and inspection results.

Computer and Mobile Forensic Boot Camp

InfoSec Institute's Authorized Computer and Mobile Forensics Boot Camp (IACRB) offers IACRB Certified Computer Forensic Examiner (CCFE) and IACRB Certified Mobile Forensic Examiner (CMFE) certification examinations by imparting necessary skills to investigate computer and mobile threats and computer crimes. It provides hands-on training with data from real forensic cases.

POINTS TO REMEMBER

- Forensic tools are used by forensic investigators to acquire digital evidence from ECDs, index them, and perform a detailed analysis.
- Forensic tools are either proprietary or open-source and freely available.
- Cyber forensic tools can be classified into two: (a)

- hardware forensic tools, and (b) software forensic tools.
- Computer forensic tools can be categorized according to their working as tools for acquisition, integrity preservation, validation and discrimination, extraction, reconstruction, and reporting.
- A forensic suite has specialized tools for RAM analysis, hard disk imaging and analysis, verification and validation, registry analysis, data recovery and reconstruction, password cracking utility, and a lot more as a collection apart from tools for reporting.
- Some of the free and open-source forensic suites include OSForensics, Nirsoft, Windows Sysinternals Live, Autopsy, Computer Aided Investigative Environment (CAINE), PALADIN, Hiren's Boot CD and Digital Evidence and Forensic Toolkit (DEFT).
- Some of the proprietary forensic suites include Paraben Tools, Computer Online Forensic Evidence Extractor, and Vogon.
- Drive imaging and validation tools include Intelligent Computer Solutions Image Master Solo, Forensics Hard Drive Duplicator, Norton Ghost, Symantec Ghost, Safeback, Encase Forensic Imager, FTK Imager, ProDiscover Incident response (IR), X-Ways Forensic (XWF), DriveSpy, Forensic Replicator, SMART Acquisition Workshop (SAW), SMART, and WinHex.
- The tools used for integrity verification and hashing are HashMyFile and HashCalc, which are free and open-source; CRCMD5, DiskSig, and MD5summer are proprietary.
- Some of the forensic tools widely used for data recovery are Recuva, Byte Back, and IsoBuster.

- The free and open-source tools used for RAM analysis are Live RAM Capturer, Volatility, DumpIt, and Magnet RAM Capture.
- Tools such as Regshot and RegRipper are used for registry analysis and are free and open-source forensic tools.
- VeraCrypt and Encrypted Disk Detector are the free and open-source forensic tools available for encryption and decryption.
- Forensic tools for password recovery include Passware Kit Forensic, ElcomSoft, and Ophcrack Forensic. Tools used for network analysis include Wireshark, Packet Tracer, OpenVPN, Network Mapper (Nmap), Firewalk, Tripwire, Snort, and NetAnalysis,
- Miscellaneous tools such as GetSlack and GetTime are used for specific purposes.
- The dd utility and The Coroners Tool Kit are widely used for UNIX system analysis.
- Forensic tools such as BitPim, Mobile Phone Examiner Plus (MPE+), MOBILedit Forensic Express, SIMCon, AFLogical, XAMN, Oxygen Forensic Detective, Paraben's Device seizure, and XRY are used for the analysis of mobile devices.
- Specific tools such as AbusePipe and Forensics Investigation Toolkit (FIT) are used for email analysis.
- Forensic computer examiners are wanted for domains such as local and federal law enforcement, in the armed forces, and in security departments.
- Possessing forensic certifications such as BCFE and CFCE will be an added advantage for forensic examiners.

KEY TERMS

Acquisition This refers to creating an exact bit stream copy of the original storage media that exists on the subject's computer.

AFLogical This is an Android forensic tool which performs logical acquisition of any Android device running either Android 1.5 or later versions.

AnaDisk This is a specialized floppy diskette analysis tool for use in security reviews to identify data storage pattern anomalies.

Autopsy This is a graphical interface to the command line digital investigation tools in The Sleuth which helps in the investigation of file systems and volumes of a computer.

BitPim This is a program that facilitates viewing and manipulating data on many CDMA phones from LG, Samsung, Sanyo, and other manufacturers.

Bit stream imaging, disk imaging, or cloning This refers to generating a bit-for-bit copy of the original media including hidden and residual data, for example, free space, slack space, swap, residue, unused space, and deleted files.

CD/DVD Inspector This is an optical media analysis tool which is used for intensive analysis and extraction of data from CD-R, CD-RW, and all types of DVD media, including HD DVD and Blu-Ray.

Cisco Packet Tracer This is a software with which the complete network can be simulated by adding and connecting different network devices.

Computer Aided Investigative Environment (CAINE) CAINE is an Italian GNU/Linux live distribution created as a digital forensics project which offers a complete forensic environment integrating existing software tools as software modules besides providing a user friendly graphical interface.

DumpIt This is a tool from Comae Technologies that can take a memory snapshot of a host when the host is suspected to be compromised or infected.

Filter_I This is an intelligent fuzzy-logic filter used to filter data that makes no sense during the analysis of ambient computer data.

Firewalk This is a tool that employs trace route-like techniques to analyse IP packet responses to determine gateway ACL filters and map networks.

GetFree This is a tool to recover files in the unallocated space which is specifically designed for law enforcement and forensic specialists to capture data in the unallocated space on computers running the Microsoft family of operating systems.

GetSlack This tool is used to capture the file slack contained on hard drives of the Microsoft family of operating systems which are further analysed with standard computer utilities or specialized NTI tools.

GetTime This program is used to document the CMOS system date and time associated with allocated files and previously deleted files immediately after the seizure of evidence.

Hash algorithm This is used as a utility to create a binary or hexadecimal number that represents the uniqueness of a data set such as a file or a disk drive.

Hiren's Boot CD This is an all-in-one bootable CD, the latest version being Hiren's BootCD 15.2 which has utilities that can be categorized as follows: antivirus tools, backup tools, BIOS/CMOS tools, browsers/file managers, cleaners, device driver tools, editors/viewers, file systems tools, hard disk tools, master boot record (MBR) tools, network tools, optimizers, other tools, partition tools, password/key tools, process tools, recovery tools, registry tools, remote control tools, startup tools, system information tools, testing tools, and tweakers.

Live RAM capturer This is a free forensic tool which is used to reliably extract the entire contents of a computer's volatile memory even if it is protected by an active anti-debugging or anti-dumping system.

Mandiant's Redline This is a handy utility that allows the detection of newly released viruses and other types of malware that are likely to be missed by standard antivirus solutions.

Message Digest Algorithm 5 (MD5) This is a mathematical algorithm that uses a cryptographic hash function to produce a 128-bit hash value.

Mobile phone examiner plus (MPE+) This is a standalone mobile device investigation solution that includes enhanced smart device acquisition and analysis capabilities.

NetAnalysis This is a Windows-based application for the forensic recovery and examination of browser history and cache data.

Nirsoft This is a website available at http://www.nirsoft.net/ and provides a unique collection of freeware utilities such as password recovery utilities, network monitoring tools, Internet-related utilities, MS-Outlook tools, command line utilities, desktop utilities, and freeware system tools.

Nmap This is a free and open-source (licence) utility for network discovery and security auditing.

OpenVPN This is a robust and highly flexible tunneling application that uses all of the encryption, authentication, and certification features of the OpenSSL library to securely tunnel IP networks over a single TCP/UDP port.

Ophcrack This is a free Windows password cracker based on a time-memory trade-off using rainbow tables.

OSForensics This tool is used for discovering the evidence, and identifying and reporting it.

PALADIN This is a modified 'live' Linux distribution based on Ubuntu which simplifies various forensic tasks in a forensically sound manner via the PALADIN Toolbox.

Passware Kit Forensic This is the complete electronic evidence discovery solution that reports all password-protected items on a computer and decrypts them.

PhotoMe This is an easy-to use utility that unlocks and organizes a photograph's metadata.

Recuva This is a user friendly recovery tool which can recover accidentally deleted files (pictures, music, document, videos, and emails), images, and data in rewriteable media (memory card, external hard drives, and USB sticks).

RegRipper This is an open-source forensic software application which is used for registry analysis in forensics examinations.

Regshot This is an open-source registry compare utility that helps to quickly take a snapshot of the registry and compare it with a second one.

SIMCon This is a program that allows the users to securely image all files on a GSM/3G SIM card to a computer file with the SIMCon forensic SIM card reader.

Snort This is an open-source network security and protection tool developed for UNIX and now applicable for Win32. It is an intrusion detection/prevention system that is capable of real-time traffic analysis and packet logging.

Tripwire This is a host-based IDS which focuses on monitoring files for changes.

Volatility This is a command line-based memory forensic analysis platform which supports memory dumps from all major 32- and 64-bit Windows versions and service packs including XP, 2003 Server, Vista, Server 2008, Server 2008 R2, and Seven.

Windows Sysinternals Live This is a service that provides a collection of Sysinternals tools to be directly executed from the web (https://live.sysinternals.com/) without manually downloading them.

Wireshark This is the most widely used network protocol analyser which captures packets in real time and displays them in human readable format.

MULTIPLE-CHOICE QUESTIONS

1. The purpose of the certified computer examiner (CCE) is to _____.
   (a) certify forensic tools
   (b) set high forensic and ethical standards for forensic computer examiners
   (c) provide forensic hardware for mobile devices
   (d) none of these
2. The functionality of Linux dd command utility is _____.
   (a) backing up and restoring an entire hard drive or a partition
   (b) creating a virtual file system to back up images on a CD and DVD as image files
   (c) backing up and restoring of MBR
   (d) all of these
3. Choose the correct representation of tools from Nirsoft under the free and open-source forensic suite.
   (a) OutlookAttachView
   (b) Windows Sysinternals Live
   (c) Both (a) and (b)
   (d) Autopsy
4. Ophcrack is a free _____ password cracker.
   (a) Linux   (c) Mac
   (b) Windows   (d) Android
5. AFLogical is a _____ forensic tool for _____.
   (a) free and open-source, computer
   (b) free and open-source, mobile devices
   (c) proprietary, computer
   (d) proprietary, mobile devices
6. XAMN allows users to view the contents of upto _____ different XRY files in one place.
   (a) 20   (c) 40
   (b) 30   (d) 50
7. _____ is the only hand-held, cellular exploitation device worldwide that requires no PC or associated phone drivers.
   (a) Cellebrite
   (b) CellDEK
   (c) Both (a) and (b)
   (d) None of these
8. ACE refers to _____.
   (a) assistant certified examiner
   (b) associate certified examiner
   (c) AccessSystem certified examiner
   (d) AccessData certified examiner
9. Which forensic training and certification has two phases (peer review phase and certification phase)?
   (a) Advanced information security (AIS)

(b) Certified computer examiner (CCE)
(c) Certified hacking forensic investigator (CHFI)
(d) Certified forensic computer examiner (CFCE)
10. _____ certifications ensure mastery in critical, specialized InfoSec domains.
   (a) Certified electronic evidence collection specialist (CEECS)
   (b) Certified computer forensics examiner (CCFE)
   (c) Computer and mobile forensic boot camp
   (d) SANS GIAC
11. Bit stream imaging is otherwise known as _____.
   (a) disk imaging (c) cloning
   (b) computer imaging (d) both (a) and (c)
12. _____ is an algorithm for computing the condensed representation of a message or a data file.
   (a) CRC32 (b) MD5
   (c) SHA (d) None of these
13. Some of the tool(s) from Windows Sysinternals Live is/are _____.
   (a) Sysmon v6.10
   (b) Process Monitor v4.51
   (c) Both (a) and (b)
   (d) Autoruns v21
14. CAINE is an _____ GNU/Linux live distribution created as a digital forensics project.
   (a) Indian (c) Asian
   (b) Italian (d) African
15. _____ remote service mode is included with every version of PALADIN.
   (a) SIMURI (c) SEMURI
   (b) SAMURI (d) SUMURI

REVIEW QUESTIONS

1. List out and define the types of cyber forensics tools.
2. Explain the categories of cyber forensics tools.
3. Mention some of the free and open-source forensic suites.
4. What are the forensics tools for integrity verification and hashing and data recovery?
5. Discuss in detail the forensic utility for metadata processing.
6. Define and differentiate PIN and PUK code.
7. Explain briefly the need and career prospects for computer forensic investigators.
8. Give a short note on proprietary forensic suites.
9. Explain briefly about the forensic analysis tools for mobile devices.
10. List out the forensics tools for email analysis.
11. Discuss in detail the miscellaneous tools.

APPLICATION EXERCISES

1. A popular company where employee A was working witnessed an internal threat— confidential data from the company had been leaked. The mode of leakage may have been either confidential data copied to a removable storage or sent via email. As a forensic examiner you are entrusted with the task of gathering evidence about the data being copied from the system that contained the confidential data. The hard disk containing the confidential data has to be examined after data acquisition and analysis for any deleted files, email sent with date and time, etc. In addition, any access to the system over a network should be ruled out. Suggest the forensic tools you would use for this scenario as well as the reason for choosing the specific tool.
2. As a forensic examiner you are endorsed with the task of retrieving data from a hard disk in a laptop whose volume containing the necessary data is encrypted. Moreover, the laptop itself is password protected. What forensic tools would you use to gain access to the laptop and gather data from the encrypted volume in the hard disk? Consider both the scenarios: (a) when the laptop is in ON state, and (b) when the laptop is in OFF state.
3. Which among the mobile forensic tools discussed in this chapter is the best? Give justifications.

BIBLIOGRAPHY

1. LinOxide (2011), *Learn Linux DD Command – 15 Examples with All Options*, available at: https://linoxide.com/linux-command/linux-dd-command-create-1gb-file/ (Accessed 01 October 2017)
2. OS Forensics, *OSForensics V5*, available at: https://www.osforensics.com/ (Accessed 01 October 2017)
3. Mapt, 'Chapter 2. Working with FTK Imager', available at: https://www.packtpub.com/mapt/book/hardware_and_creative/9781783559022/2 (Accessed 01 October 2017)
4. Access Data, *Access Data: Product Downloads*, available at: http://www.accessdata.com/product-download#digital-forever (Accessed 01 October 2017)
5. Belkasoft, *Capture Live RAM Contents with Free Tool from Belkasoft*, available at: https://belkasoft.com/ram-capturer (Accessed 01 October 2017)
6. Nirsoft, *NirSoft*, available at: http://www.nirsoft.net/ (Accessed 01 October 2017)
7. Kali Tools (2014), *Volatility Package Description*, available at: https://tools.kali.org/forensics/volatility (Accessed 01 October 2017)
8. Passmark Software, *Volatility Workbench Windows GUI for Volatility memory analysis*, available at: https://www.osforensics.com/tools/volatility-workbench.html (Accessed 02 October 2017)
9. Microsoft (2017), *Windows Sysinternals*, available at: https://docs.microsoft.com/en-us/sysinternals/ (Accessed 02 October 2017)
10. Lenny Zeltser (2017), *One-Click Windows Memory Acquisition with DumpIt*, available at: https://zeltser.com/memory-acquisition-with-dumpit-for-dfir-2/ (Accessed 02 October 2017)
11. Hacking Tools, *EnCase Forensic*, available at: http://www.hackingtools.in/free-download-encase-forensic/ (Accessed 02 October 2017)
12. Jamie McQuaid (2015), *Acquiring Memory with Magnet RAM Capture*, available at: https://www.magnetforensics.com/computer-forensics/acquiring-memory-with-magnet-ram-capture/ (Accessed 02 October 2017)
13. Magnet Forensics, *Encrypted Disk Detector*, available at: https://www.magnetforensics.com/free-tool-encrypted-disk-detector/ (Accessed 02 October 2017)
14. CCleaner, *Recuva*, available at: https://www.piriform.com/recuva (Accessed 02 October 2017)
15. Craig Wright (2009), *A Step-by-Step Introduction to Using the AUTOPSY Forensic Browser*, available at: https://digital-forensics.sans.org/blog/2009/05/11/a-step-by-step-introduction-to-using-the-autopsy-forensic-browser (Accessed 02 October 2017)
16. John Lehr, *CAINE*, available at: http://www.caine-live.net/ (Accessed 03 October 2017)
17. Sumuri, *PALADIN Forensic Suite*, available at: https://sumuri.com/software/paladin/ (Accessed 03 October 2017)
18. Hiren, *Hiren BootCD 15.2*, available at: https://www.hiren.info/pages/bootcd (Accessed 03 October 2017)
19. Softpedia, *DEFT*, available at: http://linux.softpedia.com/get/System/Operating-Systems/Linux-Distributions/DEFT-27206.shtml (Accessed 03 October 2017)
20. Chris Hoffman (2017), *How to Use Wireshark to Capture, Filter and Inspect Packets*, available at: https://www.howtogeek.com/104278/how-to-use-wireshark-to-capture-filter-and-inspect-packets/ (Accessed 03 October 2017)
21. GitHub, *OpenVPN / openvpn-gui*, available at: https://github.com/OpenVPN/ openvpn-gui (Accessed 03 October 2017)
22. Nmap, *Nmap Security Scanner*, available at: https://nmap.org/ (Accessed 03 October 2017)
23. Hashmyfiles, *HashMyFiles 2.25*, available at: https://hashmyfiles.soft112.com/ (Accessed 03 October 2017)
24. Hashcalc, *HashCalc 2.0*, available at: http://hashcalc.software.informer. com/2.0/ (Accessed 03 October 2017)
25. Softpedia (2014), *MD5Summer*, available at: http://www.softpedia.com/get/System/File-Management/MD5summer.shtml (Accessed 03 October 2017)
26. Sourceforge, *Regshot*, available at: https://sourceforge.net/projects/regshot /files/regshot/1.9.0/ (Accessed 03 October 2017)
27. Forensicswiki (2017), *RegRipper*, available at: http://www.forensicswiki.org/wiki/Regripper

(Accessed 04 October 2017)

28. FireEye, *Redline – Free Download*, available at: https://www2.fireeye.com/PPC-mandiant-redline-download-analyze-malware-ioc.html (Accessed 04 October 2017)
29. Softpedia, *Mandiant Redline*, available at: http://www.softpedia.com/get/Security/Security-Related/Mandiant-Redline.shtml (Accessed 04 October 2017)
30. PCWorld (2009), *PhotoME*, available at: https://www.pcworld.com/article/234048/photome.html (Accessed 04 October 2017)
31. Mobiledit, *MOBILedit Forensic Express*, available at: http://www.mobiledit.com/forensic-express/ (Accessed 04 October 2017)
32. Forensicswiki (2009), *Paraben Device Seizure*, available at: http://www.forensicswiki.org/wiki/Paraben_Device_Seizure (Accessed 04 October 2017)
33. Aimtech, *XAMN*, available at: http://aimtech.ru/en/catalog/43 (Accessed 04 October 2017)
34. Mobile Forensics Central, *CellDEK*, available at: http://www.mobileforensicscentral.com/mfc/products/celldek.asp?pg=d&pid=&prid=347&return=undefined (Accessed 04 October 2017)
35. Mike D. Schiffman (1998), *Firewalk*, available at: https://packetfactory.net/Projects/Firewalk/ (Accessed 04 October 2017)
36. Ric Messier, *Network Forensics*, Wiley Online Library
37. Albert Marcella, Jr., Doug Menendez, *Cyber Forensics: A Field Manual for Collecting, Examining, and Preserving Evidence of Computer Crimes*, CRC Press (Accessed 04 October 2017)
38. Security Wizardry (2003), *The Coroners Toolkit*, available at: http://www.securitywizardry.com/index.php/products/forensic-solutions/forensic-toolkits/the-coroners-toolkit.html (Accessed 05 October 2017)
39. Bruce Middleton, *Cyber Crime Investigator's Field Guide*, 2nd edition, CRC Press (Accessed 05 October 2017)
40. John Rittinghouse and William M. Hancock (2004), *Cybersecurity Operations Handbook*, Elsevier (Accessed 05 October 2017)
41. Micah Solomon, Diane Barrett, Neil Broom, *Computer Forensics JumpStart*, Wiley Publishers (Accessed 05 October 2017)
42. Marie-Helen Maras (2014), *Computer Forensics*, Jones & Bartlett Learning
43. CyberPunk: The Best Open Source CyberSecurity Tools, available at: https://n0where.net/best-digital-forensics-tools/ (Accessed 05 October 2017)
44. Digital Defective, *Net Analysis*, available at: http://www.digital-detective.net/digital-forensic-software/netanalysis/ (Accessed 06 October 2017)
45. SC Magazine (2012), *Paraben Device Seizure v4.6*, available at: https://www.scmagazine.com/paraben-device-seizure-v46/review/6597/ (Accessed 06 October 2017)
46. GoCertify, *AccessData Certified Examiner*, available at: http://www.gocertify.com/certifications/accessdata/accessdata-examiner.html (Accessed 06 October 2017)
47. The International Society of Forensic Computer Examiners, *CCE Certification*, available at: https://www.isfce.com/certification.htm (Accessed 06 October 2017)
48. EC-Council, *Computer Hacking Forensic Investigator Certification*, available at: https://www.eccouncil.org/programs/computer-hacking-forensic-investigator-chfi/ (Accessed 06 October 2017)
49. The International Association of Computer Investigative Specialists, Certification – 2, available at: https://www.iacis.com/certification-2/ (Accessed 06 October 2017)
50. Safari, *Certified ProDiscover Examiner (CPE)*, available at: https://www.safaribooksonline.com/library/view/computer-forensics-jumpstart/9781118067659/bapp03-anchor-7.xhtml (Accessed 06 October 2017)
51. Guidance Software, *Certifications*, available at: https://www.guidancesoftware.com/training/certifications (Accessed 06 October 2017)
52. GIAC, *GIAC Certified Forensic Examiner (GCFE)*, available at: https://www.giac.org/certification/certified-forensic-examiner-gcfe (Accessed 07 October 2017)
53. Proexamvault, *Professional Certified Investigator (PCI)*, available at: https://proexamvault.com/issuers/asis-international/professional-certified-investigator-pci. (Accessed 07 October 2017)
54. Sans, *SANS*, available at: https://www.sans.org/ (Accessed 07 October 2017)

55. Darknet, *Want some COFEE? Microsoft Computer Online Forensic Evidence Extractor*, available at: https://www.darknet.org.uk/2008/05/want-some-cofee-microsoft-computer-online-forensic-evidence-extractor/ (Accessed 07 October 2017)
56. Vogon, *Forensic Software*, available at: http://www.vogon-computer-forensics.com/evidential_systems-03.htm (Accessed 07 October 2017)
57. Vogon, *Company Profile*, available at: http://www.vogon-computer-forensics.com/company_profile.htm (Accessed 07 October 2017)
58. Digital Intelligence, *Software for Investigations*, available at: https://www.digitalintelligence.com/software/passware/passwarekitforensic/ (Accessed 07 October 2017)
59. Isobuster, *IsoBuster*, available at: https://www.isobuster.com/ (Accessed 07 October 2017)
60. Free Download Manager, *CD/DVD Inspector*, available at: https://en.freedownloadmanager.org/Windows-PC/CD-DVD-Inspector.html (Accessed 07 October 2017)
61. Payne Group, *Metadata Assistant*, available at: http://www.thepaynegroup.com/products/metadata/ (Accessed 07 October 2017)
62. CFI, *XRY Forensic – Mobile Phone Forensic Tool*, available at: http://www.cfi.co.th/xry-forensic.html (Accessed 07 October 2017)
63. Infosec Institute, *Hands-on Practice: Practice with Data from Real Forensics Cases*, available at: https://www.infosecinstitute.com/courses/computer-forensics-boot-camp (Accessed 07 October 2017)

Answers to Multiple-choice Questions

1. (b)	2. (d)	3. (a)	4. (b)	5. (b)
6. (d)	7. (a)	8. (d)	9. (d)	10. (d)
11. (d)	12. (c)	13. (a)	14. (b)	15. (d)

Acquisition and Handling of Digital Evidence

Learning Objectives

This chapter gives an overview of the acquisition and handling of digital evidence. The objective of this chapter is to provide the preliminaries of electronic evidence, an overview on how to acquire evidence, a detailed insight into the seizure process, and a deeper insight into acquiring evidence from computers, email, the Internet, and mobile devices. Besides these topics, the chapter also explains the process involved in acquiring evidence from other devices and media, as well as from third-party organizations. Finally, it explains the handling of digital evidence. The reader will be familiar with the following after studying the chapter:

- Preliminaries of electronic evidence
- Search and seizure of digital evidence
- Acquiring evidences from electronic communication devices (ECDs)
- Handling of digital evidence

8.1 PRELIMINARIES OF ELECTRONIC OR DIGITAL EVIDENCE

Evidence is the lifeline of any trial, enquiry, and quasi-judicial enquiries, and is primarily of two types, namely oral evidence and documentary evidence. Section 60 of the Indian Evidence Act (IEA) states that oral evidences must always be direct, that is, those evidences that are personally seen or heard by the witness giving them and not heard or told by someone else. The statements that the witness makes in court, in person, regarding the truth or the facts of the case, are called *oral evidences*.

Section 3 of the IEA defines all those documents that are presented in court for inspection regarding a case. These are referred to as *documentary evidences*. Documentary evidence may be either primary evidence or secondary evidence (Box 8.1).

> ### Box 8.1 Definitions of Primary and Secondary Evidence
>
> *Primary Evidence*
>
> Section 62 of IEA:
>
> *Primary evidence means the document itself is produced for the inspection of the court.*
>
> *Explanation 1: Where a document is executed in several parts, each part is a primary evidence of the document.*
>
> *Where a document is executed in counterparts, each counterpart being executed by one or some of the parties only, each counterpart is primary evidence as against the parties executing it.*
>
> *Explanation 2: Where a number of documents are all made by one uniform process, as in the case of printing, lithography, or photography, each is primary evidence of the contents of the rest; but, where they are copies of a common original, they are not primary evidence of the contents of the original.*

(Contd)

Box 8.1 (Contd)

> *Secondary Evidence*
>
> Section 63 of IEA:
>
> *Secondary evidence means and includes*
>
> 1. *certified copies given under the provisions hereinafter contained;*
> 2. *copies made from the original by mechanical processes which in themselves ensure the accuracy of the copy and copies compared with such copies;*
> 3. *copies made from or compared with the original;*
> 4. *counterparts of documents as against the parties who did not execute them;*
> 5. *oral accounts of the contents of a document given by some person who has himself seen it.*

Digital evidence is defined as information or data stored on, transmitted, or received by an electronic device/electronic communication device (ECD) in binary form, which is of value to a crime investigation and is relied upon in court. Digital evidence is unique and is neither primary nor secondary evidence. Consider the hard copies of a word document. It is not possible for a human observer to judge which copy is primary or secondary. Therefore, the integrity of the collected evidence has to be guaranteed at all times. Digital evidence is acquired when an electronic communication device (ECD) reported or suspected to be involved in a crime is seized and is examined by a cyber forensic examiner (CFE).

8.1.1 Categorization of Source of Digital Evidence

The sources from where evidences are collected are categorized as follows:

1. Standalone computers or devices—offline
2. Devices connected to the Internet—live forensics
3. Mobile devices

Henseler (2000) has categorized the sources of digital evidence as (a) open computer systems, which include computers—laptops, desktops, and servers; (b) communication systems, which include telecommunication systems that transfer short messaging service (SMS)/multimedia messaging service (MMS) messages and the Internet that makes emailing possible; and (c) embedded computer systems, which include mobile devices and smart cards (Box 8.2).

Box 8.2 Definition of Electronic Evidence

Section 3(2) of IEA:

"*electronic evidence is documentary evidence*

All documents including electronic records produced for the inspection of the Court, such documents are called documentary evidence"

Section 2(t) of IT Act 2000:

"*'electronic record' means data, record or data generated, image or sound stored, received or sent in electronic form or micro film or computer generated micro fiche*"

8.1.2 Locality of Digital Evidence

Digital evidence is usually collected from ECDs, namely computer hard drives, mobile phones, personal digital assistants (PDAs), CDs, or DVDs, flash cards in a digital camera, memory sticks, and memory cards, removable disks, iPads, tabs, in-built memories of telephones or phones with caller ID, FAX machines, and printers. Thus from open systems, the digital evidence is collected from its storage, for example, a file which may provide information as to who created it and when. A communication system may provide information such as the time the message was sent, the person who sent it, what was contained in the message, and so on. Embedded systems may provide evidence in the form of photographs, videos, and other personal data.

The presence of cybercriminals and attackers can be ascertained from multiple traces throughout the environment, such as in file systems, registries, system logs, and network-level logs. For example, consider an email harassment case, wherein threatening messages are sent through a web-based email service. From the sender's hard drive, the files and links, along with date- and time-related information stored by the web browser while sending these messages, can be traced. Besides this, from the email's service provider, web server access logs, IP addresses, and the entire message in the sent mail folder of the offender's email account can be obtained. During operation, operating systems and applications create files in the background that include automatic backup files, globally unique identifiers, Internet browser files, Internet history files, metadata, power saver features, temporary files, temporary Internet files, spooler files, virtual memory, and swap files. These are the areas where a CFE might find the best evidence.

It is therefore necessary for an investigating officer (IO) to identify what evidence is present, where it is stored, and how it is stored, so as to acquire it. Besides this, an IO has to identify the type of information stored in the ECD, its format, and the appropriate technology to be used to acquire it.

8.1.3 Roles played by Digital Evidence

From the perspective of digital evidence, the ECD may serve as any one of the following: (a) object/victim, when the device is affected as a result of execution of crime, say, it is stolen, destroyed, or has witnessed an attack by a cybercriminal; (b) subject, when the device is in the environment where the crime is committed, say, a virus on a device which causes inconvenience to the users who use the device, (c) witness, for example a CCTV camera employed to record the events on a hard disk; (d) tool, for example a software on the device used to commit the crime when the device is used to forge a document; (e) accomplice, for example, a weakness in the design of hardware or software used to commit the crime.

8.1.4 Characteristics of Digital Evidence

The following are the characteristics of digital evidence:
1. It is latent or is invisible and hence must be developed using special tools and equipment.
2. It is highly fragile and volatile.
3. It may be available in a number of devices, locations, and in various formats.
4. It is easily altered or destroyed and hence requires precautions to prevent alteration.
5. It requires specialized training and expert testimony for handling.
6. It is portable and can be spread across borders with ease.
7. It is time-sensitive.

8.1.5 Physical versus Digital Evidence

Digital evidence is very different from physical evidence in that the latter can be tampered with, but it is usually much harder to do and leaves discernible marks. On the contrary, digital evidence has the potential for unauthorized copies to be made without leaving behind any trace of the copy having been made. So as to be admissible in court, digital evidence must be collected in a manner where any copies made must be exactly the same, bit for bit. In other words, the checksum of the original media and the forensic copy on the target media must be identical.

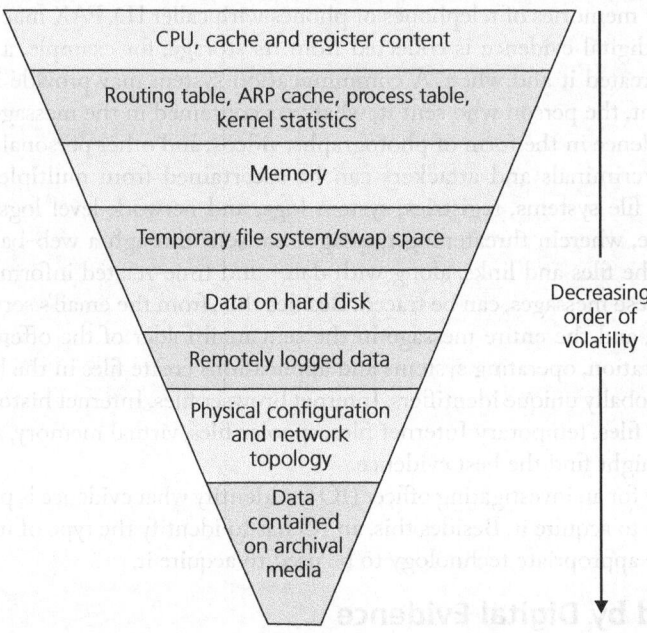

Fig.8.1 Order of volatility of digital evidence

8.1.6 Order of Volatility of Digital Evidence

Figure 8.1 presents the order of volatility of digital evidence in decreasing order from the top. Hence a CFE pays due attention and collects digital evidence without violating the order of volatility as otherwise the data is likely to disappear. He/She collects the most volatile evidence first and continues to collect data until nothing is left. The order is as follows:

CPU, cache, and register content These contents are volatile, in the order of nano seconds, and are likely to change in no time. Therefore, the CFE should collect them before they are lost.

Routing table, ARP cache, process table, kernel statistics These may refer to data located on network devices which are subject to change while they are in operation. Kernel statistics correspond to data moved back and forth between the main memory and cache. Hence, this category of evidence should be gathered quickly.

Memory The data in random access memory (RAM) will be lost in the event of either the power going off or a power spike. Hence the data from RAM must be obtained quickly.

Temporary file system or swap space This class of data and information are not very volatile and can be available for a while if not erased intentionally. However, such contents have the potential to be part of future legal proceedings.

Data on hard disk Data on hard disk is available as evidence unless otherwise the data is deleted, erased, or wiped intentionally. Forensic tools can retrieve deleted files.

Remotely logged data This category of data is more prone to change than the data on a hard disk but is not as important as the data on a hard disk.

Physical configuration and network topology These data (network cache and remote logs on firewalls, intrusion detection systems, proxy servers, etc.) are also more prone to change like remotely logged data but is not as important as data on a hard disk. Though these types of data could help in investigation, they are not very volatile and are not vital.

Data contained on archival media This refers to data on a DVD or a tape that qualifies as evidence but is non-volatile.

8.1.7 List of Crimes and Probable Location of Evidence

Offences may be classified into three as follows: (a) offences against individual, (b) offences against nation, and (c) offences against property. The following is the list of offences (shown in Fig. 8.2) which are categorized accordingly and involve the use of computer or electronic media along with the potential evidence that can be recovered from them.

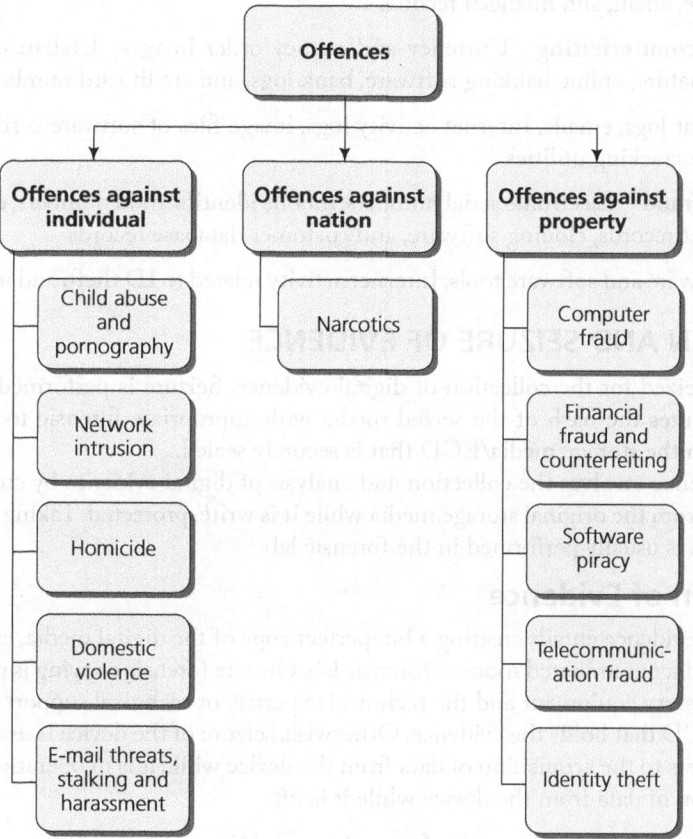

Fig. 8.2 Classification of offences

Offences against Individual

Child abuse and pornography Chat logs, images, digital camera software, Internet activity logs, email, mobile files, games, graphic editing, and viewing software

Network intrusion Address books, configuration files, email, executable programs, Internet activity logs, source code, text files with usernames, and passwords

Homicide Address books, email, financial records, Internet activity logs, legal documents and wills, medical records, telephone records, diaries, maps, and photo

Domestic violence Address books, diaries, email, financial asset records, and telephone records

Email threats, stalking, and harassment Address books, diaries, email, financial asset records, images, Internet activity logs, legal documents, telephone records

Offences against Nation

Narcotics Address books, calendar, databases, drug recipes, emails, financial asset records, and Internet activity logs

Offences against Property

Computer fraud Account data available online, address books, calendar, chat logs, credit card data, databases, digital camera software, email, and financial records

Financial fraud and counterfeiting Currency and money order images, databases, email, financial asset records, images of signature, online banking software, bank logs, and credit card numbers

Software piracy Chat logs, emails, Internet activity logs, image files of software certificates, software serial numbers, and software cracking utilities

Telecommunication fraud Electronic serial numbers, mobile identification numbers, emails, Internet activity logs, and financial asset records, cloning software, and customer database records

Identity theft Hardware and software tools, Internet activity related to ID theft, and identification templates

8.2 ACQUISITION AND SEIZURE OF EVIDENCE

The ECD has to be seized for the collection of digital evidence. Seizure is performed at the scene of crime (SOC). The IO computes the hash of the seized media with appropriate forensic tools and sends it to the forensic lab along with the storage media/ECD that is securely sealed.

Acquisition of evidence involves the collection and analysis of digital evidence by creating an exact copy of the original evidence from the original storage media while it is write-protected. Taking possession of evidence from the source media is usually performed in the forensic lab.

8.2.1 Acquisition of Evidence

Acquisition of digital evidence entails creating a bit-perfect copy of the digital media, either on-site where the ECD is kept, or in an uncontaminated room or forensic lab. On-site forensic copying is planned and performed if the IO has the necessary equipment and the technical expertise or technical support to take a forensic copy of the media in the ECD that holds the evidence. Otherwise, seizure of the device is usually planned. The term *hot data acquisition* refers to the acquisition of data from the device while it is in operation. *Cold data acquisition* refers to the acquisition of data from the device while it is off.

8.2.2 Precautionary Measures before Acquisition

Digital evidence is usually acquired from the scene of offence which could be a house with one or more computers, a cyber cafe, or an organization/company where systems are networked. The IO looking into the crime will analyse the technical details about the systems, electronic devices, and the network before proceeding with the seizure. Since digital evidence is perishable, appropriate precautionary measures have to be taken by the IO. For example, if systems are 'off' they should not be turned 'on' and vice versa.

In case the scene of offence is a house, the IO may gather information such as the type of Internet connection (wired/wireless), the number of systems, and whether they are connected to the Internet, details about the storage media (both permanent and removable), and other peripheral devices.

If the scene of offence is a cyber cafe/organization, the IO may gather information from CCTV clippings or any other management software, in addition to those mentioned earlier. A preservation notice is usually served by the IO to prevent digital evidence from being tampered with. The preservation notice may ask for stopping further access to the ECD for preserving the log information, stopping access to email to prevent deletion of emails, etc.

According to the standard operating procedure (SOP), during the investigation of the crime scene, the following should be done:

- The ECD should be quarantined so that no one can tamper with the data and the data is free from corruption and damage.
- The status of the ECD has to be ascertained and if it is 'live', its status should be recorded using photographs. Turning on a system that is turned off may result in changes being made by the operating system in the background which will change the evidence.
- The device should be disconnected from the Internet or network at the earliest. However, care should be taken to ensure that this does not result in any loss of information. This should especially be taken care of in mobile phones, as it may bring about changes to its internal data, which would otherwise have been useful to the investigation.
- Other than portable devices, all electronic devices should be powered off and safely shut down.
- All electronic gadgets should be seized along with their power chords.

8.2.3 Search and Seizure

Searches can only be carried out by the IO, who is a competent authority. The Information Technology (IT) Act 2000 allows any police officer who is not below the rank of deputy superintendent of police to investigate any offence under this Act. Section 80 of the IT Act 2000 (amended 2008) states that any police officer, not below the rank of a police inspector, or any other officer of the central or state government authorized by the central government may enter any public place and search and arrest without warrant any person who is reasonably suspected to having committed, of committing, or of being about to commit any offence or crime under this Act.

The competent authority shall call upon two witnesses to attend and witness the search and may issue an order in writing for the same. The witnesses should preferably be computer literates in the case that the evidence to be acquired is digital in nature. The person-in-charge of the premises where the offence occurred shall be permitted to witness the search.

Any seizure should be justified, appropriate, and proportionate[Marshall, 2008] which means that the ECD should have evidence related to the crime and that the value of information in it outweighs the seizure. The seizure process starts with the preservation of digital evidence. It involves seizing the ECD or taking custody of it.

Thus, after the filing of a complaint to the cybercrime cell, investigation begins with search and seizure by a team headed by the IO in the presence of two witnesses. The IO may seek the support and technical expertise of the forensic examiner depending on the complexity prevailing during the seizure of evidence.

The IO should perform the sequence of steps shown in Fig. 8.3 during crime scene investigation. These are listed here:

1. Identification and securing of the crime scene—the IO locates the crime scene once a complaint is received and takes control of it.
2. Documentation of the crime scene—the IO prepares a report that exactly reflects the crime scene.
3. Collection of evidence—the IO gathers evidences that exist physically such as user manuals, passwords, or other login credentials available as hard copy and electronic communication devices that hold digital evidences.

4. Labelling and documentation of evidence—all the evidences gathered are labelled and details of it are documented by the IO.
5. Packaging and transportation of evidence to court—all the collected evidences are packed so as to be safe during transit to court and are produced before court.

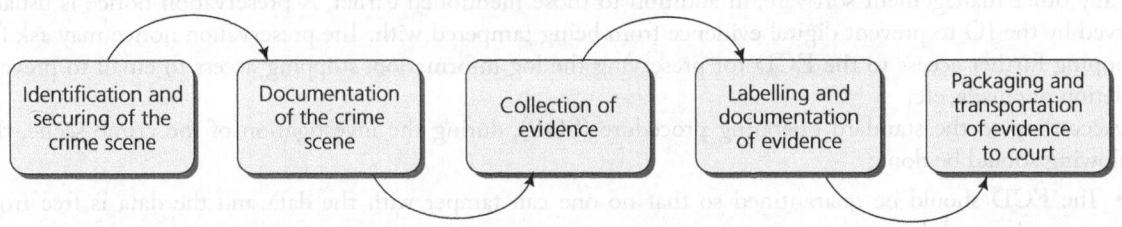

Fig. 8.3 Steps in crime scene investigation

8.2.4 Seizure Memo

Seizure Memo
(To be sent to court without any delay)

A search was undertaken on _____ (date) at _____ (place) and the following items were seized:

S. no.	Description of the items seized	Quantity	Remarks

The items seized were sealed and packed as under:

Name and signature of the investigating officer

Witnesses:
(Signature, date, name, and address)
1.
2.

(Signature, name, and address of the person-in-charge of the premises searched, who receives this letter after the search)

Fig. 8.4 Format of seizure memo

The seizure memo should include the following recorded by the IO:
1. Reasons as to why the search is necessary from the facts and circumstances of the case or the offence
2. Description of the place and location of occurrence of the offence
3. Names of the witnesses
4. Data and time when the search commenced and closed
5. Owner of the premises or the ECD suspected of holding the evidence
6. Unique identification details of the ECD such as its serial number, make, and model
7. Reasons for seizure
8. Date and time of seizure
9. Photograph/Video of the entire process of seizure in the presence of witnesses

The seizure memo is prepared in the presence of two independent witnesses and the evidences seized are labelled, sealed, and packed in their presence, after which the seizure memo is signed by the witnesses. The format of this memo is shown in Fig. 8.4.

8.3 CHAIN OF CUSTODY AND DIGITAL EVIDENCE COLLECTION FORM

Both chain of custody and digital evidence collection form provide information about the evidence collected. While the former provides information on who possessed the evidence during the course of the case right from acquisition from the crime scene to disposition in the court, the latter provides information on the actual evidence gathered from the electronic communication devices.

8.3.1 Chain of Custody

Chain of custody refers to the documentation containing the names of the people who were in possession of the physical and electronic evidence during the seizure, custody, control, transfer, and disposition (final movement of evidence when a case is closed). These persons should appear as witnesses to testify that the digital evidence has not been tampered with and that it is genuine and authentic. These persons may be any one of the following: (a) a competent authority who seized the equipment, (b) a person who transferred the evidence from the SOC to court, (c) a person who transferred the evidence from the court to the forensic lab, (d) a person who analysed the evidence, (e) a person who presented the report after analysis along with the evidence to the court, and (f) the IO of the case.

All the evidence that is acquired should be maintained in a chain-of-custody record, which can be used to trace the location of the evidence from the moment it is collected until it is presented for judicial proceeding. A typical chain of custody document should include the following:

1. Reason for the seizure and collection of evidence
2. Date, time, and location from where the evidence was collected
3. Name of the IO
4. Name of the owner of the media or the ECD
5. Details of the media collected such as type of media, its serial number, make and model, and capacity
6. Details of the ECD, for example, its physical description (i.e., 'on' or 'off')
7. Method of data acquisition and the tools used
8. Name of the forensic copy that was collected
9. Hash values of the original media and the forensic copy for verification
10. Details of issues encountered, if any
11. Name and signature of persons relinquishing control over or receiving control over the evidence with date and time

8.3.2 Digital Evidence Collection Form

The digital evidence collection (DEC) form is part of the documentation containing information related to the collection of evidence from every ECD, usually in the forensic lab. It includes information about the media that serves as the original evidence and the media that holds the forensic copy. For example, it preserves information such as the kind of software and the version used, and the opening and closing time of evidence collection. Besides this, the hash value computed by the IO during seizure is also recorded in the DEC form. The process, the tool, and the hashing algorithm used are also recorded in the DEC in case of on-site acquisition. A DEC form is shown in Fig. 8.5.

Digital evidence collection (DEC) form	
Standard details	
Crime number	
Applicable section of the law	
Date when the device was seized	
Name of the IO	
Place from where the evidence was acquired	
System information	
Device type which is used to collect evidence	
Manufacturer information of the device	
Model number of the device	
Serial number of the device	
Whether acquisition was done at the crime scene or in a forensic lab (Yes / No)	
If yes, actual date/time of acquisition of evidence	Time zone
BIOS information about the device such as BIOS date/time.	
Evidence number (unique number for every device with respect to a crime)	
Evidence drive information	
Type of media (HDD/USB/floppy/CD/DVD)	
Type of HDD (SATA, IDE/SCSI HDD)	
Name of the manufacturer	
Model number	
Serial number of the media	
Number of logical partitions	
General information	
Forensic software used for acquisition (Cyber Check Suite, Encase, FTK)	
Write-protect device type	
Drives information (original and image)	
Name of the image file and its format	
Other details of acquisition	

Fig. 8.5 DEC form

The following two factors are closely associated with the collection of evidence:

Authentication of evidence Authentication mandates that the person who collected the evidence should testify during examination that the information is what the proponent claims. Two points must be followed: (a) a record of the person who collected the evidence, and (b) the way in which the evidence was collected has to be recorded and documented.

Evidence safe Evidence custodians are specially trained personnel who can make an inventory of the evidence and ensure that it is stored safely. The best evidence, usually the original or the first copy, has to be stored in a safe storage room and should not be accessed by anyone except the evidence custodians. This storage area is referred to as evidence safe.

8.4 Fourth Amendment and Seizure

Twelve amendments were submitted by the US Congress in 1790 to safeguard the rights of individuals from interference from the federal government (Box 8.3).

Accordingly, seizure may be effected either with or without a search warrant depending on the crime and the items to be seized. In the Indian context, Article 19 (Right to Freedom) and Article 20 (Protection in Respect of Conviction for Offences) offer protection from unreasonable searches and Section 93 of the Criminal Procedure Code (CrPC) provides the same conditions of 'probable cause' as in US law. However, Section 165 of the CrPC provides exceptions as granted in the Fourth amendment, whereby if the police believe that waiting for a search warrant will compromise the evidence, they can go ahead without any warrant.

8.4.1 Search and Seizure with Search Warrant

A search warrant gives IOs the authority to enter the premises of the crime scene and search for objects and devices named in the warrant, and seize them. Search warrants must be supported by *probable cause* (There exists known facts and circumstances sufficient for the issue of warrant with the belief that the evidence for the crime will be found) and *particularity* (The warrant should specify the location where the search will be made and what will be seized).

The warrant will only be valid if it states the crime being investigated, the location where the search will be conducted, and the items that will be seized. For example, if the crime being investigated is child pornography, the warrant which authorizes the search and seizure of computers and electronic storage devices containing the images of minors engaging in sexual activity as dictated by child pornography is sufficiently particular. Any deviation from the warrant specification will make the seizure inadmissible in court. Thus, a search warrant is a process involving a neutral and detached magistrate, an oath or affirmation, establishment of probable cause and particularity, and the time and manner of execution.

8.4.2 Warrantless Searches

A search without a warrant is justified for the following: stop-and-frisk procedures, open fields, automobile exceptions, search incident to arrest, exigent circumstances, plain view, consent searches, and border searches. The first three happens during general search and seizure, whereas the others are apt for cybercrime and digital evidence.

Box 8.3 The Fourth Amendment

The Fourth Amendment to the US Constitution encompasses an individual's right to privacy and provides that

The right of the people to be secure in their persons, houses, papers and effects against unreasonable searches and seizures, shall not be violated and no Warrants shall issue, but upon probable cause, supported by Oath or affirmation, and particularly describing the place to be searched, and the persons or things to be seized.

Stop-and-frisk procedure   A law enforcement agent may attempt a search if he/she believes that the individual (suspect) is armed.

Open field   A warrantless search may occur in an open field as it is exposed and accessible to the public.

Automobile   A search without a warrant may occur if the law enforcement agent has probable cause to believe that the vehicle holds evidence of a crime.

Subpeona   This is the first option and it compels the individual or the organization that owns the computer system to surrender it. This is usually done when it is sure that a notification will not result in evidence being destroyed by the equipment owner.

Search incident to arrest   This encompasses the searches that occur upon the arrest of the individuals (suspect) and the areas under their control. This is to prevent the evidence from being destroyed by the individual and also ensures that the individual does not escape with them. However, the primary objective is to prevent the destruction of evidence. For example, with pagers, there is a finite memory for the pages and so any incoming pages may destroy the ones that are already in memory. There is a chance of the content from some of the pages getting destroyed by turning off the power or touching a button.

Exigent circumstances   Warrantless searches may occur with exigent circumstances if the destruction of evidence is imminent and there is probable cause that the item seized holds evidence of some criminal activity. Exigent circumstances arise when acquiring evidence from computers and electronic media as the suspect may attempt to physically damage the device/equipment that holds the evidence or may delete the files using computer commands or special programs (so as to delete the evidence).

Consent   Searches may occur without warrants and without a probable cause if the individual who has authority over the place and the items to be searched has given the consent to search. It is a part of the standard investigation technique followed by law enforcement agencies. Consent search may occur in a person's home or office, on roads and under informal and unstructured conditions. Consent search becomes invalid and illegal if the individual is tricked or coerced into consenting to the search. A warrant is not required either if the suspect himself/herself consents to the search or a third party (employer/parent/spouse/relative) consents to the search provided that the item under search is not locked. The best practice is to get the consent in writing in order to serve as evidence that the consent was given voluntarily.

Border searches   The US Congress has authorized custom searches at borders to regulate the entry and exit of individuals at its borders as well as the conditions for custom searches as border search doctrine. Routine border searches do not require a probable cause or a warrant, whereas non-routine searches do. Any non-routine search is preceded by reasonable suspicion and the search should not exceed what is necessary to prove a crime. For example, if child pornography is strictly prohibited in a state, and if during entry into the state suspicion is raised over an individual possessing offensive content in his/her laptop, or secondary or removable storage, the individual may be subjected to border search and the items can be seized.

Plain view   The plain view doctrine allows IOs to seize evidence not specified in a search warrant during a search. It is especially applicable to cybercrimes, wherein the search of a computer may give some clue about other items or incriminating information not specified in the warrant. Thus, the plain view exception to the warrant requirement only gives legal authority to seize a computer, hardware, software, and electronic media, but does not give legal authority to conduct a search of the electronic media.

8.5 ACQUISITION OF COMPUTER AND ELECTRONIC EVIDENCE

The primary sources of evidence in computer systems are hard drives and external storage devices, data residing in volatile memory, running processes, cache data, DNS entries, ARP, navigation, recycling bin, active sessions and users, input/output traffic.

Box 8.4 Forensic Boot Disk

It is used to boot the suspect system safely. It contains a file system and statically linked utilities such as ls, fdisk, ps, nc, dd, and ifconfig. It places the suspect media in a locked or read-only state. It does not swap any data to the suspect media. Some of the open source bootable images include FIRE (http://biatchux.dmzs.com/?section=main), Linuxcare Bootable Business Cards (http://lbt.linuxcare.com/index.epl), and Trinux (http://trinux/sourceforge.net/).

Power-up the system to access the data or remove the physical storage to collect the data. Removal of disk ensures integrity of the data. However, there are chances of compatibility issues which may prevent evidence from being copied from the drive. Search warrants may require on-site collection of digital evidence from the crime scene if the suspect's system is highly sensitive. However, there should not be any alteration of data.

8.5.1 Acquisition of Configuration Information through Controlled Boots

The following steps are performed to retrieve configuration information from the suspect's system through controlled boots (Box 8.4).
1. Perform a controlled boot to capture CMOS/BIOS information and test functionality.
   (a) Change the boot sequence to ensure that the system boots from the floppy or CD-ROM drive and acquire time, date, and power on passwords.
2. Perform a second controlled boot to test the computer's functionality and the forensic boot disk.
   (a) Ensure the power and data cables to the floppy or CDROM drive are properly connected, but those to the storage devices are still disconnected.
   (b) Place the forensic boot disk into the floppy or CD-ROM drive. Boot the computer and ensure the computer will boot from the forensic boot disk.
3. To capture the drive configuration information from the CMOS/BIOS, reconnect the storage devices and perform a third controlled boot.
   (a) Ensure there is a forensic boot disk in the floppy or CD-ROM drive to prevent the computer from accidentally booting from the storage devices.
   (b) Acquire drive configuration information which includes logical block addressing (LBA), large disks, cylinders, heads, and sectors (CHS), or auto-detect.
   (c) This is illustrated in Fig. 8.6.

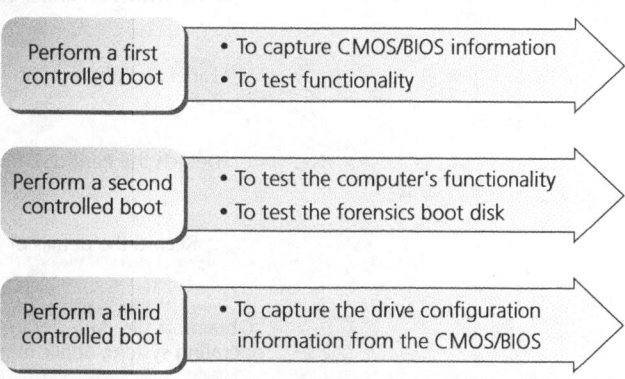

Fig. 8.6 Acquisition of configuration information through controlled boots

8.5.2 Acquisition of Evidence from Switched-off Systems

The procedure for acquiring evidences from switched-off systems is illustrated in Fig. 8.7 and is as follows:
1. Disable the modem and the network connections.
2. Make sure that the subject system is switched off. Ensure this with the hard drive and monitor activity lights. In case of laptops, remove the battery as, at times, it may power on while opening the lid.
3. Unplug the power cable and never switch on under any circumstances.
4. Label and photograph the components (ports, cables).
5. Detach the hard disk from the motherboard by disconnecting the data transfer cable and power cable.

6. Record the details of the hard disk such as the unique identifier, make, model, and serial number.
7. Obtain off-site data storage, if any, and details such as the operating system, application packages, and various users of the computer system.
8. After the removal of the hard disk, switch on the system, run BIOS, and record its date and time.
9. Complete the DEC form, documenting all the actions related to the acquisition of evidence.
10. The hard disk is now connected to the IO's computer through a write-block device to obtain the forensic copy from which forensic examination begins. The steps are as follows:
    (a) Connect the target drive to the forensic computer.
    (b) Connect the hard disk (evidence) to the forensic computer through the use of a write-blocker.
    (c) Use a forensic software tool to obtain an initial hash value for the evidence.
    (d) Use a forensic software tool to transfer the evidence onto the target drive.
    (e) Verify that the acquisition has been completed successfully and that the initial hash value for the evidence matches the acquisition hash value for the forensic copy.

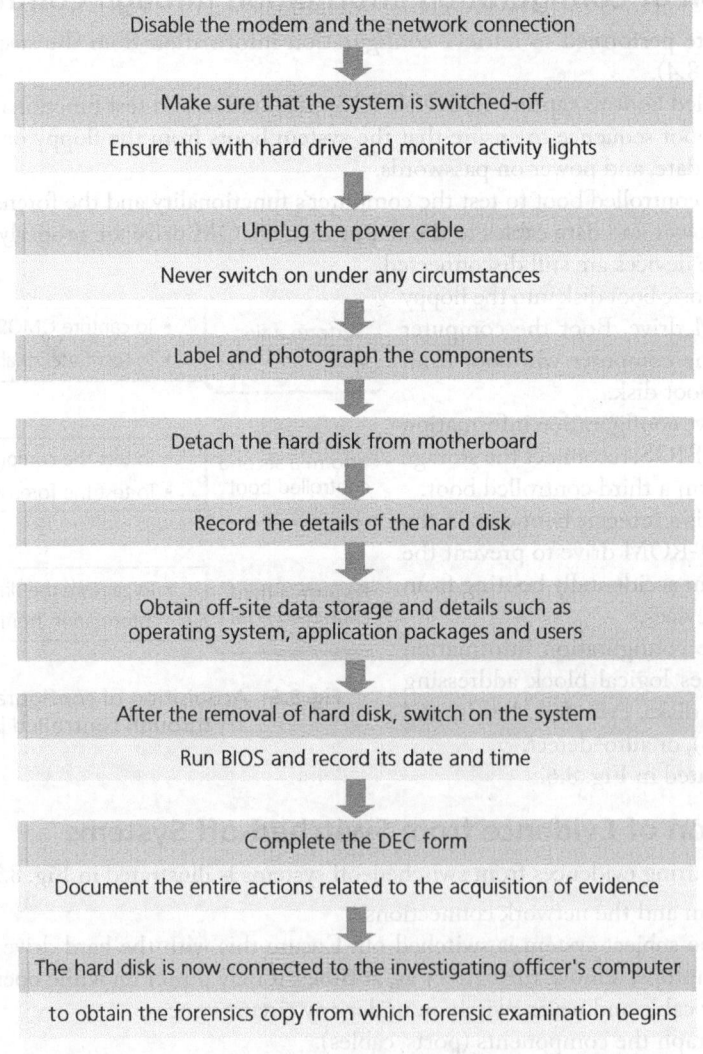

Fig. 8.7 Acquisition of evidence from switched off systems

8.5.3 Collection of Volatile Data

The first information to be captured is from the address resolution protocol (ARP) cache. It contains the mapping of the physical address and the network address. This information is useful in determining to which other computers in the network the suspect computer is connected to. The following information has to be collected from live systems before unplugging the computer. These include the following: (a) system date and time, (b) network connections, (c) list of open ports and the applications listening to those ports, and (d) applications that are running at the moment. The trace route command can be used to track the path taken by a packet from the source to destination. In other words, it reveals the routers through which the data has travelled. The netstat utility is used to list all the active connections, list of protocols running, and the list of open ports. Live acquisition facilitates the acquisition of data.

Live acquisition is possible in two ways: (a) saving the information to a removable disk, and (b) saving the information in a remote forensic system. Netcat tool (a free tool) is used to create a reliable connection between the target system and the remote forensic computer. In this way, the data is moved to the target in a relatively short span of time for possible analysis at a later point. Crptcat tool is an encrypted version of netcat which facilitates the encryption of data being exchanged between the target system and the forensic computer, thereby preventing data contamination.

8.5.4 Acquisition of Evidence from Live Systems

The process involved in obtaining evidence from live systems is illustrated in Fig. 8.8 and is as follows:

1. Disconnect the modem, if attached, and if networked, obtain expert opinion before disconnecting it.
2. Label and photograph all the components (ports, cables).

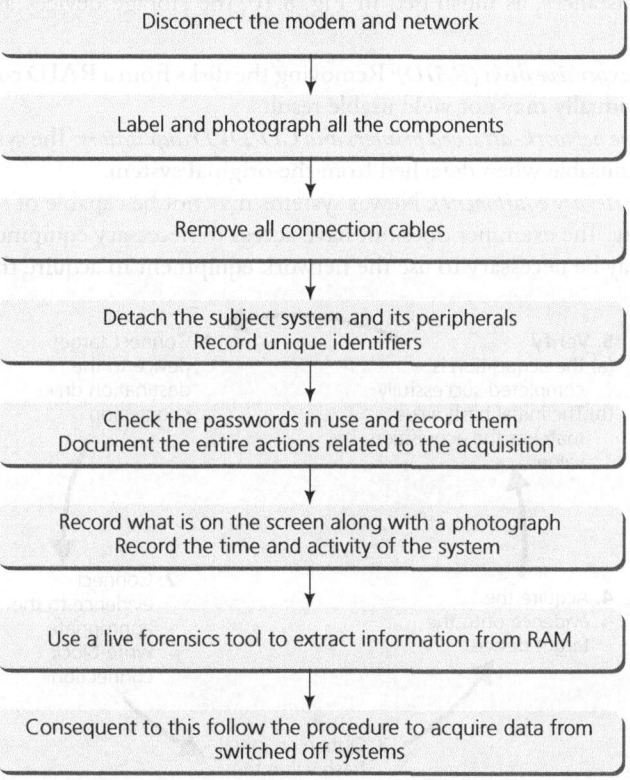

Fig. 8.8 Acquisition of evidence from live systems

3. Remove all connection cables leading from and to the subject system.
4. Detach the subject system and its peripherals, record their unique identifiers, and allow them to cool.
5. Check for passwords in use and record them along with documenting all the actions related to the acquisition of evidence.
6. Record what is on the screen along with a photograph of the same. Check whether the system is password-protected or not. Record the time and activity of the system.
7. Use a live forensics tool to extract information from RAM. In the absence of technical expertise, fit an uninterruptible power protection device and remove the end of the power supply cable attached to the system. Doing so will avoid any data being written to the hard drive.
8. Consequent to this, follow the procedure to acquire data from switched off systems.

8.5.5 Acquisition of Evidence from Standalone Hardware Device

The procedure for taking possession of evidence from standalone hardware devices is illustrated in Fig. 8.9, and is listed here:

1. Connect the target drive to the destination drive connection on the hardware device.
2. Connect the evidence to the appropriate write-block connection on the hardware device.
3. Obtain an initial hash value for the evidence.
4. Use the hardware device to transfer the evidence onto the target drive.
5. Verify that the acquisition is completed successfully and that the initial hash value for the evidence matches the acquisition hash value for the forensic copy.

8.5.6 Acquisition of Evidence from Non-detachable Hard Disk Drive

Under exceptional circumstances, as illustrated in Fig. 8.10, the storage devices are not removed from the subject system.

1. *Redundant array of inexpensive disks (RAID)*: Removing the disks from a RAID configuration and acquiring data from them individually may not yield usable results.
2. *Laptop systems and some network-attached printers and CD/DVD duplicators*: The system drive may be difficult to access or may be unusable when detached from the original system.
3. *Hardware dependency (legacy equipment)*: Newer systems may not be capable of reading older drives.
4. *Equipment availability*: The examiner does not have access to necessary equipment.
5. *Network storage*: It may be necessary to use the network equipment to acquire the data.

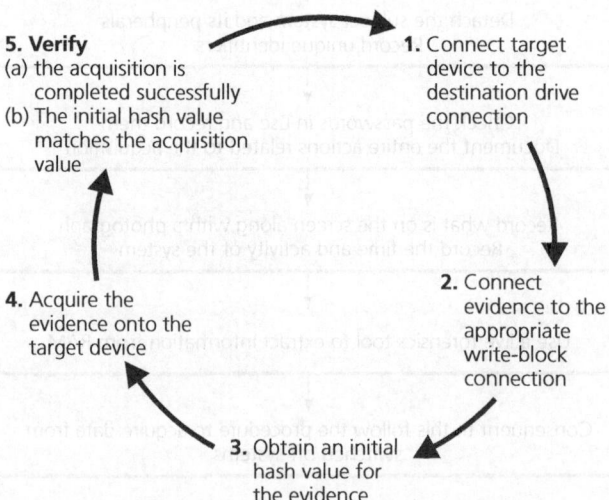

Fig. 8.9 Acquisition of evidence from standalone hardware device

In such a situation, the entire device becomes evidence. Forensic copying network acquisition, wherein the evidence computer is connected to the forensic computer using a special ethernet cable called cross cable or network crossover cable. On a crossover cable, on one end only the positive and negative 'receive' pair are switched with the positive and negative 'transmit' pair respectively to maintain polarity. This facilitates machines to talk with each other over the network crossover cable.

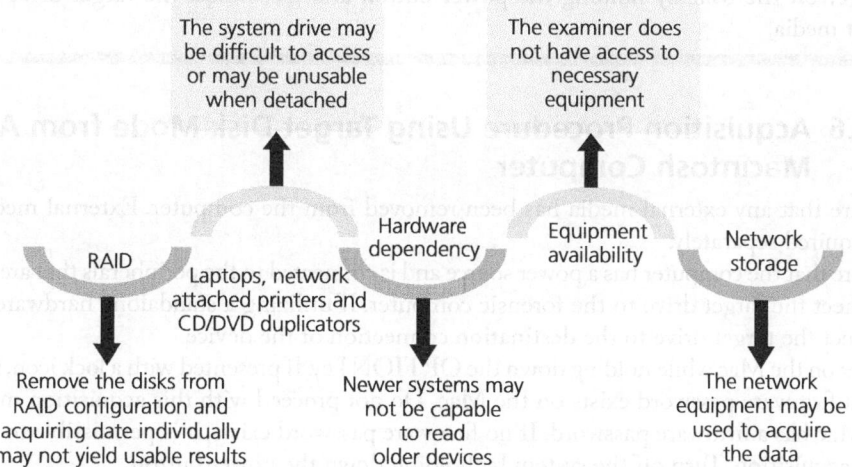

Fig. 8.10 Acquisition of evidence from non-detachable hard disk drive

Network Drives Imaging and Logical File Collection

It is not possible to shut down the machine and take the hard drive out or turn off the evidence machine if digital evidence has to be acquired from file servers or database servers serving business critical applications. Data is copied to external drives using forensic tools such as Cyber Check Suite, Encase Logical File Collection, or Robocopy (Boxes 8.5 and 8.6).

Box 8.5 Acquisition Procedure Using Forensic Boot Media from Apple Macintosh Computer (Adopted if Mac has more than one hard drive)

1. Ensure that any external media has been removed from the computer. External media shall be acquired separately.
2. Confirm that the computer has a power source and is connected to the peripherals that are needed.
3. Insert or connect the boot media into the Mac computer while powered off.
4. Connect the target drive to the Mac computer.
5. Power on the Mac while holding down the OPTION key. If presented with a lock icon, then a firmware password exists on the Mac. Do not proceed with this acquisition method if the Mac has a firmware password. If no firmware password exists, it is permissible to proceed with acquisition. The available bootable devices will appear on the screen.
6. Select the boot media. Be aware that the boot media may be listed as a 'Windows' disk as this is the default naming convention for non-Mac media. Be aware that some Mac computers have a Windows partition.
7. Obtain an initial hash value of the evidence Mac.

(Contd)

Box 8.5 (Contd)

8. Transfer the evidence from Mac onto the target drive.
9. Verify that the acquisition has been completed successfully and that the initial hash value for the evidence matches the acquisition hash value for the forensic copy.
10. Power off the Mac by holding the power button and disconnect the target drive and the boot media.

Box 8.6 Acquisition Procedure Using Target Disk Mode from Apple Macintosh Computer

1. Ensure that any external media has been removed from the computer. External media shall be acquired separately.
2. Ensure that the computer has a power source and is connected to the peripherals that are needed.
3. Connect the target drive to the forensic computer. If utilizing a standalone hardware device, connect the target drive to the destination connection of the device.
4. Power on the Mac while holding down the OPTION key. If presented with a lock icon, it means that a firmware password exists on the Mac. Do not proceed with this acquisition method if the Mac has a firmware password. If no firmware password exists, it is permissible to proceed with acquisition. Turn off the system by holding down the power button.
5. Power on the Mac by holding down the 'T' key to enter Target Disk Mode.
6. When the Target Disk Mode symbol appears on the screen, connect the evidence Mac to a hardware write-blocker attached to the forensic computer. If utilizing a standalone hardware device, connect the evidence Mac to the write-block connection of the device.
7. Obtain an initial hash value of the evidence Mac.
8. Acquire the evidence Mac onto the target drive.
9. Verify that the acquisition is completed successfully and that the initial hash value of the evidence matches the acquisition hash value of the forensic copy.
10. Disconnect the Mac and power off by holding down the power button.

Another method of getting data from hard drives is with a network cable between the machine containing the target media booted with a forensic tool for DOS and a second machine with a forensic tool for Windows. It provides the best offering—the advantages of DOS boot combined with Windows.

Network cable acquisition is useful when there are geometry mismatches in BIOS. It is also useful with RAID configurations because RAID can be booted to DOS using its native hardware configuration to mount the logical physical device. The forensic imaging tool views RAID as a mounted physical device, thus enabling acquisition and preview via the network cable connection to the forensic tool in Windows.

Once the computers are connected, the evidence computer is booted from a forensic distribution such as Helix or Linen and data can be acquired like a regular hard drive acquisition, from the evidence computer to the forensic computer, using a forensic tool like Encase.

8.6 ACQUISITION OF EVIDENCE FROM EMAIL AND INTERNET

Digital evidence has to be acquired from email in case of incidents such as email abusing, email phishing, and email scams. Information that can be obtained as evidence from email include the actual sender and recipient of the concerned emails, timestamp of the email transmission, intention of the mail, and record of the complete set of email transactions. Figure 8.11 shows the potential sources for the acquisition of evidence from email and the Internet, and is explained in detail in the following subsections.

Header

Information about the sender and/or the path along which the message has traversed can be obtained from the metadata (envelope and headers, including headers in the message body) of the email message. Besides this, the header also includes information such as IP address of the sender and the messaging initiation protocol (HTTP or SMTP). In some cases, the header is spoofed to conceal the identity of the sender. The header is analysed to acquire information about spoofing and the metadata is used as evidence.

Bait Tactics

This is done to confirm the sender of an email address and to track him/her which could then be used as evidence. In bait tactics, an email with http: 'Error! Filename not specified.' tag having an image source at a computer monitored by the IO is sent to the sender of the email under investigation containing a real (genuine) email address. When the email is opened by the sender of the email under investigation, a log entry containing the IP address of the recipient is recorded on the http server hosting the image and thus the sender is tracked. However, if the recipient (sender of the email under investigation) is using a proxy server, the IP address

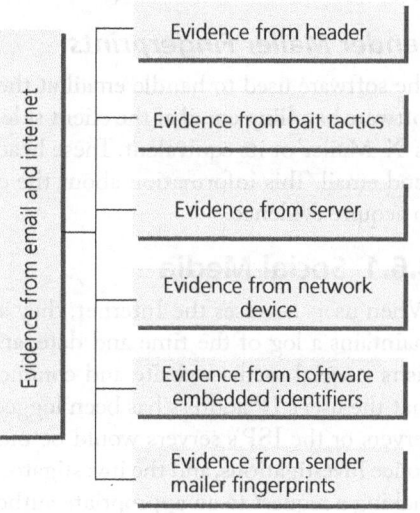

Fig. 8.11 Potential sources of evidence from email and Internet

of the proxy server is recorded. The log on the proxy server can be used to track the sender of the email under investigation. If the proxy server's log is unavailable for some reason, the IO may send a tactic email containing (a) an embedded Java applet that runs on the receiver's computer or (b) an HTML page with Active X Object to extract the IP address of the receiver's computer and email it to him/her.

Server

Server logs provide information about the source of an email message. Emails purged from the clients (senders or receivers), whose recovery is impossible, may be requested from servers (proxy or ISP) as most of them store a copy of all the emails after their deliveries. Further, the address of the computer responsible for making the email transaction can be traced from the logs maintained by servers. However, servers store the copies of emails and server logs only for a limited period of time and some may not co-operate with the investigators. Time is very important as HTTP and SMTP logs are archived frequently, especially by large ISPs. If a log is archived, it could take time and effort to retrieve and decompress the log files needed to trace the emails. Further, SMTP servers which store data such as credit card number and other data pertaining to the owner of a mailbox can be used to identify the person behind an email address.

Network Device

Digital evidence pertaining to email can be obtained from logs maintained by network devices such as routers, firewalls, and switches and is used to investigate the source of an email message. This is hard to acquire and is attempted when the logs of servers (proxy or ISP) are unavailable due to some reason, for example, when ISP or proxy does not maintain a log, lack of cooperation by the ISP, or failure to maintain the chain of evidence.

Software Embedded Identifiers

Information about the creator of an email, attached files, or documents may be included with the message by the email software used by the sender for composing the email. This information may be included in the form of custom headers or in the form of MIME content as a transport neutral encapsulation format (TNEF). Some vital information about the sender's email preferences and options could help in client-side evidence gathering.

It can reveal PST filenames, Windows logon usernames, MAC addresses, etc., of the client computer that was used to send the email messages.

Sender Mailer Fingerprints

The software used to handle email at the server can be revealed from the received header field. Similarly, the software handling email at the client side can be identified and ascertained using a different set of headers such as 'X-Mailer' or its equivalent. These headers describe the applications and their versions used by the clients to send email. This information about the client computer of the sender can be used to devise an effective plan to acquire evidence.

8.6.1 Social Media

When users accesses the Internet, they are allocated a unique address known as an IP address and their ISP maintains a log of the time and date, and the identity of the user allocated to any IP address. When a user visits a social media website and conducts some activity, for example, logs on or posts a message, it is likely that the user's IP address has been logged by the website. However, digital data resident on social networking servers or the ISP's servers would be more problematic to access. Access to such data would be restricted to police investigations, and the investigators involved would have to apply to the social network services provider, making a request to an appropriate authority.

Misuse of a social networking application can be ascertained using standard computer forensic procedures from the data resident on the hard drive of a computer, for example, in the web cache, Internet history, log-ins, username, and password relating to the social networking application. On mobile phones, a database related to, say, the Facebook application is stored in the phone's memory. The database stores the data of each friend in the list including their names, ID numbers, and phone number. Twitter uses directories to store information about Twitter account data, attachments sent with Tweets, usernames, and date and time values. MySpace uses an SQLite file to store the username of the MySpace application, as well as comments that the user has posted along with timestamps.

Digital evidence relating to social media usage can be acquired by either a physical or logical method. However, with logical acquisitions, there is the possibility that data stored in slack space may be missed.

Printouts of social media communications are considered documents, which may contain relevant evidence.

8.7 ACQUISITION OF EVIDENCE FROM MOBILE PHONE AND PDA

Mobile phones play a vital role in communication and hold digital evidence of value for an investigation. With Internet capabilities added to mobile phones, their significance in crime investigations has become of utmost importance. The evidence acquired from mobile phones (the device and SIM card) include user-created information (SMS, MMS, images, video, stored files, stored voice mail, connected computers, contacts), phone-created information (call history), Internet-related information (online accounts, email, social networking information), etc. Besides this, mobile devices, if tethered, can back up the phone's data onto a local workstation. GPS-enabled devices provide information on locations visited and maps. Mobile phones may also be synced with computer systems at home and at the workplace, and also to a cloud. Besides these, mobile phones may also be paired with music players in cars in order to play music and accept calls where the contact list in the mobile phones may be available. EXIF data embedded in photos can be a source of evidence as it provides information such as date and time when the photograph was captured, the device type used to create it, and the GPS coordinates to where the photo was taken. The primary sources of evidence from mobile devices include data captured from tablets and smartphones, supporting most operating systems, such as iOS, Android, RIM and Windows Phone, from SIM cards, or backup files. The other source of evidence is the cloud.

There are some unique points to be considered when preserving mobile devices as a source of evidence. Most mobile devices are networked devices, and they send and receive data through telecommunication systems, Wi-Fi access points, and Bluetooth piconets (an adhoc network formed using Bluetooth technology that can support a maximum of eight devices). Hence there is a possibility that the information which can serve as digital evidence in mobile devices is susceptible to being overwritten by new data or experience remote destruction commands from wireless networks and is lost completely. For example, Apple provides a Web-based service to remotely wipe a lost or stolen iPhone, and organizations that centrally manage BlackBerry devices can remotely wipe a specific device from the BlackBerry Enterprise Server. However, from a forensic perspective the major advantage of mobile devices is that they can retain deleted information even after an individual has attempted to render it unrecoverable. The underlying reason for this persistence of deleted data on mobile devices is in the use of flash memory chips to store data. Flash memory is physically durable against impact, high temperature, and pressure, making it more difficult to destroy. In addition, flash memory has a limited number of writes and can only be erased block-by-block, and mobile devices generally wait until a block is full before erasing data. Further, mobile devices use proprietary wear leveling algorithms to spread write/erase across flash memory blocks. This ensures that the deleted data remains for some time while new data is written to lesser used portions of the memory. In order to access and recover older/deleted copies of the data, it is necessary to acquire a full copy of the physical memory. Figure 8.12 presents the sources of acquiring digital evidence from mobile devices.

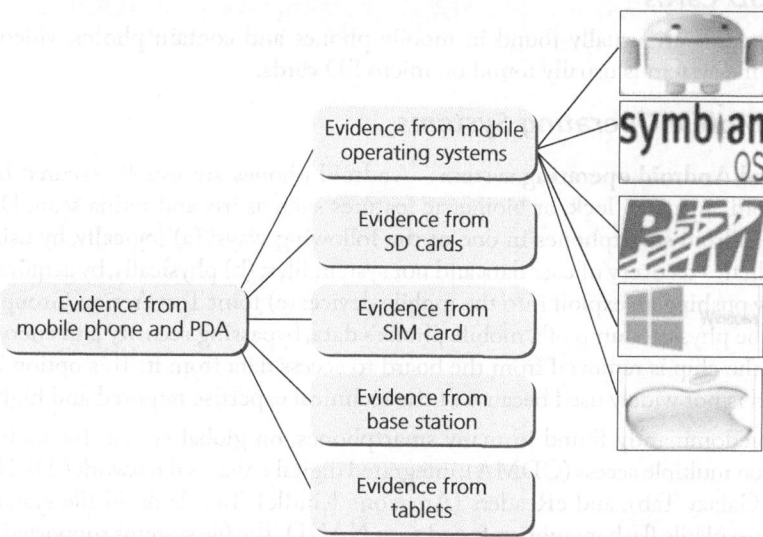

Fig. 8.12 Potential sources of digital evidence from mobile devices

Hardware imaging devices are used to acquire information from the mobile phone and its SIM card. Many forensic software and hardware are available for almost all the popular brands of mobile phones. However, it is not sufficient to address any specific issues as many new brands of mobile phones enter the market every year. Mobile operating systems are also more in number than that predominantly available for computer systems. Some of them are Symbian, iOS, Windows, Android, and RIM.

Evidence from Base Station

The first level of information on the call type, calls made and received, its duration, location (as the first and the last cell ID) of the user may be obtained from the call detail records (CDR) maintained by the service provider, usually for billing, upon request by an IO. Every mobile phone can be uniquely traced from its subscriber identity module (SIM) number in case of a global system for mobile communications (GSM) network and international mobile equipment identity (IMEI) number. Every SIM card contains an international mobile subscriber identity (IMSI) which is an internationally unique number to identify a user on a network. Besides this, tracking information can be obtained from connected cell towers over time and the location at different times. Details pertaining to Internet/data usage can also be obtained.

Evidence from SIM Card

A SIM holds an electronically erasable programmable read only memory (EEPROM) which holds the operating system, file system, user authentication information, and so on. Exploring the hierarchical file system, which comprises the master file (MF), dedicated files (DF), and elementary files (EF), helps in acquiring information. For instance the contacts on a mobile phone are located in the folder EF_ADN (extended file_abbreviated dialing numbers), and the list of outgoing calls made from EF_LND (extended file_last numbers dialled). Acquiring can be complicated if the SIM is PIN-protected. PIN Unlock Key (PUK) can also be changed by the user online and hence an IO can seek the support of the service provider. Consequently, the SIM card is cloned, similar to hard disk cloning by forensic tools, at the onset of investigation.

The type of evidence obtained from the SIM may be call details, as discussed earlier, SMS, MMS, photos, videos, apps, and maps. SMS may be stored on either the mobile device or the SIM. On the SIM it is present in DF_TELECOM. The status flag indicates if a message has been read, deleted, or sent. MMS is used for sending multimedia content such as image, audio, and video.

Evidence from SD Cards

Secure digital (SD) cards are usually found in mobile phones and contain photos, videos, apps, and maps. Microsoft's FAT32 file system is usually found on micro SD cards.

Evidence from Mobile Operating Systems

Android phones and Android operating system Android phones are usually secured by users using PIN protection, a password, a pattern lock, or biometric features such as iris and retina scan. Digital evidence can be extracted from Android smartphones in one of the following ways: (a) logically, by using mobile forensic software which facilitates recovery of user data and not system files; (b) physically, by acquiring a backup from a physical image or by pushing an exploit into the mobile device; (c) Joint Test Action Group (JTAG), an IEEE standard to obtain the physical dump of a mobile phone's data, bypassing security and encryption features; and (d) chip-off, where the chip is removed from the board to access data from it. This option is chosen when the chip is damaged and is not widely used because of the technical expertise required and high cost to conduct it.

Android OS is predominantly found in many smartphones, on global system for mobile communication (GSM), code-division multiple access (CDMA), integrated digital enhanced network (iDEN) cellular networks, tablets (Samsung's Galaxy Tab), and eReaders (Amazon's Kindle). The Android file system is located in the Android device's non-volatile flash memory referred to as NAND. The file systems supported by Android devices include Ext4, FAT32, and YAFFS2. YAFFS2 is an open source file system developed for use with NAND flash memory. Besides the file system, evidence is present in libraries, especially, the SQLite database which is an open source relational database standard predominantly found in mobile devices. Table 8.1 presents some of the sources of evidence on Android devices and their use.

Symbian OS This mobile operating system is usually found in Nokia, Sony Ericsson, Samsung, and Hitachi handsets but is not very popular nowadays.

Research in Motion—RIM This operating system is used on BlackBerry smartphones and tablets. Its files have .ipd extension. An IPD backup file is available on a synched computer which may serve as evidence even in the absence of the handset. Just as tools and sources of evidence are available for Android, tools exist to acquire evidence from IPD files available on computer systems, to extract email, SMS, MMS, call logs, photos, and so on.

Windows Phone This operating system by Microsoft is found in mobile phones of HTC, Samsung, and Nokia, and tablets. JTAG is used to acquire data from a Windows Phone and is in new technology file system (NTFS). Therefore, it does not require any conversion.

Table 8.1 Some of the sources of evidence on Android devices

Source of evidence	Use of evidence
Cache.WiFi	This is helpful in locating the user as it records the Wi-Fi hotspot, when the user walks past, regardless of whether the user connects to it or not.
Fb.db	This SQLite database contains contacts, chat logs, photos, and searches made by a user
com.google.android.gm	This can be used as a source to retrieve the Gmail login ID and password of the suspect if he/she has synced the gmail account on his/her mobile device.
/data/data/com.android.providers.telephony	This can be used as cellular telephone evidence of SMS.
/data/data/com.android.mms	This can be used as cellular telephone evidence of MMS.

Reference: Hayes, 2014.

iOS Evidence is acquired from an iOS device by placing it in 'airplane mode' so as to avoid remote wipe-off of the device or tampering of evidence by an outside entity. There are four approaches [Proffitt 2012] to acquiring evidence from iOS. They are as follows:

Acquisition via iTunes backup iTunes performs an automated backup when the mobile device synchronizes with any Windows or Mac OS for updates or syncing movies, music, or application. The backup folder is the source of evidence where the root contains the files presented in Table 8.2. The backup folder is stored in different locations depending on the operating system available in the synced systems as presented in Table 8.3 and forms the evidence. iPhoneAnalyzer by Crypticbit, a free Java-based multiplatform tool, is used to acquire data from the backup folder of iOS.

Table 8.2 Sources of evidence on iOS devices in the backup folder

Sources of evidence	Use
Status.plist file	Provides data about the latest backup
Info.plist file	Provides confirmation as to whether the backup matches the device from the IMEI number
Manifest.plist	Provides metadata about the backed up files

Acquisition via Logical methods This approach is adopted to acquire evidence from active files on iOS, especially SMS, call logs, calendar events, contacts, photos, web history, and email accounts. To acquire evidence, the iOS device has to be cabled with the forensic workstation, wherein a tool like iPhone Explorer is pre-installed.

Acquisition via physical methods Physical acquisition of device is essential to acquire evidence from the slack space and also of deleted items in the unallocated space. Further, to obtain a bit-by-bit copy of the original media, physical acquisition is necessary. Tools such as Lantern 2 forensic suite and iXAM are used to extract an image of the iOS device.

Acquisition by jail breaking Jail breaking refers to replacing the firmware partition in the iOS device with a hacked version so that tools such as SSH and Terminal can be installed by a forensic examiner to acquire evidence.

Evidence from tablets Most of the tablets available in the market rely on iOS and Android as the operating system. Some of them have variants compatible to run on cellular networks. To acquire evidence from them, Device Seizure and Cellebrite are the widely used forensic tools.

Table 8.3 Location of evidence (backup folder) from iOS devices on synced systems

Operating system at synced system	Location of backup of folder
Windows XP	%systempartition%\documents and settings\ %username%\Application Data\Apple Computer\MobileSync\Backup
Windows 7	%systempartition%\Users\%username%\AppData\Roaming\Apple Computer\MobileSync\Backup\
Mac OS	Users/%username%/Library/Application Support/MobileSync/Backup

8.7.1 Procedure for Acquiring Evidence from Mobile Phones

Acquisition principles that apply to any computing device are applicable to mobile devices. The procedure is as follows:

1. If the mobile is *off*, do not turn it *on*. If it is in *on* state do not turn it *off* as it would enable the password option, thus preventing evidence from being acquired.
2. Allowing the device to receive data through wireless networks might bring new evidence but might also overwrite the existing data. Hence it is advisable to isolate the mobile device from the networks while it is powered on.
3. Photograph the device and its display.
4. Label and collect all the cables pertaining to it.
5. Keep the device charged. Forensic analysis should begin immediately prior to complete battery discharge, if no data could be captured while charging the device.
6. Seize additional storage media such as memory card and flash memory. Faraday bags are used with seizure of mobile devices to prevent any manipulation of digital evidence during covert acquisition. It actually prevents any change that may take place on a mobile device upon receiving a signal.
7. Complete the DEC form.

8.8 ACQUISITION OF EVIDENCE FROM OPTICAL AND REMOVABLE MEDIA, DIGITAL CAMERAS

The following subsections briefly introduce how acquisition of digital evidence is possible from optical media, USB drives, and digital cameras.

8.8.1 Evidence from Optical Media

The following is the technique involved in securing evidence from an optical media:

1. Connect the target drive to the forensic computer.
2. Insert the optical media evidence into a read-only optical drive.
3. Use forensic software tools to obtain an initial hash value for the evidence.
4. Use forensic software tool to acquire the evidence onto the target drive.
5. Verify that the acquisition has been completed successfully and that the initial hash value of the evidence matches the acquisition hash value of the forensic copy.

8.8.2 Evidence from USB Drives

USB drives can be imaged using drive imaging tools such as FTK Imager, Encase, and Safeback, as discussed in Chapter 7. Prior to that, software or hardware-based write-blockers were used to connect it to the forensic machine.

8.8.3 Evidence from Digital Cameras

The memory card and the internal memory in the camera can be acquired using the technique used for USB drives.

1. Memory cards shall be removed from the digital camera and acquired separately.
2. Determine if the digital camera has internal memory storage. If an adapter cable for the digital camera is available, the internal memory of the camera shall be examined.
3. Connect the target drive to the forensic computer or forensic hardware device.
4. Connect the evidence to the forensic computer through the use of a write-blocker or forensic hardware device.
5. Use forensic software tool or hardware device to obtain an initial hash value for the evidence.
6. Use forensic software tool or hardware device to acquire the evidence onto the target drive.
7. Verify that the acquisition has been completed successfully and that the initial hash value of the evidence matches the acquisition hash value of the forensic copy.

8.9 ACQUISITION OF EVIDENCE FROM THIRD PARTY, EXTERNAL AGENCY, OR ORGANIZATION

The list of potential sources of evidence, especially third party organizations, is as follows:

Time zone conversion Time zone plays an important role in linking the acts and incidents to the local time so that offences can be mapped to offenders. Time zone computes local time as an offset from UTC (Greenwich Mean Time). For example, the website at http://www.timeanddata.com/worldclock.meeting.html helps in converting time to Indian standard time (IST).

Internet service providers The following information can be obtained by law enforcement officers upon request: (a) username, (b) personal details, (c) day-wise activity (usage history), (d) physical address, and (e) telephone number in case of DSL/CDMA/3G.

Social networking sites The following can be obtained from social networking sites such as Facebook, Orkut, and so on upon request by law enforcement officers: (a) username; (b) personal details updated in the profile, especially updated email IDs, if any; (c) IP address from where the profile is accessed; (d) user activity; and (e) friends and group-related information, etc.

Financial/Banking institutions The following can be obtained from financial institutions upon request by law enforcement officers: (a) personal details updated in the profile of the account holder, (b) transaction details, (c) CAF and other supporting documents submitted by the customer along with the introducer details, and (d) IP address from where an Internet banking transaction was initiated.

Website domain hosting providers The following can be obtained from financial institutions from website domain hosting providers: (a) registration details of the website, (b) access details, (c) FTP logs, (d) payment details, (e) technical/administrative/owner of the domain, (f) details of the website developer.

8.10 CHALLENGES TO ACQUISITION OF DIGITAL EVIDENCE

Some of the challenges to digital evidence acquisition are summarized here:

1. An increase in the number and models of ECDs, each containing immense volumes of data, poses challenges to evidence acquisition as it requires a forensic examiner who is involved in acquisition to be proficient enough and skilled to gather the right evidence from the ECDs.
2. The advancement in technology as well as the offenders becoming skilled at concealing evidences within the ECDs makes it difficult to acquire evidence if not done in a timely manner.

3. Software for wiping data from hard drives can be procured online, which may erase the evidence from the storage media before the acquisition of evidence.
4. Offenders can adopt counter forensic measures such as encryption and steganography, as they are freely available on the Internet.
5. The IO who is not a trained computer examiner usually seizes the evidence from ECDs and packs the evidence and sends it to a forensic lab so as to be examined by a trained computer forensic examiner. At times, the examination process takes several days to years and may raise questions on acquisition at a later period rather than providing the solution for examination.
6. Besides basic file system challenges, updates made to the operating system with every version brings obstacles to digital forensic investigations as what worked on one version may not work on the other.

8.11 HANDLING OF DIGITAL EVIDENCE

The following are the processes in the handling of digital evidence acquired from various sources:

Shut Down Computer and Take Digital Photographs of Original System and/or Media

The computer system is shut down using proper commands. Photographs of the system, from all angles, are necessary to document the system hardware components and how they are connected. If the system is in the *on* state, the status on the screen, the list of programs running in the background, list of files open, and system statistics are all recorded and photographed, and the computer is shut down at the earliest.

Document Hardware Configuration of System

1. Disconnect storage devices (using the power connector or data cable from the back of the drive or from the motherboard) to prevent destruction, damage, or alteration of data.
2. Disassemble the case of the computer to be examined to permit physical access to the storage devices. Identify storage devices that need to be acquired. These devices can be internal, external, or both.
3. Document internal storage devices and hardware configuration, drive condition (e.g., make, model, geometry, size, jumper settings, location, drive interface), internal components [e.g., sound card, video card, network card (including media access control, MAC address), personal computer memory card international association (PCMCIA) cards].

Chain of Custody

A good way of ensuring that acquired evidence data remains uncorrupted is to keep a chain of custody. This is a detailed list of what was done with the original copies once they were collected. It includes documentation of everything that happened with the evidence—who found the data, when and where it was transported (and how), who had access to it, and what they did with it. It plays a major role during the admissibility of evidence in court.

Packaging and Labelling of Evidence

Packaging and labelling refers to acquiring evidences and numbering them so that it becomes easy to locate and retrieve at a later point of time. Different types of evidence need special packaging, such as evidence envelopes, bags, and containers. Each piece of evidence should be packaged separately, properly sealed, labelled, and documented.

Labelling helps the system to be easily reconnected when the system configuration is restored to its original condition at a secure location or in the forensics lab. Label all media appropriately with evidence labels.

Transportation of Evidences

The evidence is transported to a secure place where the chain of custody is ensured and is safe. Diskettes have fragile magnetic media which could be damaged and data may be lost if it is loosely packed and disturbed during transit.

Box 8.7 Forensic Duplication

Forensic duplication retains every bit on the source media including deleted files (except host protected area, HPA, discussed in Chapter 6) so as to make the evidence admissible in court proceedings. It is also called forensic copying or forensic imaging. Forensic image is of three types, namely complete disk, partiton and logical (copies selected files/folders, active data and not deleted files). Hashing is used to ensure that the data copied from the source is unaltered but accurate. Write blockers (hardware or software) are used to ensure that the source media is not modified during forensic copying. Forensic duplication is done with tools such as Encase, FTK Imager, Safeback, dd utility (discussed in Chapters 7 and 9) such that no changes are made to the original storage medium during the process. Further, the results generated by such tools should be repeatable and verifiable by a third party. The tools have logs which provide details about the actions taken with it on the source media, as well as errors encountered, if any, during the duplication process.

The computer components, especially integrated circuits (memory, processor, and expansion cards), should be guarded against electrostatic discharge. During transportation, the electromagnetic field created by magnets and radio transmitters can alter or destroy data. Hence the CPU, devices, and media should not be exposed to drastic changes in temperature (extreme heat or cold), high humidity, and moisture and should not be placed in a vehicle trunk. The computer system should be moved to a secure location where the chain of custody can be maintained. The computers and the media should never be left unattended unless it is locked in a secure location (Box 8.7).

Make Bit Stream Backups of Storage Media

The original evidence is left untouched and forensic examination is performed only on the working copy or the bit stream copy. Fill out an evidence tag for the original media, or for the forensic copy or forensic duplication, and store the best evidence in an evidence-safe location.

Evidence Validation

Evidence validation is done to ensure that the acquired evidence is identical to the one presented in court. The forensic copy is validated with the original evidence using hashing algorithms. This is usually done by comparing the MD5 hash of the original media and the forensic copy.

Examine Contents of Hard Drive Currently Placed within Computer, Record Information about Computer System under Examination

Document system date and time The system date and time setting, at the time the computer system is acquired as evidence, should be documented to serve as proof in court.

Make list of key search words Instead of looking at every aspect of the file, keywords are used to make a search on all computer hard disk drives and floppy diskettes using automated software to find some evidence. When relevant evidence is identified, the fact should be noted. When new keywords are identified, they should be added to the list, and a new search should be conducted using the text search utility.

Evaluate Windows swap file and file slack, and unallocated space or erased files The Windows swap file is a valuable source of evidence. Automated tools take only a few minutes to do this in contrast to several days with other tools. File slack is a data storage area. The data dumped from the memory ends up being stored at the end of allocated files, beyond the reach or view of the computer user. Specialized forensic tools for file slack can provide a wealth of information and investigative leads. The hard disk drive retains information about deleted files and unallocated space should be evaluated using specialized and automated forensic tools. Unallocated space is typically a good source of data that was previously associated with word processing temporary files and other temporary files created by various computer applications. The list of deleted files is analysed and the files are sorted based on the filename, file size, file content, creation date, and last modified date and time. Such information can provide a timeline of computer usage.

Deal with file, program, and storage anomalies Encrypted files should be handled separately as they cannot be handled by generalized tools. After decrypting, a keyword search has to be carried out to look for evidence. Any hidden partition in the hard disk should be identified and added in the documentation.

Document and retain copies of used software All the software used for forensic evaluation of the evidence, including the version number of the programs, is documented. A copy of the software used has to be included with the output of the forensic tool or the documentation, to eliminate confusion, when the software version is upgraded (which will result in a change in the output). Retaining the original version would help.

Evidence Log and Role of Evidence Custodian

An evidence custodian enters a record of the best evidence in the evidence log. For each piece of best evidence, there will be a corresponding entry in the evidence log. An evidence custodian ensures that backup copies of the best evidence are created. He/She will create tape backups once the principal investigator for the case states that the data will no longer be needed in an expeditious manner. An evidence custodian performs a monthly audit to ensure all of the best evidences are present, properly stored, and labelled.

Post Seizure of Evidence

Once the digital evidence is seized, orders of the competent court should be obtained to retain them under the custody of the IO. Orders should also be obtained to perform forensic analysis and seek expert opinion. If the release of seized evidence is claimed by the accused, the IO should prepare appropriate objections, and can only return the forensic copy of the original evidences if directed by the court. On no account is the original evidence seized returned.

Preservation of Evidence

To preserve digital evidence for later producing it before the court during trial, the following may be done:
1. Keep the digital media in an anti-static cover that is labelled and sealed properly.
2. Maintain an inventory list of all the media seized, with case number.
3. Preserve and store the media in a cool and dry place.
4. Preserve and store the media in a fire-proof and tamper-proof storage device.
5. Update the chain of custody if the media is taken out for some reason.

8.12 PRECAUTIONS INVOLVED IN HANDLING DIGITAL EVIDENCE

Digital evidence is sensitive to humidity, shock, extreme temperatures, static electricity, magnetic fields, etc., besides being fragile. Hence the IO should adopt precautionary measures while photographing, packaging, transporting, and documenting the digital evidence so as to avoid alteration, damage, and destruction. Some of the precautionary measures while handling digital evidence are as follows:
1. It should be packed in anti-static packaging and not in plastic bags as it produces static electricity, in addition to humidity and condensation, which may destroy the evidence.
2. It should be packed correctly so as to prevent it from getting scratched and to avoid bending and deformation.
3. Mobile devices should be packed in Faraday bags, aluminium foil, or radio-frequency shielding material so as to prevent exchange of data to or from the devices with the intention to alter or destroy the evidence.

Acquisition and Handling of Digital Evidence

4. It should be kept away from magnetic fields so as to protect it from induced static electricity.
5. It should not be held in vehicles for a long time as it may be prone to heat, cold, or humidity which may destroy the evidence.
6. It should be stored in a secure place, free from the effects of extreme weather conditions and any other radiations that could destroy it.

POINTS TO REMEMBER

- Evidence may be oral or documentary, and documentary evidence may be primary or secondary.
- Digital evidence is unique and can be judged as neither primary nor secondary evidence.
- Digital evidence is defined as information or data stored on, transmitted, or received by an electronic device in binary form that is of value in a criminal investigation and that is relied upon in court.
- The sources of digital evidence include open computer systems, communication systems, and embedded computer systems.
- Digital evidence is usually collected from a computer's hard drive, mobile phone, personal digital assistant (PDA), CD or DVD, flash card in a digital camera, memory stick and memory card, removable disk, ipad, tab, and so on.
- The characteristic of digital evidence is that it is latent or is invisible, highly fragile and volatile, is time-sensitive, can be easily altered or destroyed, and requires specialized training for handling.
- The crime scene is usually a house with one or more computers, a cyber café, or an organization/company with systems being networked.
- Acquisition of digital evidence entails creating a bit-perfect copy of digital media, either on-site where the electronic device is kept, or, if the device can be transported, in a clean room or a forensic lab.
- The order of volatility of digital evidence is as follows: CPU, cache and register content, routing table, ARP cache, process table, kernel statistics, memory, temporary file system/swap space, data on hard disk, remotely logged data, physical configuration and network topology, and data contained on archival media.
- Any seizure should be justified, appropriate, and proportionate.
- Seizure may be effected either with or without a search warrant depending on the crime and the items to be seized.
- Forensic copying refers to creating an exact, bit stream copy of the original storage media that exists on the subject computer.
- A write-blocker is used to ensure that the backup or imaging process does not alter the data on the original media.
- After backup or imaging, the message digest is computed so as to verify integrity.
- Digital evidence can be obtained from computer systems from volatile data, boot configuration, switched-off systems, live systems, and non-detachable hard drives by following certain procedures and using the appropriate tools.
- Digital evidence can be obtained from emails and headers (using bait tactics), servers, network devices, software embedded identifiers, and sender mailer fingerprints using the respective tools.
- Digital evidence can be obtained from mobile devices and PDA with help from the service provider, base station, SIM, mobile operating system, and SD card by following appropriate procedures and using the respective tools.
- Digital evidence can be obtained from optical media, camera, and also from third-party organizations.
- While handling digital evidence authentication, the chain of custody and evidence-safe have to be ensured.

KEY TERMS

Chain of custody This refers to the documentation containing a list of the people who were in possession of the evidence.

Cold data acquisition This refers to acquiring data from a device while it is off.

Digital evidence collection or DEC form This

refers to a part of the documentation containing information related to the collection of evidence from every electronic device.

Digital evidence This refers to information or data stored on, transmitted, or received by an electronic device in binary form that is of value in a criminal investigation and that is relied upon in court.

Evidence safe This refers to the storage area where the best evidence, usually the original or the first copy, has to be stored and should not be accessed by anyone except the evidence custodians.

Forensic copying This refers to creating an exact, bit stream copy of the original storage media that exists on the subject computer.

Hot data acquisition This refers to acquiring data from a device while it is in operation.

MULTIPLE-CHOICE QUESTIONS

1. Evidence is the lifeline of any _____.
   (a) trial
   (b) trial and enquiry
   (c) trial, enquiry, and quasi-judicial enquiries
   (d) none of these

2. _____ of Indian Evidence Act (IEA) includes that oral evidences must always be direct, that is, those evidences which are personally seen or heard by the witness giving them and not heard or told by someone else.
   (a) Section 20   (b) Section 40
   (c) Section 50   (d) Section 60

3. _____ of IEA includes all those documents which are presented in court for inspection regarding a case, known as documentary evidences.
   (a) Section 3   (c) Section 5
   (b) Section 4   (d) Section 6

4. Henseler has categorized the sources of digital evidence as _____.
   (a) open computer systems and communication systems
   (b) open computer systems, communication systems, and embedded computer systems
   (c) open computer systems and embedded computer systems
   (d) communication systems and embedded computer systems

5. The electronic evidence may serve as any of the following:
   (a) object, subject, witness, tool, accomplice
   (b) victim, subject, tool
   (c) object/victim, subject, witness, tool, accomplice
   (d) victim, tool, accomplice

6. The characteristic of digital evidence is _____.
   (a) visible   (c) time sensitive
   (b) non-volatile   (d) unhandy

7. The checksum of the _____ and the forensic copy on the _____ must be exactly the same.
   (a) original media, any media
   (b) original media, target media
   (c) target media, original media
   (d) target media, any media

8. _____ is an example of offences against individual.
   (a) Computer fraud   (c) Identity theft
   (b) Software piracy   (d) Homicide

9. On-site forensic copying is planned and performed if _____.
   (a) the investigating officer (IO) has the necessary equipment
   (b) the investigating officer (IO) has the necessary technical expertise
   (c) the investigating officer (IO) has the necessary equipment and technical expertise
   (d) the investigating officer (IO) has the necessary equipment and technical expertise or the technical support to take a forensic copy of the media in the ECD that holds the evidence.

10. According to the standard operating procedure (SOP), during investigation of a crime scene, the following should be done:
    (a). Status of the electronic device should be quarantined.
    (b) Electronic device has to be ascertained.
    (c) The electronic device should be disconnected from the Internet.
    (d) Except portable devices, all the electronic devices should be powered off.

11. _____ of IT Act 2000 (amended 2008) states that any police officer not below the rank of a police inspector or any other officer of the central or state government authorized by the central gov-

ernment in this behalf may enter any public place and search and arrest without warrant any person who is reasonably suspected, having committed, of committing, or of being about to commit any offence or crime under this Act.
(a) Section 40 (c) Section 80
(b) Section 60 (d) Section 20

12. _____ compels the individual or the organization that owns the computer system to surrender it.
(a) Stop-and-frisk procedure
(b) Subpeona
(c) Open field
(d) Search incident to arrest

13. _____ acquisition is useful when there are geometry mismatches in BIOS.
(a) Network cable (c) Cable
(b) Network (d) None of these

14. MySpace uses an _____ file to store the username of the MySpace applications as well as comments.
(a) DB (c) SQLite
(b) SQL (d) XL

15. _____ plays an important role in linking the acts and incidents to the local time.
(a) Time zone conversion
(b) Internet service providers
(c) Social networking sites
(d) Financial/Banking institutions

REVIEW QUESTIONS

1. List the types of cybercrime and their probable location of digital evidence.
2. What are the steps to be followed by an IO during a crime scene investigation?
3. What is the information provided by computer hardware to identify that an individual has used a computer?
4. Discuss in detail the acquisition of computer and electronic evidence.
5. What are the major concerns in the handling of digital evidence? Explain in detail the process of handling of digital evidence.
6. Mention and explain in detail the sources of acquiring evidence other than from the hard disk drive.
7. Define and distinguish a warrant search from a warrantless search.
8. Explain the precautionary measures to be taken before acquisition of digital evidence.

APPLICATION EXERCISES

1. Imagine that you are the forensic examiner for a crime reported as 'theft of intellectual property' by an organization, say, X. The IO had reported after a thorough investigation that an employee, say, Y had become disgruntled in recent months and left the organization. On the day Y left the organization, he was spotted by co-workers as downloading some data from the computer to a removable storage media which is against the organization's security policy. Employee Y was associated with a research team in organization X where the research findings were on the production line and was worth huge money. The IO had already confirmed with employee Y about the usage of removable storage on the day he left the organization, for which he had given a justification to the effect that he had copied only the personal content which he didn't want to leave behind with the organization. He also added that most of the content on the removable storage had already been deleted. The removable storage has now been seized for forensic examination. Answer the following questions from the given context:
(a) Explain in detail the documentation that has to be done for the seizure and acquisition of evidence.
(b) How will you describe the physical characteristics of the evidence in the report?
(c) What will you do to secure and preserve the crime scene?

2. Imagine that you are on-site during a search warrant and the suspect's computer is powered on and you observe one of the following icon's in the status bar (shown in Fig. 8.13). What could this imply and what actions would you take as a result of your observation while on scene?

3. Imagine that you are an IO. What are the types of digital evidence that could reveal a cybercriminal when you are handling a case that involves emails and the Internet?

Fig. 8.13 Icon on computer

BIBLIOGRAPHY

1. Darren Hayes, R. (2014), *A Practical Guide to Computer Forensics Investigations*, Pearson IT Certification
2. John Vacca, R. (1995), *Computer Forensics Computer Crime Scene Investigation*, 2nd edn: Charles River Media
3. Marjie Britz, T. (2013), *Computer Forensics and Cyber Crime - An Introduction*, 3rd edn: Pearson Education India
4. Robert Newman, C. (2007), *Computer Forensics Evidence Collection and Management* 1st edn: Auerbach Publications
5. Eoghan Casey (2011), *Digital Evidence and Computer Crime Forensic Science, Computers and the Internet*. 3rd edn: Academic Press
6. Michael Solomon, G, Ed Tittel, Neil Broom and Diane Barrett (2011), *Computer Forensics JumpStart*. 2nd edn: Sybex
7. Michael Sheetz (2007), *Computer Forensics An Essential Guide for Accountants, Lawyers and Managers*, Wiley Publications
8. Angus Marshall, M. (2008), *Digital Forensics Digital Evidence in Criminal Investigation*. Chichester, UK; Hoboken, NJ: Wiley-Blackwell
9. *Cyber Crime Investigation Manual*. NASSCOM
10. Tariq, M. and Banday, P.G. (November 2011), *Techniques and Tools for Forensic Investigation of E-mail*. International Journal of Network Security and its Applications (IJNSA), Vol.3, No.6
11. Charalambous, Elisavet Bratskas, Romaios Koutras, Nikolaos, Karkas George and Anastasiades (2016), '*Email Forensic Tools: A Roadmap to Email Header Analysis through a Cybercrime Use Case*'. Journal of Polish Safety and Reliability Association Summer Safety and Reliability Seminars, Volume (7), available at: http://jpsra.am.gdynia.pl/upload/SSARS2016PDF/Vol1/SSARS2016-Charalambous.pdf (Accessed 05 December 2017)
12. U.S. Department of Homeland Security, United States Secret Service (2007), *Best Practices For Seizing Electronic Evidence v.3 A Pocket Guide for First Responders*, available at www.crime-scene-investigator.net/SeizingElectronicEvidence.pdf (Accessed 05 December 2017)
13. Tim Proffitt (05 November 2012), *Forensic Analysis on iOS Devices*. SANS Information Security Training, available at: https://www.sans.org/reading-room/whitepapers/forensics/forensic-analysis-ios-devices-34092 (Accessed 05 December 2017)
14. Henseler J. (2000), *Computer Crime and Computer Forensics*, Encyclopedia of Forensic Science, London: Academic Press
15. Peter Coons, (2005), *How to Document Your Chain of Custody and Why It's Important*, available at: http://d4discovery.com/discover-more/how-to-document-your-chain-of-custody-and-why-its-important#sthash.D8scM4JR.dpbs (Accessed 10 February 2018)

Answers to Multiple-choice Questions

1. (c)	2. (d)	3. (a)	4. (b)	5. (c)
6. (c)	7. (b)	8. (d)	9. (d)	10. (d)
11. (c)	12. (b)	13. (a)	14. (c)	15. (a)

Analysis of Digital Evidence

9

Learning Objectives

This chapter provides an overview of the analysis of digital evidence. The objective of this chapter is to provide an insight into forensic copying, computation of hash, analysis of files stored in storage media, identification and retrieval of deleted files, etc. The chapter also presents the steps associated with live forensics. It introduces the working of various forensic tools such as FTK Imager, Autopsy, Volatility, and WinHex. Further, it explains email tracking and tracing with necessary tools. Finally, it explains the role of a forensic analyst along with report preparation. The reader will be familiar with the following after studying the chapter:

- Analysis of digital evidence
- Working with forensic tools used for analysis
- Role of the analyst and report preparation

9.1 INTRODUCTION TO ANALYSIS OF DIGITAL EVIDENCE

Analysis of digital evidence is the process of identifying, preserving, interpreting, and documenting the evidence recovered for presentation in a civil or criminal court. Evidence may usually be present in the form of computer documents, emails, texts and instant messages, transactions, images, and Internet histories on electronic communication devices (ECDs) which may be either mobile or standalone. Analysis of evidence involves the seizure of the ECD and acquisition of evidence. This process is also called forensic analysis of digital media, digital discovery, electronic discovery, and forensic examination.

Analysis of evidence provides assistance to forensic investigators in a variety of ways—for example, to determine evidence associated with a financial crime,, to reconstruct evidence by analysing emails, documents, and correspondence on an employee's computer so as to prove wrongdoing (e.g., theft of data), to acquire evidence from email, SMS, and text messages when they are deleted intentionally in harassment cases, etc.

Analysis of digital evidence helps in recovering deleted files, searching the slack space and unallocated space on a hard drive where potential evidence usually resides, examining Windows artifacts (explained in Chapter 6), etc.

Digital evidence should be examined by only those trained specifically for that purpose (forensic examiner or forensic analyst) and only a good analyst knows, for example, how to process hidden files that may contain past usage information that may be essential for use in court. Examination of ECDs should be made in an isolated chamber so as to prevent connections to any networks and interference from other mobile devices.

Once the digital evidence is sent to the laboratory for analysis, the following are the steps a qualified examiner performs during the analysis of evidentiary digital media:

1. He/She verifies if the physical seal of the digital media/ECD is intact and then removes it from the storage media.
2. After opening the physical seal, he/she computes the hash value of the content available in the storage media/ECD.
3. He/She compares the computed hash value with the hash value computed and specified by the investigating officer (IO) at the scene of crime.

4. He/She creates the forensic copy (or work copy) of the original storage device prior to analysis of digital evidence.
5. He/She computes the hash value of the forensic copy and compares it with the original (as explained in step 3). All these hash values should be the same if the chain of custody and the integrity of evidence has been preserved.
6. He/She creates a forensic archive from the forensic copy on another form of media so as to keep the original pristine. Analysts should use a clean storage media, preferably wiped, to prevent contamination (ensured with a hash value of zero which is an indication that the storage media is clean) and use write-blocker to avoid introduction of data from some other source.
7. He/She determines the make and model of the storage media/ECD once a working copy is created to select appropriate extraction software that can completely parse data and view its contents. He/She also looks at the request made by the IO to choose the appropriate tool to analyse the forensic copy.
8. The forensic examiner examines the files on the drive, hidden files, deleted files, etc., apart from evidence from the Internet, such as chat rooms, messaging applications, websites, Internet addresses, and email headers, and exports potential probative digital data related to the investigation.
9. After gathering the relevant digital evidence from the forensic copy, he/she applies the same tool and the method on the original media and obtains the digital evidence.
10. He/She then documents all the aforementioned steps and prepares a report to be produced in court.

All of these steps are critical to the overall success of the investigation and eventual prosecution of the culprit. Analysis also looks for information from individuals associated with the case to determine the forensic tools required to look for relevant evidence. Forensic tools are required to look for keywords and deleted files, to evaluate the file slack and unallocated space on the disk, and to evaluate encrypted files. A forensic analyst should document the date and time at which the ECD was taken into possession. He/She should document the filename, and its date of creation and last modification, which are relevant to the case. The documentation process should also include the software and its version that was used for forensic evaluation and retain its copy. Documentation should contain the findings, the issues that are identified as well the evidence found relevant to the case.

This chapter presents how the evidence present in various media and ECDs are analysed with the respective forensic tools. In the ensuing subsections, the following are explained with screenshots that were obtained by performing tests on an i3 machine with 8GB RAM and 500 GB HDD:

1. Live forensics or RAM analysis with Volatility
2. Acquisition of a forensic copy with FTK Imager
3. Acquisition of a forensic copy, computing of hash, and analysis of the hard disk with WinHex
4. Email tracking and tracing with emailTrackerPro and OnlineEmailTracer
5. Analysis of deleted files with Autopsy

9.2 CAPTURING OF FORENSIC COPY OF MEMORY AND HARD DRIVE WITH FORENSIC TOOLKIT IMAGER

Potential evidence during analysis is the memory and the hard drive and so all the data from them has to be retrieved as explained in the following sub-sections.

Capturing Main Memory

The runtime version of the Forensic Toolkit (FTK) Imager from Access Data can be used to access the memory and capture the contents of the drive, especially passwords, before the contents get encrypted. Upon running FTK Imager and clicking on *Capture Memory* (as shown in Fig. 9.1) and specifying the destination path and filename [as shown in Figs 9.2(a) and (b)], the memory contents are captured in the respective file as shown in Fig. 9.4. The Windows system file, that is, swap file, pagefile.sys contains useful information. This is shown in Fig. 9.3.

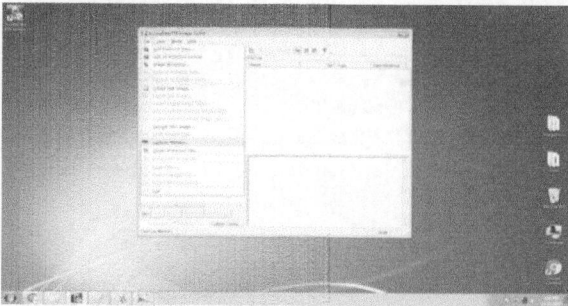

Fig. 9.1 Capturing memory using FTK Imager

Registry files cannot be acquired for analysis while the system is running. Rather, the registry files can be extracted from the forensic copy of the hard drive. To do so, the *Obtain Protected Files* option from the File menu of FTK Imager has to be selected to enable copying of live registry files. This provides information stored within the registry files, for example, passwords, if the user has allowed the operating system to remember his/her passwords. The *Minimum Files for Login Recovery* option retrieves users, system, and SAM files which can recover the user's account information. The *Password Recovery and All Registry Files* option can retrieve users, system, SAM, NTUSER.DAT, default, security, and software files, which contain both account information and passwords to other files.

Capturing Hard Drive

To create a forensic copy, *Create Disk Image* from the File menu is selected. In order to get the forensic copy of the entire device, the required *Physical Drive* is chosen from among the choices. After selecting the source drive, choose *Finish*. The type of image file to be created can be specified. Raw is a bit-by-bit uncompressed copy of the original, whereas the other three alternatives (SMART, E01, and AFF) are designed for use with a specific forensics program. For example, E01 is an EnCase forensic copy file format. The destination location to hold the forensic copy is specified (as shown in Fig. 9.6) where the forensic copy will be created. Forensic copies can be verified once they are created. When forensic copying is complete and the *Image Summary* button is clicked, a pop-up window (shown in Fig. 9.5) appears, to show the name of the forensic copy, the sector count, computed (before forensic copy creation) and reported (after forensic copy creation) MD5 and SHA1 hash values with a confirmation that they match, and a list of bad sectors (if any). This information is also saved as a text file. The directory listings of all the files in the forensic copy can also be created as shown in Fig. 9.7.

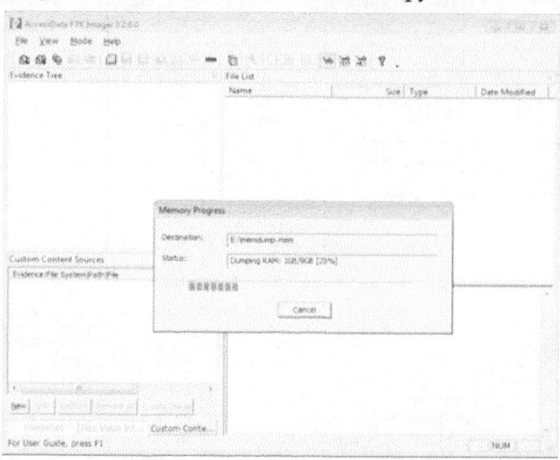

Fig. 9.2 (a) Acquiring memory dump

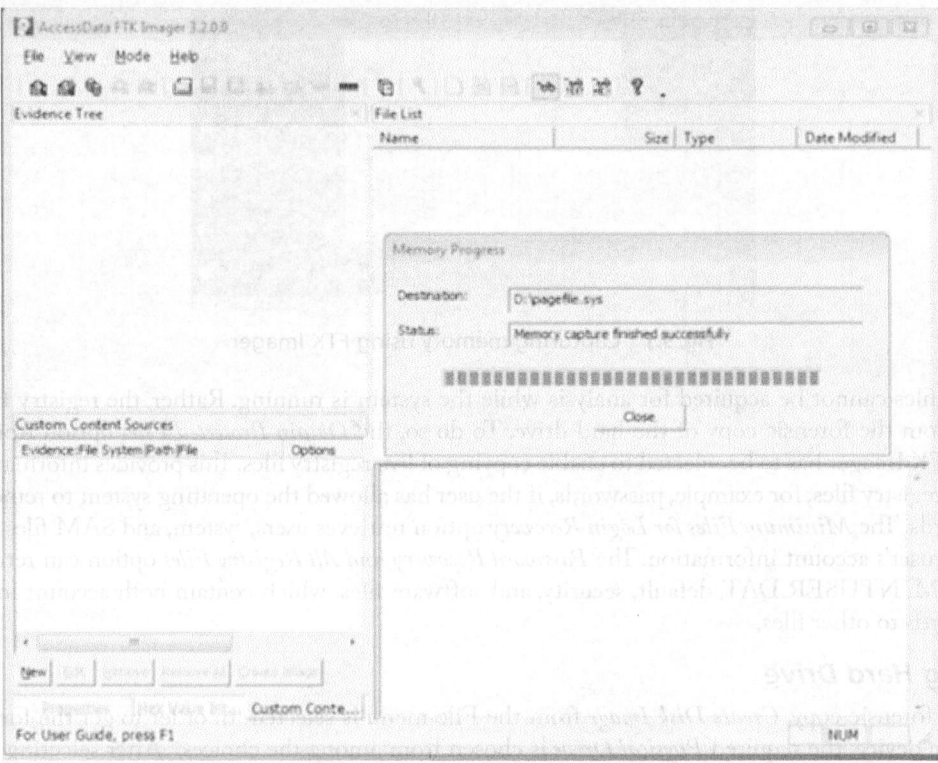

(b) Acquiring memdump.mem in D:

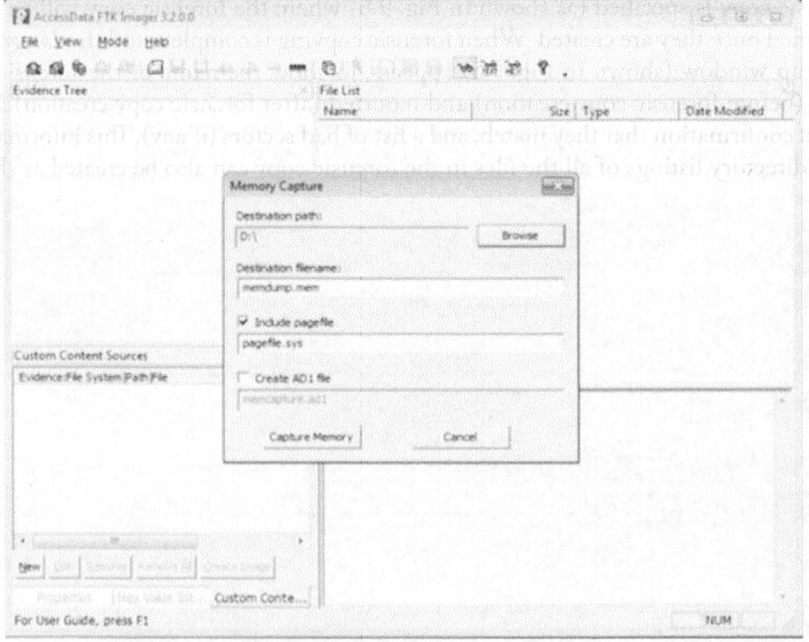

Fig. 9.3 Acquiring pagefile.sys in D:

Analysis of Digital Evidence

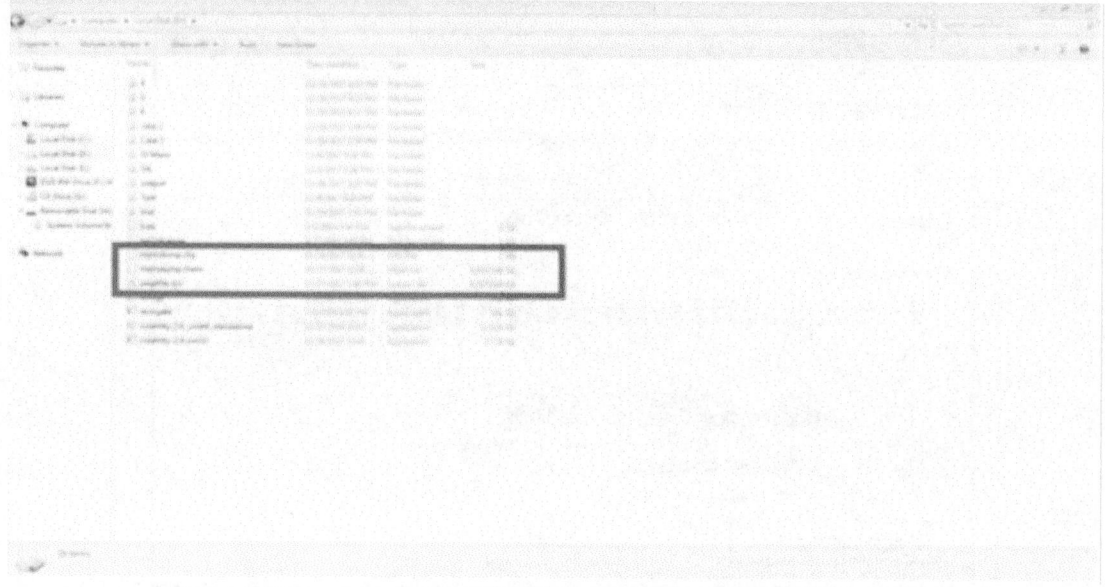

Fig. 9.4 Acquiring memory in D:

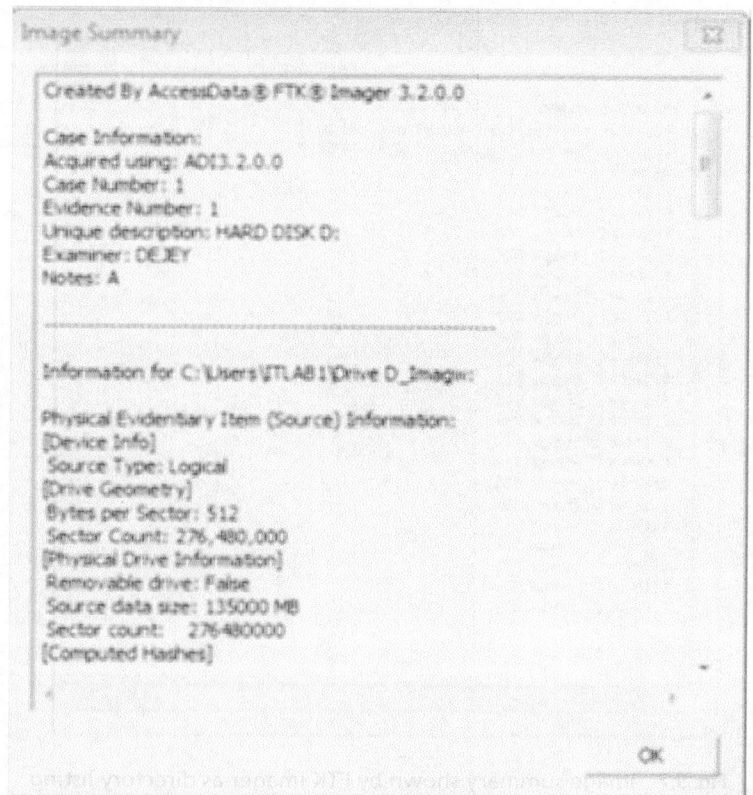

Fig. 9.5 Image summary

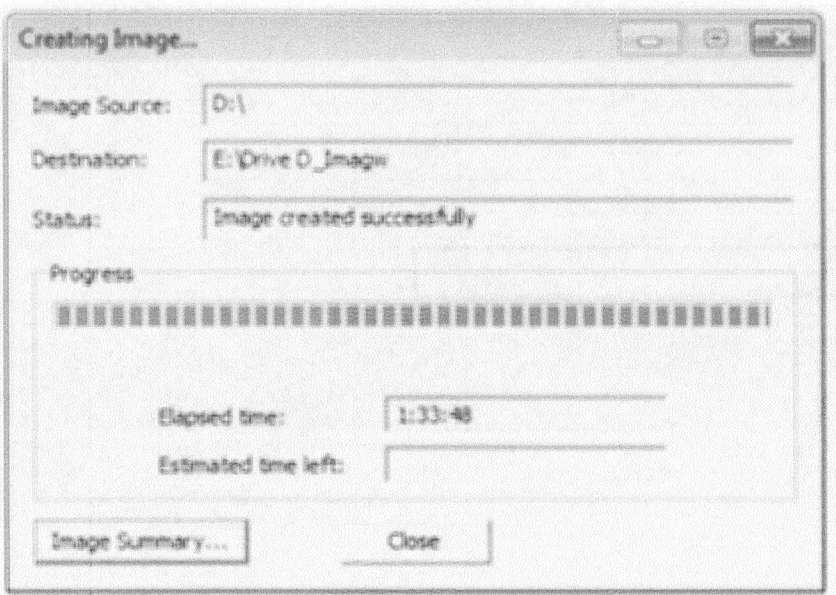

Fig. 9.6 Acquiring forensic copy of hard drive in E:

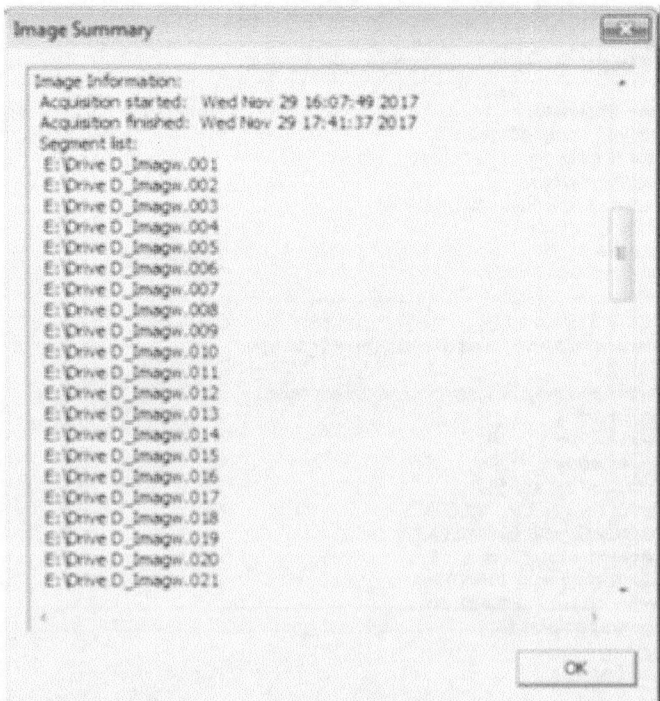

Fig. 9.7 Image summary shown by FTK Imager as directory listing

9.3 RAM ANALYSIS WITH VOLATILITY

RAM artifacts include the data used by a software application or a hardware device. RAM analysis helps to identify the running processes, malicious processes, network activities, open connections, etc., in a compromised system. Information about network connections such as remote IP address and port number helps during analysis while gathering information pertaining to computer intrusion and identifying remote destination where the malware is communicating (hence RAM analysis is also called malware analysis). RAM analysis lists dynamic link libraries (DLLs) that help the FA to locate any malicious DLL that has injected itself into a process. Open registry keys for a process that were responsible for the malware surviving a reboot as well as open files associated with a process can be identified with RAM analysis. The list of running processes provide information to the FA on how the system is being used. Authentication information (username and passwords) for social networking accounts, email accounts, etc., can be gathered from RAM analysis.

Volatility Framework is an open collection of tools, implemented in Python under the General Public License (GNU). It is used for the extraction of digital artifacts from volatile memory (RAM) samples. It is implemented in Python scripting language and can be used for the analysis of both 32/64 bit systems that run on Windows, Linux, Mac, and Android. Volatility Framework is also used to analyse crash dumps, raw dumps, VMware, and VirtualBox dumps.

Some of the essential commands of Volatility are discussed in this section. *imageinfo* command is used to identify the profiles supported for the forensic copy of the dumped memory (Fig. 9.8). *pslist* command is used to get a list of running process in the memory dump (Fig. 9.9). It helps to locate any running process that is irrelevant to the respective operating system and that may be malicious. *pstree* command is used to figure out the parent–child relationships between processes with their process identifier (PID) and parent process identifier (PPID) and to identify the parent process of the malicious program. The output of *pstree* command is shown in Fig. 9.10. *malfind* plugin of Volatility is used to detect the malicious DLLs in the process that can be dumped into an output directory (Fig. 9.11) where no DLL is actually found by the plugin. However, for the process with PID1652, the output of *malfind* is shown in Fig. 9.12. The respective folder contains the dumps of malicious programs which can be scanned using antivirus or anti-malware software. *cmdscan* plugin of Volatility shows the history of commands that are run on the machine on the command prompt (Fig. 9.13) show an empty list when no commands are run from the command prompt. The *netscan* plugin is used to find the network connections from the forensic copy of the memory (Fig. 9.14). Thus, memory analysis can be used to identify whether the system is compromised by any malware.

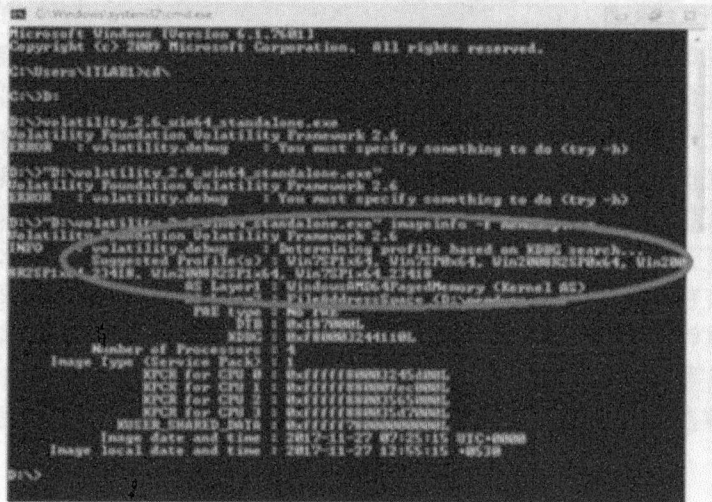

Fig. 9.8 Output of imageinfo command

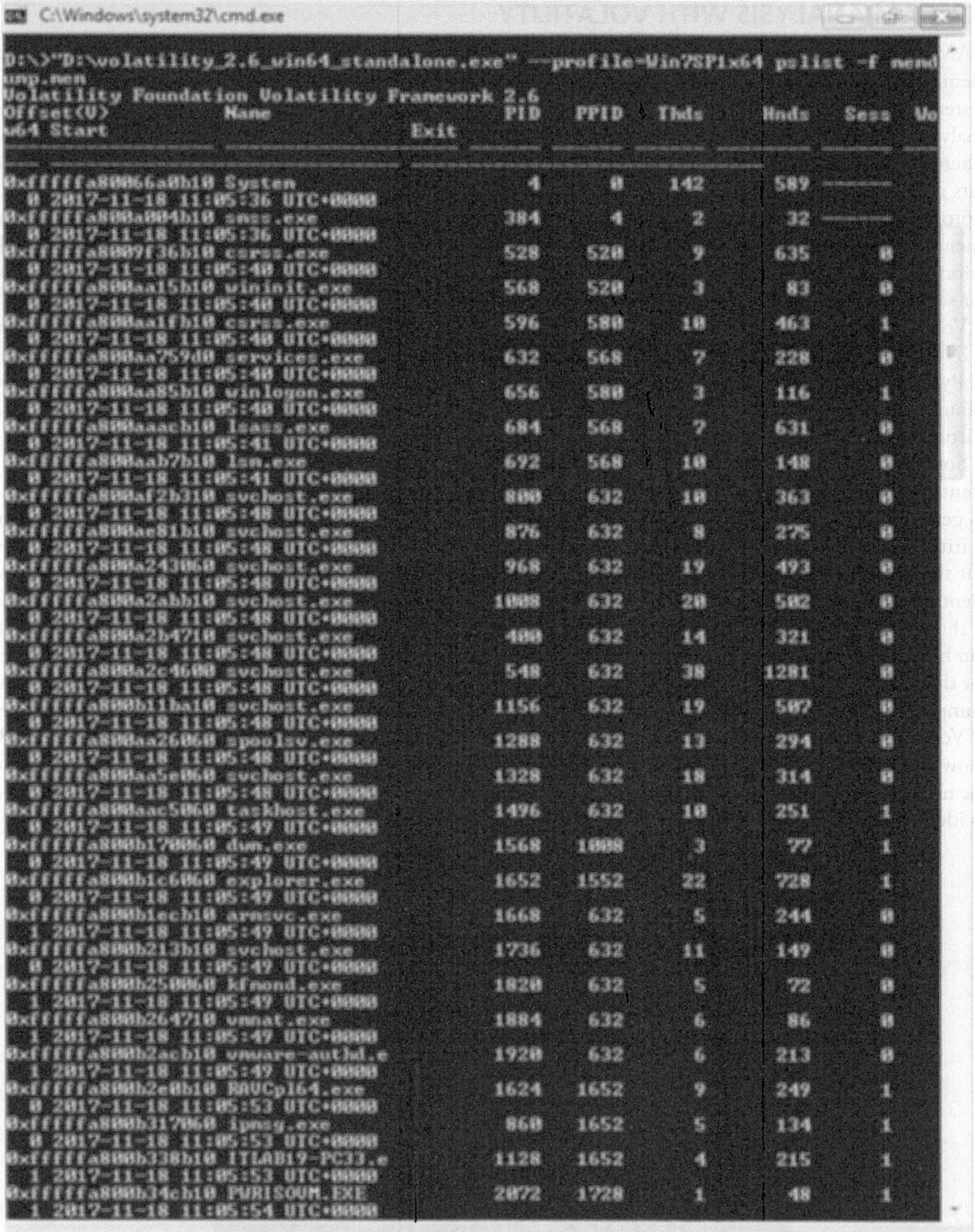

Fig. 9.9 Output of pslist command

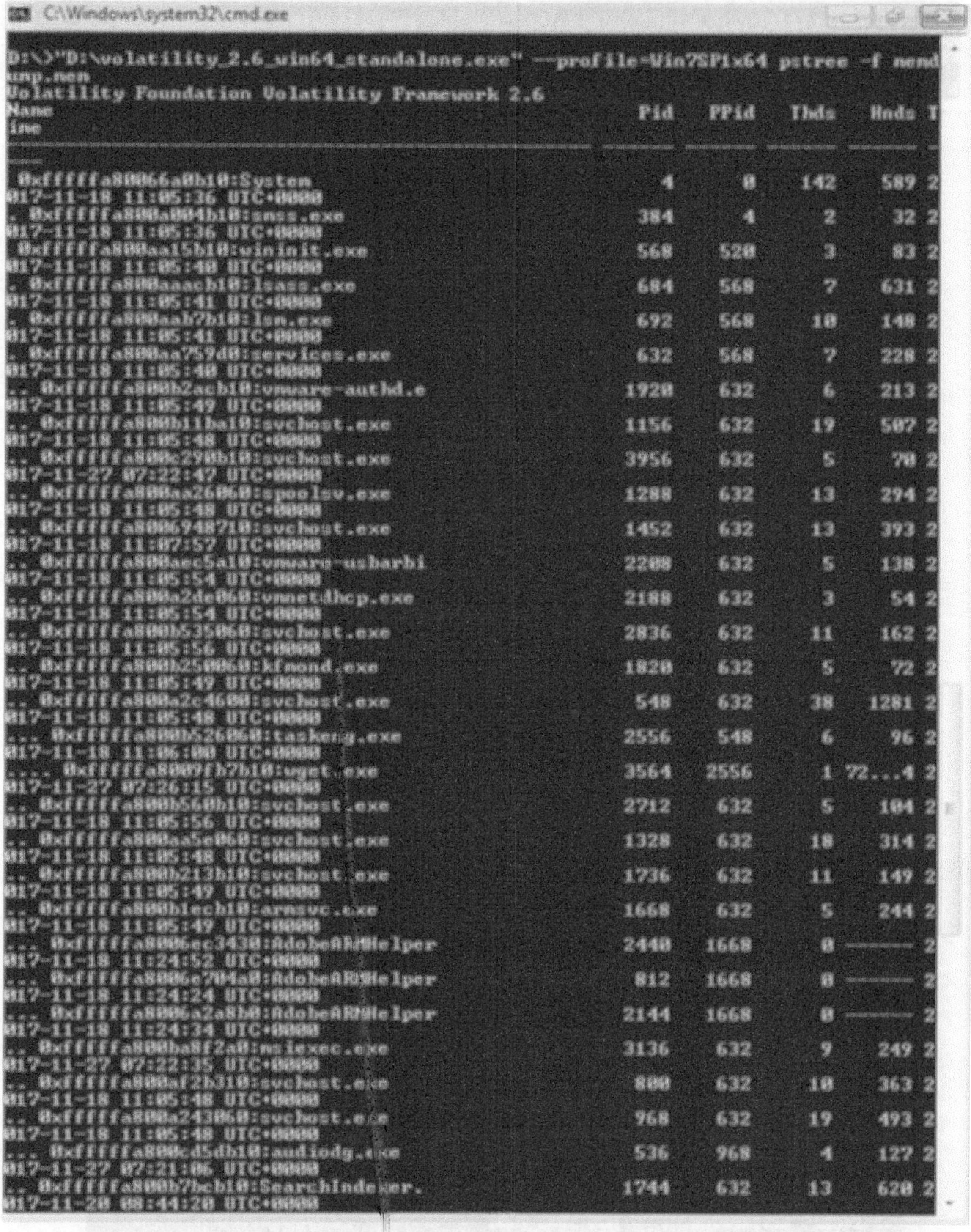

```
C:\Windows\system32\cmd.exe

D:\>"D:\volatility_2.6_win64_standalone.exe" --profile=Win7SP1x64 pstree -f memd
ump.mem
Volatility Foundation Volatility Framework 2.6
Name                                           Pid     PPid    Thds    Hnds  T
ime
----------------------------------------------------------------------------------
 0xfffffa80066a0b10:System                       4       0     142     589  2
017-11-18 11:05:36 UTC+0000
. 0xfffffa800a004b10:smss.exe                  384       4       2      32  2
017-11-18 11:05:36 UTC+0000
 0xfffffa800aa15b10:wininit.exe                568     520       3      83  2
017-11-18 11:05:40 UTC+0000
. 0xfffffa800aaacb10:lsass.exe                 684     568       7     631  2
017-11-18 11:05:41 UTC+0000
. 0xfffffa800aab7b10:lsm.exe                   692     568      10     148  2
017-11-18 11:05:41 UTC+0000
. 0xfffffa800aa759d0:services.exe              632     568       7     228  2
017-11-18 11:05:40 UTC+0000
.. 0xfffffa800b2acb10:vmware-authd.e          1920     632       6     213  2
017-11-18 11:05:49 UTC+0000
.. 0xfffffa800b11ba10:svchost.exe             1156     632      19     507  2
017-11-18 11:05:48 UTC+0000
.. 0xfffffa800c290b10:svchost.exe             3956     632       5      70  2
017-11-27 07:22:47 UTC+0000
.. 0xfffffa800aa26060:spoolsv.exe             1288     632      13     294  2
017-11-18 11:05:48 UTC+0000
.. 0xfffffa8006948710:svchost.exe             1452     632      13     393  2
017-11-18 11:07:57 UTC+0000
.. 0xfffffa800aec5a10:vmware-usbarbi          2208     632       5     138  2
017-11-18 11:05:54 UTC+0000
.. 0xfffffa800a2de060:vmnetdhcp.exe           2180     632       3      54  2
017-11-18 11:05:54 UTC+0000
.. 0xfffffa800b635060:svchost.exe             2036     632      11     162  2
017-11-18 11:05:56 UTC+0000
.. 0xfffffa800b250060:KMSond.exe              1820     632       5      72  2
017-11-18 11:05:49 UTC+0000
.. 0xfffffa800a2c4600:svchost.exe              548     632      38    1281  2
017-11-18 11:05:48 UTC+0000
... 0xfffffa800b526060:taskeng.exe            2556     548       6      96  2
017-11-18 11:06:00 UTC+0000
.... 0xfffffa8009fb7b10:wget.exe              3564    2556       1   72...4 2
017-11-27 07:26:15 UTC+0000
.. 0xfffffa800b560b10:svchost.exe             2712     632       5     104  2
017-11-18 11:05:56 UTC+0000
.. 0xfffffa800aa5e060:svchost.exe             1328     632      18     314  2
017-11-18 11:05:48 UTC+0000
.. 0xfffffa800b213b10:svchost.exe             1736     632      11     149  2
017-11-18 11:05:49 UTC+0000
.. 0xfffffa800b1ecb10:armsvc.exe              1668     632       5     244  2
017-11-18 11:05:49 UTC+0000
... 0xfffffa8006ec3430:AdobeARMHelper         2440    1668       0    ----  2
017-11-18 11:24:52 UTC+0000
... 0xfffffa8006e704a0:AdobeARMHelper          812    1668       0    ----  2
017-11-18 11:24:24 UTC+0000
... 0xfffffa8006a2a8b0:AdobeARMHelper         2144    1668       0    ----  2
017-11-18 11:24:34 UTC+0000
.. 0xfffffa800ba8f2a0:mslexec.exe             3136     632       9     249  2
017-11-27 07:22:35 UTC+0000
.. 0xfffffa800af2b310:svchost.exe              800     632      10     363  2
017-11-18 11:05:48 UTC+0000
.. 0xfffffa800a243060:svchost.exe              968     632      19     493  2
017-11-18 11:05:48 UTC+0000
... 0xfffffa800cd5db10:audiodg.exe             536     968       4     127  2
017-11-27 07:21:06 UTC+0000
.. 0xfffffa800b7bcb10:SearchIndexer.          1744     632      13     620  2
017-11-20 08:44:20 UTC+0000
```

Fig. 9.10 Output of pstree command

Fig. 9.11 Output of *malfind* command for process with PID 2712

Fig. 9.12 Output of *malfind plug*in for process with PID 1652

Fig. 9.13 Output of *cmdscan* plugin

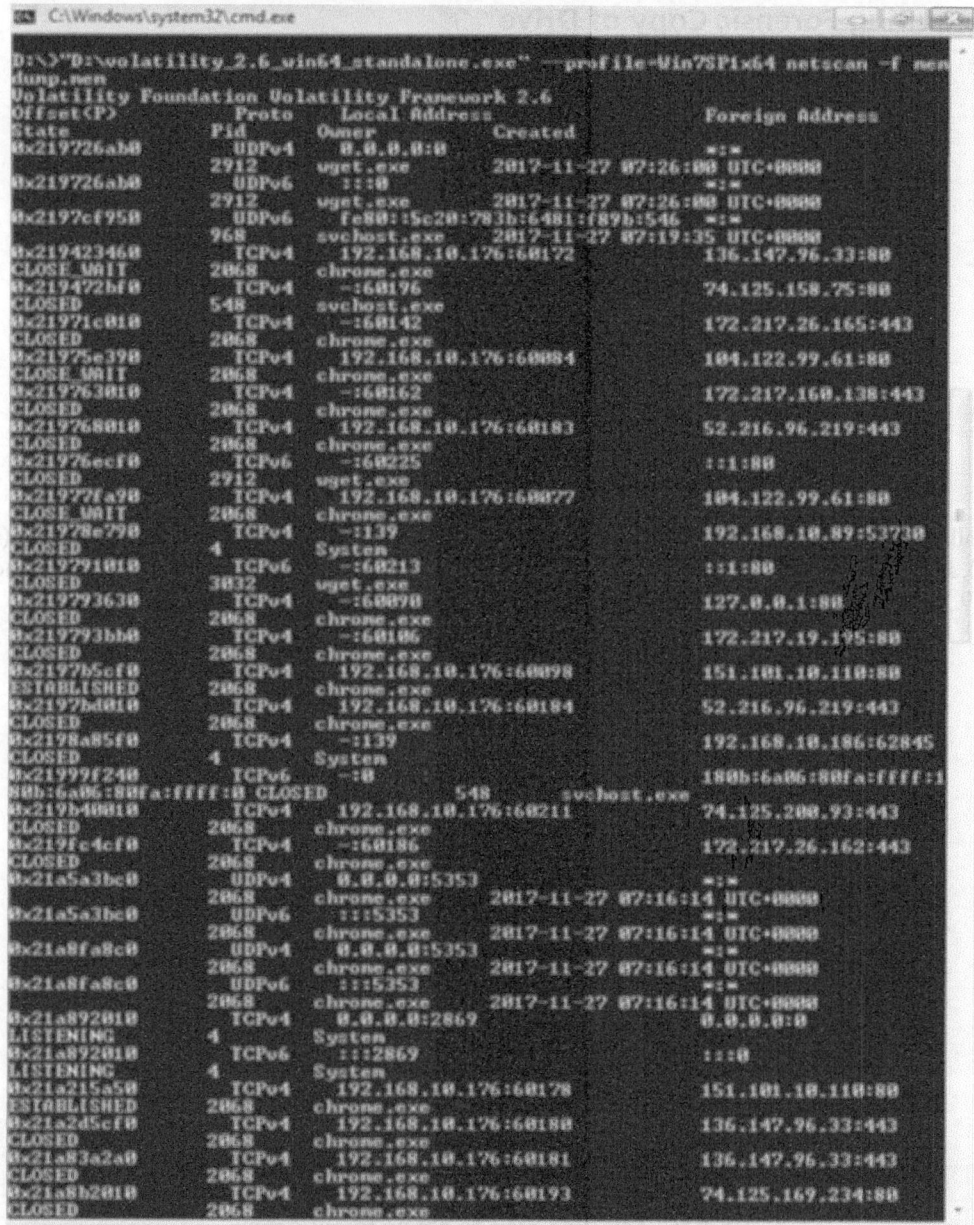

Fig. 9.14 Output of *netscan* plugin

9.4 ANALYSING HARD DRIVE WITH WINHEX

WinHex is a proprietary tool from X-waysForensics. It is a disk editor primarily used for data recovery and digital forensics. It can be used to acquire a forensic copy of the drive; inspect files from storage media such as hard disks, optical storage media, and memory cards/sticks; recover deleted files; analyse and compare files; and many other features that come with licensing. This section presents some of the results obtained with WinHex version 16.7.

9.4.1 Acquiring Forensic Copy of Drive

Fig. 9.15 Open Disk from Tools menu

After installing WinHex 16.7 and opening it, to get the forensic copy of the drive, under *Tools* menu (from the menu bar), select *Open Disk* (Fig. 9.15) and choose the drive for which a forensic copy is to be obtained. Click OK (Fig. 9.16). Figure 9.17 shows the output after choosing a drive.

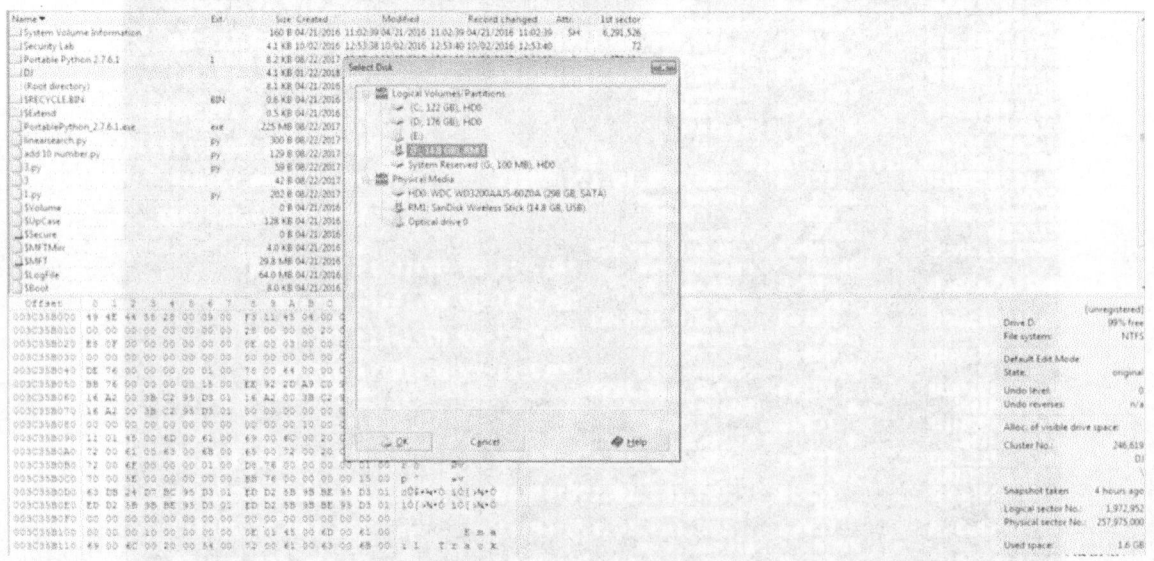

Fig. 9.16 Selecting drive to obtain forensic copy

Analysis of Digital Evidence   313

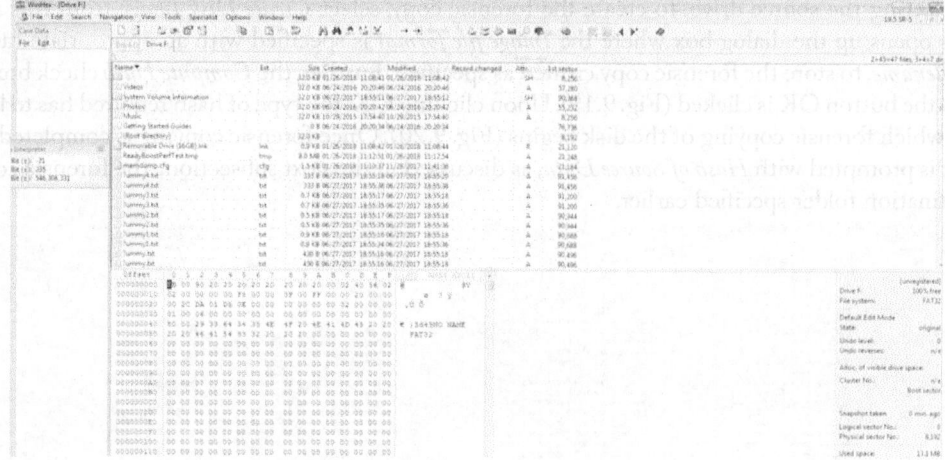

Fig. 9.17   Output of Open Disk

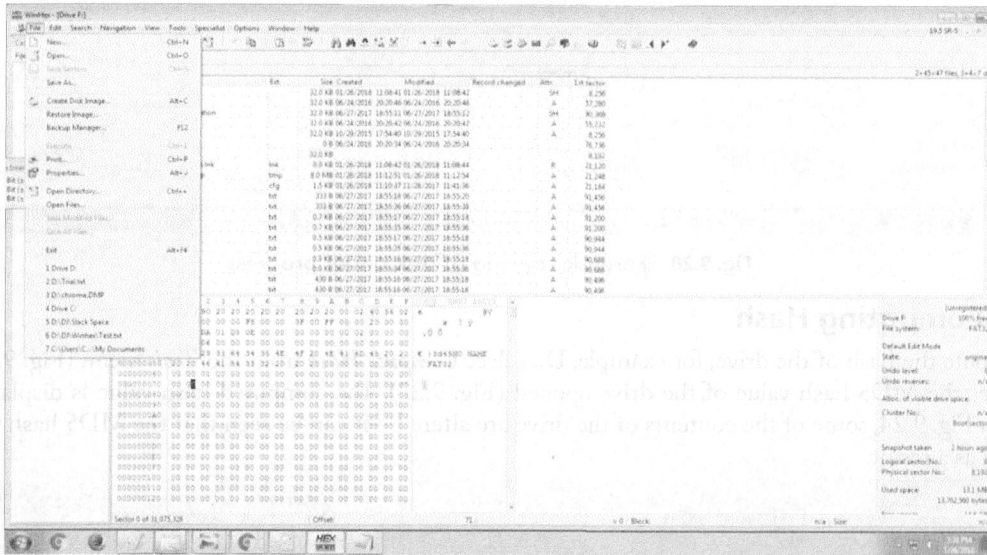

Fig. 9.18   Choosing Create Disk Image from File menu

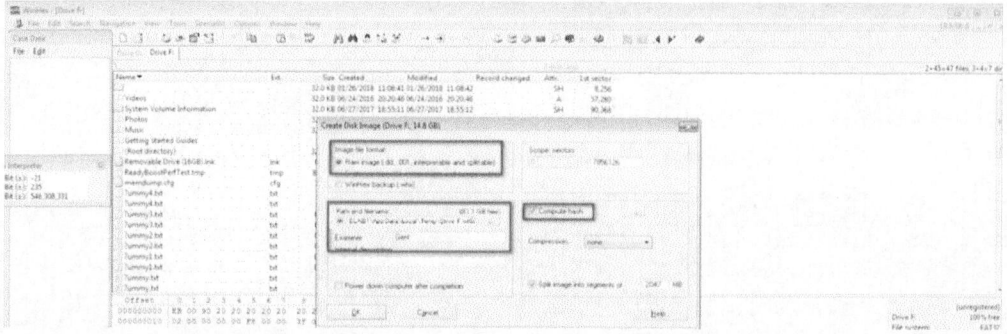

Fig. 9.19   Specifying file format of forensic copy and destination

After selecting the source drive, to create the forensic copy, select *Create Disk Image* from *File* menu (Fig. 9.18). This opens up the dialog box where the *Image file format* is specified with an appropriate radio button, *Path and filename*, to store the forensic copy created as specified. Further, the *Compute Hash* check box is enabled and finally the button OK is clicked (Fig. 9.19). Upon clicking OK, the type of hash required has to be specified following which forensic copying of the disk begins (Fig. 9.20). Once forensic copying is completed, a message dialog box is prompted with *Hash of Source Data*, as discussed in the next subsection. The forensic copy appears in the destination folder specified earlier.

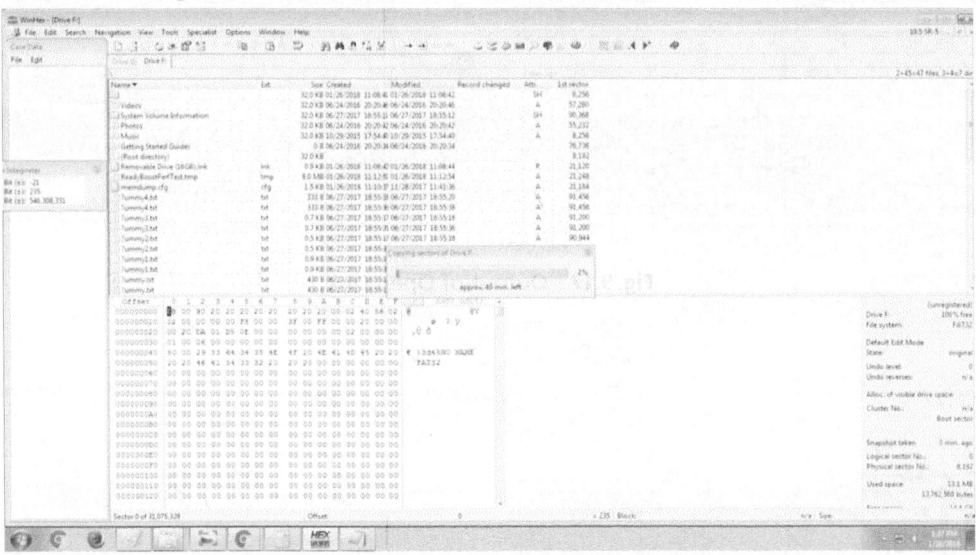

Fig. 9.20 Forensic copying of disk under progress

9.4.2 Computing Hash

To compute the hash of the drive, for example, D:, select *Compute Hash* from the *Tools* menu (Fig. 9.21). This computes the MD5 hash value of the drive opened (Fig. 9.22). The computed hash value is displayed (Fig. 9.23). In Fig. 9.24, some of the contents of the drive are altered—it can be seen that the MD5 hash value has changed as a result.

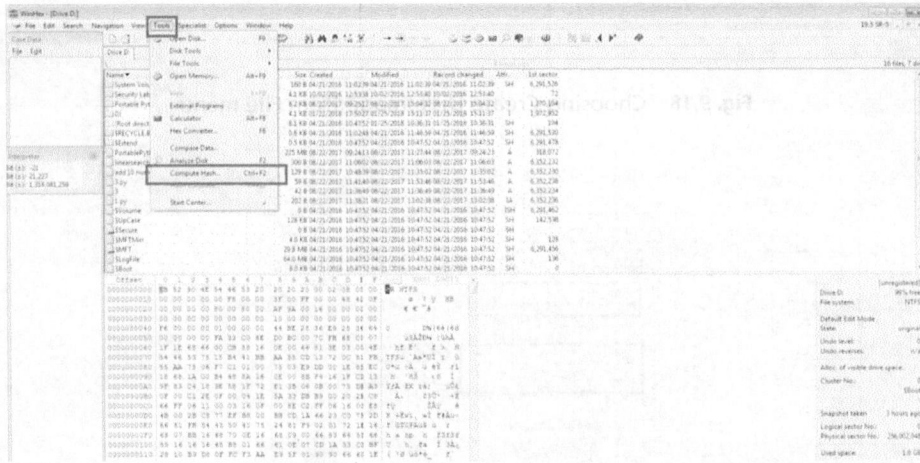

Fig. 9.21 Selecting *Compute Hash*

Analysis of Digital Evidence 315

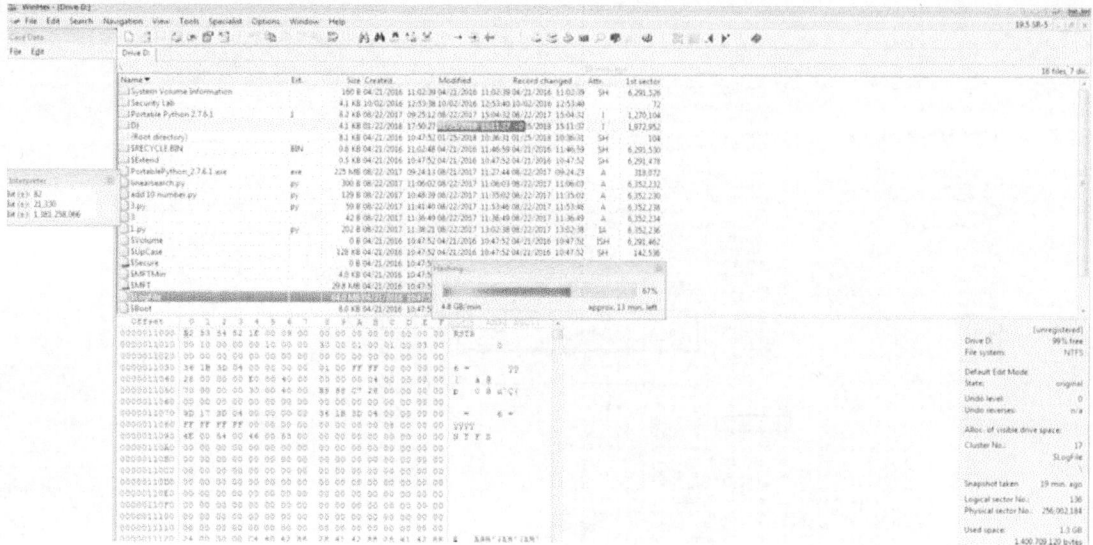

Fig. 9.22 Hashing in progress

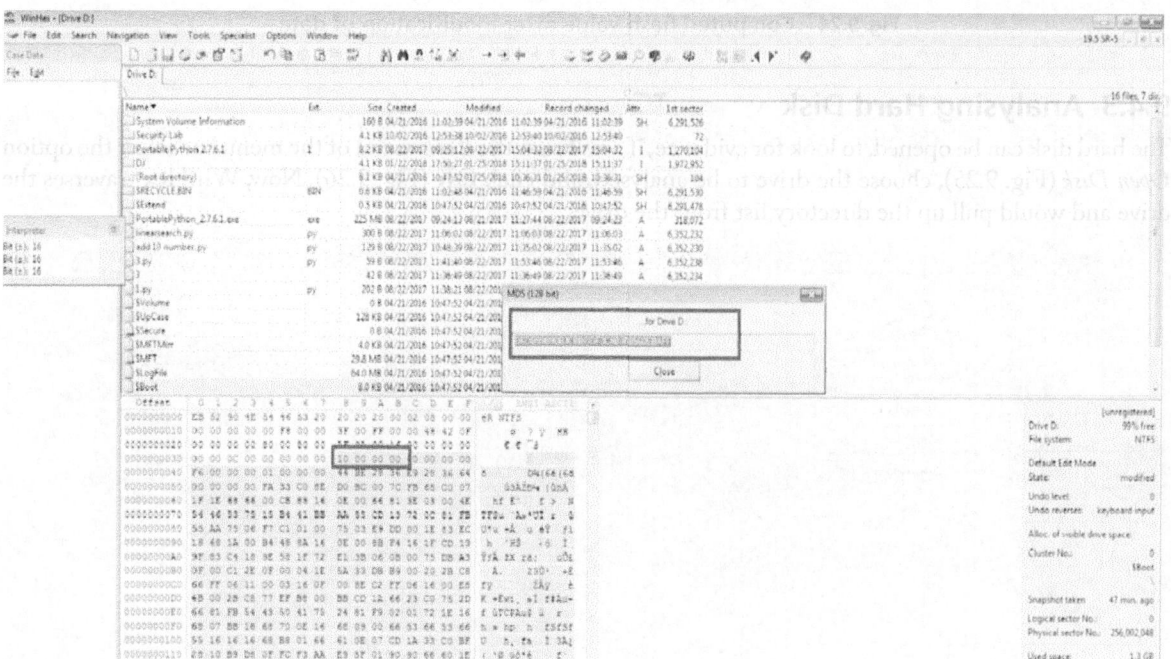

Fig. 9.23 Computing hash value for drive

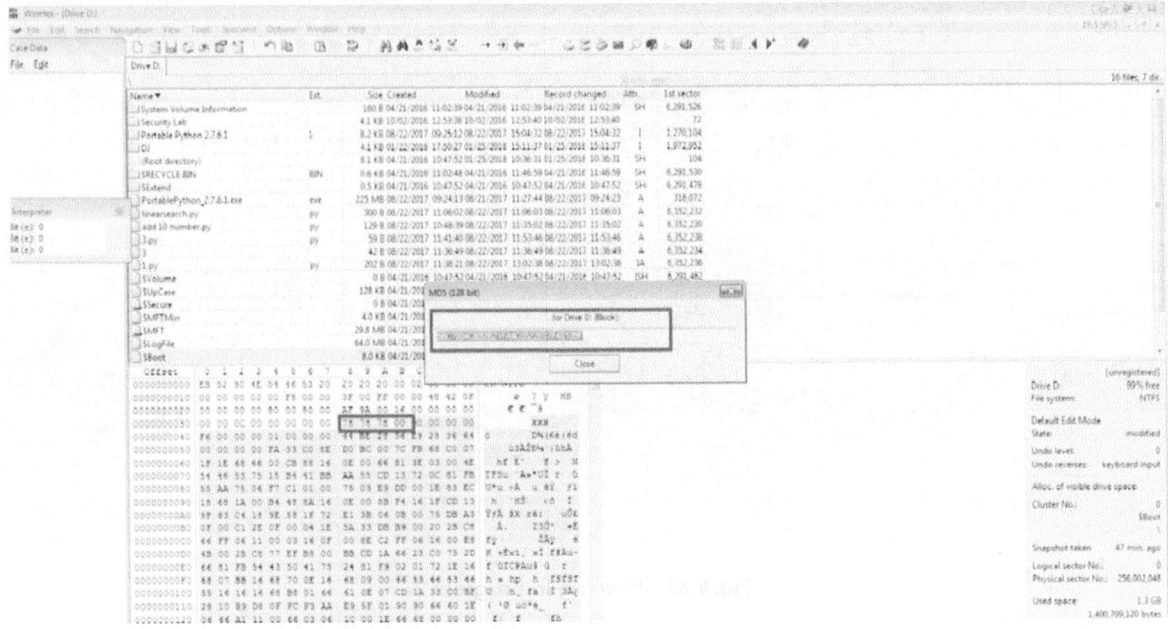

Fig. 9.24 Computed hash value after modification of data

9.4.3 Analysing Hard Disk

The hard disk can be opened, to look for evidence, if any. From the *Tools* menu of the menu bar select the option *Open Disk* (Fig. 9.25), choose the drive to be analysed, and click OK (Fig. 9.26). Now, WinHex traverses the drive and would pull up the directory list from the drive.

Fig. 9.25 Selecting Open Disk

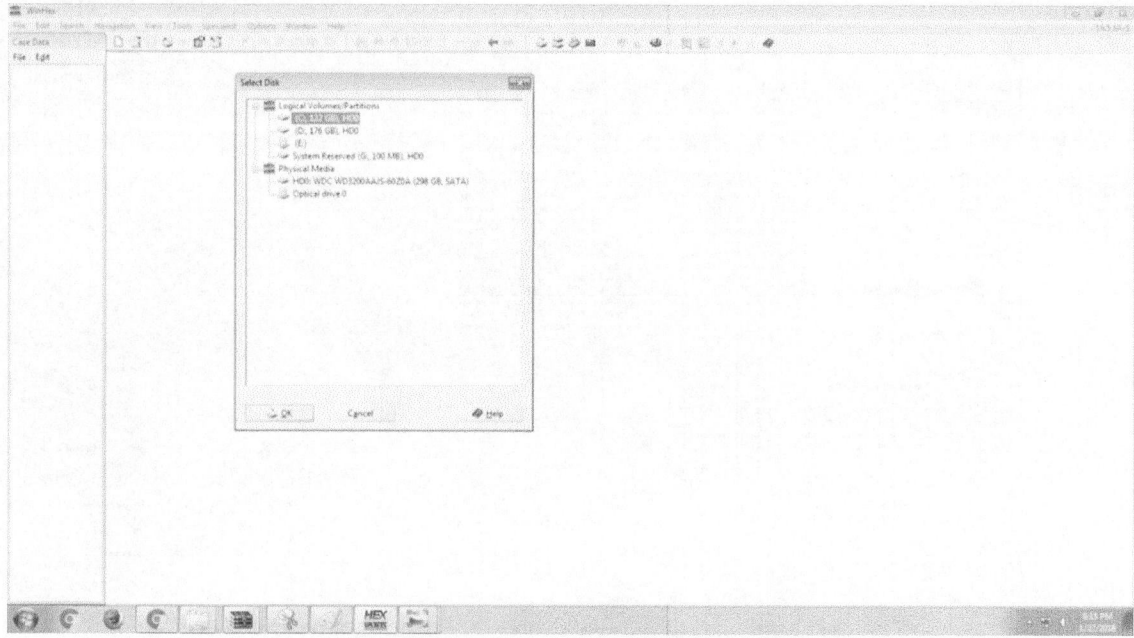

Fig. 9.26 Choosing drive to be analysed

Under the *Tools* menu from the menu bar, choosing *Disk Tools* and then choosing *Take Volume Snapshot* from the pop-up menu helps in obtaining a snapshot of the drive to work with.

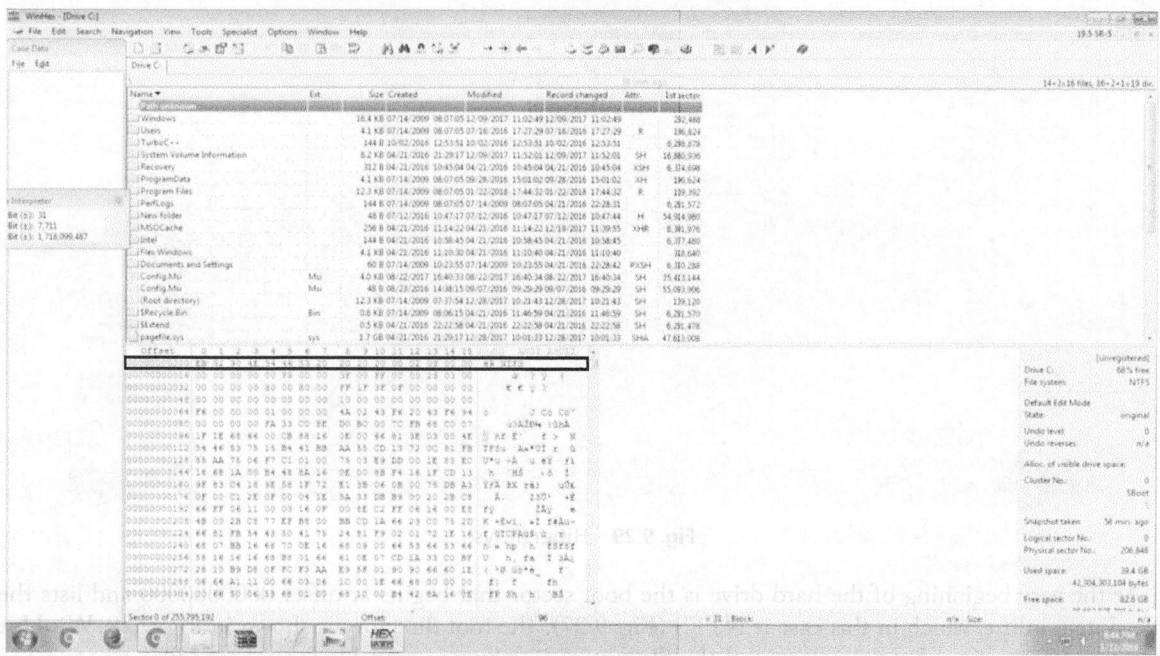

Fig. 9.27 Boot sector

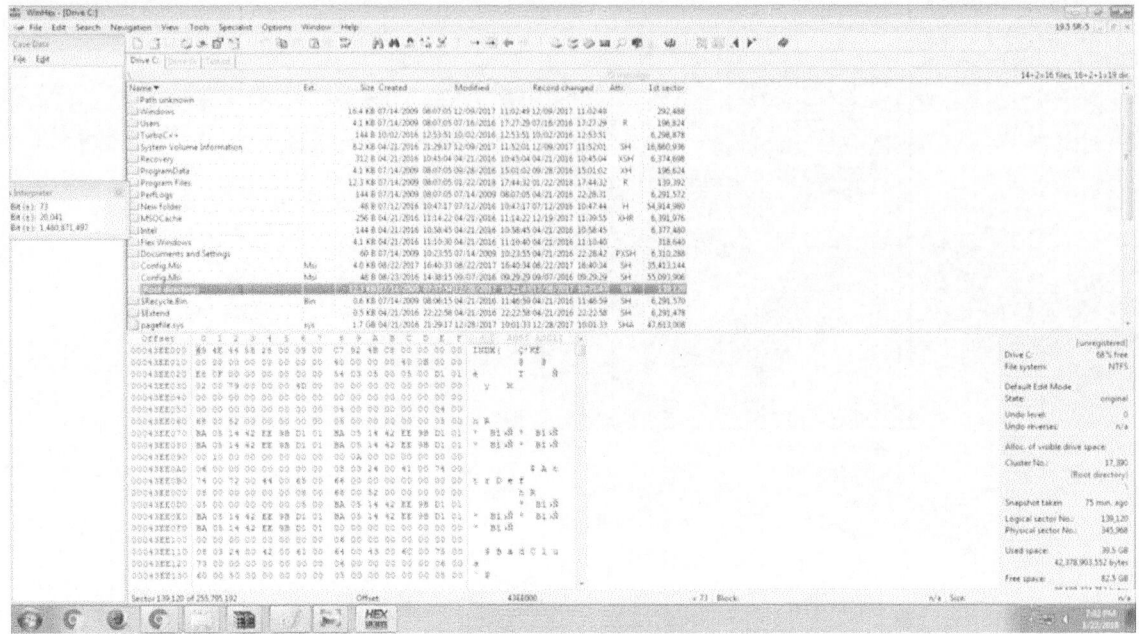

Fig. 9.28 Root directory

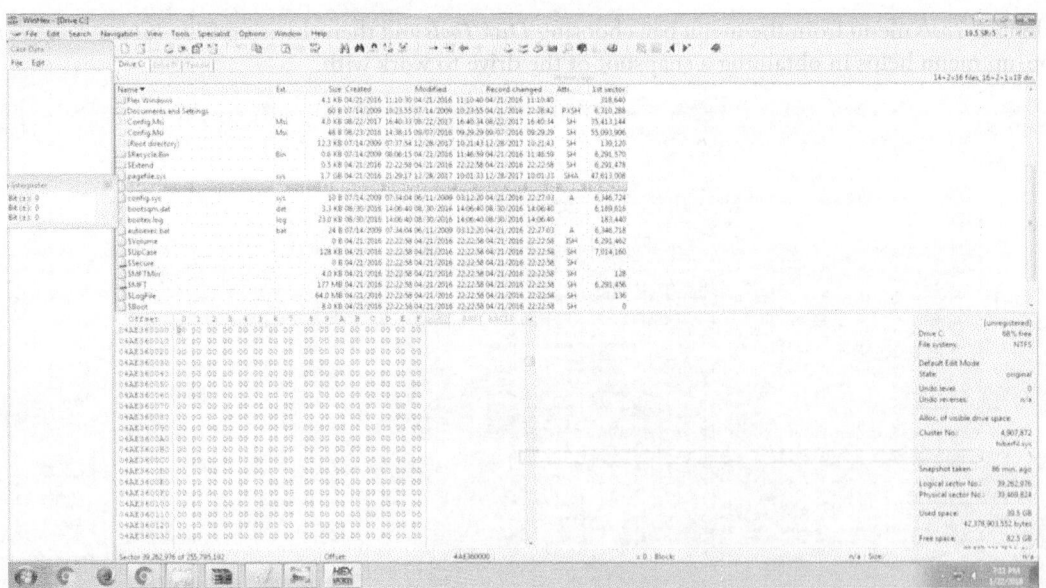

Fig. 9.29 Hiberfil.sys

At the very beginning of the hard drive is the boot sector that begins at offset 0000000000 and lists the file system in use, which in this case is NTFS (Fig. 9.27). The root directory is shown in Fig. 9.28. WinHex presents information about the drive and the file system in a human readable form. Some files are stored in

Analysis of Digital Evidence 319

the root of the hard drive. For example, *hiberfil.sys* is allocated (Fig. 9.29). *Pagefile.sys* and *config.sys* are shown in Figs 9.30 and 9.31.

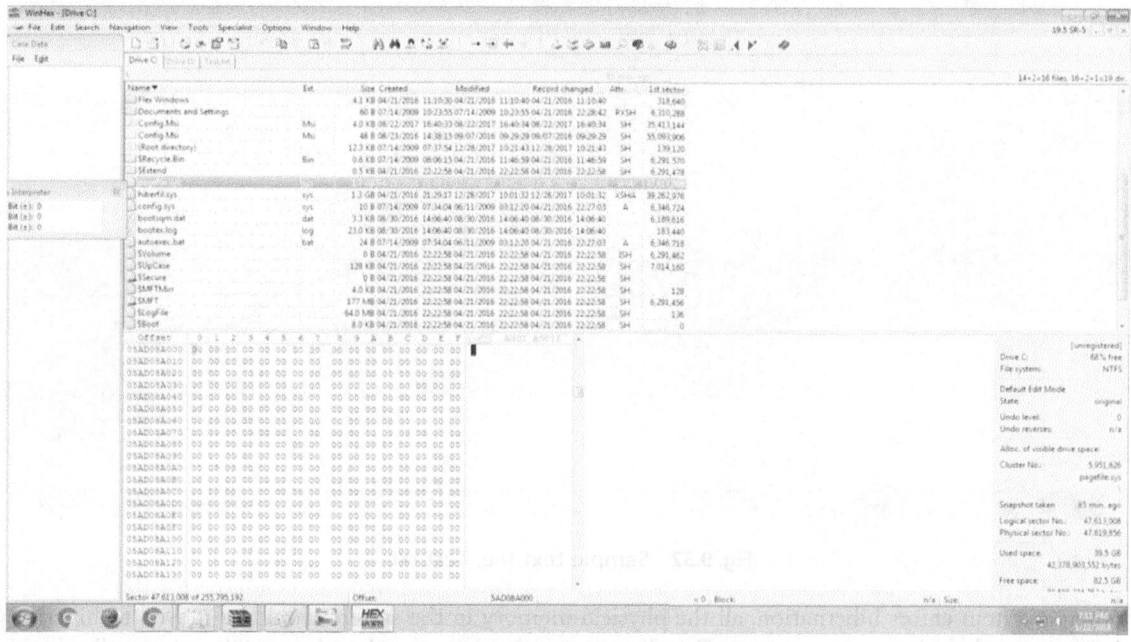

Fig. 9.30 Pagefile.sys

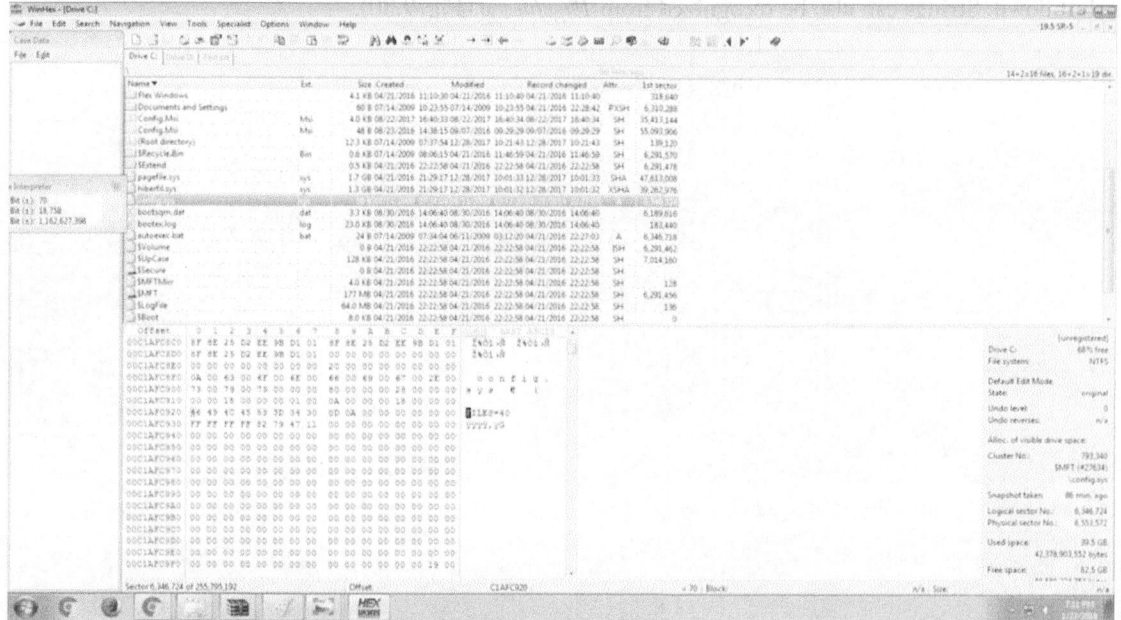

Fig. 9.31 Config.sys

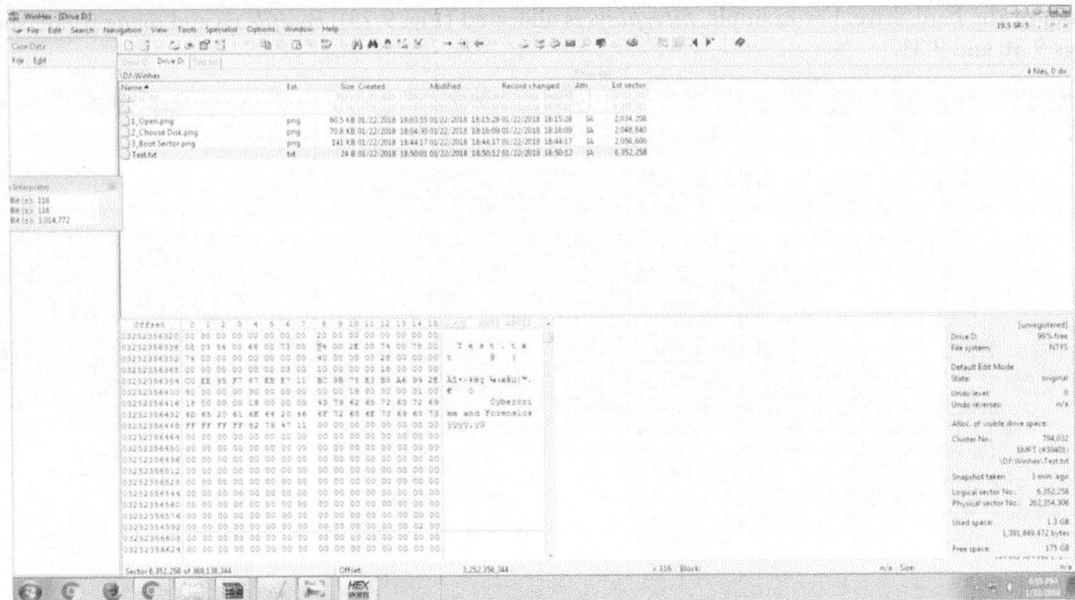

Fig. 9.32 Sample text file, Test.txt

When a system enters hibernation, all the physical memory in the system would be moved to hiberfil.sys and vice versa when the system resumes. *Pagefile.sys* corresponds to the virtual memory and actually contains information to control swapping. This file contains information such as programs that are stored, used, and running in the memory. It also helps to look at artifacts from programs that were running when the system was shut down. Strings can also be recognized from *Pagefile.sys* (Fig. 9.30).

Fig. 9.33 File header of the sample text file

Analysis of Digital Evidence **321**

For example, a text file (Fig. 9.32) can be analysed by merely clicking on it where WinHex lists the metadata about the file on the right and the file header (Fig. 9.33). Other metadata files of New Technology File System NTFS that were discussed in Chapter 6 are shown in Figs 9.34 through 9.44.

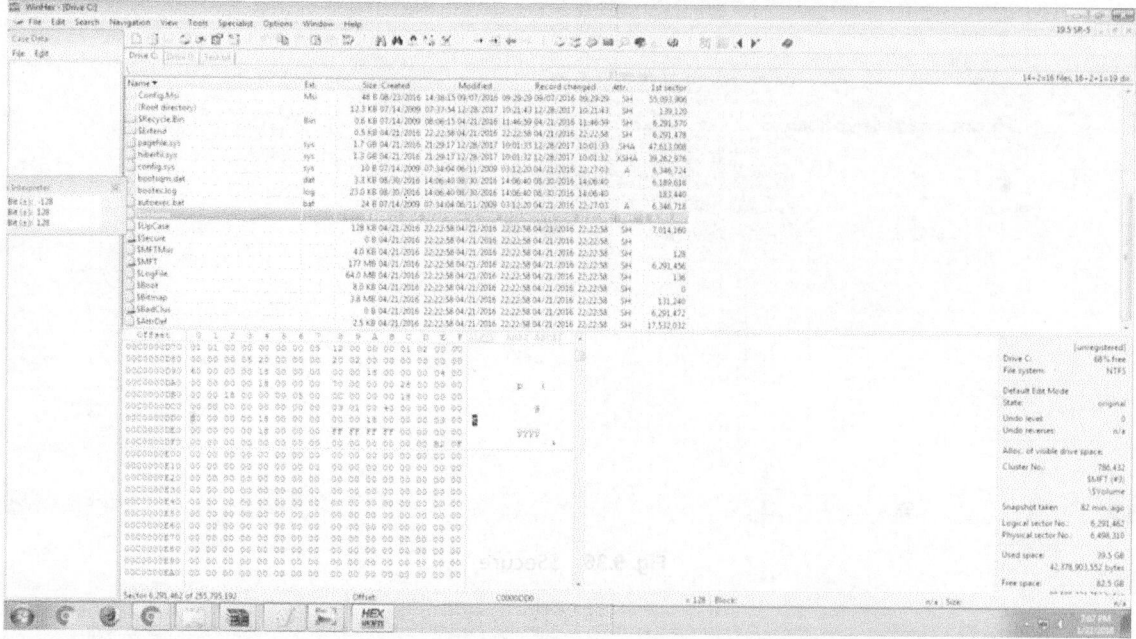

Fig. 9.34 $Volume

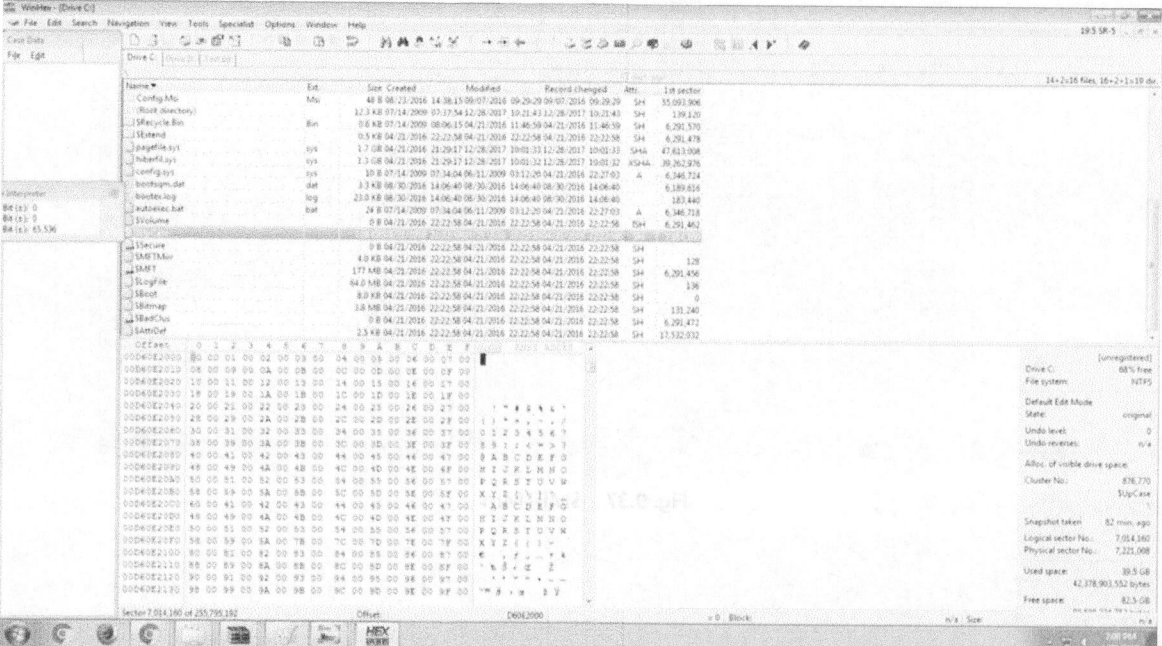

Fig. 9.35 $UpCase

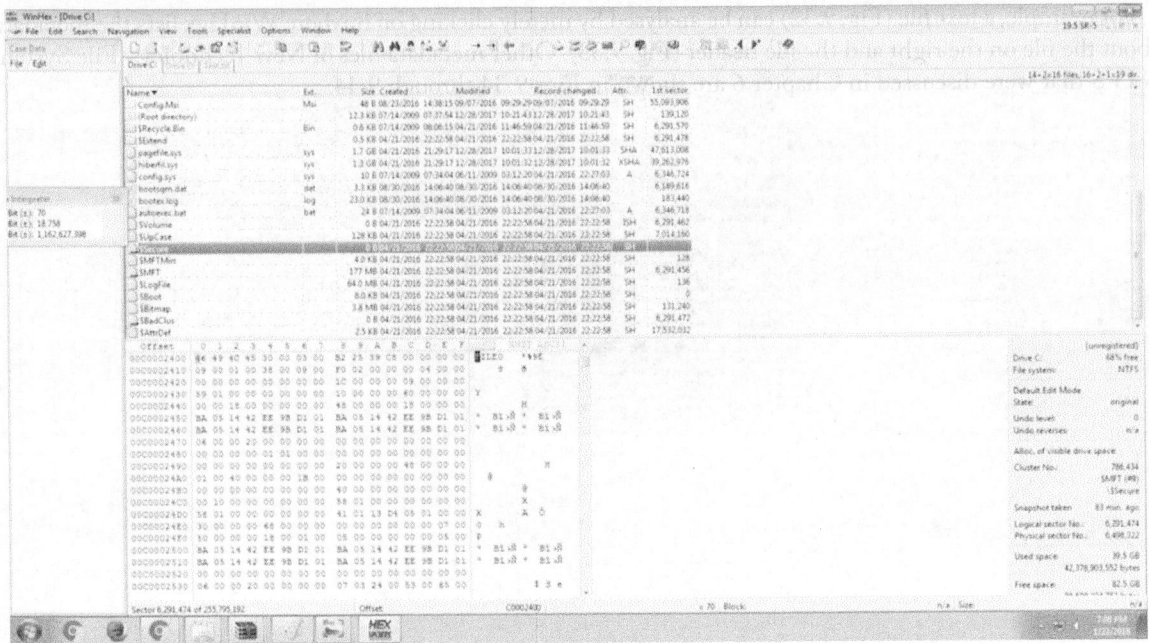

Fig. 9.36 $Secure

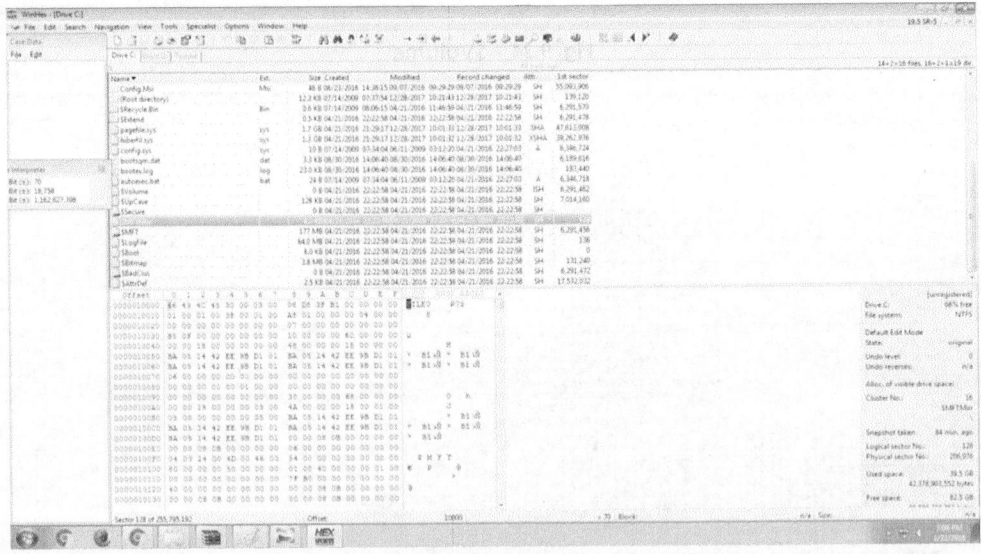

Fig. 9.37 $MFTMirr

Analysis of Digital Evidence 323

Fig. 9.38 $MFT

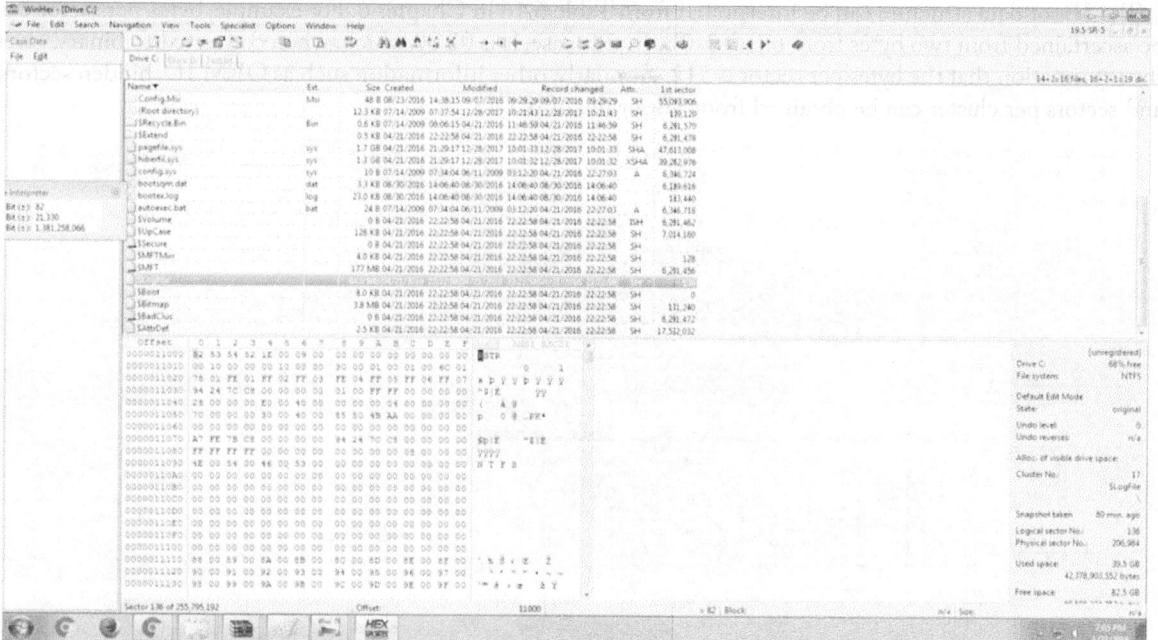

Fig. 9.39 $LogFile

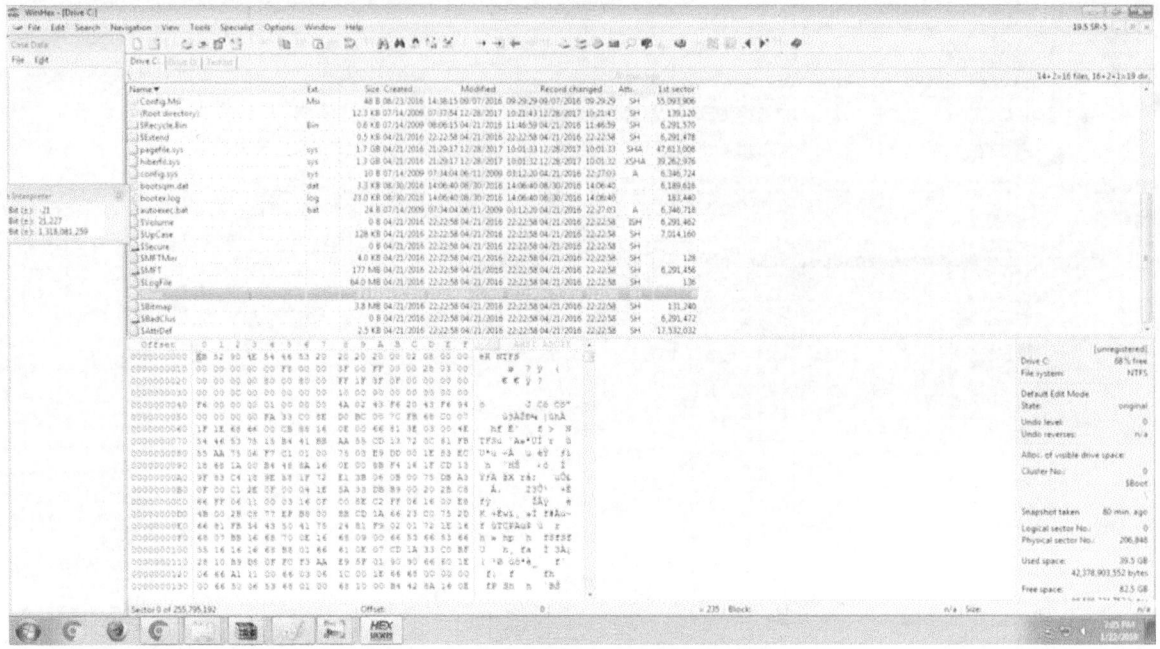

Fig. 9.40 $Boot

The $Boot data structure can be interpreted from Table 6.16 in Chapter 6. For example, bytes per sector can be ascertained from two bytes from offset 0x0B. In this case, it is 02 00 which, when converted to binary, gives the information that the bytes per sector is 512. Similarly, other information such as OEM ID, hidden sectors, and sectors per cluster can be obtained from $Boot file of NTFS.

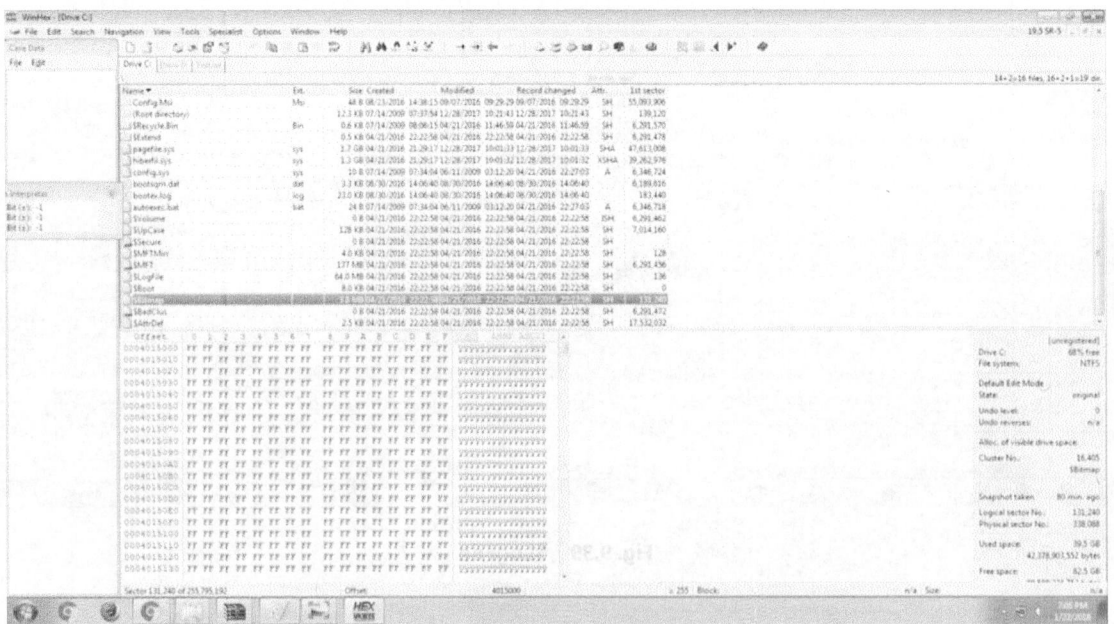

Fig. 9.41 $Bitmap

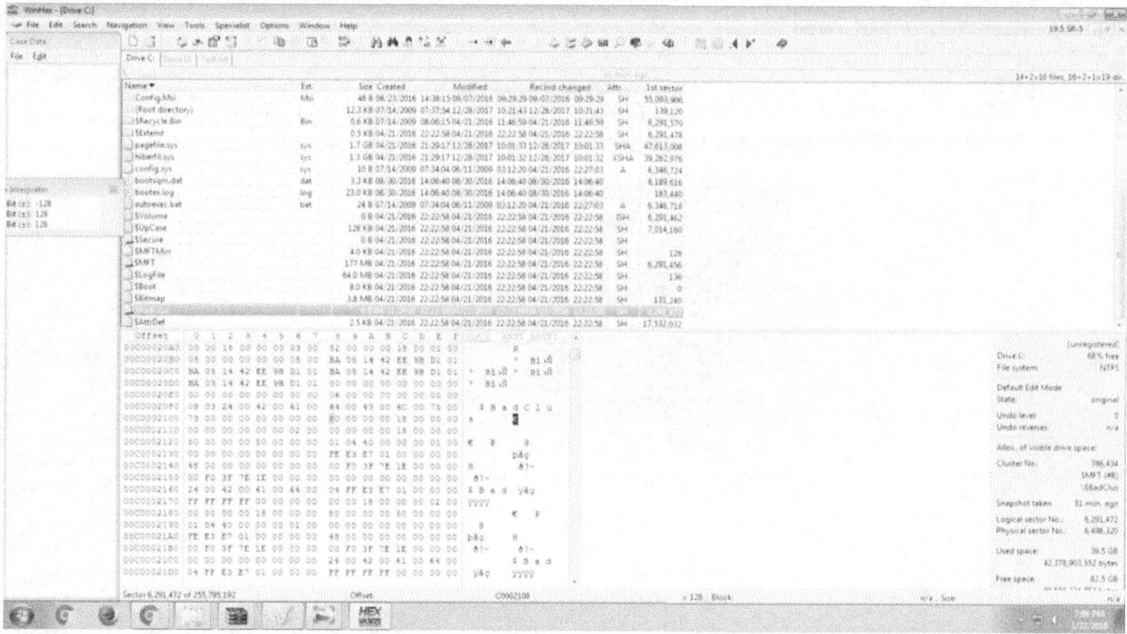

Fig. 9.42 $BadClus

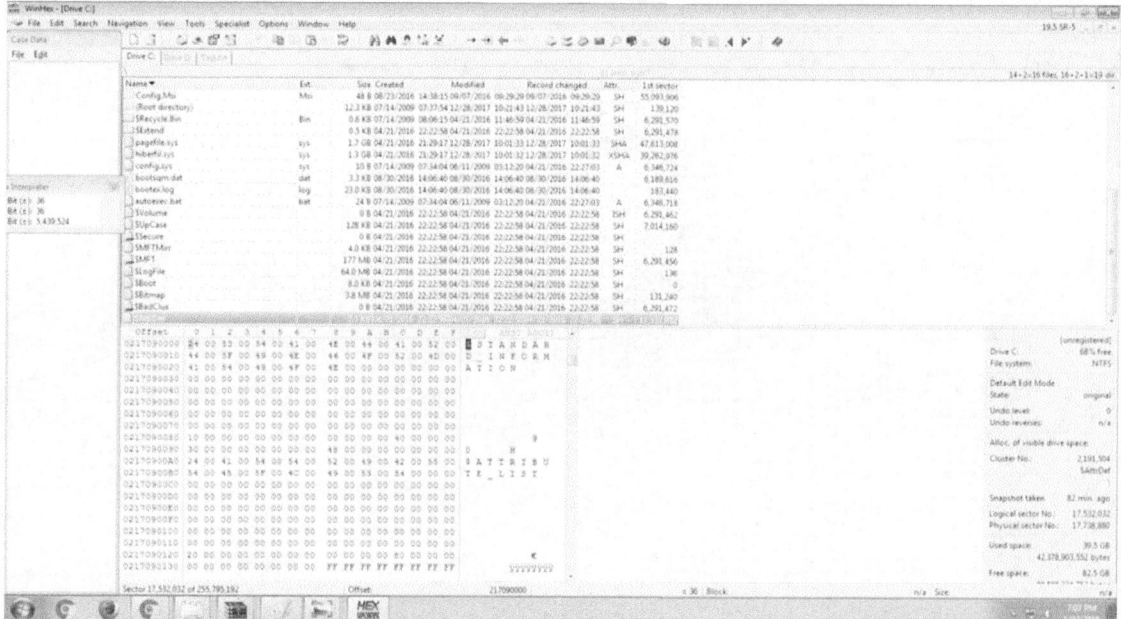

Fig. 9.43 $AttrDef

326   Cyber Forensics

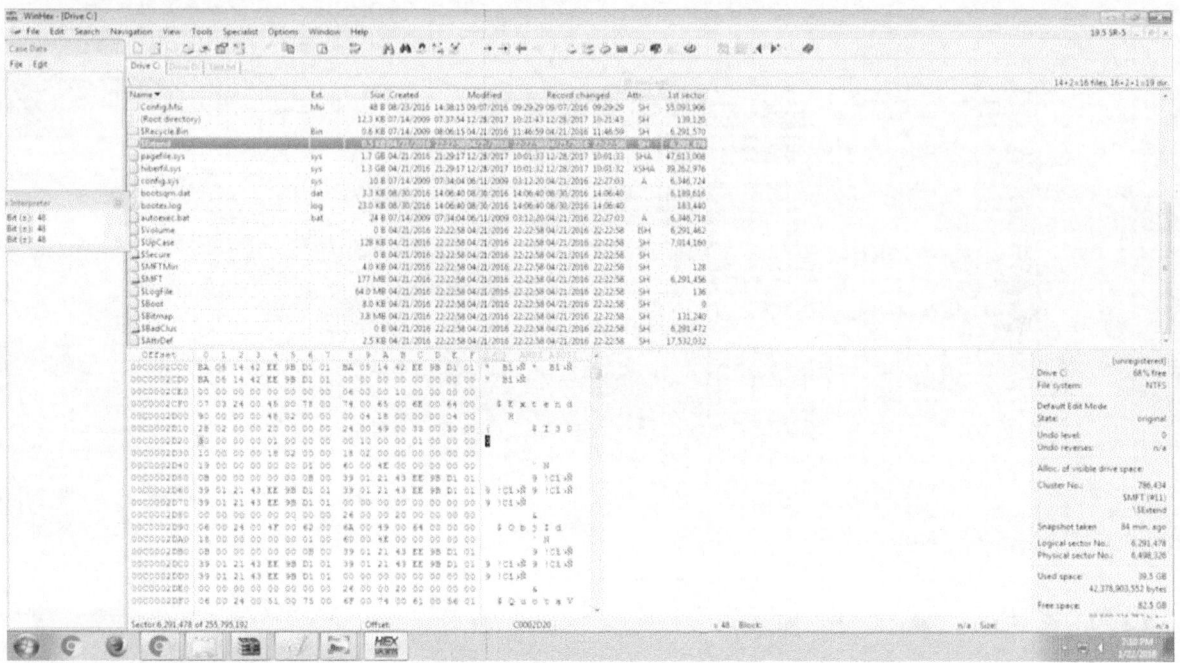

Fig. 9.44 $Extend

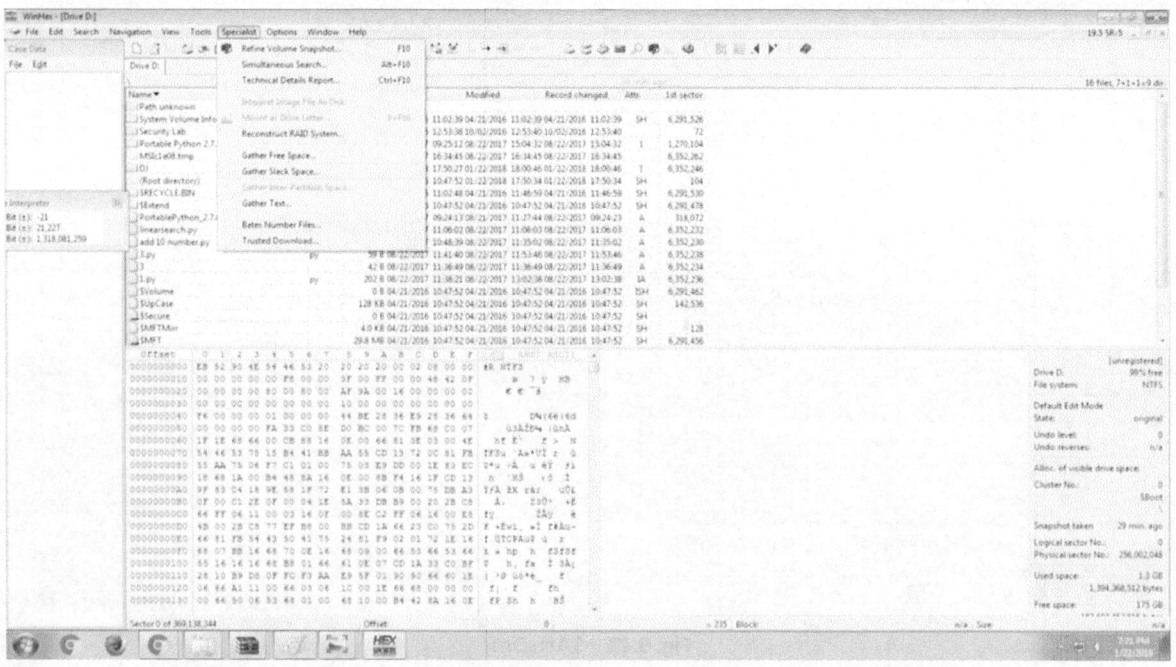

Fig. 9.45 Gather Slack Space

9.4.4 Analysing Slack Space and Free Space

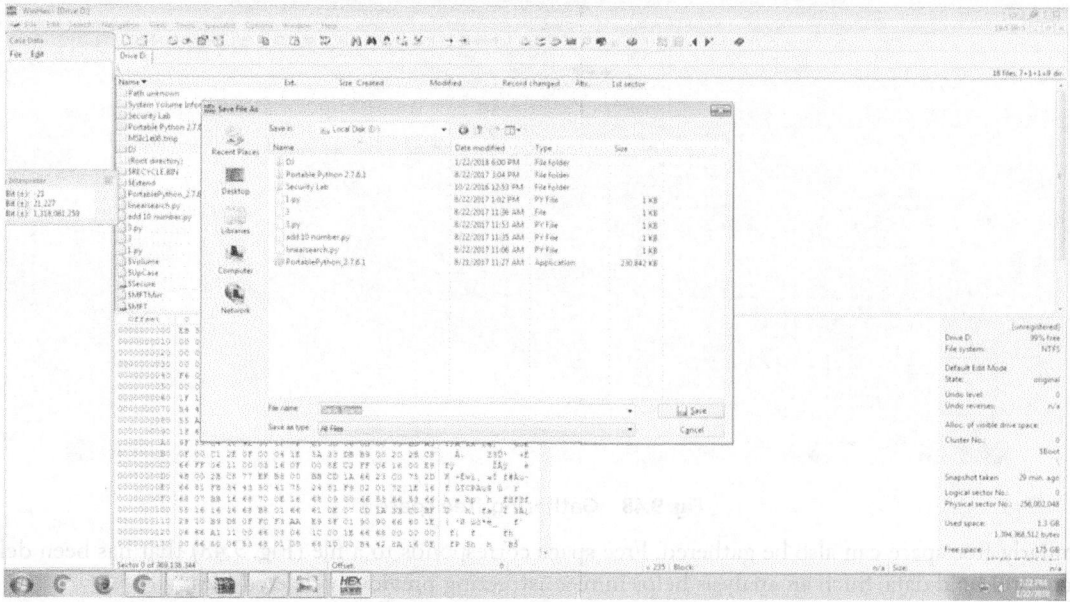

Fig. 9.46 Location of Slack Space

Some bytes in the memory allocated to the file are free and not in use; these actually correspond to the slack space. To analyse the slack space, choose the *Specialist* menu from the menu bar and click the *Gather Slack Space* option which gathers all the slack space (Fig. 9.45) and stores them in a specified location (Fig. 9.46). Usually a large amount of storage is required to acquire slack space. The file containing the data and all the slack space can be analysed exclusively for the presence of evidence (Fig. 9.47).

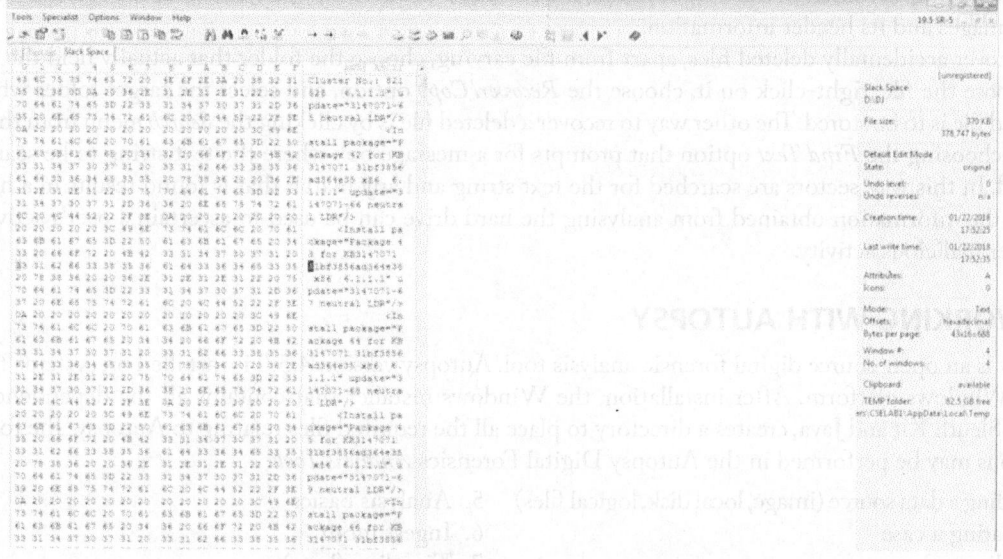

Fig. 9.47 Slack Space

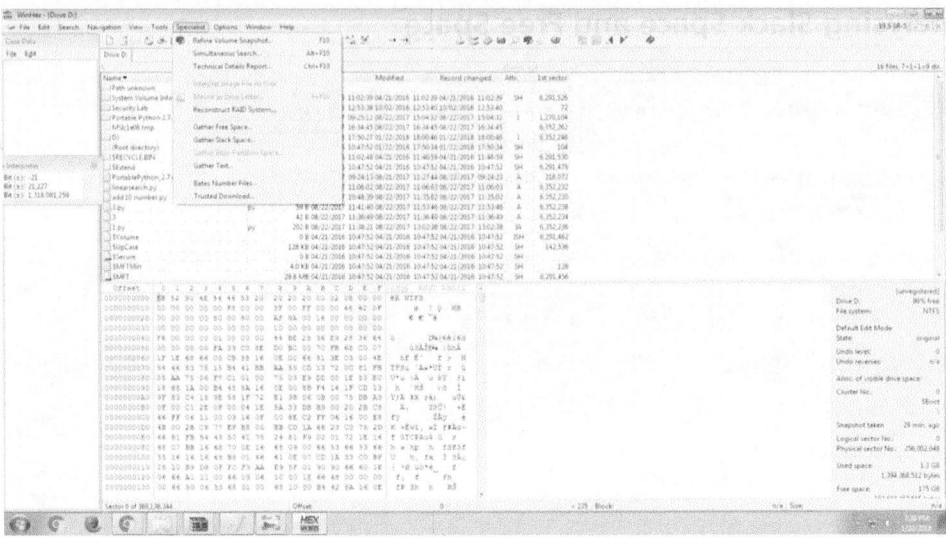

Fig. 9.48 Gathering free space

Similarly, free space can also be gathered. Free space corresponds to a file (Fig. 9.48) that has been deleted but may still hold data. Such an analysis helps in reconstructing previously deleted files.

9.4.5 File Carving

File carving is done by selecting the *Tools* menu from the menu bar, choosing *Disk Tools*, and then *File Recovery by Type* from the pop-up menu. A pop-up window then prompts and lists the file types that can be carved, such as *Pictures*, *Documents*, *E-mail*, *Internet*, *OSArtifacts*, *Music/Video*, *Programs*, *Application*, etc. For example, to carve JPEG images, expand *Pictures* and check *JPEG* from the hierarchy of file types. The output folder has to be specified. The option *Look for Headers in Free Cluster Only* and *Search All Sector Boundaries* have to be chosen from the combo box. A message box prompts once recovery is complete. Now the carved images are available in the specified output folder. In the output folder, the text file *File Recovery by Type* lists all the sector boundaries for the images and its header information.

To recover accidentally deleted files, apart from file carving, choose the folder that actually held the deleted file. Choose the file, right-click on it, choose the *Recover/Copy* option, and select the target folder where the recovered file is to be stored. The other way to recover a deleted file is by clicking the *Search* menu from the menu bar and choosing the *Find Text* option that prompts for a message box where the text string to be searched is specified. In this way, sectors are searched for the text string and any match that is found results in a hit.

Thus, the information obtained from analysing the hard drive can be used to ascertain a user's activity and any other malicious activity.

9.5 WORKING WITH AUTOPSY

Autopsy is an open-source digital forensic analysis tool. Autopsy version 4.5.0 is installed on systems that run on the Windows platform. After installation, the Windows installer that contains all the dependencies, including Sleuth Kit and Java, creates a directory to place all the required files relating to Autopsy. The following operations may be performed in the Autopsy Digital Forensics *Analysis Tool*:

1. Adding a data source (image, local disk, logical files)
2. Creating a case
3. Adding a data source
4. Ingest modules
5. Analysis basics
6. Ingest inbox
7. Timeline (beta)
8. Reporting

Adding Data Source—Image, Local Disk, Logical File

The data source is the location of the data, and can be a file, spreadsheet, database, etc. The data source may be either single or multiple. The data sources are added to a case which can be either a single data source or multiple data sources. A single report is generated for an entire case. The report can also be created for individual data sources by creating one data source for each case.

Creating Case

A new case is created using either the *Create New Case* option on the home screen of the Autopsy tool or from the file menu. Then, a *New Case Wizard* appears where the name of the case, the directory to store all the case results, case numbers, and other details (e.g., name of the examiner) have to be entered.

Adding Data Source

The input data source is then added to a case. After creating a case, the *Add Data Source Wizard* will start automatically. Data source can also be added to the case by manually selecting the *Add Data Source* option from the file menu or toolbar. Then, the type of input data source (an image, a local disk, or logical files and folders) has to be specified to add to the case. The input data source has to be supplied with the location of the source to add.

1. For a disk image, only the first file in the data set is browsed and Autopsy identifies the rest of the files in the data set. Autopsy supports E01 and raw (dd) files.
2. For a local disk, one of the detected disks is selected. Autopsy will add the current view of the disk which is a snapshot of the metadata. Though the content of the individual files gets updated with the changes made to the disk, Autopsy has to be run on Administrator mode to detect all the files.
3. For logical files, the *Add* button is used to add one or more files or folders to the case. A logical file may be a single file or a folder of files. Folders are added recursively to the case.

Options are available in the *Adding Data Source* wizard to make the ingest process faster so as to deal with deleted files. It takes longer time to analyse the unallocated space and the entire device is searched for the deleted files. Recovery steps are performed in some situations. Autopsy will begin analysis once the data sources are added to the case and internal database.

Ingest Modules

Ingest modules run in the background and perform specific tasks. They analyse files in a prioritized order so that files in the user's directory are analysed before the files in other folders. Ingest modules are developed by third parties. Some of the ingest modules in Autopsy are as follows:

1. Recent activity
2. Hash lookup
3. Keyword search
4. Archive extractor
5. Exif image parser
6. Thunderbird parser

Recent activity Recent activity extracts the activities of the user from the web browser and the operating system. It also runs regripper on the registry hive.

Hash lookup Hash lookup uses hash database to ignore known files (which may be known software, file signatures, etc) from the repository, National Institute of Standards Technology - National Software Reference Library (NIST NSRL) which is used predominantly by law enforcing authority and organization involved in forensic investigations to flag known bad files. The advanced button is used to add and configure the hash database that needs to be used. As ingest occurs, an update on known bad files is obtained. The hash databases can be added later by selecting *Options* from the Tool menu.

Keyword search Keyword search uses a list of keywords to identify files with specific words in the list. The keyword lists to search can be selected automatically. A new list of keywords can be created using the advanced option. Keyword search performs search activity only after the ingest process has been completed. The keyword lists that are searched during ingest will be searched at periodic intervals and the results are generated in real time. There is no need to wait for all the files to be indexed.

Autopsy provides options to change the settings of an ingest module. The change settings options are available while selecting the ingest module. For example, it is possible to configure which keyword search lists and hash databases are to be used.

The progress bar is displayed in the lower right-hand side while the ingest modules are running in the background. The graphical user interface (GUI) can be used to review incoming results and perform other tasks.

9.5.1 Analysis Basics

The analysis techniques will start from the tree on the left and can be seen in Fig. 9.52. The *Data Source* root node shows all the data in the case. The individual image nodes show the structure of the forensic copy of the disk or local disks in the case. The *Logical File Set* nodes show the logical files in the case. The *Views* node shows the same data from a file type or timeline perspective. The *Results* node shows the output from the ingest modules.

When a node is selected from the tree on the left, a list of files is shown on the upper right-hand side. The Thumbnail view on the upper right-hand side can also be used to view the pictures. When a file from the upper right-hand side is selected, the contents of the file are shown on the lower right-hand side. The tabs are used in the lower right-hand side to view the text of the file, an image, or the hex data. If the files are viewed from the *Views and Results* node, it is possible to get the location of the file by right-clicking on the file. This feature is very useful to see what else the user has stored in the same folder for the current file. The file can be extracted to the local system by right-clicking on it. To search a single keyword, a search box on the upper right-hand side can be used. The results are shown in a table on the upper right-hand side. Arbitrary files can be tagged or bookmarked so as to find them more quickly or include them specifically in the report.

Ingest Inbox

The ingest modules will run in the background and the results are shown in the tree as soon as the ingest modules find them and report them. The Ingest Inbox receives messages from the ingest modules as they find results. The inbox can be opened to see what has been recently found by the ingest modules. It keeps track of what messages have been read. The intended use is to let the FA focus on some data for a while and then check back on the ingest inbox at a convenient time. Further, it is possible to see what else was found while focus was on the previous task. Upon selecting a message, a jump is made to the *Results* tree where more details can be found or a jump is made to the location of the file in the file system.

9.5.2 Timeline

Autopsy has a basic timeline view which can be accessed by selecting the *Make Timeline* feature from the Tools menu.

Reporting

A final report is generated which contains all the analysis results. The *Generate Report* button can be used to create the report. Autopsy creates an HTML or XLS report in the Reports folder of the case folder. The location of the case folder can be determined using the *Case Properties* option in the File menu. Autopsy provides an option to export the report to a separate folder outside the case folder.

9.5.3 Example Use Cases

The following are some of the examples to do common analysis tasks:

1. Web artifacts
2. Known bad hash files
3. Media—images and videos

Web Artifacts

To view the recent activity of a user, the *Recent Activity* ingest module is enabled. Then, go to the *Results* node in the tree on the left and then into the *Extracted Data* node to find bookmarks, cookies, downloads, and history.

Known Bad Hash Files

To see if the data source has known bad files, *Hash Lookup* is enabled in ingest module. Then the *Hashset Hits* section in the Results area of the tree on the left is viewed. The hash lookup can take a long time. Therefore, it is updated as long as the ingest process is occurring. The Ingest Inbox is used to keep track of what known bad files have been found recently. If a known bad file is found, right-click on the file to view the original location of the file. All the relevant additional files are stored in the same folder of this file.

Media—Images and Videos

All the images and videos on the forensic copy can be viewed under the *File Types* feature of the *Views* section on the left of the tree. The Thumbnail option on the upper right-hand side can be used to view thumbnails of all the images. The selected image or video can be viewed on the lower right-hand side.

9.5.4 Analysis of Deleted Files with Autopsy

To perform analysis of deleted files, some files are deleted in the system. The Autopsy is run in Administrator mode in the system. *Create New Case* is clicked in the Welcome wizard (Fig. 9.49). Under *New Case Information* wizard, case name and browse base directory under case info section are entered (Fig. 9.50). Case number and examiner are entered under the additional information section (Fig. 9.51). Under *Add Data Source* wizard, four steps are performed: *Select Type of Data Source to Add*, *Select Data Source*, *Configure Ingest Modules*, and *Add Data Sources*. Among the four types of data sources, the type of add source to add as logical disk is selected in the example shown in Fig. 9.53. The data source is selected as Removable Disk (H:) which is available under the *Select Data Source* section of the *Add Data Source* wizard (Fig. 9.52). Out of the many ingest modules, only the recent activity ingest module is selected (Fig. 9.54). The progress of the analysis process is shown under the *Add Data Source* section (Fig. 9.55). The status of adding a data source is displayed (Fig. 9.56). After the data source has been added successfully, the logical disk to be analysed is displayed in the table on the right-hand side of the Autopsy UI (Fig. 9.57). Selecting the logical disk displays the contents in the form of a tree on the left-hand side of the Autopsy UI (Fig. 9.58) from which the deleted files (marked with ×) can be viewed by expanding the corresponding nodes in the tree.

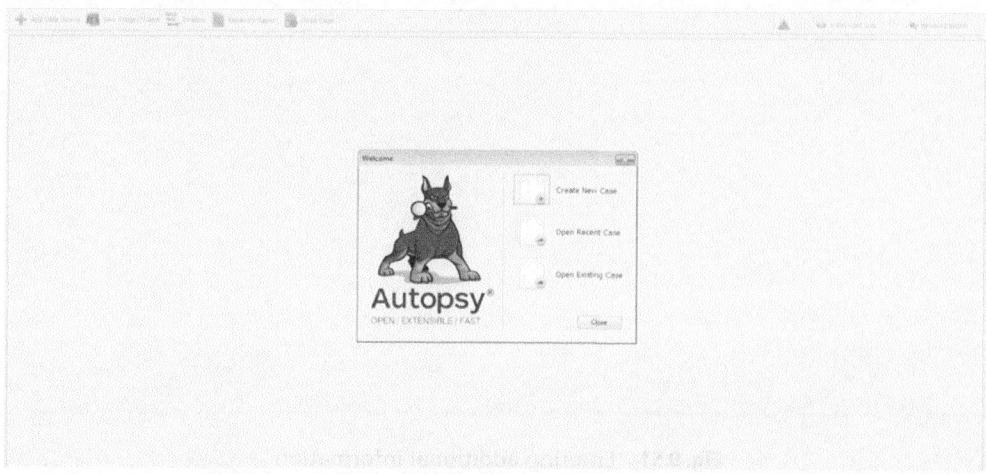

Fig. 9.49 Creating new case

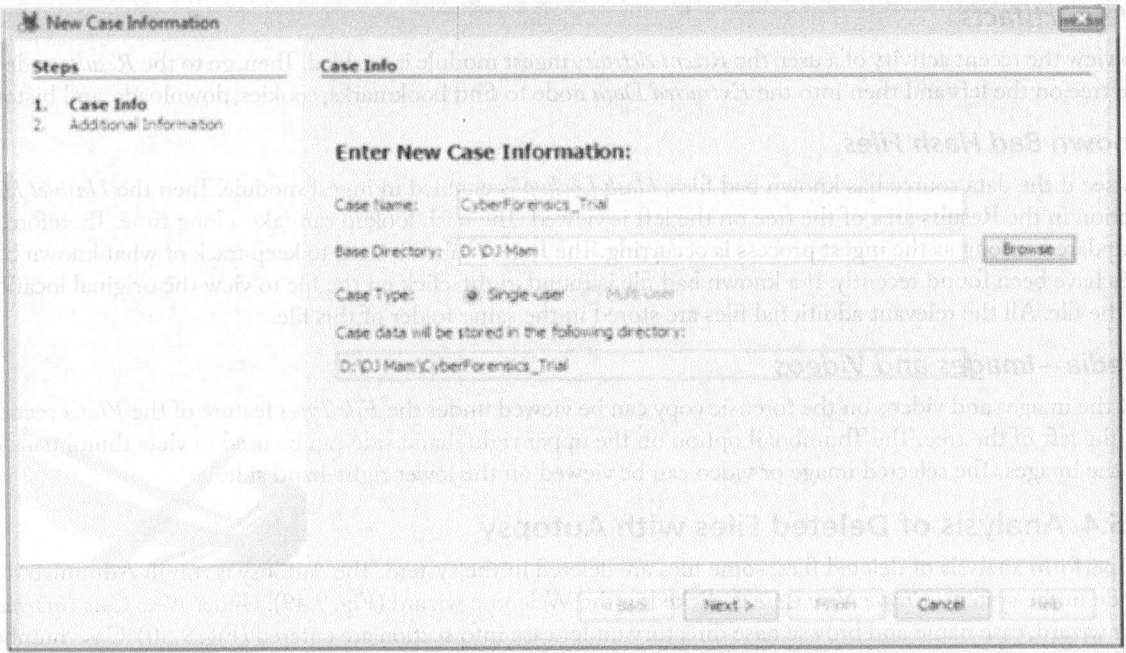

Fig. 9.50 Entering new case information

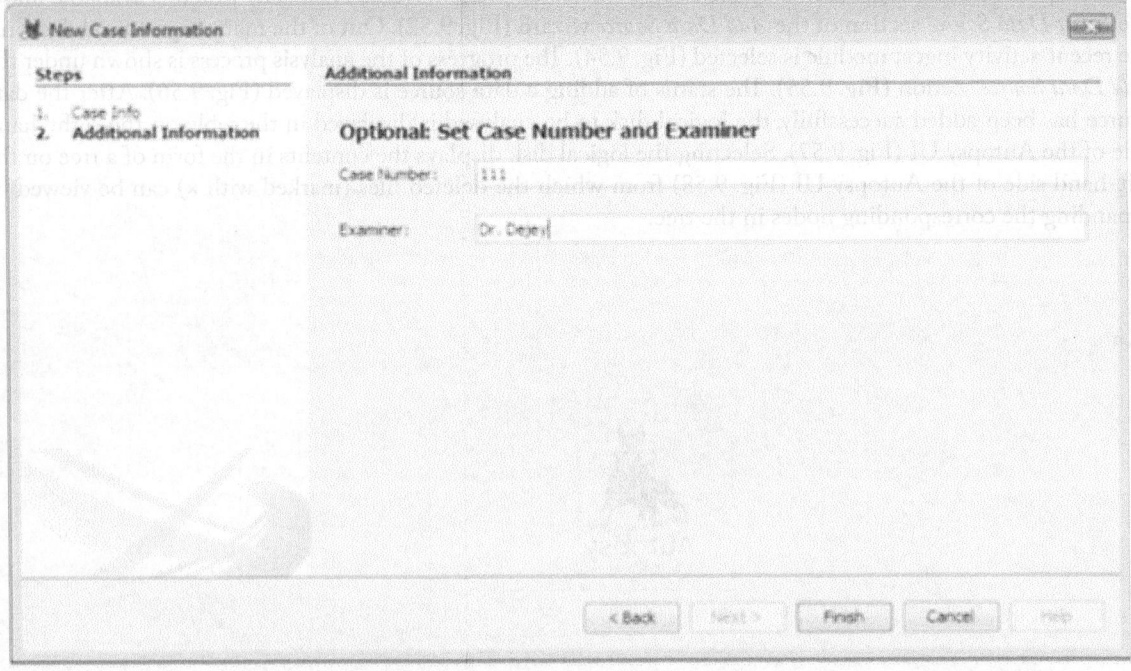

Fig. 9.51 Entering additional information

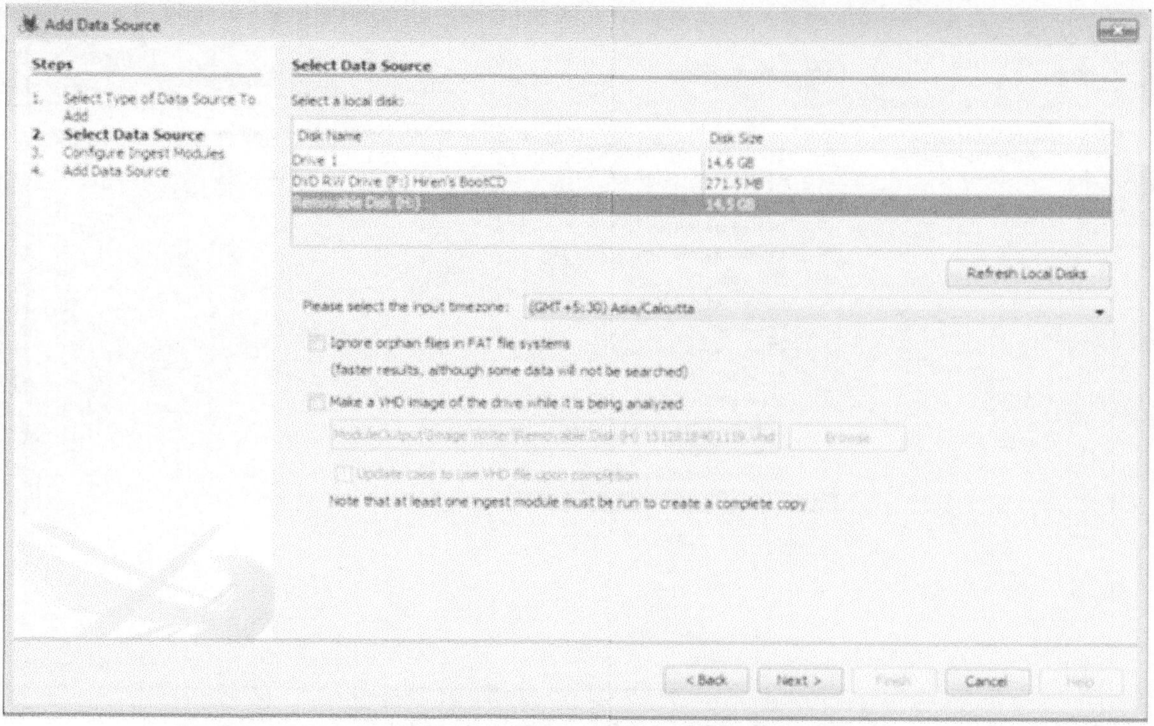

Analysis of Digital Evidence 333

Fig. 9.52 Selecting type of data source to add

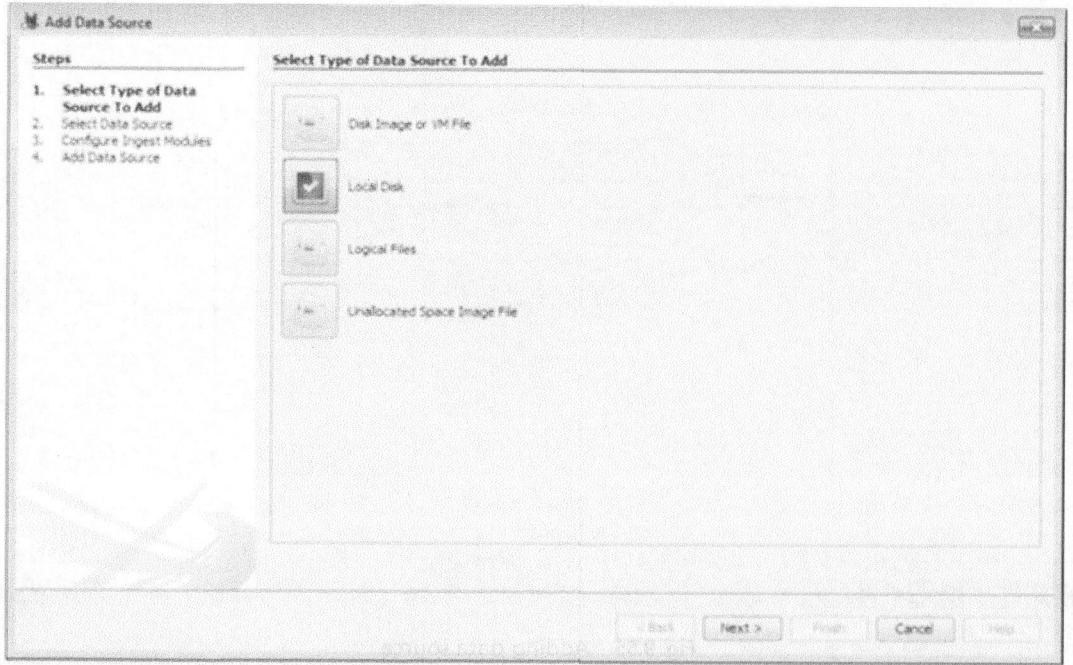

Fig. 9.53 Selecting data source

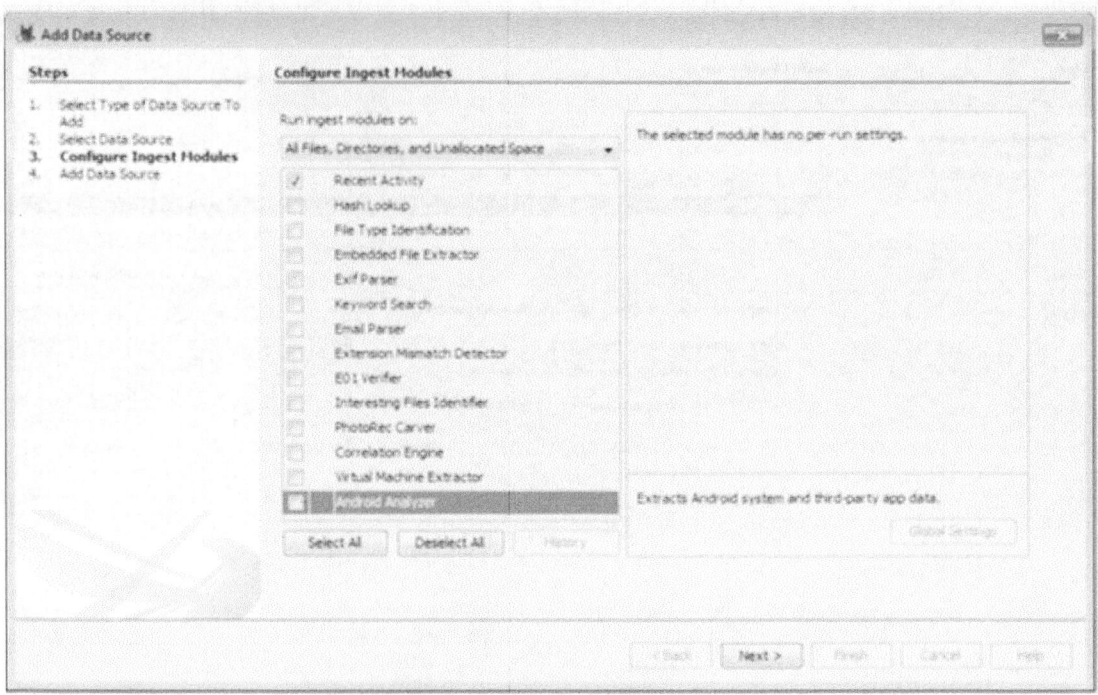

Fig. 9.54 Configuring ingest modules

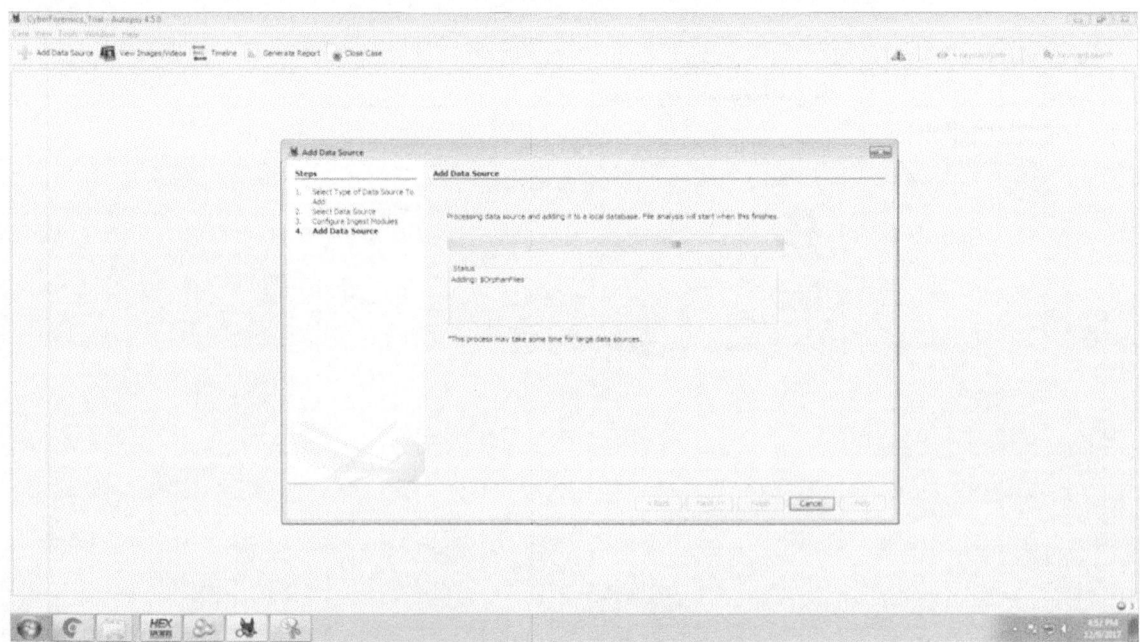

Fig. 9.55 Adding data source

Analysis of Digital Evidence 335

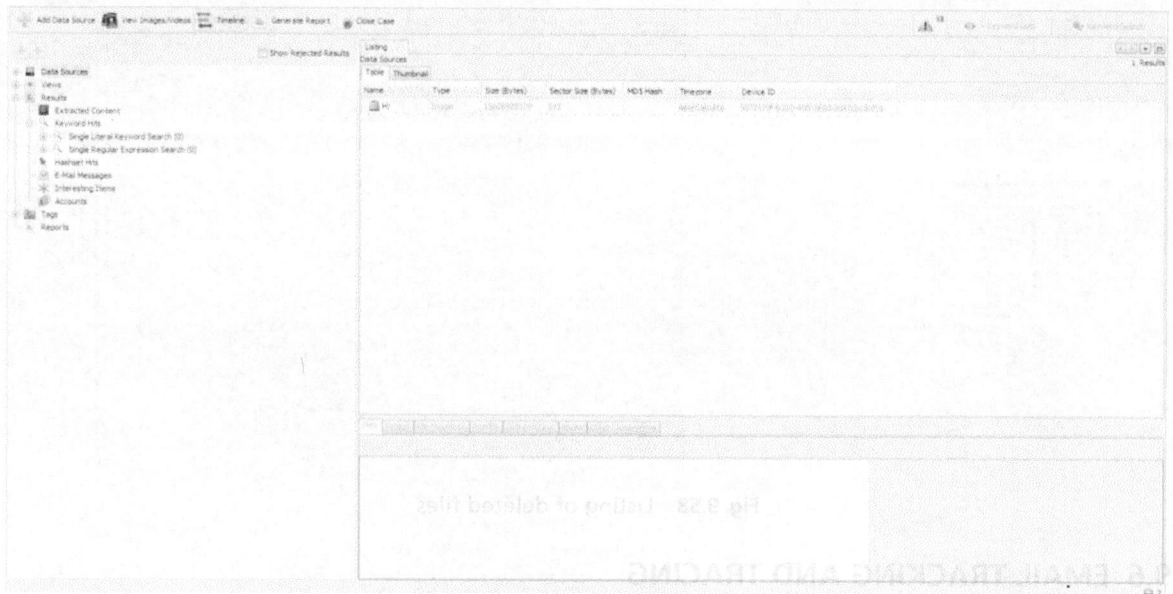

Fig. 9.56 Status of add data source

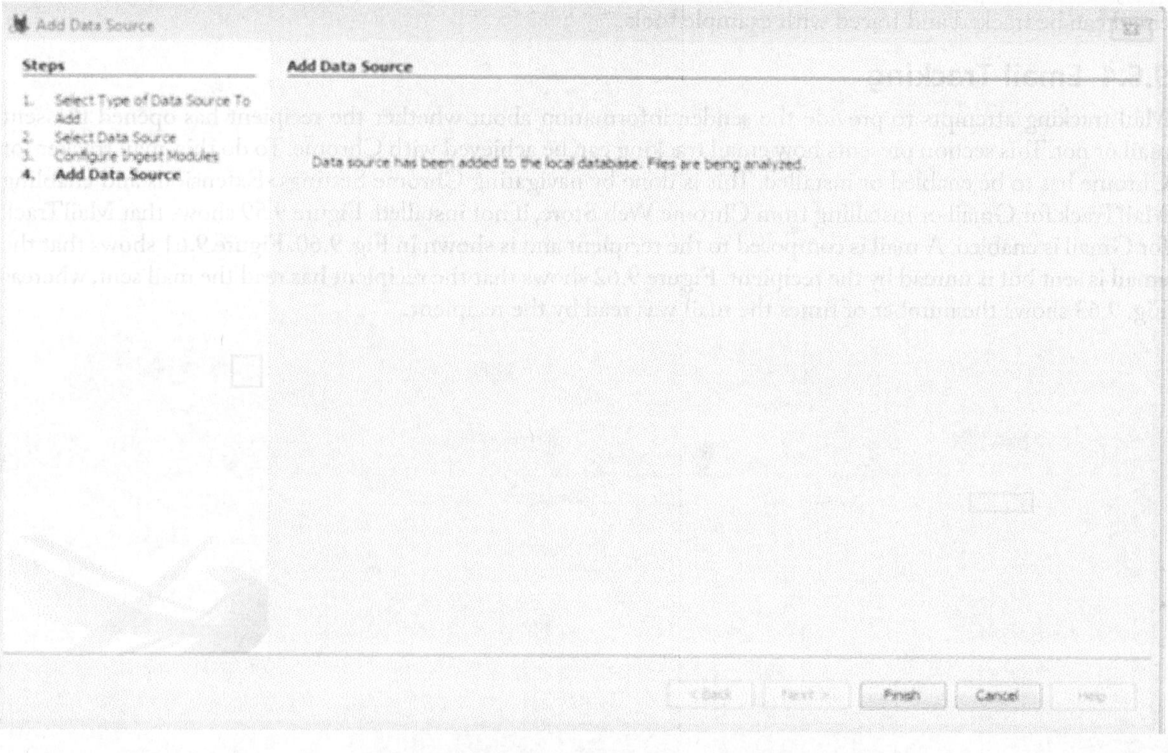

Fig. 9.57 Display of logical disk in table

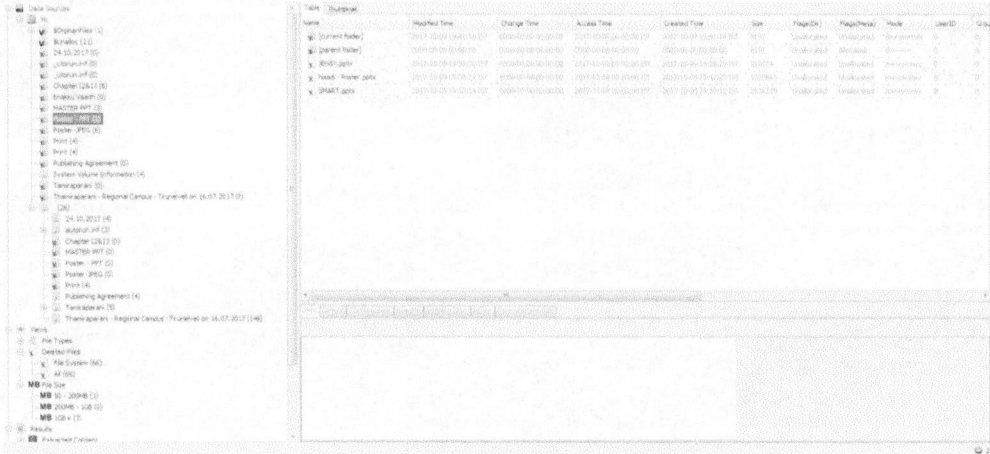

Fig. 9.58 Listing of deleted files

9.6 EMAIL TRACKING AND TRACING

Email tracking is done primarily to monitor delivery of mail. On the other hand, tracing an email provides a way for the recipient of the mail to know who the sender is. The following subsections briefly explain how email can be tracked and traced with example tools.

9.6.1 Email Tracking

Mail tracking attempts to provide the sender, information about whether the recipient has opened the sent mail or not. This section presents how email tracking can be achieved with Chrome. To do this, mail tracker for Chrome has to be enabled or installed. This is done by navigating Chrome Settings>Extensions and enabling MailTrack for Gmail or installing from Chrome Web Store, if not installed. Figure 9.59 shows that MailTrack for Gmail is enabled. A mail is composed to the recipient and is shown in Fig. 9.60. Figure 9.61 shows that the email is sent but is unread by the recipient. Figure 9.62 shows that the recipient has read the mail sent, whereas Fig. 9.63 shows the number of times the mail was read by the recipient.

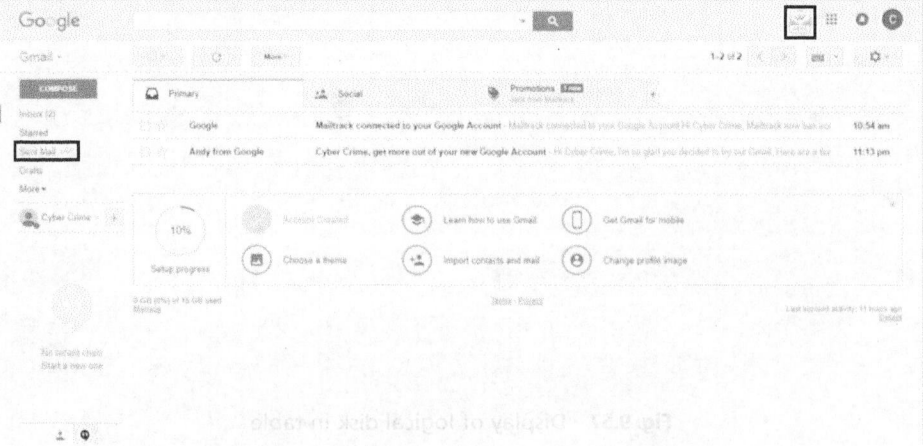

Fig. 9.59 Adding Mail Tracker to Gmail account

Analysis of Digital Evidence **337**

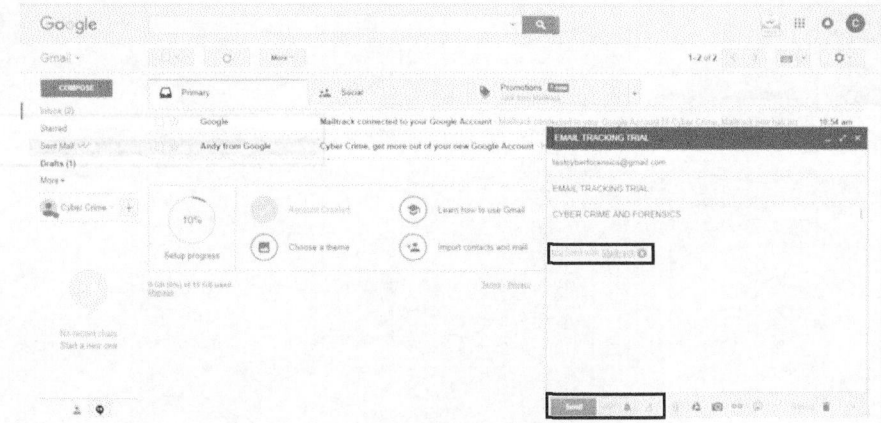

Fig. 9.60 Composing email for recipient

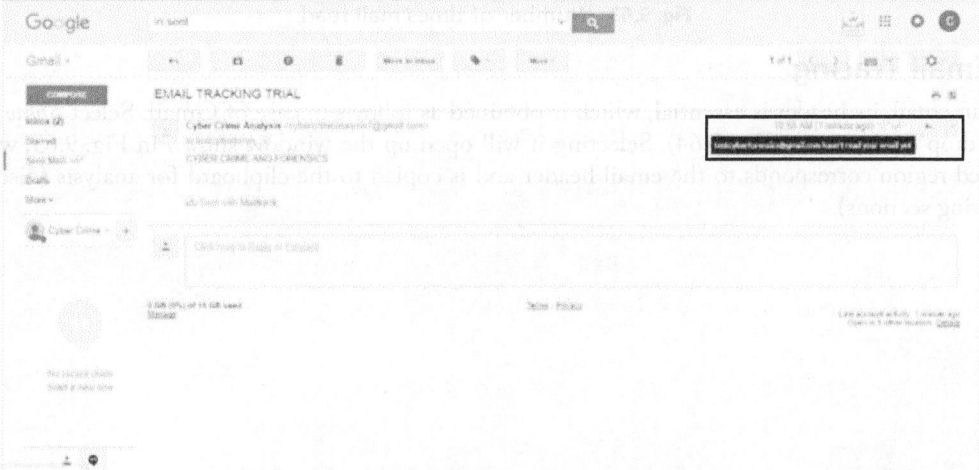

Fig. 9.61 Email sent but unread by recipient

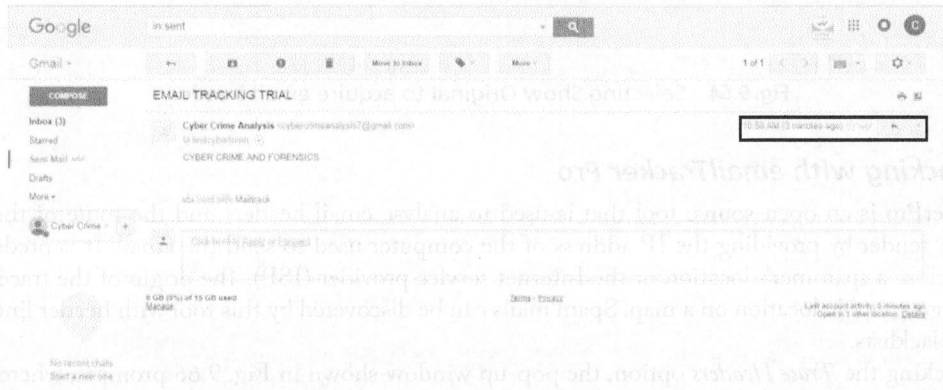

Fig. 9.62 Sent email status

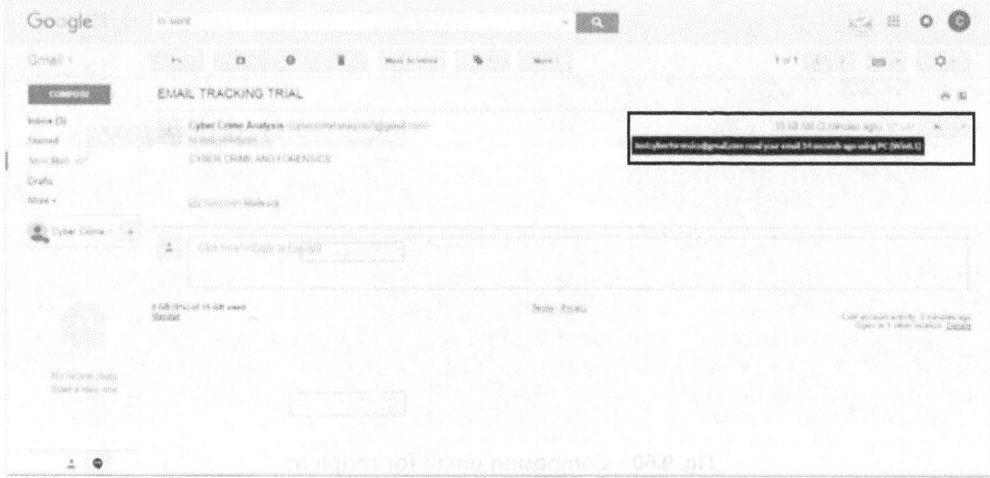

Fig. 9.63   Number of times mail read

9.6.2 Email Tracing

To trace an email, its header is essential, which is obtained as follows in case of Gmail: Select *Show Original* from the drop down menu (Fig. 9.64). Selecting it will open up the window shown in Fig. 9.65, where the highlighted region corresponds to the email header and is copied to the clipboard for analysis (discussed in the following sections).

Fig. 9.64   Selecting Show Original to acquire email header

Email Tracking with emailTracker Pro

emailTrackerPro is an open-source tool that is used to analyse email headers and the route of the email to disclose the sender by providing the IP address of the computer used to send the email. It is predominantly used to disclose a spammer's location or the Internet service provider (ISP). The origin of the traced email is shown as a geographic location on a map. Spam mails can be discovered by this tool with header line mistakes and DNS blacklists.

Upon clicking the *Trace Headers* option, the pop-up window shown in Fig. 9.66 prompts, where the radio button *Trace an Email I have Received* is chosen. The email header is copy–pasted and the *Trace* button is clicked (Fig. 9.67). The result of Trace Header is shown in Fig. 9.68. If the radio button *Look Up Network Responsible for an Email Address* is chosen, an email address is given (Fig. 9.69) and the resultant network addresses are displayed (Fig. 9.70).

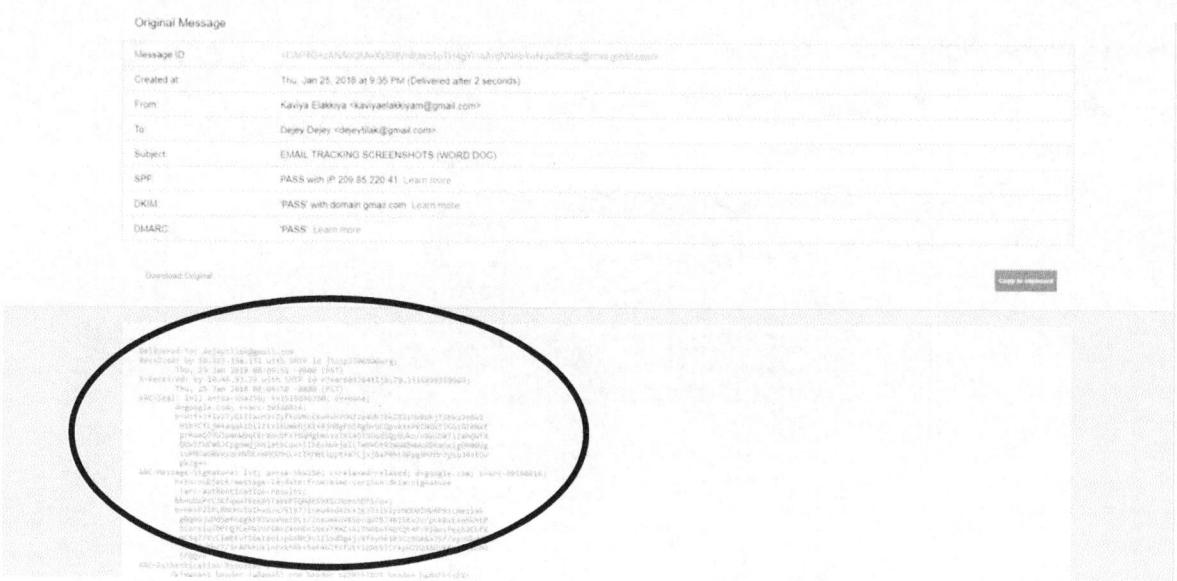

Fig. 9.65 Sample email header

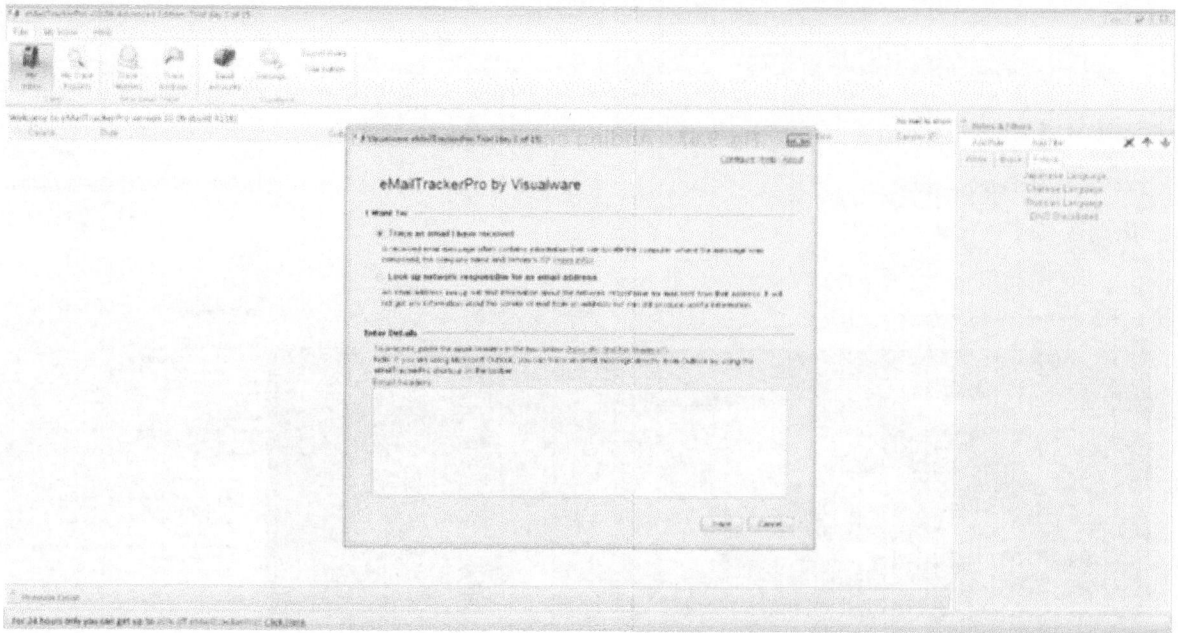

Fig. 9.66 Adding email header with Trace Header

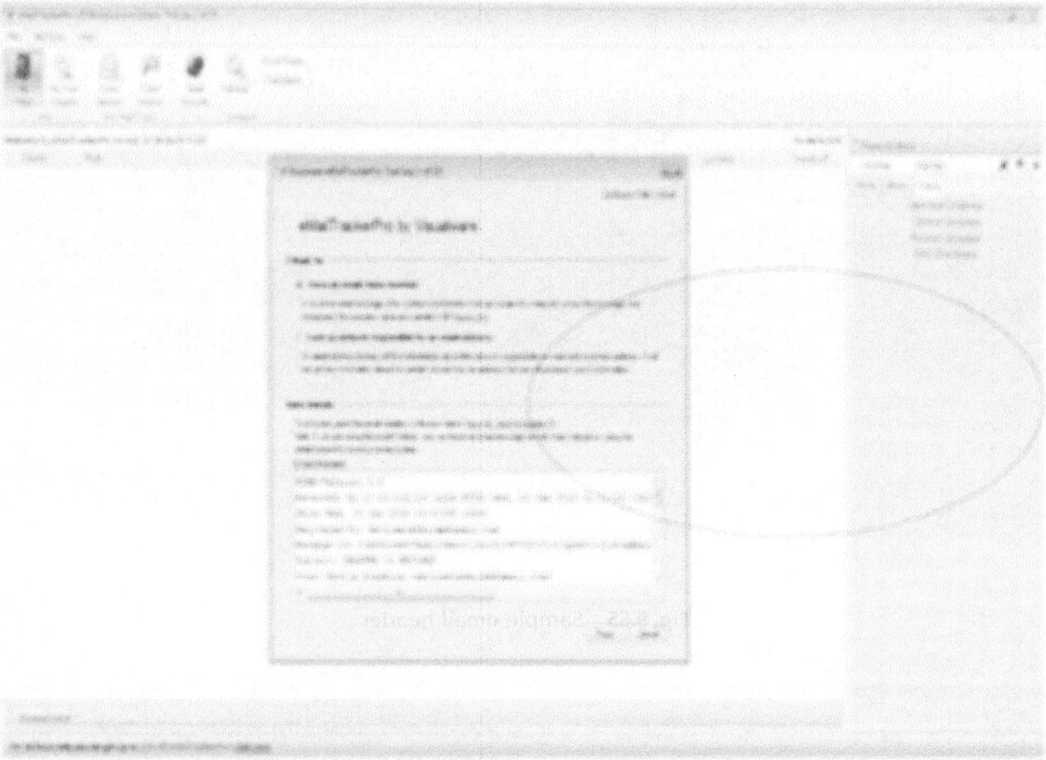

Fig. 9.67 Adding email header

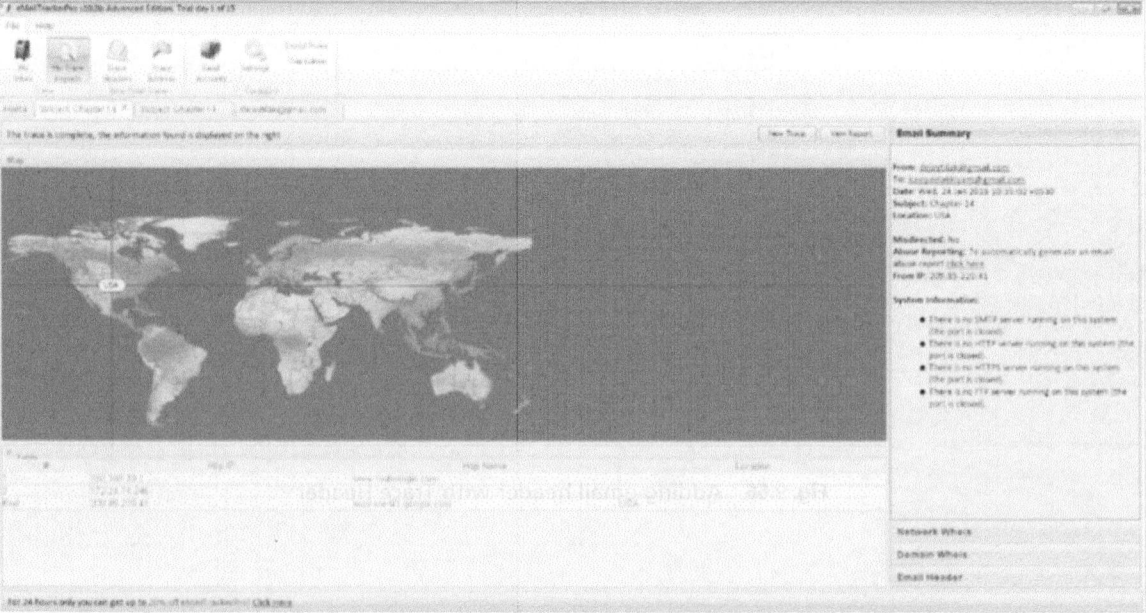

Fig. 9.68 Result of Trace Header

Analysis of Digital Evidence 341

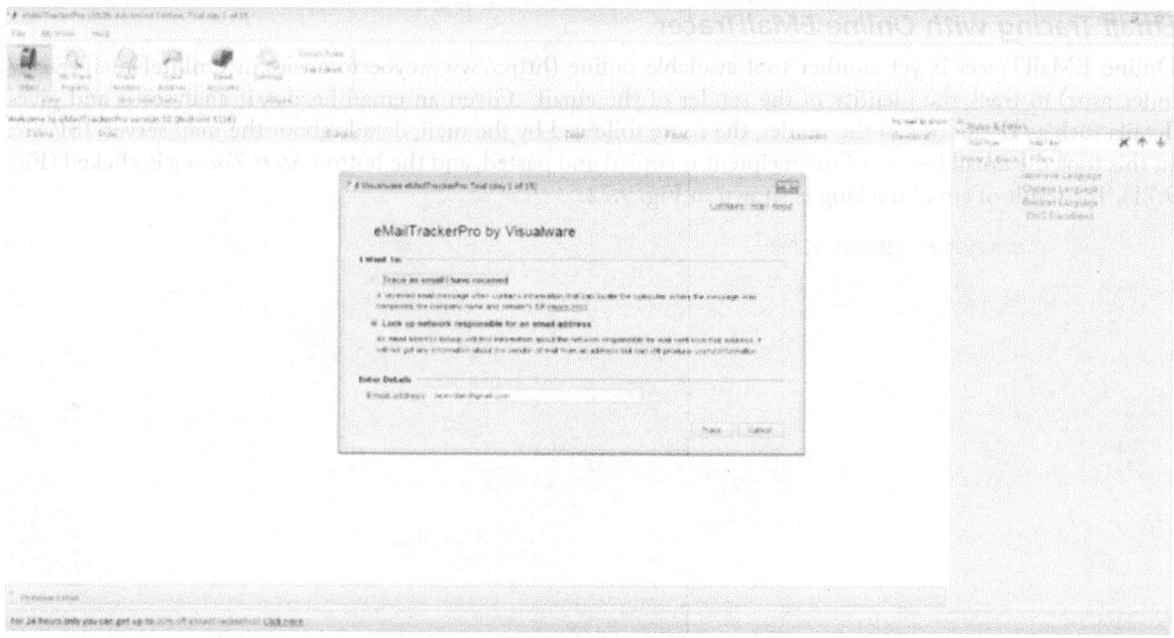

Fig. 9.69 Tracing network with email address

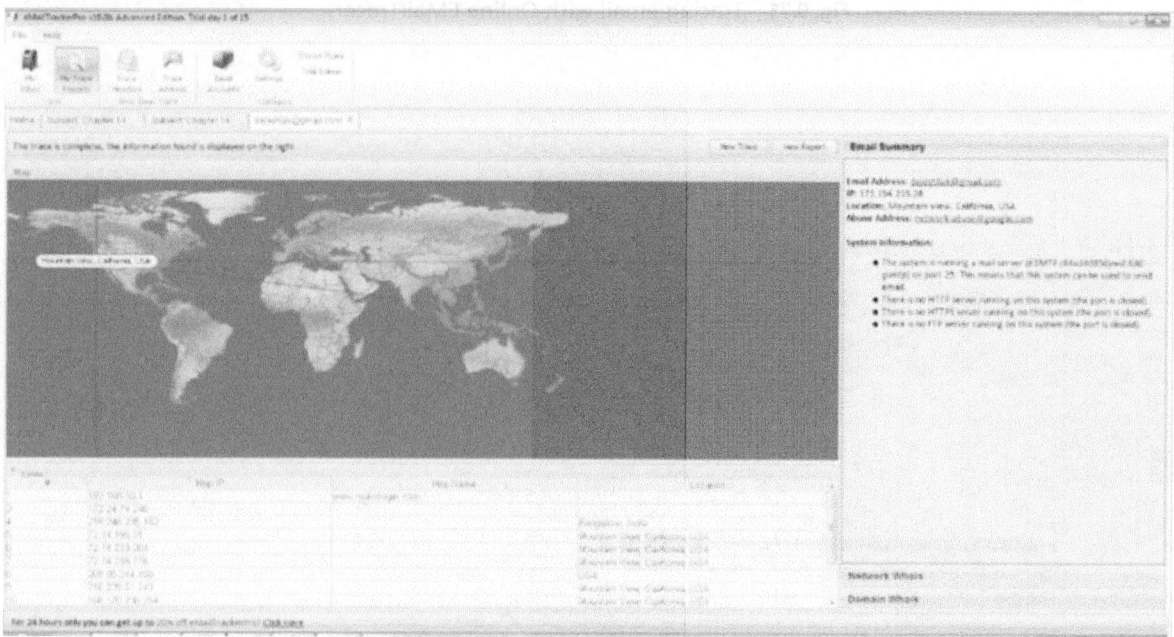

Fig. 9.70 Result of Trace Address

Email Tracing with Online EMailTracer

Online EMailTracer is yet another tool available online (http://www.cyberforensics.in/OnlineEmailTracer/index.aspx) to track the identity of the sender of the email. Given an email header, it analyses it and gives details such as IP address of the sender, the route followed by the mail, details about the mail server, ISP, etc. In this tool, the email header of the recipient is copied and pasted, and the button *Start Tracing* is clicked (Fig. 9.71). The result of email tracking is shown in Fig. 9.72.

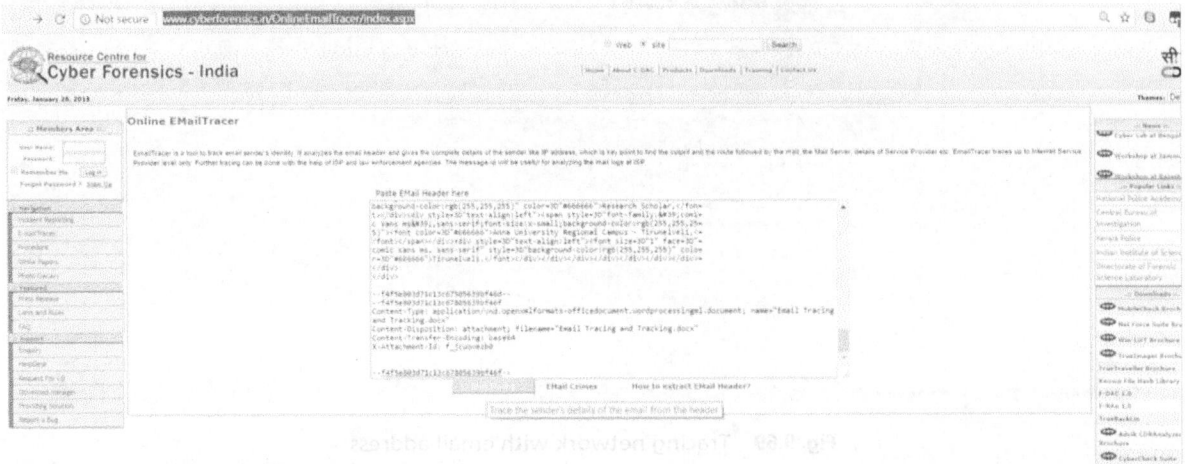

Fig. 9.71 Tracing email with Online EMailTracer

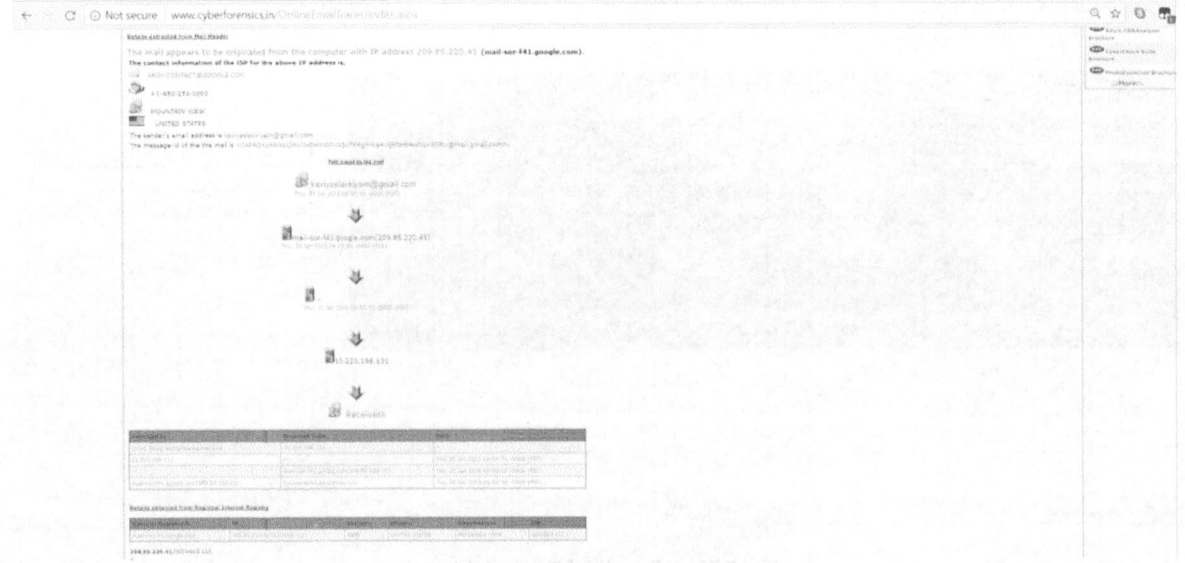

Fig. 9.72 Results of email tracing

Format of requisition to forensic lab

Crime no.				Police station		
Section of the law				District		
Date				State		
I. Nature of Crime						
Nature of crime						
Brief case history						
Any other relevant details						
II. List of exhibits for examination						
S. no./Barcode	Description of exhibits		How, when, and by whom found		Source of exhibits	Remarks
III. Nature of examination required						
S. no./Barcode	Description of exhibits		Nature of examination required		Date or any keyword or filter	Remarks
S. no.	Full name	Occupation		Sex	Date and time of arrest	Whether bailed, court, or police custody
Seal				Rank and sign of IO		
O/W no.				Date		
Forwarded to the director						
Specimen seal/s impression/s on exhibits or parcel/s				Sign and designation of forwarding officer		

Fig. 9.73 Format of requisition sent to forensic lab for analysis

9.7 ROLE OF FORENSIC ANALYST IN ANALYSIS

To begin forensic examination, the IO sends a request to the forensic lab in the format shown in Fig. 9.73. The IO should furnish the following information to the forensic analyst: facts about the case, description of the objects seized that carries the evidence, name of the person, organization or location from where it was seized, etc. Examples of how the description of seized objects (e.g., hard disk, credit card, mobile phone) is relevant to the case, is shown in Box 9.1.

> **Box 9.1 Description of Seized Objects in Cases**
>
> Hard disk
>
> One hard disk (Seagate MODEL ST310212A P/N: 7T6789-123 S/N: wer14ty 10.2 Gbytes) kept in a sealed paper parcel marked HARD DISK SEAGATE....B.No.: xxx/xx, Cr. No.: xxx/xx,...A2 <Name of the accused>

(*Contd*)

Box 9.1 (Contd)

> **Credit card**
>
> The following information should be produced in case of credit/debit cards:
>
> Item description : ICICI Bank VISA Card HP Gold Colour
>
> Name as found printed on the card : xxxxxxxxxxxxxxxx
>
> Number as found printed on the card : xxxx xxxx xxxx xxxx
>
> Validity as printed on the card : 05/15–05/20
>
> **Mobile phone**
>
> Item No. X: One black/silver mobile phone (Sony Ericsson W205 TY 1234-5678 S/N: FG789GHRY7 12345678-123456-1), a battery (Sony Ericsson Standard Battery TYU-34), and a memory card (Sony M2 1 GB 6T789Q) and with a SIM Card (Airtel 12345 00000 45678 890] kept in a sealed paper parcel marked, CCB Cr. No.: xxx/xx B.No. xxx/xx, SONY Ericsson W205 IMEI No: xxxxxxxxxxxxxxxx

A forensic analyst performs analysis on the evidence, which varies depending on the case. In some cases, he/she may be specifically asked to examine devices for some keywords or files, account details, passwords, etc. Table 9.1 presents some example cases and the examination that must be performed by the forensic analyst.

Table 9.1 Information to be provided by forensic analyst for example cases

Example cases	Information to be extracted and examined by forensic analyst
Computer that is a target of some crime	Operating system installed with time of installation, usernames in the system, list of software installed on the system, MAC address and IP address, information related to email address, presence of any suspicious malware, Internet history, log files, etc.
Sending threatening email	Presence of sender's email address in victim's hard disk and vice versa, presence of contents of email in the hard disk, email tracing for obtaining information related to IP addresses
Obscene content	Presence of obscene information on the hard disk, Internet history, log files, etc.
Computer as a repository or an instrument to commit crime	Operating system installed with time of installation, usernames in the system, list of software installed in the system, MAC address and IP address, information related to email address, passwords, date/time change, Internet history, log files, etc.

A forensic analyst has to perform any one of the following for providing information required for every case: acquire a forensic copy of the suspected storage media, compute hash, perform live forensics, examine slack space/free space, check for deleted files, look for artifacts, trace email, etc., as discussed in earlier sections. He/She then has to prepare the report of the forensic examination including item-wise details of examination with reference to the points raised by the IO. An example report prepared by a forensic analyst after examining a hard disk is shown in Box 9.2. Finally, he/she represents his/her opinions and draws a conclusion of the case.

Box 9.2 Report Prepared by Forensic Analyst after Examining Hard Disk

The hard disk was examined using the software EnCase v 6.0. The disk report of the hard disk is given as Annexure x. A file named 'xxxx.doc' was found stored with email correspondence details. The contents of the mail are as follows:

xxxxxxxxxxxxxxxxxxxxxxxxxxxxxxxxxxxxxxxxxxxxxxxxxxxxxxxxxxxxxxxxxxxxxxxxxxxxxx
xxxxxxxxxxxxxxxxxxxxxxxxxxxxxxxxxxxxxxxxxxxxxxxxxxxxxxxxxxxxxxxxxxxxxxxxxxxxxx
xxxxxxxxxxxxxxxxxxxxxxxxxxxxxxxxxxxxxxxxxxxxxxxxxxxxxxxxxxxxxxxxxxxxxxxxxxxxxx
xxxxxxxxxx

The above findings revealed that email correspondence had occurred between person X and the accused A2.

Information about bank account, IFSC code, and money transfer were found stored in the hard disk. The hard copy of the documents is given as Annexure x.

POINTS TO REMEMBER

- Analysis of digital evidence is the process of identifying, preserving, interpreting, and documenting the evidence recovered for presentation in a civil or criminal court.
- Analysis of evidence follows seizure of the ECD and acquisition of evidence.
- Forensic tools are required to look for keywords and deleted files, to evaluate the file slack and unallocated space on disk, and to evaluate encrypted files.
- The runtime version of the FTK Imager from Access Data can be used to access the memory and capture the contents of the drive, especially passwords, before the contents get encrypted.
- Registry files cannot be acquired for analysis while the system is running.
- Information about network connections such as remote IP address and port number helps during analysis in gathering information pertaining to computer intrusion, identify remote destinations where the malware is communicating, etc.
- The Volatility framework is a completely open collection of tools, implemented in Python under the GNU General Public License, for the extraction of digital artifacts from volatile memory (RAM) samples.
- The Volatility framework is also used to analyse crash dumps, raw dumps, VMware, and VirtualBox dumps.
- Some of the essential commands of Volatility are *imageinfo*, *pslist*, *pstree*, *malfind*, *cmdscan*, and *netscan*.
- Memory analysis can be used to identify whether the system is compromised by any malware.
- WinHex is a proprietary tool from X-ways Forensics which is primarily used for data recovery and digital forensics. It can also be used to acquire a forensic copy of the drive; inspect files from storage media such as hard disks, optical storage media, and memory cards/sticks; recover deleted files; analyse and compare files; and many other features that come with licensing.
- WinHex presents information about the drive and the file system in a human readable form.
- The information obtained from analysing the hard drive can be used to ascertain user activity and any other malicious activity.
- Autopsy is an open-source digital forensic analysis tool that performs several operations such as adding a data source (image, local disk, logical files), creating a case, ingest modules, analysis basics, ingest inbox, timeline (beta), and reporting.
- In Autopsy, the ingest modules analyse files in a prioritized order so that files in the user's directory are analysed before the files in other folders.
- Some of the ingest modules in Autopsy are Recent Activity, Hash Lookup, Keyword Search, Archive Extractor, Exif Image Parser, and Thunderbird Parser.

- Email tracking is primarily done to monitor delivery of mail.
- Tracing an email provides a way for the recipient of the mail to know who the sender is.
- emailTrackerPro is an open-source tool that is used to analyse email headers and the route of an email to disclose the sender by providing the IP address of the computer used to send the email.
- Online EMailTracer is a tool available online (http://www.cyberforensics.in/OnlineEmailTracer/index.aspx) to track the identity of the sender of the email.

KEY TERMS

Analysis of digital evidence This is defined as the process of identifying, preserving, interpreting, and documenting the evidence recovered for presentation in a civil or criminal court.

Autopsy This is an open-source digital forensic analysis tool that performs several operations such as adding a data source (image, local disk, logical files), creating a case, ingest modules, analysis basics, ingest inbox, timeline (beta), and reporting.

emailTrackerPro This is an open-source tool that is used to analyse email headers and the route of an email to disclose the sender by providing the IP address of the computer used to send the email.

Online EMailTracer This is a tool available online (http://www.cyberforensics.in/OnlineEmailTracer/index.aspx) to track the identity of the sender of the email.

RAM analysis This refers to identifying the running processes, malicious processes, network activities, open connections, etc., in a compromised system.

RAM artifacts This refers to data used by a software application or a hardware device.

Volatility framework This refers to a completely open collection of tools, implemented in Python under the GNU General Public License, for the extraction of digital artifacts from volatile memory (RAM) samples.

WinHex This is a proprietary tool from X-ways Forensics which is primarily used for data recovery and digital forensics.

MULTIPLE-CHOICE QUESTIONS

1. Analysis of evidence follows _____.
   (a) seizure of electronic communication device
   (b) acquisition of evidence
   (c) both (a) and (b)
   (d) none of these
2. Analysis of evidence is also called _____.
   (a) forensics analysis of digital media
   (b) forensics analysis of digital discovery
   (c) forensics analysis of electronic discovery and forensic examination
   (d) forensics analysis of digital media, digital discovery, electronic discovery, and forensic examination
3. Examination of ECDs should be made in an isolated chamber so as to prevent _____.
   (a) connections to any networks and interference from other mobile devices
   (b) interference from networks
   (c) interference from other mobile devices
   (d) connection to any networks
4. A forensic copy is also known as _____.
   (a) digital image   (c) digital copy
   (b) work copy   (d) work image
5. Analysts should use _____ to avoid introduction of data from some other source.
   (a) clean storage media
   (b) write-blocker
   (c) both (a) and (b)
   (d) none of these
6. Forensics tools are required _____.
   (a) to look for keywords and deleted files
   (b) to evaluate the file slack and unallocated space on the disk
   (c) to evaluate encrypted files
   (d) all of these
7. _____ option retrieves users, system, and SAM files in FTK Imager.
   (a) Obtain Protected Files

(b) Minimum Files for Login Recovery
(c) Password Recovery and all Registry Files
(d) Obtain Protected Files and Password Recovery and all Registry Files

8. _____ is a bit-by-bit uncompressed copy of the original.
(a) SMART   (c) Raw
(b) E01   (d) AFF

9. The Volatility framework runs on _____.
(a) Windows, Linux, Mac, and Android
(b) Linux
(c) Linux and Mac
(d) Mac and Android

10. *imageinfo* command is used to _____.
(a) get a list of running processes in the memory dump
(b) figure out the parent–child relationships between processes
(c) identify the profiles supported for the dump memory image
(d) to show the history of commands

11. In WinHex, boot sector begins at offset _____.
(a) 00000000   (c) 000000000000
(b) 0000000000   (d) 0000000000000000

12. In WinHex, *Gather Slack Space* option is available under _____ menu.
(a) Tools   (c) Options
(b) Specialist   (d) Search

13. The ingest modules analyse files in _____ order so that files in the user's directory are analysed before the files in other folders.
(a) First-in first-out   (c) Prioritized
(b) Last-in first-out   (d) Linear

14. emailTrackerPro is used to analyse _____.
(a) email headers   (c) routes of network
(b) routes of email   (d) both (a) and (b)

15. To begin a forensic examination, the IO sends a request to the _____.
(a) forensic lab   (c) forensic examiner
(b) forensic officer   (d) forensic analyst

REVIEW QUESTIONS

1. Define analysis of evidence.
2. What are the five essential tasks that a qualified examiner performs during the analysis of evidentiary digital media?
3. Discuss in detail about capturing forensic copying with Forensic Toolkit (FTK) Imager.
4. Explain RAM analysis with volatility.
5. Give a brief note on analysing hard drives with WinHex.
6. Briefly discuss autopsy.
7. Explain in detail the analysis of deleted files with autopsy.
8. Give a detailed note on email tracking and tracing.
9. What kind of information must be extracted and examined by a forensic analyst in cases where threatening emails are sent?
10. Mention the role of the forensic analyst in analysis.

APPLICATION EXERCISES

1. How can a forensic analyst determine that the contents of the SD card have been tampered with, since being seized?
2. Analyse the email header of any of the emails in your inbox and determine the IP address of the sender and the ISP.
3. Refer to Fig. 9.34. Prepare a report on $volume as the forensic evidence file.
   (a) What is the volume label for this file?
   (b) What is the name of the record where it is stored?
   (c) In what cluster is it stored?
4. How will you prove that a particular file is part of an active file? (*Hint*: $Data Attribute of NTFS may be referred.)
5. With the data structure shown in Table 9.2 for $MFT File Record data, interpret the $MFT shown in Fig. 9.38.

Table 9.2 $MFT file record data

Offset in hex	Length in bytes	Data
0x00	4	Signature
0x04	2	Offset to fixup array
0x06	2	Number of entries in the fixup array
0x08	8	$LogFile sequence number
0x10	2	Incremental sequence count
0x12	2	Link count
0x14	2	Offset to the first attribute
0x16	2	Allocation status (deleted file/directory, allocated file/directory)
0x18	4	Logical size of the $MFT record
0x1C	4	Physical size of the $MFT record
0x20	8	File reference to the base record
0x28	2	The next attribute identification (shows an account of the attributes in the record)
0x2A	2	Fixup codes and attributes
0x2C	4	$MFT file record number

6. Prepare a forensic examination report for a hard disk containing obscene content that is determined after analysis, with the following information:
   1. Date when the evidence was received and from whom
   2. Purpose of examination
   3. Dates of examination
   4. Description about the evidence examined
   5. Tools used
   6. Finding
   7. Examination conclusions
   8. Recommendations (if any)

Analyse in both the aspects as if the drive contains offensive content and also it does not contain such offensive content.

BIBLIOGRAPHY

1. John Barbara, J. (2014), *Digital Evidence Analysis*, available at: https://www.forensicmag.com/article/2014/01/digital-evidence-analysis (Accessed 24 January 2018)
2. Forensic Science Simplified, *A Simplified Guide to Digital Evidence*, available at: http://www.forensicsciencesimplified.org/digital/how.html (Accessed 24 January 2018)
3. Rakesh Kumar Mishra (2017), *Analysis of Digital Evidence*, available at: https://www.slideshare.net/MISHRA8931/analysis-of-digital-evidence (Accessed 24 January 2018)
4. Computer Forensics Associates, *Electronic Evidence Analysis*, available at: http://www.computerforensicsassociates.com/forensics_services/electronic_evidence_analysis.html (Accessed 24 January 2018)
5. Mark Wade (2011), *Memory Forensics: Where to Start*, available at: https://www.forensicmag.com/article/2011/06/memory-forensics-where-start (Accessed 25 January 2018)
6. Tools (2014), *Volatility*, available at: https://tools.kali.org/forensics/volatility (Accessed 25 January 2018)

7. Ahmad (2017), *How to Install and Use Volatility Memory Forensic Tool*, available at: https://www.howtoforge.com/tutorial/how-to-install-and-use-volatility-memory-forensic-tool/ (Accessed 25 January 2018)
8. Doug Austin (2013), *How to Create an Image Using FTKImager – eDiscoveryBest Practices*, available at: https://www.ediscovery.co/ediscoverydaily/how-to-create-an-image-using-ftk-imager-ediscovery-best-practices/ (Accessed 25 January 2018)
9. Heinz Tschabitscher (2017), *eMailTrackerPro 10 Review – Email Tracking Tool*, available at: https://www.lifewire.com/emailtrackerpro-review-1174405 (Accessed 26 January 2018)
10. Cyberforensics, *Online EMailTracer*, available at: http://www.cyberforensics.in/OnlineEmailTracer/index.aspx (Accessed 26 January 2018)
11. X-ways, *WinHex: Additional Features of Specialist Licenses*, available at: https://www.x-ways.net/winhex/specialist_tools.html (Accessed 26 January 2018)
12. EmailTracer, *Email Tracing vs. Email Tracking*, available at: http://www.emailtracer.com/content/email_search/types_of_online_email_search/email_trace/email_tracing_vs_email_tracking.html (Accessed 26 January 2018)

Answers to Multiple-choice Questions

1. (c)	2. (d)	3. (a)	4. (b)	5. (b)
6. (d)	7. (b)	8. (c)	9. (a)	10. (c)
11. (b)	12. (b)	13. (c)	14. (a)	15. (d)

10
Admissibility of Digital Evidence

Learning Objectives

This chapter provides an overview of the admissibility of digital evidence. The objective of this chapter is to provide a deeper insight into the basics of electronic records supported by law, admissibility of electronic records in accordance with the rules, categorization of evidence, and to discuss some examples. The chapter also discusses the steps involved in presenting the evidence, namely reporting and testimony, with guidelines and challenges. The court presentation system and a summary of the investigation process is also given. The reader will be familiar with the following after studying the chapter:

- Admissibility
- Electronic records
- Admissibility of electronic records
- Pre-trial preparation
- Presentation of evidence
- Summary of digital investigation

10.1 INTRODUCTION

After the acquisition and analysis of evidence, admissibility is the next step in the process of any cybercrime investigation. *Evidence* is produced before a judge or a jury to prove a fact in a case. Admissibility of evidence refers to the application of a set of legal rules by a judge to determine whether or not the *evidence* may be proffered. The rules are extensive. Figure 10.1 illustrates the stages in the investigation of cases involving digital evidence. To be admissible in court, the evidence should possess the following characteristics:

Admissible The evidence must be usable in court. Following the standard operating procedure (SOP) during the acquisition and handling of evidence is the basic requirement to make evidence admissible in court. Failure to stick to the SOP during the acquisition of evidence and improper documentation and reporting can make the evidence unusable (Box 10.1).

> ### Box 10.1 Case Laws–Evidence Inadmissibility Cases
>
> *Anvar P.V vs P.K. Basheer and Ors. on 18 September 2014 [Kurian, 2014]*
>
> The Supreme Court of India concluded that electronic evidence without certificate U/s 65B cannot be proved by oral evidence. Further, the opinion of the expert U/s 45A Evidence Act cannot be resorted to, to make such electronic evidence admissible.
>
> *Jagdeo Singh vs The State and Ors. on 11 February 2015 [Neeraj, 2015]*
>
> In a judgment pronounced by the Hon'ble High Court of Delhi, while dealing with the admissibility of an intercepted telephone call in a CD and CDR which were without a certificate U/s 65B Evidence Act, the court observed that the secondary electronic evidence without the certificate U/s 65B Evidence Act is inadmissible and cannot be looked into by the court for any purpose whatsoever.

Authentic The evidence should leave no doubt that it was acquired from a specific location/electronic communication device and that it is a complete and accurate copy of the acquired digital evidence and the integrity of the evidence is assured. Maintaining a proper *chain of custody* helps in maintaining consistency and prevents contamination of evidence. It reveals from where the evidence was acquired and all those who had control over it, right from acquisition. Maintaining *integrity documentation* ensures its integrity. It guarantees that the digital evidence has not been altered since its acquisition. In addition to this, evidence must be reliable, genuine, and relevant to the incident so as to be able to prove something.

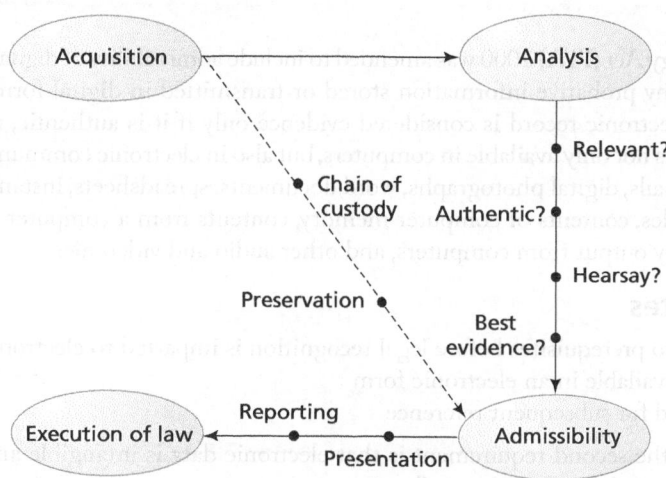

Fig. 10.1 Three stages in investigation of digital evidence

Complete The evidence should capture all dimensions of the incident, that is, it should be collected in such a way that it is sufficient to prove or disprove an action.

Reliable The evidence has to be assessed for reliability while authenticating it. There are two approaches: (a) to determine whether the electronic communication device from which the evidence was collected is functioning normally, and (b) to examine whether the digital evidence collected from the electronic communication device is damaged, contaminated, or tampered. The acquisition and analysis of evidence should not raise any doubt on the authenticity and reliability of the evidence. Evidence should be free from interference and contamination.

Believable The evidence presented before the court should be understandable and believable to the jury. Digital evidence, especially, should be conveyed to the jury in simple terms, as opposed to presenting it in the form of a binary dump to a jury who is computer-illiterate.

10.2 DIGITAL EVIDENCE—ELECTRONIC RECORD

An electronic record refers to information recorded by a computer that is produced or received in the initiation, conduct, or completion of an agency's or individual's activity. Examples of electronic records include email messages, word-processed documents, electronic spreadsheets, digital images, and databases [Michigan]. An electronic record is in machine readable form and so, appropriate hardware and software are required to interpret its meaning. Table 10.1 presents the provisions with respect to electronic records in the Information Technology Act (ITA) 2000 [IT Act 2000, Information Technology Act 2000, Bare Act, Information Technology Law, 2013] (Boxes 10.2–10.5). Electronic evidence is secondary evidence as per the Indian Evidence Act (IEA), meaning that it does not have the same persuasive value.

Table 10.1 Provisions with respect to electronic records in ITA 2000

Section	Scope
4	Legal recognition of electronic records
5	Legal recognition of digital signatures
6	Use of electronic records and digital signatures by the government and its agencies
7	Retention of electronic records

Information Technology Act (ITA) 2000 was amended to include admissibility of digital evidence. Accordingly, an electronic record is any probative information stored or transmitted in digital form and that can be used during trial in court. Electronic record is considered evidence only if it is authentic, relevant, and allowable under hearsay. Evidence is not only available in computers, but also in electronic communication devices. It takes various forms such as emails, digital photographs, word documents, spreadsheets, instant messages, transaction logs, browsing history files, contents of computer memory, contents from a computer moved as backup onto different media, hardcopy output from computers, and other audio and video files.

10.2.1 Prerequisites
The following are the two prerequisites before legal recognition is imparted to electronic records:
1. Rendered or made available in an electronic form
2. Accessible to be used for subsequent reference

The rationale behind the second requirement is that electronic data is intangible and by its very nature of being transient, makes it available for future reference.

10.2.2 Retention of Electronic Records
It refers to the duration of the electronic record's life. Since electronic data is intangible, it is highly susceptible to partial or complete loss of accessibility, integrity, or/and identification. For instance, if digital evidence is stored in a CD-ROM, its life expectancy or the retention period is three to five years.

Box 10.2 Section 4 of ITA 2000—Legal Recognition of Electronic Records

Where any law provides that information or any other matter shall be in writing or in the typewritten or printed form, then, notwithstanding anything contained in such law, such requirement shall be deemed to have been satisfied if such information or matter is

(a) rendered or made available in an electronic form; and
(b) accessible so as to be usable for a subsequent reference.

Box 10.3 Section 5 of ITA 2000—Legal Recognition of Digital Signatures

Where any law provides that information or any other matter shall be authenticated by affixing the signature or any document shall be signed or bear the signature of any person (then, notwithstanding anything contained in such law, such requirement shall be deemed to have been satisfied, if such information or matter is authenticated by means of digital signature affixed in such manner as may be prescribed by the Central Government.

Box 10.4 Section 6 of ITA 2000—Use of Electronic Records and Digital Signatures in Government and its Agencies

(1) Where any law provides for
   (a) the filing of any application form or any other document with any office, authority, body or agency owned or controlled by the appropriate Government in a particular manner;
   (b) the issue or grant of any licence, permit, sanction or approval by whatever name called in a particular manner;
   the receipt or payment of money in a particular manner, then, notwithstanding anything contained in any other law for the time being in force, such requirement shall be deemed to have been satisfied if such filing, issue, grant, receipt or payment, as the case may be, is effected by means of such electronic form as may be prescribed by the appropriate Government.
(2) The appropriate Government may, for the purposes of sub-section (1), by rules, prescribe
   (a) the manner and format in which such electronic records shall be filed, created or issued;
   (b) the manner or method of payment of any fee or charges for filing, creation or issue any electronic record under clause (a)

Box 10.5 Section 7 of ITA 2000—Retention of Electronic Records

(1) Where any law provides that documents, records or information are required to be retained for any specific period, then, that requirement shall be deemed to have been satisfied if the same is retained in electronic form, if
   (a) the information contained therein remains accessible so as to be usable for a subsequent reference;
   (b) the electronic record is retained in the format in which it was originally generated, sent or received or in a format which can be demonstrated to represent accurately the information originally generated, sent or received;
   (c) the details which will facilitate the identification of the origin, destination, date and time of despatch or receipt of such electronic record are available in the electronic record:
   Provided that this clause does not apply to any information which is automatically generated solely for the purpose of enabling an electronic record to be despatched or received.
(2) Nothing in this section shall apply to any law that expressly provides for the retention of documents, records or information in the form of electronic records.

The ITA 2000 under Section 7 sets forth the rules governing the retention of electronic records and is applicable to both, that is, the records which originally exist in electronic form as well as electronic retention of records that originally exist in paper form or other tangible media. It does not specify any definite retention period for electronic records but provides minimum standards as contained in Section 7(1). They are explained here (Boxes 10.6–10.8).

Accessibility All records that are archived should be accessible. However, the legal implications are ease of alteration and physical deterioration.

Box 10.6 Section 7(1)(a) of ITA 2000

… for making records, "accessible so as to be usable for a subsequent reference"

> **Box 10.7 Section 7(1)(b) of ITA 2000**
> ... even if there a change in the format, the electronic record shall be "in a format which can be demonstrated to represent accurately the information originally generated, sent or received"

> **Box 10.8 Section 7(1)(b) of ITA 2000**
> ... the details which will facilitate the identification of the origin, destination, date and time of dispatch or receipt of such electronic record are available in the electronic record."

Format integrity Any change in the format can affect the material characteristic of the electronic record, and so the integrity of the format in which the information was originally generated is important. For instance, the same version of the software used for analysis should be used for demonstration, even if several versions are available.

Identification Any form of record retention should enable identification.

10.3 ADMISSIBILITY OF ELECTRONIC RECORDS

The Indian Evidence Act 1872 is the law governing the relevance and admissibility of evidence in the courts of law in India. Advances in technology have led to the use of various digital techniques in the presentation of evidence to the courts. The Indian Evidence Act has been amended by virtue of Section 92 of the Information Technology Act 2000 (before amendment) so as to include digital evidence. Section 3 of the Act was amended and the phrase *All documents produced for the inspection of the Court* was substituted by *All documents including electronic records produced for the inspection of the Court*. Thus, electronic records are documentary evidence under Section 3 of the Evidence Act. A movie stored in MPEG format on a DVD, an email sent through a laptop, footage recorded via a CCTV are all examples of electronic records under the IT Act (Box 10.9).

> **Box 10.9 Section 65 A of IEA—Contents of Electronic Records**
> *The contents of the electronic records may be proved in accordance with the provisions of Section 65B.*

> **Box 10.10 Section 65 B of IEA—Admissibility of Electronic Records**
> **Section 65B of IEA—Admissibility of Electronic Records**
>
> **Section 65B (1)**
>
> Notwithstanding anything contained in this Act, any information contained in an electronic record which is printed on a paper, stored, recorded or copied in optical or magnetic media produced by a computer shall be deemed to be also a document, if the conditions mentioned in this section are satisfied in relation to the information and computer in question and shall be admissible in any proceedings, without further proof or production of the original, as evidence of any contents of the original or of any fact stated therein of which direct evidence would be admissible.

(Contd)

Box 10.10 (Contd)

Section 65B(2)
- The computer from which the record is generated was regularly used to store or process information in respect of activity regularly carried on by a person having lawful control over the period, and relates to the period over which the computer was regularly used;
- Information was fed in computer in the ordinary course of the activities of the person having lawful control over the computer;
- The computer was operating properly, and if not, was not such as to affect the electronic record or its accuracy;
- Information reproduced is such as is fed into computer in the ordinary course of activity.

Section 65B(3)
The following computers shall constitute a single computer –
- by a combination of computers operating over that period; or
- by different computers operating in succession over that period; or
- by different combinations of computers operating in succession over that period; or
- in any other manner involving the successive operation over that period, in whatever order, of one or more computers and one or more combinations of computers,

Section 65B(4)
Regarding the person who can issue the certificate and contents of certificate, it provides the certificate doing any of the following things:
- identifying the electronic record containing the statement and describing the manner in which it was produced;
- giving the particulars of device
- dealing with any of the matters to which the conditions mentioned in sub-section (2) relate, and purporting to be signed by a person occupying a responsible official position in relation to the operation of the relevant device or the management of the relevant activities (whichever is appropriate) shall be evidence of any matter stated in the certificate; and for the purpose of this sub-section it shall be sufficient for a matter to be stated to be the best of the knowledge and belief of the person stating it.

Similarly, as regards documentary evidence, in Section 59, for the words *Content of documents* the words *Content of documents or electronic records* have been substituted to prove the contents of electronic records through primary evidence or through secondary evidence like traditional documents. Primary evidence envisages the existence of a single original and this is impossible with electronic evidence as multiple copies of it could be generated with today's advancement in replication technology. The general law on secondary evidence under Section 63 read with Section 65 of the Evidence Act has no application in case of secondary evidence by way of electronic records. Sections 65A and 65B were inserted to incorporate the admissibility of electronic evidence which introduces safeguards in the form of mandatory conditions and a certificate to ensure the authenticity of the source and the contents of the record itself. If the genuineness of the electronic record is still questioned, the Act allows for an expert's opinion to verify the electronic record produced in terms of Section 65B (Box 10.10). Figure 10.2 shows the format of the certificate issued under Section 65B of the Indian Evidence Act. Under the provisions of Section 45A (Box 10.11), the opinion of an examiner can be sought to ascertain the relevancy of a fact that is either stored or generated from a computer.

> **Box 10.11   Section 45 A of IEA—Opinion of Examiner of Electronic Evidence**
>
> When in a proceeding, the court has to form an opinion on any matter relating to any information transmitted or stored in any computer resource or any other electronic or digital form, the opinion of the Examiner of Electronic Evidence referred to in Section 79 A of the Information Technology Act, 2000 (21 of 2000) is a relevant fact.

CERTIFICATE U/S 65B INDIAN EVIDENCE ACT

I, ... s/o, w/o, d/o .... r/o .... certify that I am submitting you the .... in relation with the case FIR no. .... and it fulfills the following conditions:

a) The computer output containing the information was produced by the computer during the period over which the computer was used regularly to store or process information for the purposes of any activities regularly carried on over that period by the person having lawful control over the use of the computer;

b) During the said period, information of the kind contained in the electronic record or of the kind from which the information so contained is derived was regularly fed into the computer in the ordinary course of the said activities;

c) Throughout the material part of the said period, the computer was operating properly or, if not, then in respect of any period in which it was not operating properly or was out of operation during that part of the period, was not such as to affect the electronic record or the accuracy of its contents; and

d) The information contained in the electronic record reproduces or is derived from such information fed into the computer in the ordinary course of the said activities.

Name of the Person Issuing Certificate:

Address: Signature:

Fig. 10.2   Format of certificate under Section 65 of Indian Evidence Act

Source: Data Security Council of India, 2011

Any documentary evidence by way of an electronic record under the Evidence Act, in view of Sections 59 and 65A, can be proved only in accordance with the procedure prescribed under Section 65B. Section 65A provides that the contents of electronic records may be proved in accordance with the provisions of Section 65B. Section 65B deals with the admissibility of the electronic record. The purpose of these provisions is to sanctify secondary evidence in electronic form, generated by a computer. Figure 10.3 provides a glimpse of the provisions for proof of evidence.

> **Box 10.12   Section 22A of IEA—Relevancy of Oral Evidence Regarding the Contents of the Electronic Records**
>
> Oral admissions regarding the contents of electronic records are not relevant unless the genuineness of the electronic records produced is in question.

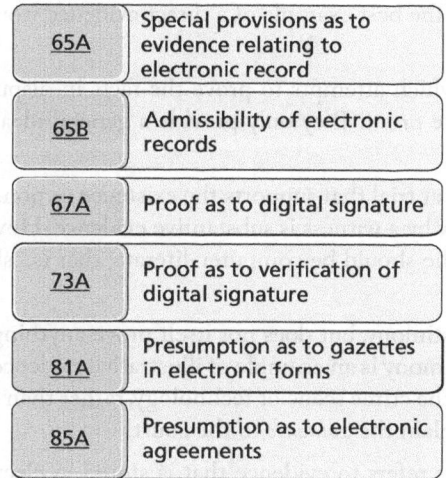
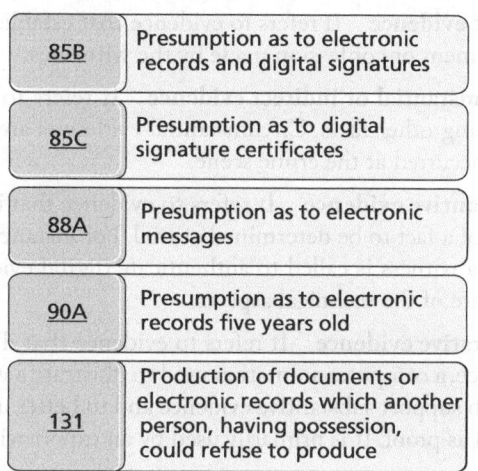

Fig. 10.3 Glimpse of provisions for proof of evidence

10.3.1 Rules of Admissibility of Electronic Evidence

In order to be admissible in a court of law, the following are the rules that any evidence is expected to satisfy. This is illustrated in Fig.10.4.

Relevant It reveals whether one fact is connected to the other either directly or proves/disproves the fact in any way as provided in Section 5(55) of the Evidence Act. Section 5 of the Evidence Act provides that evidence can be given only for facts that are at the core of the issue or of relevance. Since electronic records are susceptible to tampering, hiding, destruction, malicious viral and malware attack, and getting corrupted, metadata information such as the title of the document, date of creation, and location saved in the computer are used by the computer forensic examiner (CFE) to determine its relevance during analysis. However, since such metadata information can be easily embedded with technology, other means should be used to establish the facts. Section 22A has been inserted into the Evidence Act to provide for the relevancy of oral evidence regarding the contents of electronic records (Box 10.12).

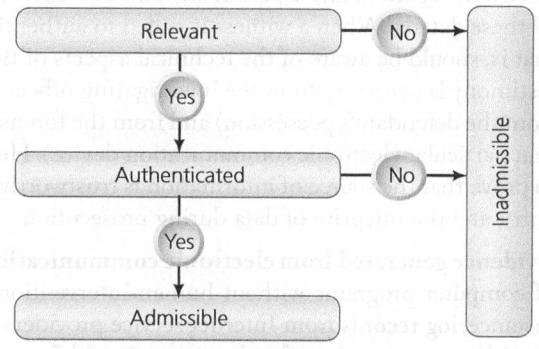

Fig. 10.4 Rules of admissibility of electronic evidence

Authenticated Authentication means satisfying the court that (a) the contents of the record have remained unchanged, (b) that the information in the record does in fact originate from its purported source, whether human or machine, and (c) that extraneous information such as the apparent date of the record is accurate. It also refers to establishing facts such as testimony of witnesses, expert witness with specimens, and distinctive characteristics (e.g., appearance, content, substance, pattern, or public records) where the electronic document contains electronic or digital signatures. Metadata from electronic data is an important source of authentication.

10.3.2 Categorization and Characteristics of Evidence with Respect to Law

Evidence is categorized into different types based on nature and use during the legal process. Some of the common forms of evidence are presented here:

Direct evidence It refers to evidence that establishes a fact. The best example of a direct evidence would be a statement or confession made by the witnesses.

Circumstantial or indirect evidence It refers to evidence which attempts to prove the facts in dispute by providing other facts. Circumstantial evidences are not definite proof. They only provide a general idea as to what occurred at the crime scene.

Substantive evidence It refers to evidence that is produced at trial that supports the existence or non-existence of a fact to be determined at trial. For instance, testimony by a witness is substantive evidence. However, when a witness is called to authenticate digital evidence, he/she should be computer-literate, that is, should be aware of the technical aspects.

Illustrative evidence It refers to evidence that illustrates testimony, but does not itself prove anything. For instance, a computer animation used to illustrate a witness' testimony is an example of illustrative evidence. It is used to support substantive evidence and to better understand the crime scene or technology, rather than being served as proof. It is primarily used by an expert witness to explain the *evidence* to the court.

Evidence stored in electronic communication devices This refers to evidence that is stored in electronic form and human-generated. These are treated as hearsay evidence. For instance, word documents, spreadsheets, email, and Internet chat messages are all electronic records stored in electronic communication devices. These evidences come with authentication and hearsay issues. Authentication is an issue because of the need to ascertain who the owner of the electronic record is (i.e., who holds possession), and that there is no change in the electronic record in any aspect at any point of time with respect to the case. Chain of custody, of course, reveals all these details. When a witness is called to authenticate digital evidence, he/she should be computer-literate, that is, should be aware of the technical aspects of the electronic record stored. Besides this, authentication by testimony is necessary from the investigating officer, IO (that the electronic communication device was seized from the defendant's possession) and from the forensic examiner (that the electronic record was recovered from that particular electronic communication device). Hearsay is an exception to business records, and it is required to prove that the source of information is trustworthy, that is, includes reliability of the record, accuracy of data entry, and the integrity of data during prosecution.

Evidence generated from electronic communication devices This refers to evidence that are direct outputs of computer programs without human intervention. It refers to electronic records and not hard copies. For instance, log records from Internet service providers (ISPs) and call detail records (CDR) from mobile service providers are examples of evidences generated from electronic communication devices. This type of evidence is also termed as real evidence. These evidences also come with authentication issues, but not hearsay. In case of Internet-related crimes, for instance, say child pornography, authentication depends much on circumstantial evidence. However, the log record obtained from the defendant's electronic communication device and information from ISP are sufficient to establish authorship.

Besides this, there are chances of evidence being a combination of both—the electronic communication device where it is stored, as well as the electronic communication device where it is generated—and may raise both the aforementioned issues. This type of evidence is usually referred to as *derived evidence* (Box 10.13).

Box 10.13 Examples of Cases on Admissibility of Evidence with Respect to Offences

Case 1: Child pornography

Consider an offence such as publishing of child pornography. In this case, first the server in which it is published has to be identified. This is done by determining the IP address. The cyber police have to collect information about the IP address, following which the police will search the physical

(Contd)

Box 10.13 (Contd)

address and trap the culprit. The next step is seizure and acquisition of evidence. The hard disk is seized and hard copies from the computer of the accused are also collected. The cyber forensic examiner can make an analysis about the authenticity of the collected evidence. Further, the Internet service provider may be approached to gather information, for example, if the computer was in regular usage and operated by the accused. The ISP usually supplies log information which forms a major source of evidence. Besides this, CCTV footage, Internet browsing history, and the hard disk itself will act as evidence. All of them should be in compliance with Section 65B(2). Further, statements produced as evidence should be accompanied by a certificate and should have been signed by a person holding a responsible official position according to Section 65B(4).

Case 2: Email-related offence

While deciding the admissibility of email, as per Sections 65B and 88 of IEA, the email downloaded and printed from an email account can serve as evidence. Any testimony of the witness who has carried out the process of downloading and printing is sufficient to prove the electronic communication. The original electronic media serves as primary evidence. Secondary evidence, if any, should accompany the certificate in terms of Section 65B so as to be admissible.

Case 3: Image forgery

Image forgery is a crime wherein an original image is falsely altered with the intention to produce a fake image, usually by way of image retouching, image splicing, and copy–move attack. When an incident on image forgery is reported, for instance on Facebook, then the log is obtained from the Facebook server. Whether the photo or the image is forged or not is determined based on the findings and report from the cyber lab which relies on special software. If it is deemed to be forged, then the IP address obtained from the log given by Facebook is used to trace the culprit behind the image forgery. The evidence which is the log file as well as the report from the cyber lab that are produced in court should be in compliance with Section 65B(2) of IEA. Further, it should be accompanied by a certificate and should have been signed by a person holding a responsible official position according to Section 65B(4) of IEA.

Limitations on Admissibility of Digital Evidence The very nature of digital evidence raises concerns over its admissibility. First, acquisition of volatile data (live forensics) is tough as it is very difficult to recover this from storage media. Being invisible, digital evidence is more susceptible to tampering and alteration. This can be avoided by employing a write blocker. Even then, digital evidence is invalid without the Section 65B IEA certificate. Thus, strict compliance with Section 65B IEA is mandatory to rely upon websites, emails, and any other electronic records for both civil and criminal trials before the courts in India. Second, if the chain of custody is not properly maintained, digital evidence is deemed as tampered evidence. Though hashing is used to prove the integrity of digital evidence, admissibility of computer-generated electronic records cannot be relied upon solely and can be used as corroborative evidence.

Besides this, the prosecutor and the court should be in touch with the ongoing legal and technological advancements as electronic evidence has substantial impact during trial. This necessitates that admissibility issues have to be discussed with the court prior to commencement of trial.

The need of the hour in India is to devise a mechanism for ensuring the veracity of contents of electronic records but there is a long way to go.

10.4 PRE-TRIAL PREPARATION

The ability to convince the court that digital evidence is worthy to the case is dependent on the qualifications and competence of the IO and the FE, the skill and knowledge of the public prosecutor (PP) in leading such evidence, and the quality of the digital evidence itself. Hence, in civil cases where digital evidences are involved, pre-trial preparation should involve the IO, forensic examiner, and the PP. These three should conduct a pre-trial meeting, well in advance of the trial, and plan for the presentation of the case. The PP has to go through the reports prepared by the IO and FE before the meeting to clarify doubts with them. The PP has to review the scope of investigation and prepare an understandable theory of the case for presentation while obtaining clarifications on technological issues from the FE. The scope and limitations of the evidence should also be analysed. Besides this, any anticipated defence by the opposing party should be analysed. The chances of direct examination and cross-examination should also be discussed with the IO and FE by the PP. Any substantive evidence required for the case should be planned for. If illustrative evidence has to be presented for the case, the mode and medium of presentation has to be analysed.

The following are essential for trial:
1. Thorough understanding of the case
2. Thorough knowledge of the strategy used for analysis
3. Precise presentation of the evidence collected and its relevance to the case
4. Good listening skills, with respect to the questions raised by the jury
5. Good presentation aid

10.5 PRESENTING DIGITAL EVIDENCE

Qualified and competent experts (consulting expert, court's expert, and CFE) play a significant role in maintaining credible reputation of digital evidence. The role played by them during various stages of the investigation process is illustrated in Fig. 10.6. The SOP should be adopted when digital technology is used for the examination and presentation of evidence.

Presentation of facts after analysing the evidence should include four points, namely (a) what was done with the evidence; (b) why it was done; (c) how it was done; and (d) what was found as a result. For instance, it should include, how the evidence, say, a file, is stored in the media, what a forensic copy is, how was the deleted data recovered from the media, what data was contained in the Internet history, and so on, depending on the case.

Presenting digital evidence in the court includes the following two important aspects:

Reporting It aims at providing a transparent view of the investigative process wherein the final reports contain important details from each step, including reference to protocols followed and methods used to seize, document, acquire, preserve, recover, reconstruct, organize, and search for key evidence. The purpose of reporting is to document and detail each process followed by the CFE, including even alternative theories, thus demonstrating the independence and objectivity of the CFE, and his/her investigation.

Testimony It is necessary that the IO presents his/her findings to a court or other forum. He/She will have to be able to convey his/her findings and report, first to the court, by giving testimony viva voce, and second, to stand the scrutiny of cross-examination on his/her report, findings, and testimony. The issue with digital evidence for the investigator is that he/she may be well versed in his/her field, and translate facts that are technical in nature, to a court which is not technically sound. This necessitates that the evidence be kept concise and understandable.

10.5.1 Reporting—Expert Report

Cyber forensic examiners produce a written summary of their findings after analysing the digital evidence, usually in the form of an affidavit or *expert report* (ER). This report should present solid arguments with supporting evidence to convince the opposition to settle the case out of court, whereas lack of foundation in evidence and a weak report would lead the opposition to proceed to trial. Thus, poor work during reporting and a poorly presented ER can undermine the gains that have been made thus far in a case. Reliability of digital evidence is of utmost importance in criminal prosecutions. An ER should typically include the following:

Introduction It should give an overview of the case, the evidence analysed, who requested the analysis, and what was asked to be analysed. It also includes the profile of the CFE.

Evidence summary It must provide details about the physical characteristics of the device (make, model, and serial number), description of the evidence (when, where, and from whom it was obtained), what was analysed from it, method opted for analysis, and the tools used.

Examination summary It should present the findings after the examination, and summarize the tools used for examination and recovery, filtering out irrelevant files and data.

File system examination It should present the data related to the device that had undergone forensic examination—date timestamps, hash values, path to the file, and its physical location. Any traces of deletion, wiping, and reformatting should also be highlighted.

Forensic analysis and findings It must present a detailed description of the forensic analysis, findings, and supporting evidence.

Conclusions It should give a logical conclusion, drawn from the facts, and interpret the digital evidence.

Besides this, the following should be considered while preparing an expert report:
1. Details on where the evidence was found and how it was analysed and fetched should be included in the ER for the PP to interpret the report and for any other peer to verify the results.
2. Assertions should be included by the CFE along with multiple independent sources of evidences, so that the weakness of any particular evidence will not prevent one from arriving at the conclusion.
3. The evidences should be ordered so that there is coherence among them when presented before a judge which otherwise would make the process complex.
4. Highlighting important aspects of evidence in the form of figures and illustrations will justify the findings and strengthen the conclusions.
5. Presenting alternative scenarios and demonstrating why such a scenario is not feasible will reinforce the conclusions.
6. Any alteration in the evidence observed during analysis should be mentioned in the ER.

10.5.2 Guidelines for Cyber Forensic Examiners

The following are some of the guidelines for cyber forensic examiners to preserve the evidence so that it is admissible:
1. The CFE should create the master copy of the original data and leave the original and the master copy in the evidence store. He/She should always work with secondary copies. This is done to minimize the handling of the original data as any change to the original would affect the outcome of any analysis done to the copies at a later stage.
2. Any alterations in evidence should be accounted for and a detailed log of actions taken on the evidence should be maintained and documented.
3. The CFE should capture an accurate forensic copy of the evidence acquired and minimize the corruption of the original evidence. If there is any deviation in the forensic copy from that of the original, the CFE should account for the differences.
4. The CFE should ensure that his/her actions are repeatable. The conclusions and interpretations will be believable only if the actions are replicated and the same results are reached. Any error in the processes would eventually rule out a trial.

5. Besides this, the CFE is expected to work fast, as volatile evidence may vanish entirely if not acquired on time. Further, this will ensure that there is lesser likelihood of the data changing.
6. The CFE should not shut down the electronic communication device before acquiring the evidence. This is because there is a possibility of loss of volatile evidence. Moreover, while such a device is in the *On* mode, there are chances of an attacker using the Trojan to target the startup of the computer, shutdown scripts, or use a plug-and-play device to alter the system configuration and wipe out the temporary file systems.
7. The CFE should not attempt to run any programs on the affected system as it would inadvertently trigger a response that could change or destroy evidence.

10.5.3 Testimony

The public prosecutor facing trial can have a meeting with the IO and the CFE to discuss the review findings, clarify doubts, and deliberate on how information has to be presented. The CFE has to be accepted as an expert by the court. This decision is however made by the court based on his/her qualification, experience, and other credentials.

When the case appears before a judge in the court, the IO/CFE will undergo two phases of trial, namely (a) direct examination and (b) cross-examination. In direct examination, the judge poses questions to the witness for which testimony has to be provided in return. Cross-examination attempts to weaken the testimony and involves questioning by the opposing counsel.

During cross-examination, a response has to be given by the CFE with proper explanation and with necessary evidence. Further, the CFE may even give a brief about how the evidence was handled and analysed, and demonstrate the chain of custody.

During direct examination, the PP refers to the digital evidence and makes it available to the judge. All notes related to the work of the CFE are documented.

The FE can provide technical testimony in civil and criminal litigation. However, digital investigations and cybercrime offences restrict it to civil litigations. The forensic team should be aware of the laws that apply to the case and the evidence. The legal team should work in line with the forensic team to convince the judge while offering forensic evidence and testimony. Detailed analysis of the ER and a good level of knowledge-sharing on the case among the IO, FE, and the PP can discredit any testimony by the opposing team. In general, public prosecutors have limited knowledge on electronic communication devices and they require technical guidance from the FE during the course of a case. It will help the PP to decide how to approach the case. Public prosecutors may opt for visual aids to give a live demonstration of the case if the court permits. The FE will not only assist the PP on what can be done, but also act as an expert witness during cross-examination. The FE is much like an IO with respect to digital evidence and has to work in coherence with the case when new facts about the case are revealed, which requires evidence to be reviewed again to ascertain whether something more could be discovered and that it is applicable to the case.

The FE might be a technical witness or expert witness during testimony. A technical witness presents what the evidence is and how it was obtained, thus revealing only facts and not any conclusions. On the other hand, an expert witness presents opinions and observations additionally. Hence an expert should present the technical matters in such a way that even a non-expert would be able to understand it. Besides this, during cross-examination, he/she is expected to justify the professional opinion offered. Thus, the witnesses offered either as a technical or as an expert witness may be applicable to the legal system in any one of the following four ways [Robert, 2007]:

1. Consulting an expert who advises the legal team on the technology and the strategy to be used
2. Court's expert who gives valuable input to judges while the opposing parties present divergent views
3. Testifying expert where a technologist serve as an expert witness
4. Expert as witness where the expert serves as witness without offering his/her expertise

The PP should consider the following:
1. He/She has to file a counter affidavit and argument to establish that the claim made by the defence counsel is incorrect. In other words, he/she has to disprove claims that are meaningful and reasonable. However, what to disprove depends on the issue involved in the case.
2. It is good to maintain a community of qualified technical experts with practical exposure who could assist investigations.
3. He/She should ensure that the rules of evidence and admissibility are followed by the forensic expert.
4. He/She should prepare the witness to testify at pre-trial hearings, for chief as well as cross-examination.

10.5.4 Courtroom Presentation System

The digital evidence for presentation in a court takes the form of audio, video, image, or text. A courtroom evidence presentation system consists of two parts: the sources of the evidence and the technology by which the evidence is displayed or heard. The following equipment are essential for a courtroom [Derek Miller, n.d.],[Martin Gruen, 2003].

1. Courtrooms should be equipped for electronic presentation. A computer with a video adapter, sound card, speakers for audio and video functioning, monitor, projector, and networking facility are required.
2. Monitors are similar to television sets, but provide much higher resolution making them a better choice for courtroom display systems. It helps the lawyer to display information about the evidence to the judge and jury.
3. The trial presentation computer can be equipped with the complete copy of the case data along with evidence management software. The computer should be equipped with tools to prepare documents for trial. A generic presentation tool like Microsoft PowerPoint is essential. Corel Presentations offer slide show creation programs.
4. Evidence in the form of video, an image, or text can be presented on a larger projection screen with a ceiling-installed projector.
5. The audio system in a courtroom includes microphones, an audio processor, audio amplifiers, and an audio control system. If videoconferencing is part of the proceedings, the audio system must also include an echo-cancellation system.
6. A desktop videoconferencing system includes cameras and microphones connected to the presentation computer which enables individuals or groups of people in different locations to communicate through the use of audio and video equipment.
7. A document camera consists of a vertically mounted TV camera aimed down at a flat surface, where a photo or document is placed for the camera to instantly display the image on the projector or monitor(s) to which it is attached.
8. Illustrative evidence like animations need to be presented during a trial with the presentation computer.
9. Notebook computers can be used to present documents, photographs, graphs, and animations in the courtroom.

The courtroom evidence presentation system comes in two forms: the *central podium concept* is one where all the presentation devices such as document camera, presentation computer, and video and audio players are mounted onto a customized podium. Thus it provides an attorney with all the equipment needed to present a case. On the other hand, in the *on a cart concept*, presentation devices are placed in a cart on wheels and can be moved from courtroom to courtroom, thus offering the flexibility of servicing several courtrooms with a single cart.

10.5.5 Challenges with Admissibility and Presentation of Digital Evidence

Some of the challenges that exist with digital evidence and remain a hindrance to admissibility in the court are as follows:

Improper handling Handling of digital evidence involves a sequence of steps, as discussed in Chapter 8. Any deviation from this process can make the evidence inadmissible.

Illegal search and seizure Searches have to be made with warrants, but in some situations warrantless searches are also permitted. More details on search with warrants and warrantless searches are discussed in Chapter 8. Any deviation from the constraints can make the acquired evidence inadmissible.

Expert opinion Unlike other types of evidence, electronic evidence is available in its original form in some storage medium which is known as the primary evidence of electronic records. Thus hard disks, cell phones containing call records, memory cards, and CDs are examples of primary evidence. They are presented in the court as copies of original documents made through mechanical processes and are referred to as secondary evidences. Thus, a printout of the contents of a hard disk, memory card, or CD, and call records become secondary evidence. Section 65B of the Indian Evidence Act (IEA) lays down the procedures for the admissibility of secondary evidence. Besides this, the court accepts it under the following conditions:

(a) as a record only if it is certified by a competent officer through an affidavit or by a person who is responsible for managing or operating the device;

(b) as an electronic record if the witness can identify the signature made by the competent officer or by a person who speaks the facts based on his/her personal knowledge.

When there is apprehension on the integrity and authenticity of the documents or the contents of electronic records, expert opinion is sought for consideration by the court. Experts should possess sufficient expertise, follow principles of natural justice and fairness, and be unbiased while giving opinion on the contents of the electronic record. He/She has to inform everyone involved in the process of digital investigation as well as the court if his/her opinion differs from that contained in the secondary evidence. An expert should resist influence by any one of the parties, peers, and the lawyers involved in the case.

Inclination to preconceived theories The success of digital investigation depends on a combination of the skills and experience of the IO, CFE, and prosecutors. First and foremost, investigators should analyse if a crime has actually been committed. Every case is unique in one way or another, and an inclination to preconceived theories is possible if the IOs approach the case in the same way they have dealt with the previous cases. This may even draw them to erroneous conclusions. Every case has to be analysed from all dimensions. The acquired evidence always speaks for a case and investigators should not attempt to misdirect investigations by concealing evidence.

10.6 SUMMARY OF INVESTIGATION PROCESS INVOLVING DIGITAL EVIDENCE

The investigation process involving digital evidence is illustrated in Fig. 10.5 and the role played by various authorities is illustrated in Fig. 10.6. The process of investigation is explained as follows:

1. Complaint from a complainant is a must to set the law into motion.
2. An FIR is filed by the jurisdiction police station and the IO will be nominated.
3. The suspect is located by the IO. He/She is termed a 'suspect' and not an 'accused' until convicted by a court.
4. The suspected electronic communication device is located by the IO.
5. Seizure of the electronic communication device by the IO is performed in accordance with the SOP for the acquisition of evidence. Chain of custody and other documentation are carried out. Seizure is carried out in the presence of two independent witnesses, who must preferably be computer-literate. Live evidence, if any, is acquired with the support of the FE. The seized evidence is packed and sealed in the presence of witnesses.
6. The acquired evidence and the device are presented before the court by the IO.
7. The original evidence and the master copy are preserved in the evidence store to avoid any alteration.
8. The FE generates a forensic copy of the original evidence once it is handed over to the forensic lab.

Admissibility of Digital Evidence 365

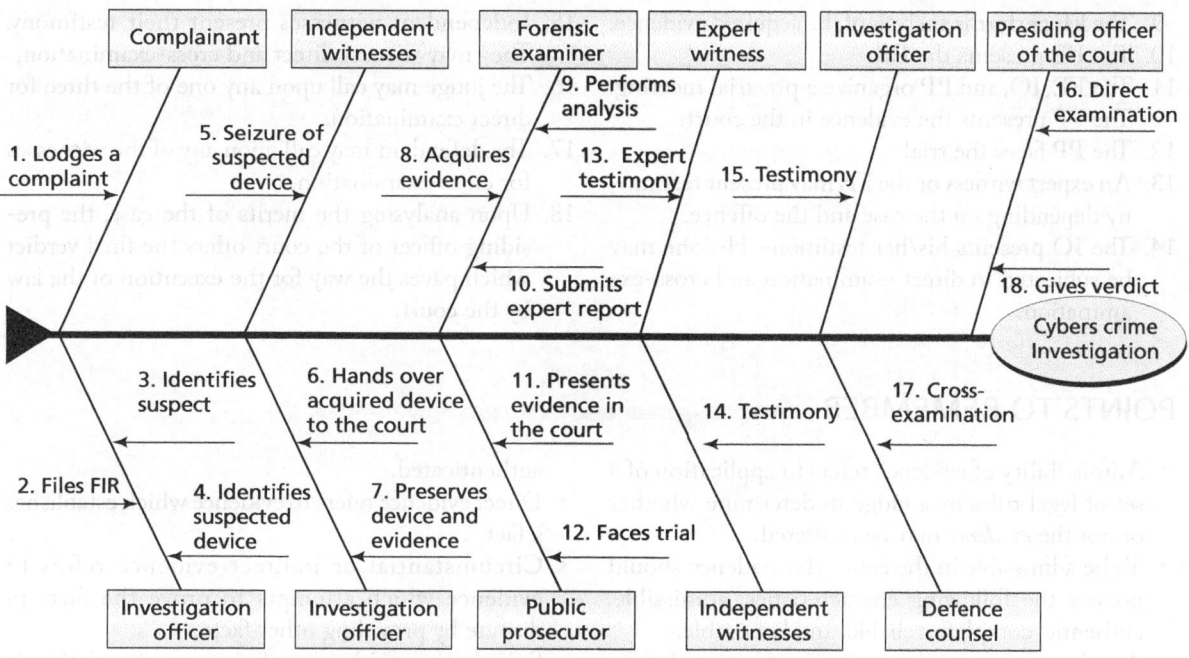

Fig. 10.5 Investigation and trial processes involving digital evidence

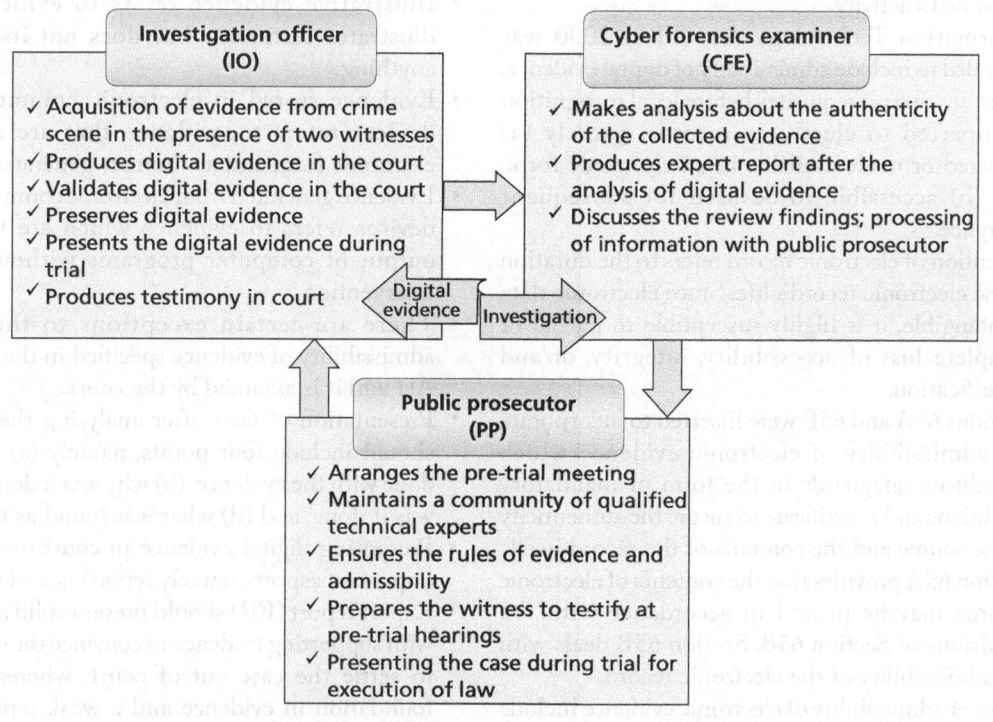

Fig. 10.6 Roles played by various authorities in investigation process

9. The FE performs analysis of the acquired evidence.
10. The FE presents the ER.
11. The FE, IO, and PP organize a pre-trial meeting. The PP presents the evidence in the court.
12. The PP faces the trial.
13. An expert witness or the FE may present testimony depending on the case and the offence.
14. The IO presents his/her testimony. He/She may be subjected to direct examination and cross-examination.
15. Independent witnesses present their testimony. They may witness direct and cross-examination.
16. The judge may call upon any one of the three for direct examination.
17. The defendant may call upon any of the witnesses for cross-examination.
18. Upon analysing the merits of the case, the presiding officer of the court offers the final verdict which paves the way for the execution of the law by the court.

POINTS TO REMEMBER

- Admissibility of evidence refers to application of a set of legal rules by a judge to determine whether or not the *evidence* may be proffered.
- To be admissible in the court, the evidence should possess the following characteristics: admissible, authentic, complete, reliable, and believable.
- An electronic record is information recorded by a computer that is produced or received in the initiation, conduct, or completion of an agency's or individual's activity.
- Information Technology Act (ITA) 2000 was amended to include admissibility of digital evidence.
- There are two prerequisites before legal recognition is imparted to electronic records, namely (a) rendered or made available in an electronic form, and (b) accessible to be used for subsequent reference.
- Retention of electronic record refers to the duration of the electronic record's life. Since electronic data is intangible, it is highly susceptible to partial or complete loss of accessibility, integrity, or/and identification.
- Sections 65A and 65B were inserted to incorporate the admissibility of electronic evidence which introduces safeguards in the form of mandatory conditions and a certificate to ensure the authenticity of the source and the contents of the record itself.
- Section 65A provides that the contents of electronic records may be proved in accordance with the provisions of Section 65B. Section 65B deals with the admissibility of the electronic record.
- Rules of admissibility of electronic evidence include the following: the evidence should be relevant and authenticated.
- Direct evidence refers to evidence which establishes a fact.
- Circumstantial or indirect evidence refers to evidence which attempts to prove the facts in dispute by providing other facts.
- Substantive evidence refers to evidence that is produced at trial and supports the existence or non-existence of a fact to be determined at trial.
- Illustrative evidence refers to evidence that illustrates testimony but does not itself prove anything.
- Evidence stored in electronic communication devices refers to evidence that are stored in electronic form and are human-generated.
- Evidence generated from electronic communication devices refers to evidence which are the direct output of computer programs without human intervention.
- There are certain exceptions to the rule of admissibility of evidence specified in the Evidence Act and it is assumed by the court.
- Presentation of facts after analysing the evidence should include four points, namely (a) what was done with the evidence; (b) why was it done; (c) how was it done; and (d) what was found as the result.
- Presenting digital evidence in court includes two important aspects, namely reporting and testimony.
- Expert report (ER) should present solid arguments with supporting evidence to convince the opposition to settle the case out of court, whereas lack of foundation in evidence and a weak report would lead the opposition to proceed to trial.

- When the case appears before a judge in the court, the investigating officer (IO)/ cyber forensic examiner (CFE) will undergo two phases of trial, namely (a) direct examination and (b) cross-examination.
- In direct examination, the judge poses questions to the witness for which testimony has to be provided in return.
- Cross-examination attempts to weaken testimony and involves questioning by the opposing counsel.
- A technical witness presents what the evidence is and how it was obtained, thus revealing only facts and not any conclusions.
- An expert witness presents opinions and observations additionally.
- An expert opinion is sought for consideration by the court. Experts should possess sufficient expertise, follow principles of natural justice and fairness, and be unbiased while giving opinion on the contents of the electronic record.
- The acquired evidence always speaks for a case and investigators should not attempt to misdirect investigations by concealing some evidence.

KEY TERMS

Admissibility of evidence It refers to the application of a set of legal rules by a judge to determine whether or not the evidence may be proffered.

Authentication It means satisfying the court that (a) the contents of the record have remained unchanged, (b) that the information in the record does in fact originate from its purported source, whether human or machine, and (c) that extraneous information such as the apparent date of the record is accurate.

Derived evidence This refers to evidence that is a combination of both, the electronic communication device where it is stored, as well as the electronic communication device where it is generated.

Electronic record It is information recorded by a computer that is produced or received in the initiation, conduct, or completion of an agency's or individual's activity.

MULTIPLE-CHOICE QUESTIONS

1. _____ refers to the application of a set of legal rules by a judge to determine whether or not the evidence may be proffered.
   (a) Acquisition of evidence
   (b) Analysis of evidence
   (c) Acquisition and analysis of evidence
   (d) Admissibility of evidence
2. Section 6 of ITA 2000 refers to _____
   (a) legal recognition of evidence records
   (b) legal recognition of digital signatures
   (c) use of electronic records and digital signatures by the government and its agencies
   (d) retention of electronic records
3. What are the prerequisites before legal recognition is imparted to electronic records?
   (a) Rendered or made available in an electronic form
   (b) Accessible to be used for subsequent reference
   (c) Both (a) and (b)
   (d) Neither (a) nor (b)
4. The minimum standards as contained in Section 7(1) of ITA 2000 are _____
   (a) accessibility
   (b) format integrity
   (c) identification
   (d) all of these
5. The rules of admissibility of electronic evidence is/are _____
   (a) relevant
   (b) authenticated
   (c) relevant and authenticated
   (d) irrelevant
6. Indirect evidence is also known as _____
   (a) substantive evidence
   (b) illustrative evidence
   (c) evidence stored in electronic communication device
   (d) circumstantial evidence

7. The consulting expert, court's expert, and cyber forensic examiner are referred to as _____
   (a) qualified experts
   (b) competent experts
   (c) qualified and competent experts
   (d) cyber experts
8. _____ aims at providing a transparent view of the investigative process.
   (a) Testimony
   (b) Reporting
   (c) Reporting and testimony
   (d) Presenting digital evidence
9. An expert report (ER) typically includes the following:
   Introduction, evidence summary, _____, _____, _____, conclusion.
   (a) examination summary, forensic analysis and findings, file system examination
   (b) file system examination, forensic analysis and findings, examination summary
   (c) examination summary, file system examination, forensic analysis and findings
   (d) examination summary, file system summary, forensic findings
10. Improper handling, illegal search and seizure, expert opinion, and inclination to preconceived theories are some of the _____
    (a) challenges with admissibility
    (b) challenges with presentation of digital evidence
    (c) challenges with admissibility and presentation of digital evidence
    (d) conflicts with admissibility
11. The role of the investigation officer is _____
    (a) producing testimony to court
    (b) discussing the review findings
    (c) presenting the case during trial
    (d) (a) and (b)
12. The role of the cyber forensic examiner is _____
    (a) producing testimony to court
    (b) discussing the review findings
    (c) presenting the case during trial
    (d) (a) an (c)
13. The role of the public prosecutor is _____
    (a) producing testimony to court
    (b) discussing the review findings
    (c) presenting the case during trial
    (d) (b) and (c)

REVIEW QUESTIONS

1. Define digital evidence. Explain the characteristics of digital evidence with respect to admissibility.
2. Discuss the rules of admissibility of electronic evidence.
3. Explain the categories and characteristics of evidence with respect to law.
4. Describe the information to be sought from the forensic examiner after performing an analysis of digital evidence.
5. Explain in detail the challenges with regard to the admissibility of digital evidence.
6. Describe the roles and responsibilities of the investigation officer (IO), cyber forensic examiner (CFE), and public prosecutor (PP).
7. Describe testimony in detail.

APPLICATION EXERCISES

1. Imagine that you are asked to ascertain the admissibility of evidence of a private website hosting unwanted data. The webmaster maintains a log of files and contents that were placed on the website at a specific time. Such information is obtained either through testimony or through documentation. What evidence would you consider to verify the admissibility of digital evidence?
2. Social networking sites are widely used by everyone today and there are increasing incidents of cybercrimes employing social engineering. What may be the methods you adopt to authenticate the postings made on social websites so as to make the evidence collected admissible?
3. Categorize the following as direct/indirect/substantive/illustrative evidence:

(a) A witness who saw a suspect fleeing the scene of a crime which is a server room.
(b) A computerized bank records in a credit card fraud case.
(c) The fact that the accused had an intense dislike of the victim.
(d) E-mails in a cyber-stalking case.
(e) A PowerPoint presentation explaining the execution of cybercrime as an animation.
(f) Image files in a child pornography case.
4. Refer to Section 5.1 and prepare an expert report for any one of the cybercrime incidents against an individual.

BIBLIOGRAPHY

1. Naggal, M. (2016), *Admissibility of Electronic Evidence under Indian Law*, available at: http://medcraveonline.com/FRCIJ/FRCIJ-02-00074.pdf (Accessed 15 December 2017)
2. Jonathan W. Hak, *The Admissibility of Digital Evidence in Criminal Prosecutions*, available at http://www.crime-scene investigator.net/admissibilitydigitaleveidencecriminal prosecutions.html (Accessed 15 December 2017)
3. Harley Kozushko (2003), *Digital Evidence*, available at: http://infohost.nmt.edu/~sfs/Students/HarleyKozushko/ Presentations/DigitalEvidence.pdf (Accessed 15 December 2017)
4. Johann Hershensohn, *I.T. Forensics: The Collection and Presentation of Digital Evidence*, available at: http://icsa.cs.up.ac.za/issa/2005/Proceedings/Full/076_Article.pdf (Accessed 15 December 2017)
5. Michigan, *Frequently Asked Questions about Electronic Records for Local Governments*, available at: https://www.michigan.gov/documents/hal_mhc_rms_electronic_records_125548_7.pdf (Accessed 15 December 2017)
6. Criminal.findlaw, *Hearsay Evidence*, available at: criminal.findlaw.com/criminal-procedure/hearsay-evidence.html (Accessed 15 December 2017)
7. Robert Newman, C. (2007), *Computer Forensics Evidence Collection and Management*, Auerbach Publications
8. Classstudio (12 March 2003), *Digital Evidence in the Courtroom: A Guide for Preparing Digital Evidence for Courtroom Presentation*, available at: http://classstudio.com/papers/grad_papers/forensics/Palmer/ digital_evidence_in_courtroom.pdf (Accessed 15 December 2017)
9. Alberto Gonzales, R., Regina Schofield, B., David Hagy, W., *Digital Evidence in the Courtroom: A Guide for Law Enforcement and Prosecutors*, available at: https://www.ncjrs.gov/pdffiles1/nij/211314.pdf (Accessed 15 December 2017)
10. Lucy Thomson, L. (2013), *Mobile Devices New Challenges for Admissibility of Electronic Evidence*, available at: https://www.americanbar.org/content/dam/aba/events/science_technology/mobiledevices_new_challenges_admissibility_of_electronic_device.authcheckdam.pdf (Accessed 15 December 2017)
11. Racolblegal (08 November 2016), *Admissibility of Electronic Evidence in Indian Courts*, available at: http://racolblegal.com/admissibility-of-electronic-evidence-in-indian-courts/ (Accessed 15 December 2017)
12. Dubey, V. (2017), *Admissibility of Electronic Evidence: An Indian Perspective*, available at: http://medcraveonline.com/FRCIJ/FRCIJ-04-00109.pdf (Accessed 16 December 2017)
13. Swati Mehita (2012), *Cyber Forensics and Admissibility of Digital Evidence*, available at: http://www.supremecourtcases.com/index2.php?option=com_content&itemid=135&do_pdf=1&id=22821 (Accessed 16 December 2017)
14. Lucy Thomas, L. and Esq (01 December 2011), *Admissibility of Electronic Documentation as Evidence in U.S. Courts*, available at: http://www.crl.edu/sites/default/files/d6/attachments/pages/Thomson-E-evidence-report.pdf (Accessed 16 December 2017)
15. Paul Grimm, W. and Kevin Brady, F., *Admissibility of Electronic Evidence*, available at: http://cyber-trail.com/wp-content/uploads/2016/05/Admissibility-of-Electronic-Evidence_-2012.pdf (Accessed 16 December 2017)
16. Inta (22 September 2009), *Board Resolutions Admissibility of Electronic Evidence*, available at:

http://www.inta.org/Advocacy/Pages/AdmissibilityofElectronicEvidence.aspx (Accessed 16 December 2017)
17. Rajendra Prasad, P., *Recent Trends in Recording and Admissibility of Evidence*, available at: http://ecourts.gov.in/sites/default/files/WORKSHOP%20IV%20NOTES%20(2).pdf (Accessed 16 December 2017)
18. Debaditya Roy (2013), *Experts Opinion and its admissibility and relevancy – Law of Evidence*, available at: http://www.legalservicesindia.com/article/article/experts-opinion-and-its-admissibility -and-relevancy-law-of-evidence-1583-1.html (Accessed 16 December 2017)
19. Lawweb (01 November 2015), *Whether CDR report is admissible in Evidence in Absence of Certificate as Per S65B of Evidence Act*, available at: http://www.lawweb.in/2015/11/whether-cdr-reportelectronic-evidence.html (Accessed 16 December 2017)
20. Alberto R. Gonzales, Regina B. Schofield and David W. Hagy (2011), *Digital Evidence in the Courtroom: A Guide for Law Enforcement and Prosecutors*, National Institute of Justice
21. Cybercrime Investigation Manual by Data Security Council of India.
22. Darren R. Hayes (2014), *A Practical Guide to Computer Forensics Investigations*, Pearson Education.
23. The Tribune (2010), *Security in Cyberspace*, available at: http://shodhganga.inflibnet.ac.in/bitstream/10603/7829/15/15_chapter%206.pdf (Accessed on 17 December 2017)
24. Gradestack, *Electronic Evidence and the Law*, available at : https://gradestack.com/Advanced-Certification-in/Electronic-evidence-and/Electronic-evidence-and/18347-3292-29744-study-wtw (Accessed 16 December 2017)
25. Derek Miller, *Best Practice Tips for Electronic Trial Presentation*, available at: http://www2.law.columbia.edu/johnson/syllabus/TrialPresentationTips.pdf (Accessed 16 December 2017)
26. Martin Gruen (2003), *The World of Courtroom Technology*, available at: http://www.legaltechcenter.net/download/whitepapers/ The%20World%20Of%20Courtroom%20Technology.pdf (Accessed 16 December 2017)
27. Neeraj Aarora (2015), *Admissibility of Electronic Evidence: Challenges for Legal Fraternity*, available at: http://www.neerajaarora.com/admissibility-of-electronic-evidence-challenges-for-legal-fraternity/ (Accessed 13 January 2018)
28. IT Act 2000, Information Technology Act 2000, Bare Act, Information Technology Law (2013), available at: http://www.itlaw.in/ (Accessed 13 January 2018)
29. Data Security Council of India (2011), *Cybercrime Investigation Manual*, available at https://uppolice.gov.in/writereaddata/uploaded-content/Web_Page/28_5_2014_17_4_36 _Cyber_Crime_Investigation_Manual.pdf (Accessed 13 January 2018)
30. Kurian (2014), *Supreme Court of India - Anvar P.V vs P.K. Basheer & Ors on 18 September, 2014*, available at: https://indiankanoon.org/doc/187283766/ (Accessed 18 January 2018)

Answers to Multiple-choice Questions

1. (d)	2. (c)	3. (c)	4. (d)	5. (c)
6. (d)	7. (c)	8. (b)	9. (c)	10. (c)
11. (a)	12. (b)	13. (c)		

Cybercrime Case Studies 11

Learning Objectives

The objective of this chapter is to provide some of the cybercrimes as case studies. A gist of the case is presented first. This is followed by an explanation on investigation and analysis along with evidence gathering. Finally, applicable sections of the law under the Indian Penal Code (IPC) and Information Technology (Amendment) Act (ITAA) 2008 are listed for every case. Figures are included to highlight the sequential flow of concepts in every case. The reader will be familiar with the following after studying the chapter:

- Role played by the investigating officer in evidence gathering
- Evidence collection procedure
- Applicable sections of the law to penalize the offenders

11.1 INTRODUCTION

This chapter presents some of the cybercrimes against individual, property, and nation. For every case, the background is presented. This is followed by the steps carried out during investigation and the conclusion drawn from the case.

11.2 CYBERCRIME AGAINST INDIVIDUAL

This section presents some of the cybercrimes reported against individuals.

11.2.1 Posting of Obscene, Defamatory, and Annoying Messages against Women Online (State of Tamil Nadu vs Suhas Katti)

This was the first case in India that was related to the posting of obscene and annoying messages on the internet.

Gist of the Case

Obscene and defamatory messages about a divorced woman (the victim) was posted in a Yahoo message group. The accused created a fake email account in the name of the victim and used it to send emails as though the victim was sending them. The posting of these messages resulted in annoying phone calls to the lady as people believed that she was soliciting clients. The victim lodged a complaint with the police and a case was registered.

Investigation—Collection of Evidence and Analysis

1. Based on a complaint, the police traced the accused and arrested him within a few days.
2. The accused was interested in marrying the victim and was a known family friend. However, the victim married another person. This marriage later ended in divorce and the accused started contacting her again. On her reluctance to marry him, the accused harassed the victim online.
3. The IP address of the computer that was used to host the profile on the Yahoo group as well as the email ID from which this profile was hosted were obtained from the email service provider. The ISP was requested to determine the physical address from the IP address of the system used by the accused during the email tracking process. The ISP then provided the physical address of the accused.

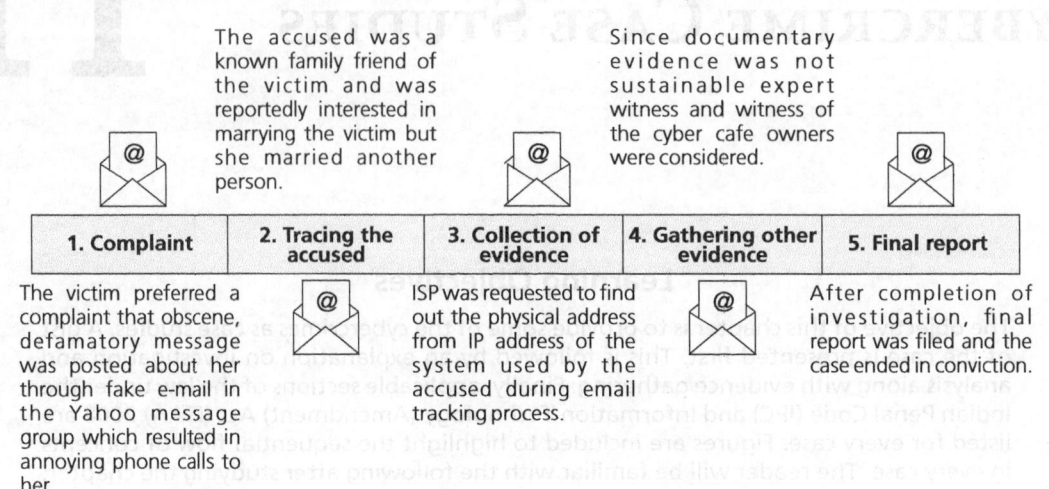

Fig. 11.1 Flow of investigation in case State of Tamil Nadu vs Suhas Katti

Conclusion

After completion of investigation, the investigating officer (IO) filed the final report before the appropriate court under the following sections:

1. Section 469 of the Indian Penal Code (IPC) (Forgery for purpose of harming reputation)
2. Section 509 IPC (Word, gesture, or act intended to insult the modesty of a woman)
3. Section 67 of the Information Technology Amendment Act (ITAA) 2008 (Punishment for publishing or transmitting obscene material in electronic form)

Case Flow

1. The Defence argued that the offending mails would have been sent either by the ex-husband of the complainant or the complainant herself to implicate the accused, as the complainant alleged to have turned down the request of the accused to marry her.
2. Further, the Defence counsel argued that some of the documentary evidences were not sustainable under Section 65 B of the Indian Evidence Act. The court considered upon expert witnesses and witnesses of the cyber cafe owners to conclude that the crime was proved.
3. The trial court, after hearing both sides, convicted the accused. This is considered to be the first case to have ended in conviction under Section 67A of the Information Technology Act 2000 in India.

The flow of investigation in this case is shown in Fig. 11.1.

11.2.2 Phishing Fraud

This case is on a fraudulent message appeared to have come from a legitimate enterprise, a bank, and had resulted in loss of money to an individual.

Gist of the Case

Mr X received a fraudulent mail disguised to be from the customer care department of the bank where he maintained an account. The mail had a link to the web page that urged him to reveal his personal information, including account number, password, and the transaction password. The link gave him the impression that it was a genuine message from the bank where he maintained his account. After some time, he noticed that some amount had been debited from his account for an online purchase, even though he had not made such a purchase. Mr X then lodged a complaint with the police.

Investigation—Collection of Evidence and Analysis

The IO gathered the following evidences:

1. Self-attested copies of the printout of the phishing email along with full headers as well as the bank account from the complainant
2. Account statement containing fraudulent transactions, the transaction IP address of the fraudulent transactions, the details of the beneficiaries (the company to which the amount was credited for the online purchase)
3. The owner of the phishing website from the website hosting company
4. From the beneficiary details provided by the bank, the IO contacted the online store and issued a notice to furnish details of the goods purchased (a laptop and a mobile phone) and the delivery address. However, further investigation revealed that the goods were not delivered to this address but were instead collected from the courier agency with fake IDs.
5. The IO requested all the mobile service providers to check whether the given IMEI number (unique) belonged to their network. One of the service providers responded in the affirmative.
6. Further, the IO obtained additional information from the mobile service provider, that is, the customer application form (CAF) to get an alternate mobile number, the address, and the call data record (CDR). The CDR revealed the people who were in touch with the accused; further enquiry helped in locating the accused.
7. The accused had purchased electronic equipment using the defrauded funds and was finally confronted.
8. The laptop that was in the possession of the accused was seized and sent to the forensic lab for examination. No useful information regarding the phishing mail or Internet banking transaction(s) were available on it.
9. The IO checked the full headers of the phishing mail received by Mr X to trace the IP address of the origin of the phishing email. The IP address was traced and found to be spoofed. Hence, it was not possible to track the original host of the phishing website. The IO notified the CERT-In and the concerned bank to pull down the phishing website.

Conclusion

After the completion of investigation, the IO filed the final report under the following sections:

1. Section 420 of IPC (cheating and dishonestly inducing delivery of property)
2. Section 66C ITAA 2008 (Punishment for identity theft)
3. Section 66D ITAA 2008 (Punishment for personation by using computer resource)

The flow of investigation in this case is shown in Fig. 11.2.

11.2.3 Hacking Using Key logger

This case is associated with the usage of key logger that captures and records a user's activity on an encrypted log file. The perpetrator had used the key logger to acquire the user's banking credentials and steal money from his account.

Gist of the Case

Mr X was a manager in an IT company. The company had a strong information security policy and restrictions regarding carrying mobile phones and browsing the Internet in office. He would visit a cyber cafe located next to his office to do online banking and other personal transactions over the Internet. One day, Mr X visited a shop and made a purchase. He attempted to make a payment for the purchase using debit card. The vendor swiped the card and found that funds were insufficient. He approached the bank, where he was informed that a huge sum had been transferred from his account to another account. After this, Mr X lodged a complaint with the police.

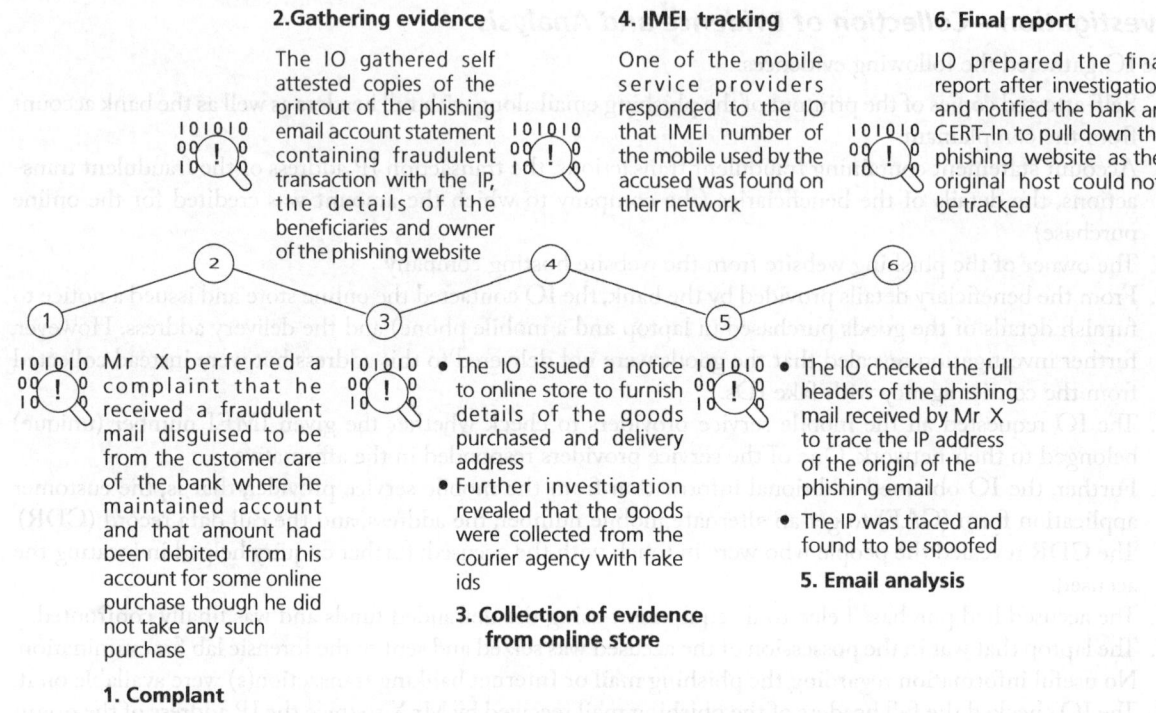

Fig. 11.2 Flow of investigation in case of phishing fraud

Investigation—Collection of Evidence and Analysis

1. The IO gathered the statement of accounts from Mr X.
2. The IO obtained the IP address from where the transaction was initiated, CAF of the beneficiary account, and other details from the bank.
3. The IO approached the ISP to provide the physical address details of the IP address given by the bank (transaction IP address).
4. The IP address was traced to the cyber cafe located next to Mr X's office.
5. The IO visited the cyber cafe and found that a proper log register had not been maintained by the owner. However, the cyber cafe had CCTV installed. CCTV logs were gathered from the cyber cafe.
6. The IO was able to locate the suspect by correlating the time of transaction gathered from the bank with the timestamp in the video footage. The same was confirmed with the cyber cafe owner.
7. The cyber cafe owner was instructed to notify the IO the next time the suspect visited the cyber cafe. He did so and the IO rushed to the spot and interrogated the suspect who confessed to having carried out the fraudulent transaction.
8. Further interrogation revealed that the accused used to visit many cyber cafes in the vicinity of IT companies and had installed the keylogger program to collect the credentials of IT employees who visited the cyber cafe. The accused had used it to transfer funds.
9. The accused had saved the keystrokes in the form of an email uploaded to his email ID rather than saving it in any removable storage media.
10. The CCTV clippings and the hard disk from the cyber cafe were seized by the IO and sent to the forensic lab for examination.
11. Analysis of video footage confirmed the suspect seen in the CCTV clippings. Analysis of the hard disk confirmed the presence of the keylogger program on the seized hard disk and the email ID which the accused had used to upload the key log file as well as the bank account number of Mr X.

Conclusion

After the completion of investigation, the IO filed the final report under the following sections:

1. Section 420 IPC (Cheating and dishonestly inducing delivery of property)
2. Section 66C, ITAA 2008 (Punishment for identity theft)
3. Section 66D, ITAA 2008 (Punishment for cheating by personation by using computer resource)

The flow of investigation in this case is shown in Fig. 11.3.

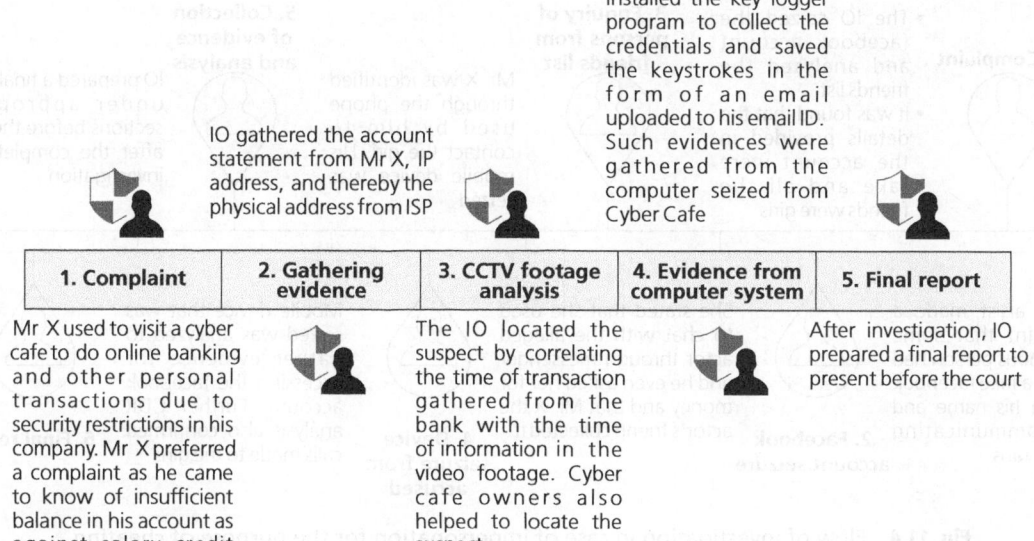

Fig. 11.3 Flow of investigation in case of hacking using Key Logger

11.2.4 Impersonation for Purpose of Cheating

This case is an example of impersonation for the purpose of cheating and commiting a fraud that had spoilt the reputation of a celebrity.

Gist of the Case

A cine artist preferred a complaint that an anonymous person had created a fake Facebook page in his name and was communicating with his fans.

Investigation—Collection of Evidence and Analysis

1. The IO seized the said Facebook account. The account was scrutinized and it was found that the details provided in the account were all fake. Then the friends list was analysed and it was found that all the friends were girls.
2. The user IP for the fake account was requested from Facebook Inc., but it was not provided by them.
3. One of the persons in the friends list was identified and she was questioned. She stated that she used to chat with the alleged actor through messenger and he even asked her for money. She further stated that one of the actor's friends, Mr X, came to her and collected the money. Mr X was identified through the phone used by him to contact the girl.
4. He was apprehended and on enquiry, admitted to having created a fake account and having cheated many girls.

5. The mobile phone used by Mr X was seized and sent for digital analysis.

Conclusion

After the completion of investigation, the IO filed the final report and a case was registered under the following sections:
1. Section 420 IPC (Cheating and dishonestly inducing delivery of property)
2. Section 66D, ITAA 2008 (Punishment for cheating by personation by using computer resource)

The flow of investigation in this case is shown in Fig. 11.4.

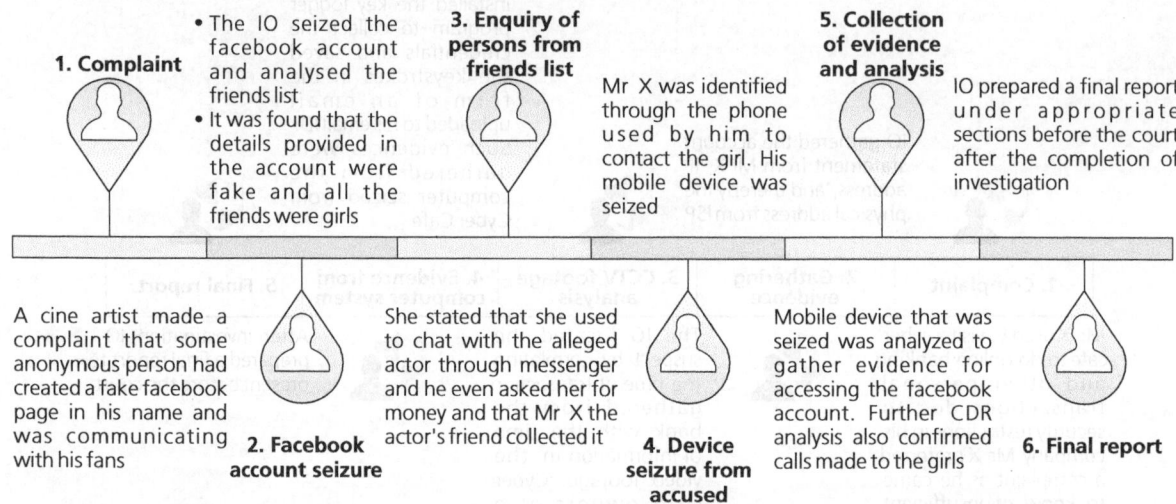

Fig. 11.4 Flow of investigation in case of impersonation for the purpose of cheating

11.2.5 Transmission of Sexually Explicit Material through Internet

This case is presented to explain how sexually explicit material that was circulated on the Internet against an individual was dealt with.

Gist of the Case

A middle-aged man gave a complaint that someone had posted his daughter's morphed nude image on a fake Facebook profile.

Investigation—Collection of Evidence and Analysis

1. The IO requested for the IP logs from Facebook Inc. and it was provided by them.
2. The IP address associated with the fake Facebook profile was obtained.
3. The user for the particular IP and the physical address were collected from the concerned ISP.
4. The physical address was traced to the complainant's maternal home.
5. On the particular date and time of posting of the obscene image, his daughter-in-law was present in her father's house, and the laptop used for creating the morphed image was seized and sent to the forensic lab for analysis.
6. Analysis confirmed the presence of the morphed image and the tools used to create them.
7. The evidences gathered were sent to court.

Conclusion

After the completion of investigation, the IO filed the final report and a case was registered under the following sections:
1. Section 354C IPC (Punishment for voyeurism)
2. Section 67A ITAA 2008 (Punishment for publishing or transmitting of material containing sexually explicit act, etc., in electronic form)

11.2.6 Criminal Intimidation and Sending Obscene Material through Internet

Gist of the Case

A lady preferred a complaint with the police stating that her daughter was married and residing in Singapore with her husband. However, her ex-boyfriend was threatening her and sending obscene pictures to her husband through email.

Investigation—Collection of Evidence and Analysis

1. The IO gathered the email ID used by the accused.
2. The email was traced and the IP address associated with it was obtained from the log.
3. The ISP was approached to obtain the physical address associated with the IP address.
4. Further, it was revealed that the IP address belonged to a software company. However, the source of the IP address was ascertained and the suspect was apprehended.
5. The laptop of the suspect was seized from his residence and sent to the forensic lab for analysis.
6. Analysis confirmed that all the obscene pictures sent by the accused to the complainant's daughter were found in his laptop.

Conclusion

After the completion of investigation, the IO filed the final report and a case was registered under the following sections:
1. Section 354C IPC (Punishment for voyeurism)
2. Section 67A ITAA 2008 (Punishment for publishing or transmitting of material containing sexually explicit act, etc., in electronic form)

The flow of investigation in this case is shown in Fig. 11.5.

11.2.7 Trolling on Social Media

Social media posts can cause social turmoil when any member of a group or a community deliberately offends another member or members. One such case is presented here.

Gist of the Crime

A famous singer and social activist used to express her views on various social events through Twitter. A group of Tamil chauvinists started stalking her on social media and posted threatening and abusive contents. The comments were extremely vulgar. Hence, she lodged a complaint with the police.

The flow of investigation in this case is shown in Fig. 11.6.

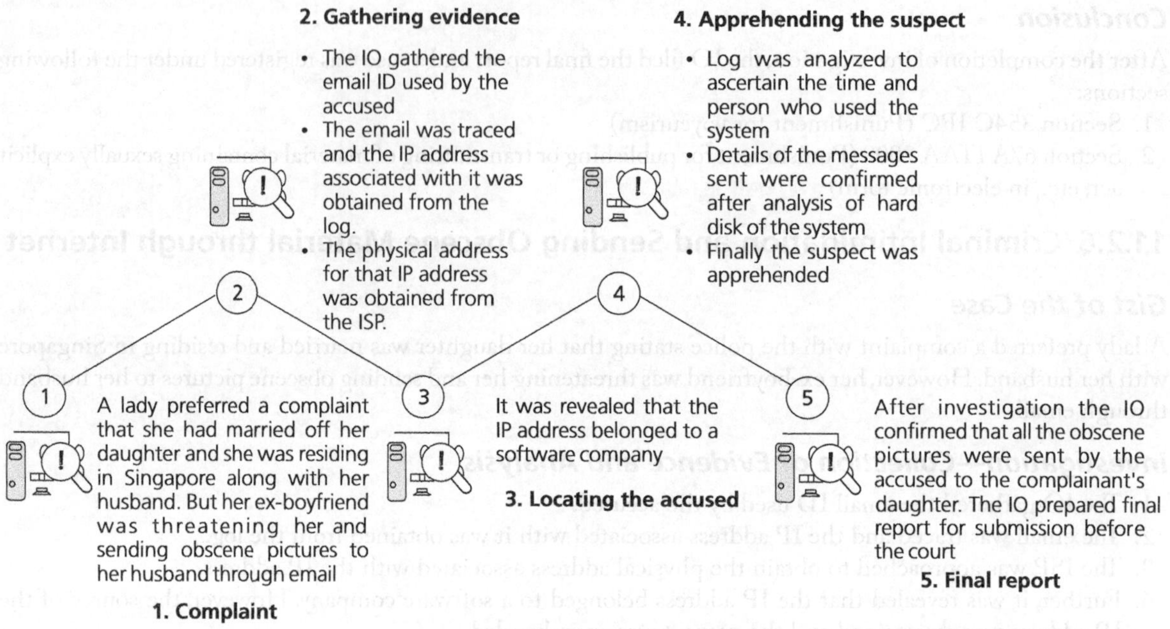

378 *Cyber Forensics*

1. Complaint
A lady preferred a complaint that she had married off her daughter and she was residing in Singapore along with her husband. But her ex-boyfriend was threatening her and sending obscene pictures to her husband through email

2. Gathering evidence
- The IO gathered the email ID used by the accused
- The email was traced and the IP address associated with it was obtained from the log.
- The physical address for that IP address was obtained from the ISP.

3. Locating the accused
It was revealed that the IP address belonged to a software company

4. Apprehending the suspect
- Log was analyzed to ascertain the time and person who used the system
- Details of the messages sent were confirmed after analysis of hard disk of the system
- Finally the suspect was apprehended

5. Final report
After investigation the IO confirmed that all the obscene pictures were sent by the accused to the complainant's daughter. The IO prepared final report for submission before the court

Fig. 11.5 Flow of investigation in case of criminal intimidation and sending obscene material through Internet

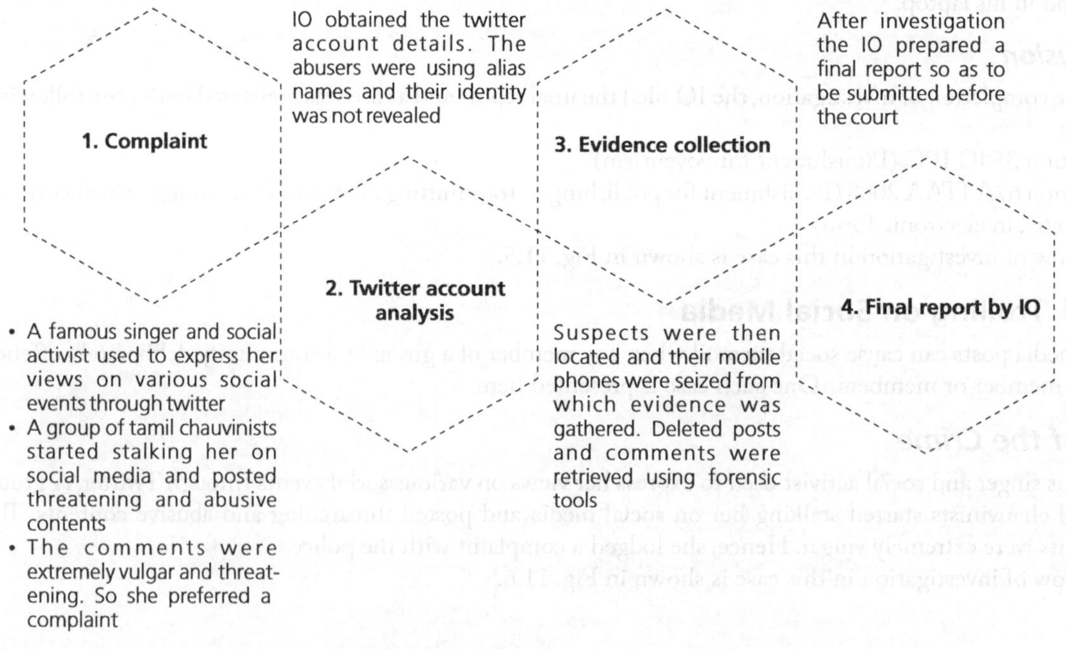

1. Complaint
- A famous singer and social activist used to express her views on various social events through twitter
- A group of tamil chauvinists started stalking her on social media and posted threatening and abusive contents
- The comments were extremely vulgar and threatening. So she preferred a complaint

2. Twitter account analysis
IO obtained the twitter account details. The abusers were using alias names and their identity was not revealed

3. Evidence collection
Suspects were then located and their mobile phones were seized from which evidence was gathered. Deleted posts and comments were retrieved using forensic tools

4. Final report by IO
After investigation the IO prepared a final report so as to be submitted before the court

Fig. 11.6 Flow of investigation in case of trolling on social media

Investigation—Collection of Evidence and Analysis
1. The IO seized the Twitter account of the complainant.
2. Twitter refused to provide any information about the suspects. The abusers were using aliases and their identities were not known initially.
3. The IO meticulously probed the account information related to the suspect and some leads were found.
4. The suspects were then located and their mobile phones were seized and sent to the forensic lab for analysis.
5. Analysis confirmed the presence of abusive threatening content. Further, even the deleted posts and comments were retrieved with appropriate forensic tools.

Conclusion
After the completion of investigation, the IO filed the final report under the following sections:
1. Section 153 A IPC (Wantonly giving provocation with intent to cause riot)
2. Section 354D IPC (Stalking)

11.3 CYBERCRIME AGAINST PROPERTY
This section presents some of the cybercrimes reported against property.

11.3.1 Online Lottery Scam
This case deals with the commission of fraud with an SMS notification that a large sum of money is won through lottery along with a command to keep the notification secret. It is an attempt to initiate a dialogue with potential victims and fraudulently extract money by pretending to be a legitimate lottery company, but sending the money to the scammers responsible.

Gist of the Case
Mr X gave a complaint stating that he had lost ₹36 lakh to online fraudsters. He was working abroad.

Investigation—Collection of Evidence and Analysis
1. The IO gathered details from the complainant. Mr X had received an SMS declaring him the winner of a lottery for $100,000. To claim the amount, he had to contact an email address and give his personal details, bank details, email ID, etc.
2. In the reply email, Mr X was instructed to keep this matter secret. He also got a confirmation email with a certificate declaring him the winner of the lottery. He then received an email asking him to deposit ₹10,000/- for registration charges, ₹25,000/- for an anti-money laundering certificate, ₹75,000/- for an anti-drug trafficking certificate, ₹2,50,000/- for RBI clearance, ₹10,00,000/- for tax and many more such fraudulent reasons.
3. Mr X deposited a total of ₹36 lakh in 16 fraudulent accounts in a total of 28 transactions.
4. The IO approached the bank and obtained the transaction log.
5. The beneficiary account details were obtained and the concerned banks were contacted. It was found that there was nil balance in the accounts.
6. The KYC was obtained and the physical address was found to be fake.
7. The withdrawals had been made in ATMs without cameras. Hence, the accused could not be identified.

The IO filed this in the report.

Conclusion
After the completion of investigation, the IO filed the final report and a case was registered under the following sections:
1. Section 420 IPC (Cheating and dishonestly inducing delivery of property)

2. Section 66C ITAA 2008 (Punishment for identity theft)
3. Section 66D ITAA 2008 (Punishment for personation by using computer resource)

The flow of investigation in this case is shown in Fig. 11.7.

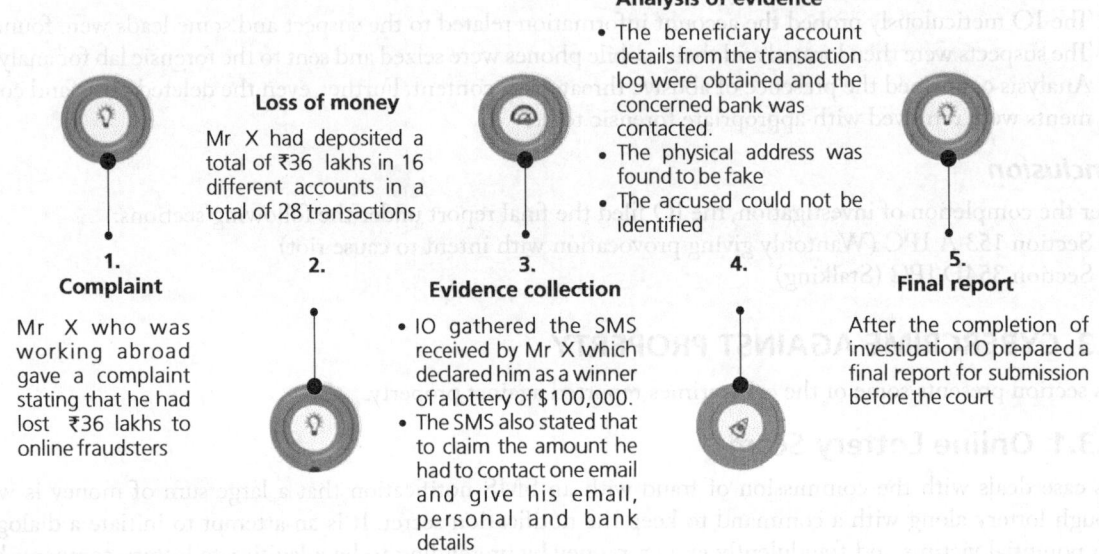

Fig. 11.7 Flow of investigation in case of online lottery scam

11.3.2 Theft in ATM

This case is associated with money being stolen by a bank employee while loading it in the ATM.

Gist of the Case

A nationalized bank lodged a complaint that a certain amount (in lakhs) was found missing from an ATM attached to the branch of the bank. This was brought to light after an audit which revealed that the cash fell short against entries in the accounts in the last six months. Loading of cash in that ATM was the responsibility of the bank manager. Two key system (passwords), one for the manager and the other for the cashier, had to be used to access the ATM machine. Every day, around 15 lakhs of rupees was being loaded. The passwords were changed every day. After reconciliation of statements, there was a shortage of money and in this connection a complaint was given to the police.

Investigation—Collection of Evidence and Analysis

1. After registering a case, the police visited the ATM and found that the CCTV camera was inside the ATM room. During analysis of the video footage, it was observed that only the cashier alone had loaded the money into the ATM and so he must have swindled the money.
2. The ATM had different cases to load currency notes of different denominations. After placing the currency in the respective cases, an entry had to be made in the computer on the count of notes of each denomination placed in their respective cases.
3. Following this, the tray holding the cases had to be closed. The cashier had placed the money in the case, made an entry in the computer, and had then taken the money before the tray was closed. Thus, the cashier of the bank was accused of swindling a part of the money which was supposed to be loaded in the ATM.

4. This was noticed when the video footage recorded by the camera present within the ATM machine was analysed.
5. The video footage further revealed that the manager had never visited the ATM though it was mandatory. The cashier had used the password that had to be used by the manager.
6. It was also confirmed that the manager was involved with some other work in his terminal when the cash was loaded by the cashier in the ATM. This was confirmed with the log files which recorded the status of the manager's terminal at the time the cash was loaded by the cashier (time as available in the video footage).
7. Thus the investigator was able to conclude that only the cashier was involved in the money theft and not the manager though there was dereliction of duty.

Conclusion

After the completion of investigation, the IO filed the final report before the court under the following sections:
1. Section 408 IPC (Criminal breach of trust by clerk or servant)
2. Section 420 IPC (Cheating and dishonestly inducing delivery of property)
3. Section 66C ITAA 2008 (Punishment for identity theft)
4. Section 66D ITAA 2008 (Punishment for personation by using computer resource)

The flow of investigation in this case is shown in Fig. 11.8.

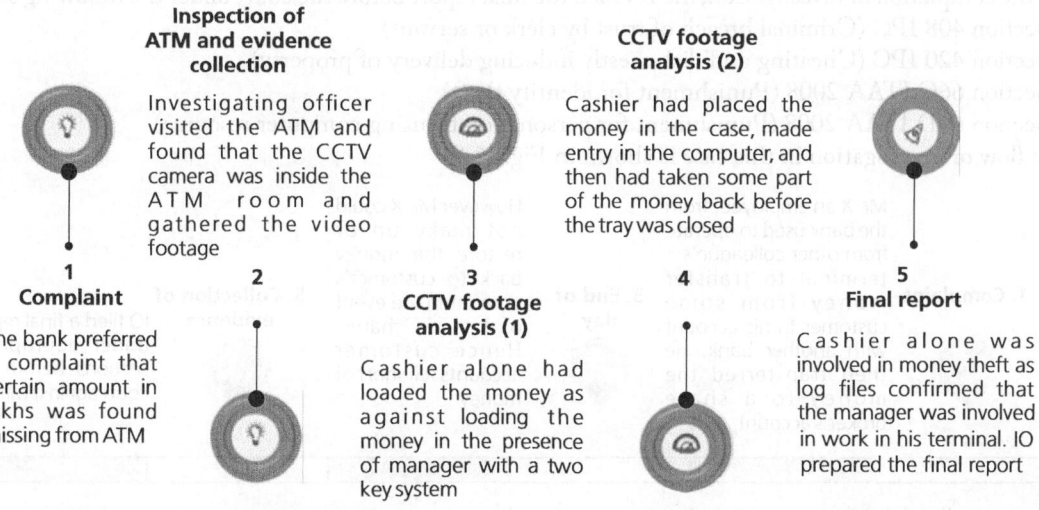

Fig. 11.8 Flow of evidence in case of theft in ATM

11.3.3 Swindling of Money by Bank Employee

This case presents a commission of fraud with bank deposits by a bank employee, which resulted in losses incurred by the bank and also earned the displeasure of customers.

Gist of the Case

From a nationalized bank a complaint was preferred that money had been debited from many customers' accounts without their knowledge. Further, a shortage of money was observed in the audit.

Investigation—Collection of Evidence and Analysis

1. Every employee in the bank was provided with login credentials and they were supposed to work from their own terminals.

2. Transfer of money to or from a customer's account could be made from any terminal once the employee had successfully logged on to the terminal. Employees were strictly instructed to log off their terminals or sign out when they left their seats, especially during lunch break and tea break. However, in practice, nobody followed this.
3. A log of transactions and the work pattern of the employees were obtained.
4. The log reflected that a high volume of transactions during that period was to a particular bank account.
5. Mr X, an employee of the bank used to operate another colleague's terminal to transfer money from a customer (preferably an account with sufficient balance and that had remained inoperative for a long time) to his bank account maintained with another bank. Then from his personal account maintained with the other bank, he used to transfer the money to a share broker's account. The share broker would then invest the money on intra-day share transactions on behalf of Mr X. Every evening the share broker would transfer the amount (proportional to the loss or gain resulted in that day on the shares) back to Mr X's account who would then deposit the money back into the respective customer's account from whom the money had been taken in the morning.
6. However, Mr X could not make up or restore the money back to the customer's account in the event of loss on shares. This sort of loss on shares accumulated over a period of time to 30 lakhs of rupees.

Conclusion

After the completion of investigation, the IO filed the final report before the court under the following sections:
1. Section 408 IPC (Criminal breach of trust by clerk or servant)
2. Section 420 IPC (Cheating and dishonestly inducing delivery of property)
3. Section 66C ITAA 2008 (Punishment for identity theft)
4. Section 66D ITAA 2008 (Punishment for personation by using computer resource)

The flow of investigation in this case is shown in Fig. 11.9.

1. Complaint

A nationalized bank preferred a complaint that money had been debited from many customers account without their knowledge

2. Fund transfer

Mr X an employee from the bank used to operate from other colleague's terminal to transfer money from some customer to his account with another bank. He then transferred the money to a share broker's account

3. End of day

Every evening the share broker would transfer the proceeds of intra day shares back to Mr X's account who then would deposit the money back into respective customer's account

4. Loss

However Mr X could not make up or restore the money back to customer's account in the event of loss on shares. Hence customer accounts fell short of money

5. Collection of evidence

The transaction log revealed that huge amount of transactions during the period were to a particular bank account

6 Final report

IO filed a final report under appropriate sections before the court against Mr X

Fig. 11.9 Flow of investigation in case of swindling of money by bank employee

11.3.4 Data Theft

This case presents theft of data that was not safeguarded with appropriate security measures. The perpetrator had managed to gain access to the system as a part of maintenance to steal data, which led to chaos.

Gist of the Case

The police arrested a celebrity who was an accused in a drug trafficking case. The accused was brought to the police station after obtaining permission from a competent court for further interrogation. The interrogation proceedings were recorded on CCTV cameras installed in different places in the interrogation room. The investigation agency was surprised to see the interrogation clippings in television news channels and video sharing websites. The police registered an FIR against the unknown person who had stolen the CCTV clippings and shared it with the media.

Investigation—Collection of Evidence and Analysis

The IO collected all relevant details such as the following:
1. Copy of the video clipping broadcasted by the media/online, original CCTV clippings for future comparison
2. Date and time of the telecast of the CCTV clippings from the television channel
3. IP address/email ID of the person who had uploaded the clippings from the video sharing website
4. The physical address of the accused from the IP address obtained from the ISP.
5. The IO seized the storage media responsible for storing the CCTV clippings in the interrogation room. It was sent to a forensic lab for analysis with a request to report the details of any removable media inserted into the computer.
6. The hardware configuration of the original storage media was documented and a duplicate of the hard drive was created in the forensic lab to preserve the evidence. The CMOS information, including the time and date, was also documented.
7. The directory and file structures, including file dates and times, were recorded.
8. A forensic tool was used to determine whether any removable media was connected to the hard disk. The report confirmed it positively, which means that the removable media was used to gather video clippings from the original storage media prior to telecast. This was also confirmed with the last accessed time of the video files.
9. To determine who inserted the removable media, the IO summoned all the personnels who had access to the interrogation room. It was observed that Mr X who belonged to the CCTV servicing company had made a log entry to check the computer system in agreement with the warranty offered at the initial stages.
10. The IO then recovered the removable media from Mr X. The make and model of the removable media matched with that specified in the forensic report.
11. Further, it was revealed that Mr X handed over the removable media to Mr Y who uploaded the video clippings onto the website.

Conclusion

After the completion of the investigation, the IO filed the final report under the following sections:
1. Section 43 ITAA 2008 (Penalty and compensation for damage to computer, computer system, etc.)
2. Section 66B of ITAA2008 (Punishment for dishonestly receiving stolen computer resource or communication device)
3. Section 66C ITAA 2008 (Punishment for identity theft)
4. Section 66D ITAA 2008 (Punishment for personation by using computer resource)

The flow of investigation in this case is shown in Fig. 11.10.

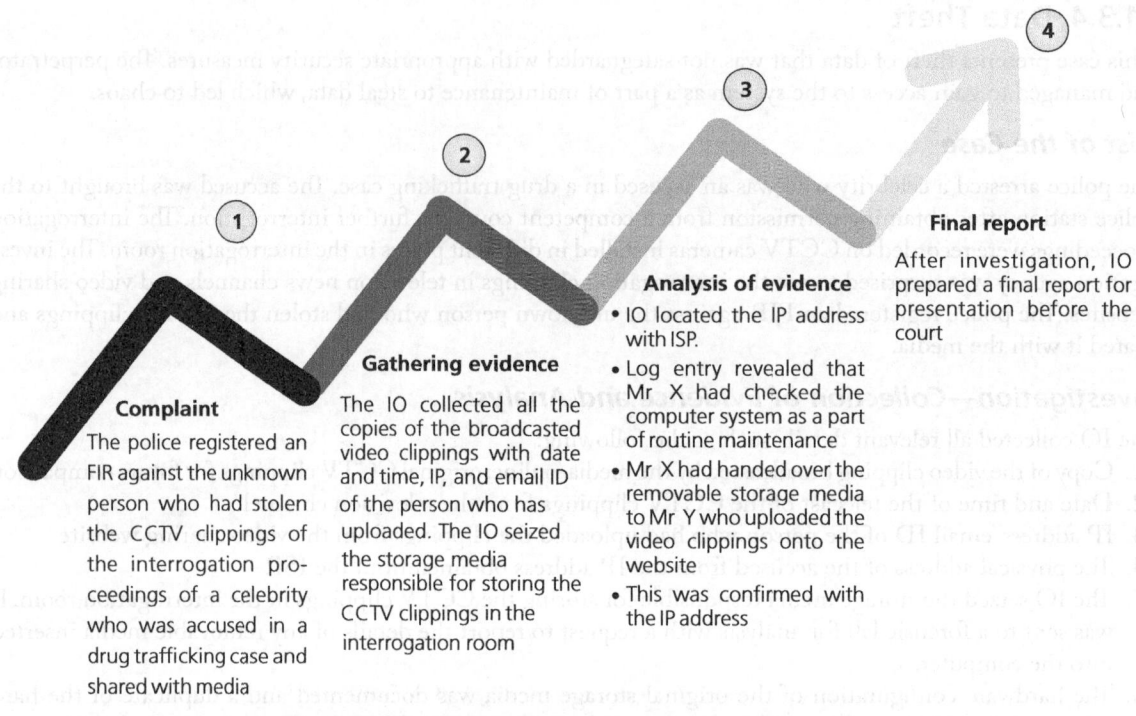

Fig. 11.10 Flow of investigation in case of data theft

11.3.5 Hacking

This case explains how the trace left by a hacker on the web server of the bank helped in cracking it.

Gist of the Case

The complainant was a bank manager of a private bank. He stated that one of their customers, who is an online merchant, alleged that someone was tampering with their website and as a result the merchant was not receiving full payments for the products delivered to the customer.

Investigation—Collection of Evidence and Analysis

1. The IO obtained the bank account details from the manager and the transaction log.
2. The IP associated with the transactions was located. It was noticed that all the disputed transactions were being done from a single IP.
3. The ISP was approached to get the user CAF associated with the IP address and his physical address.
4. The user of the IP was traced and was interrogated. On interrogation, he revealed that he used to book products on e-commerce sites and before the form was sent to the bank, he would tamper with the form and change the amount. For example, if the product cost was ₹5000/- he would change it to ₹5/-. The bank would debit ₹5/- from his account and pay to the merchant.
5. As there was a delay in settling of the amount to the merchant by the bank the products had been despatched before getting the amount. In this way, the accused had purchased goods worth ₹18 lakh.
6. The computer used by the suspect was seized and sent to the forensic lab for analysis.
7. Analysis revealed that the accused had meddled with the online purchases from the browser artifacts.

Conclusion

After the completion of investigation, the IO filed the final report and a case was registered under the following sections:
1. Section 420 IPC (Cheating and dishonestly inducing delivery of property)
2. Section 66C ITAA 2008 (Punishment for identity theft)
3. Section 66D ITAA 2008 (Punishment for personation by using computer resource.)

The flow of investigation in this case is shown in Fig. 11.11.

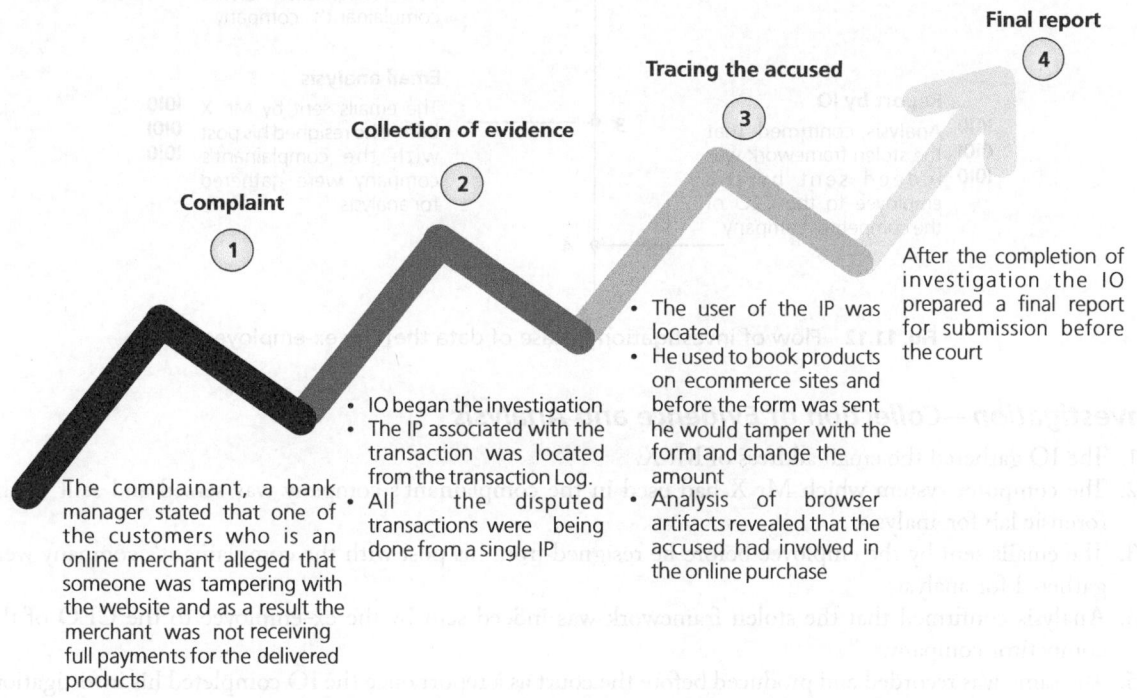

Fig. 11.11 Flow of investigation in case of hacking

11.3.6 Data Theft by Ex-employee

This case was against an ex-employee by an IT company for stealing its software, data, etc., which resulted in revenue loss to the company.

Gist of the Case

The Chief Operating Officer of a software company alleged that their software was stolen by their ex-employee, Mr X, and that he had joined a competitor company. The stolen data was misused by him and the competitor company. The alleged data was provided by a foreign client of the complainant's company as a framework on which further development had to be made. The competitor company had showed the framework on a PowerPoint presentation to the foreign client and they immediately identified their framework and informed the complainant company that there was leak of data from their company. The complainant company then lodged a police complaint.

Cyber Forensics

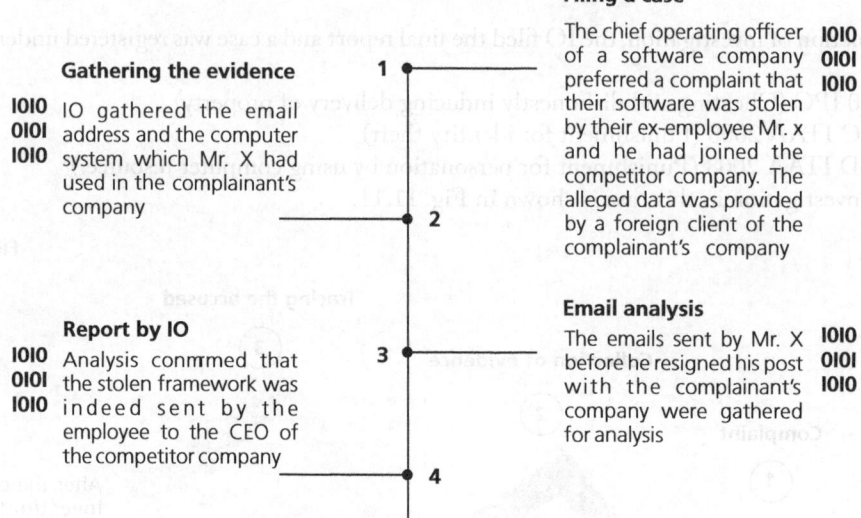

Fig. 11.12 Flow of investigation in case of data theft by ex-employee

Investigation—Collection of Evidence and Analysis

1. The IO gathered the email address of Mr X.
2. The computer system which Mr X had used in the complainant's company was seized and sent to the forensic lab for analysis.
3. The emails sent by the employee before he resigned from his post with the complainant's company were gathered for analysis.
4. Analysis confirmed that the stolen framework was indeed sent by the ex-employee to the CEO of the competitor company.
5. The same was recorded and produced before the court as a report once the IO completed his investigation. The flow of investigation in a case of data theft is shown in Fig. 11.12.

Conclusion

After the completion of investigation, the IO filed the final report and a case was registered under the following sections:

1. Section 420 IPC (Cheating and dishonestly inducing delivery of property)
2. Section 66C ITAA 2008 (Punishment for identity theft)
3. Section 66D ITAA 2008 (Punishment for personation by using computer resource)

11.4 CYBERCRIME AGAINST NATION

This section presents cybercrime cases against the nation.

11. 4.1 Preparation of Forged Counterfeits using Computers/Printers/Scanners

This case was filed by a University on a former student, for the printing of fake marksheets and utilizing it to secure a job. Such an act is forbidden by law.

Gist of the Case

Mr X was an engineering graduate and his classmate, Mr Y, had not completed his graduation because he had not passed a few subjects. They had both appeared for an interview for a job where the resume of Mr X was rejected, whereas Mr Y was selected for the job. This made Mr X suspect that Mr Y must have forged documents related to his graduation as he was very sure that Mr Y had not completed the course. He brought it to the knowledge of the university and the university obtained the records of Mr Y from the company where he had submitted forged documents of his academic qualification. Verification of records by the university confirmed that Mr Y had forged the documents of the university. The university preferred a complaint to the police against Mr Y.

Investigation—Collection of Evidence and Analysis

1. The IO gathered details of the accused and his contact information from Mr X and the university and the forged documents that were submitted by Mr Y from the company.
2. The police located Mr Y with the help of a mobile phone service provider who maintained the customer application form, the details of alternate mobile numbers in his possession, and the CDR for the specific period.
3. On preliminary enquiry Mr Y revealed that Mr Z printed the fake marksheet for him.
4. The IO conducted a search at Mr Z's premises and found computers, scanners, and printers which were used to create forged marksheets and other documents. Besides this, the IO found stationery, rubber stamps, and seals of various universities, SIM cards, and mobile phones. Further, one of the systems was switched-on and a laptop was found in a switched-off condition.
5. All the ECDs were seized as per the standard digital evidence gathering procedures. The switched-on and switched-off systems were transported for presentation before the court.
6. Now, to establish that the seized computer was responsible for creating the forged marksheet, the IO obtained permission from the court and sent them for expert opinion to the computer forensics laboratory along with sample documents and a questionnaire.
7. The forensic examiner analysed the seized media and found that the seized computers were used for preparation of the counterfeit marksheets and degrees of Mr Y. The forged documents and the printer, along with the ink, were sent to the forensic lab for examination and expert opinion revealed that the forged documents were printed from that printer.
8. Further, CDR and email exchanges between Mr Y and Mr Z confirmed the association of Mr Y with Mr Z

Conclusion

After the completion of investigation, the IO filed the final report under the following sections:
1. Section 464 IPC (Making a false document including a false electronic record)
2. Section 465 IPC (Punishment for forgery)
3. Section 468 IPC (Forgery for the purpose of cheating)
4. Section 471 IPC (Using as genuine a forged document or electronic record)
5. Section 473 IPC (Making or possessing counterfeit seal, etc.)
6. Section 66C ITAA 2008 (Punishment for identity theft)
7. Section 66D ITAA 2008 (Punishment for personation by using computer resource) before the appropriate court to prove that the cybercrime or the offence was committed using computers to create forged documents along with evidences gathered by forensic examiners.

The flow of investigation in this case is shown in Fig. 11.13.

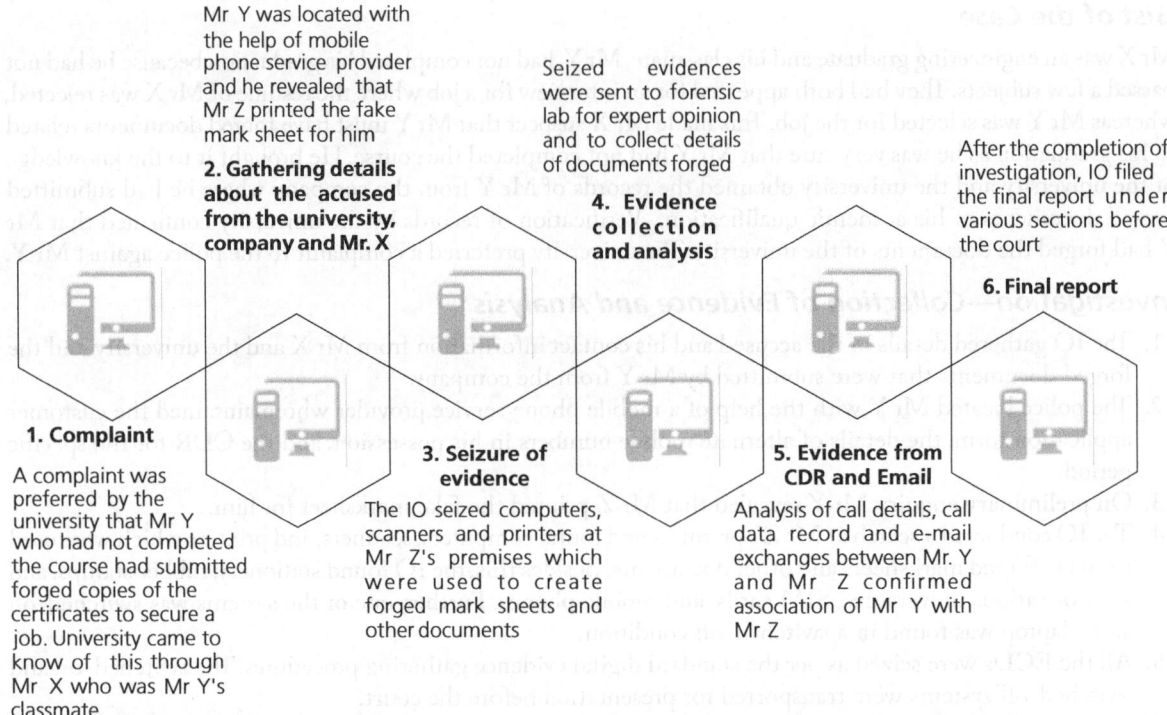

Fig. 11.13 Flow of investigation in case of preparation of forged counterfeits using computers

11.4.2 Blocking of Websites

This case concerns blocking of websites that support anti-nationals and other illegal activities, which disrupt the peace and harmony of a community.

Gist of the Case

A member of an allegedly notorious terrorist organization was shot dead by the police during an encounter. His admirers/supporters started protests on websites/blogs and had anti-Indian statements. The blogs also created communal disharmony. There was public outcry over such blatant anti-Indian propaganda. The law enforcement agency and the government felt that continuation of access to the website hosting anti-national content would threaten national security and communal harmony.

Investigation—Collection of Evidence and Analysis

1. The IO gathered the contents of the website on a storage media and took printouts of the web pages.
2. The IO filed a report to the court.

A requisition was given to the service provider to block the particular website, and access to the particular website was blocked.

Conclusion

After the completion of investigation, the IO filed the final report under the following sections:

1. Section 153 A IPC (Wantonly giving provocation with intent to cause riot)
2. Section 69 A under ITAA 2008 (Power to issue directions for blocking public access of any information through any computer resource that empowers the competent authorities to block public access of notified websites.)

The flow of investigation in this case is shown in Fig. 11.14.

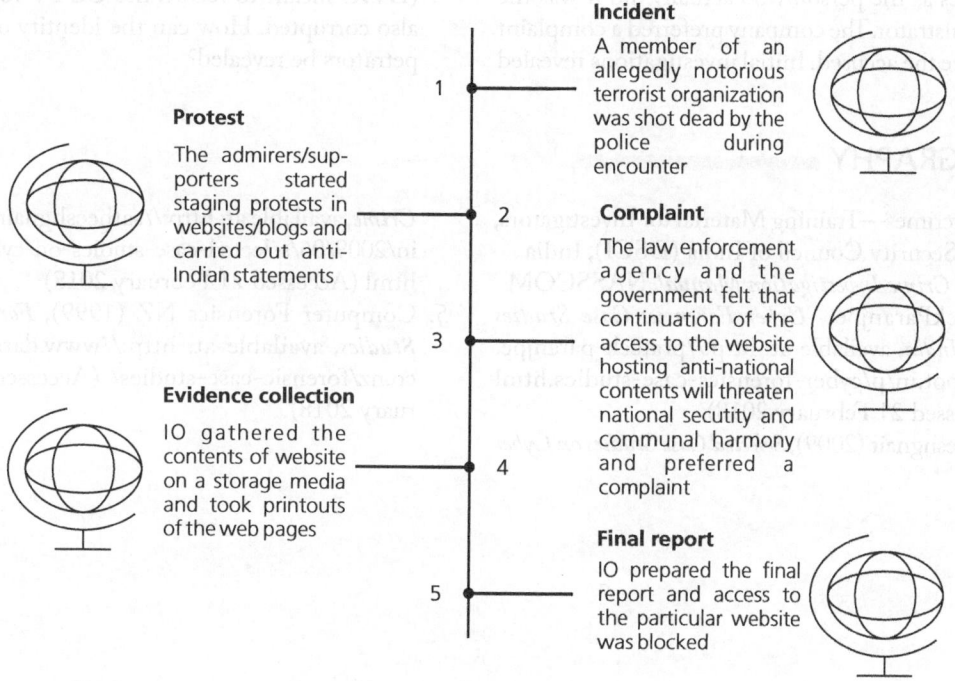

Fig. 11.14 Flow of investigation in case of blocking of websites

APPLICATION EXERCISES

The gist of cases pertaining to cybercrime is furnished below. Elucidate on the investigation process, evidence to be gathered, analysis, and the applicable sections of the law.

1. Ms X preferred a complaint that she was being harassed by telephone calls with sexual innuendos at odd hours. On interrogating the callers, she came to know that her email ID was available in chat rooms and once they sent a mail to that ID, they received an auto response with her telephone number and an invitation to call her.
2. Ms X preferred a complaint that she received emails containing her obscene morphed photographs. She was threatened that such photographs would be posted in pornographic websites.
3. Mr X who runs an online booking website preferred a complaint that a group of anonymous people had booked tickets using stolen cards through the online facility. The credit card companies in turn had charged Mr X for those fraudulent transactions which had resulted in financial loss.
4. Mr X, the owner of a software company, preferred a complaint that the source code under development had been misused by an ex-employee. Mr X had obtained copyright for the code under development.
5. Mr X, an employee of a company that worked for a bank, managed to get the credentials of the bank's customers fraudulently. He then used to transfer money from the customer's account to his

account. The bank preferred a complaint against the company.
6. Mr X was working in a company where stringent IT policies were in force. He was charged for downloading pornographic content. He denied the charges as the person who actually did it was the administrator. The company preferred a complaint to trace the accused. Initial investigations revealed an IP conflict.
7. Mr X preferred a complaint that some perpetrators had managed to enter the record room, containing CCTV footage, and had managed to steal the records. The hard drive of the digital video recorder (DVR) meant to record the CCTV footage was also corrupted. How can the identity of the perpetrators be revealed?

BIBLIOGRAPHY

1. Cybercrimes—Training Material for Investigators, Data Security Council of India (DSCI), India
2. *Cyber Crime Investigation Manual*, NASSCOM
3. PrateekParanjpe, *Cyber Forensic: Case Studies from India*, available at: http://prateek-paranjpe.blogspot.in/p/cyber-forensics-case-studies.html (Accessed 21 February 2018)
4. Satheeshgnair (2009), *Selected Case Studies on Cyber Crime*, available at: http://satheeshgnair.blogspot.in/2009/06/selected-case-studies-on-cyber-crime.html (Accessed 21 February 2018)
5. Computer Forensics NZ (1999), *Forensic Case Studies*, available at: http://www.datarecovery.co.nz/forensic-case-studies/ (Accessed 21 February 2018)

Introduction to Cyber Laws

12

Learning Objectives

This chapter provides an overview on what cyber law is. The objective of this chapter is to provide an insight into cyber laws and their need. The terms and technologies associated with cybercrime are explained in Section 2 of the Information Technology (Amendment) Act (ITAA) 2008. The laws and their legal issues have been discussed. The chapter also offers an introduction to cyber security and the strategies to be adopted. How the laws are applicable to minimize risk is discussed, along with the initiatives taken by the government. The reader will be familiar with the following after studying the chapter:

- Cyber law and its need
- Terminologies associated with cyber laws
- Cyber laws and the legal issues
- Strategies and initiatives involved in cyber security

12.1 CYBER LAWS

The virtual world encompassing the Internet is referred to as the *cyberspace* and the laws governing cyberspace are known as *cyber laws*. Any Internet user falls under cyberspace and under the ambit of cyber laws, as they carry universal jurisdiction. Thus cyber law can be defined as follows:

Cyber law is a branch of law used to describe the legal issues related to use of communications technology or internetworked information technology, particularly 'cyberspace', that is, the Internet.

In other words, cyber law is the law governing electronic communication devices, all aspects of transactions, and activities on and involving the Internet, World Wide Web, and cyberspace. It encompasses laws relating to cybercrimes, electronic and digital signatures, intellectual property, data protection, privacy, etc. All regulatory mechanisms and legal infrastructure fall under its purview. Cyber law is an attempt to integrate the challenges presented by human activities on the Internet with a legitimate system of laws applicable to the physical world. Every action and reaction in cyberspace have some legal and cyber legal perspectives.

12.2 NEED FOR CYBER LAWS

The Internet was originally developed for resource and information sharing. With the increase in the number of users, the techno-savvy environment has made the Internet highly transactional with e-business, e-commerce, e-governance, e-procurement, etc. As the users of cyberspace grow increasingly diverse, and the range of online interaction expands, there has also been steady increase in the cybercrimes being perpetrated—breach of online contracts, operation of online torts, crimes, etc. Hence arose the need to adopt a strict law by the cyber space authority to regulate criminal activities relating to cyberspace and to provide better administration of justice to the victims of cybercrimes. Crimes committed using computers, as either a tool or a target, are on the rise and all legal issues related to such crimes have to be addressed with cyber laws. Cyber laws preserve the integrity and security of information, enhance the security of government data, preserve intellectual property rights (IPR), retain the privacy and confidentiality of information, and ensure the legal status of online transactions.

The following have necessitated the need for cyber laws:
1. Though India had a well-defined legal system (Indian Penal Code—IPC, the Indian Evidence Act 1872—IEA, the Banker's Book Evidence Act 1891, the Reserve Bank of India Act 1934, the Companies Act, etc.), the advent of the Internet brought with it new and complex legal issues which necessitated the enactment of cyber laws.
2. The existing laws then were not sufficient to include all aspects of activities in cyberspace and also did not give legal validity and sanction. Activities carried out in cyberspace necessitated the enactment of relevant cyber laws.
3. The following characteristics of cyberspace has brought in the need for cyber laws:
   (a) It has intangible dimensions where conventional laws are not applicable.
   (b) It is devoid of jurisdictional boundaries.
   (c) It handles enormous amount of traffic pertaining to activities by netizens from different parts of the globe with high level of anonymity.
   (d) It offers economic efficiency, as huge revenue is exchanged over the Internet while trading software without the need for licence, shipping, etc. At the same time, the dark side is that pirated copies of the software are also distributed in a fraction of a second resulting in huge loss of revenue.
   (e) Since data is exchanged in electronic form, it is characterized by extreme mobility. However, at the same time, theft of such data is possible as multiple copies can be produced even when the original data is in the possession of the owner.
4. The following scenarios made cyber laws gain momentum due to increasing incidents of cybercrimes, such as online banking frauds, online trading and share-related frauds, theft of source code, credit and debit card fraud, attacks due to virus, cyber sabotage, phishing attacks, email hijacking, denial of service attack, hacking, pornography, etc.
   (a) The reliance of organizations, firms, departments, and companies to preserve valuable data in electronic form and the exchange of such data via a network or the Internet
   (b) The change from paper to electronic form, documents handled by the government, for example, income tax returns and the automation of all its departments
   (c) Increased use of credit and debit cards for trading
   (d) Reliance on email, mobile phones, SMS, and chats on social media for communication
   (e) Demat form of all transactions in the share market
   (f) The replacement of conventional methods of transactions in business with digital signatures and e-contracts
5. Besides cybercrimes, presence of evidence of non-cybercrime cases in electronic communication devices (e.g., in cases of divorce, murder, kidnapping, tax evasion, organized crime, terrorist operations, etc.)
Thus crimes that have some intersection with cyberspace necessitated the formulation of cyber laws.

12.3 CYBER LAWS AND LEGAL ISSUES

The following are the legal issues associated with cyber laws:

Lack of awareness Since cybercrimes are a relatively recent development, and the area and the associated technologies are new, it is the responsibility of the government and law enforcement agencies to make all the stakeholders (e.g., judicial officers, legal professionals, litigants, the public, and users at large) aware of the problems on cyberspace and the remedies at hand. This is highly essential because provisions, for example, the scope for adjudication process, is not known to even investigating officers.

Jurisdiction Jurisdiction is explicitly specified in the adjudication process. With respect to cybercrimes, jurisdiction is territory-free and borderless. However, in case of electronic records, the place of dispatch and receipt of the electronic record may fall under jurisdictional issues. This should be properly understood besides giving appropriate training to the players in the field.

Resolving conflicts while collecting evidences Evidence plays a vital role in noting that a cybercrime has been committed. Evidence may be present in the victim's computer, in the offender's computer, or with intermediaries. Irrespective of the case, measures have to be taken swiftly to capture evidence which otherwise would get destroyed. A trusted third party, in possession of sophisticated equipment, is essential to look into the suspect's system so as to collect evidence in case of dispute between the parties.

Lack of provisions for handling many cybercrimes ITA 2000 and ITAA 2008 are not sufficient to handle cybercrimes such as cybersquatting, spam mail, ISP's liability with copyright infringement, data privacy, among others. Hence new and more stringent regulations have to be developed, and the current laws have to be reinforced.

12.4 CYBER SECURITY

Cyber security is a branch of computer security specifically related to the Internet. The objective of cyber security is to establish rules and measures to counter attacks and crimes carried out on the Internet.

Cyber security offers the following advantages:
1. Defends from hacks and viruses
2. Defends from other critical attacks
3. Monitors all incoming and outgoing traffic on the computer
4. Provides a safe environment for browsing, as well as to access only trusted websites
5. Updates the databases at regular intervals so that even a new virus is deleted

12.5 STRATEGIES INVOLVED IN CYBER SECURITY

The approach to cyber security should take into account the following aspects:
1. A strong ecosystem comprising the three areas of automation, operability, and authentication can prevent cybercrimes.
2. Governments as well as organizations must be allowed to update software infrastructure in compliance with security standards.
3. Individuals and organizations should be encouraged to use open standards as it would lead to improved security against cybercrimes.
4. The government needs to work on strengthening the regulations in place to tackle cybercrimes. These include instituting agencies to handle cybercrimes and constituting cyber laws, promoting cyber security, providing education and training, and implementing new security technologies.
5. IT-related mechanisms such as end-to-end or link-oriented data encryption should be promoted to fight cybercrimes.
6. Developing e-governance so as to provide services over the Internet is an important aspect of cyber law.
7. Any outdated technology invites cybercrime. Hence, it is paramount to protect the organization's infrastructure using up-to-date software.

12.6 MINIMIZING RISK WITH CYBER LAWS

Cyber laws have come into effect so as to reduce risks on the Internet. The risk reduction strategies of cyber laws are as follows:
1. Promoting cyber security research and development
2. Gaining intelligence about threats
3. Improving firewalls
4. Developing and using protocols and algorithms
5. Introducing authentication aspects
6. Focusing on cloud and mobile security
7. Using methods of cyber forensics

Besides this, as cyberattacks are increasing, organizations should have an eye on their data supply chains and ensure that they are free from risks and the equipment are not altered, with better inspection methods. Cyber laws should focus on the supply chain, as an interruption in this process may pose a big security risk. Stringent regulatory rules are in force in many developed countries to prevent unauthorized access to networks and such an act is declared as a penal offence.

Further, to reduce risk, it is essential to realize that employees themselves may be security risks and so it is absolutely vital to promote ethical security mechanisms and promote awareness.

Minimizing information sharing can greatly reduce risk. This can be enforced by making reporting mandatory. Businesses can use a strong security framework comprising the following: (a) a core activity that allows businesses to identify, protect, detect, respond, and recover from cyber threats, (b) an implementation tier which reflects how advanced the security system adopted is, and (c) a database where businesses maintain a record of the strategies adopted.

12.7 INITIATIVES PROMOTING CYBER SECURITY

India has adopted a new initiative which is a practical approach to national cyber security. It attempts to achieve this in phases, as mentioned here:
1. Establishing a central/national cyber security body
2. Defining a national cyber security strategy
3. Establishing a national dialogue
4. Building preventive national cyber security capabilities
5. Building reactive national cyber security capabilities

Accordingly, the National Cyber Security Policy was introduced on 2 July 2013 by the Government of India. The following are the aims of this policy:
1. To monitor and protect information systems
2. To facilitate effective collaboration for public and private partnership through technical and operational cooperation
3. To emphasize and promote R&D in cyber security
4. To establish cyber security training infrastructure and to develop human resource through education and training

The following are some of the existing initiatives to ensure cyber security:
1. National Informatics Centre (NIC), which provides the network backbone and e-governance support to all government bodies across the nation
2. Indian Computer Emergency Response Team (ICERT) which ensures the security of cyber space in the country
3. The Indo–US Cyber Security Forum where high-power delegations work together
4. Alliances of cyber security with other nations such as the US, China, and North Korea to fight for the cause of cyber security as a global economy

12.8 TERMS AND TERMINOLOGIES ASSOCIATED WITH CYBER LAWS

Section 2 of the IT Act 2000 defines the key terms and concepts which help in the understanding of the jurisprudence of cyber laws. The terms are as follows:
1. *Access* with its grammatical variations and cognate expressions means gaining entry into, instructing, or communicating with the logical, arithmetical, or memory function resources of a computer, computer system, or computer network. [Sec. 2(1)(a) of the IT Act 2000]
2. *Addressee* means a person who is intended by the originator to receive the electronic record but does not include any intermediary. [Sec. 2(1)(b) of the IT Act 2000]
3. *Affixing electronic signature* with its grammatical variations and cognate expressions means adoption of any methodology or procedure by a person for the purpose of authenticating an electronic record by means of electronic signature. [Sec. 2(1)(d) of the IT Act 2000]

4. *Asymmetric crypto system* means a system of a secured key pair consisting of a private key for creating a digital signature and a public key to verify the digital signature. [Sec. 2(1)(f) of the IT Act 2000]
5. *Certifying authority* means a person who has been granted a licence to issue an electronic signature certificate under Section 24. [Sec. 2(1)(g) of the IT Act, 2000]
6. *Communication device* means cell phones, personal digital Assistance (sic), or combination of both or any other device used to communicate, send, or transmit any text, video, audio, or image. [Sec. 2(1)(ha) of the IT Act 2000]
7. *Computer* means any electronic, magnetic, optical or other high-speed data processing device or system which performs logical, arithmetic, and memory functions by manipulations of electronic, magnetic, or optical impulses, and includes all input, output, processing, storage, computer software, or communication facilities which are connected or related to the computer in a computer system or computer network. [Sec. 2(1)(i) of the IT Act 2000]
8. *Computer network* means the interconnection of one or more computers or computer systems or communication device through (i) the use of satellite, microwave, terrestrial line, wire, wireless, or other communication media; and (ii) terminals or a complex consisting of two or more interconnected computers or communication devices, whether or not the interconnection is continuously maintained. [Sec. 2(1)(j) of the IT Act 2000]
9. *Computer resource* means computer, communication device, computer system, computer network, data, computer database or software. [Sec. 2(1)(k) of the IT Act 2000]
10. *Computer system* means a device or collection of devices, including input and output support devices and excluding calculators which are not programmable and capable of being used in conjunction with external files, which contain computer programmes, electronic instructions, input data, and output data that perform logic, arithmetic, data storage and retrieval, communication control and other functions. [Sec. 2(1)(l) of the IT Act 2000]
11. *Cyber cafe* means any facility from where access to the Internet is offered by any person in the ordinary course of business to the members of the public. [Sec. 2(1)(na) of the IT Act 2000]
12. *Cyber security* means protecting information, equipment, devices, computer, computer resource, communication device and information stored therein from unauthorized access, use, disclosure, disruption, modification, or destruction. [Sec. 2(1)(nb) of the IT Act 2000]
13. *Data* means a representation of information, knowledge, facts, concepts, or instructions which are being prepared or have been prepared in a formalized manner, and is intended to be processed, is being processed, or has been processed in a computer system or computer network and may be in any form (including computer printouts, magnetic or optical storage media, punched cards, punched tapes) or stored internally in the memory of the computer. [Sec. 2(1)(o) of the IT Act 2000]
14. *Digital signature* means authentication of any electronic record by a subscriber by means of an electronic method or procedure in accordance with the provisions of Section 3. [Sec. 2(1)(p) of the IT Act 2000]
15. *Electronic form* with reference to information means any information generated, sent, received, or stored in media, magnetic, optical, computer memory, micro film, computer generated micro fiche, or similar device. [Sec. 2(1)(r) of the IT Act 2000]
16. *Electronic record* means data, record or data generated, image or sound stored, received or sent in an electronic form or micro film or computer generated micro fiche. (Sec. 2(1)(t) of the IT Act 2000]
17. *Electronic signature* means authentication of any electronic record by a subscriber by means of the electronic technique specified in the Second Schedule and includes digital signature. [Sec. 2(1)(ta) of the IT Act 2000]
18. *Function* in relation to a computer includes logic, control, arithmetical process, deletion, storage, and retrieval and communication or telecommunication from or within a computer. [Sec. 2(1)(u) of the IT Act 2000]
19. *Information* includes data, message, text, images, sound, voice, codes, computer programmes, software, and databases or micro film or computer generated micro fiche. [Sec. 2(1)(v) of the IT Act 2000]

20. *Intermediary* with respect to any particular electronic records, means any person who on behalf of another person receives, stores, or transmits that record or provides any service with respect to that record and includes telecom service providers, network service providers, Internet service providers, web hosting service providers, search engines, online payment sites, online-auction sites, online marketplaces and cyber cafes. [Sec. 2(1)(w) of the IT Act 2000]
21. *Key pair* in an asymmetric crypto system means a private key and its mathematically related public key, which are so related that the public key can verify a digital signature created by the private key. [Sec. 2(1)(x) of the IT Act 2000]
22. *Originator* means a person who sends, generates, stores or transmits any electronic message or causes any electronic message to be sent, generated, stored or transmitted to any other person but does not include an intermediary. [Sec. 2(1)(za) of the IT Act 2000]
23. *Private key* means the key of a key pair used to create a digital signature. [Sec. 2(1)(zc) of the IT Act 2000]
24. *Public key* means the key of a key pair used to verify a digital signature and listed in the digital signature certificate. [Sec. 2(1)(zd) of the IT Act 2000]
25. *Secure system* means computer hardware, software and procedure that (a) are reasonably secure from unauthorized access and misuse; (b) provide a reasonable level of reliability and correct operation; (c) are reasonably suited to performing the intended functions; and (d) adhere to generally accepted security procedures. [Sec. 2(1)(ze) of the IT Act 2000]
26. *Subscriber* means a person in whose name the electronic signature certificate is issued. [Sec. 2(1)(zg) of the IT Act 2000]

POINTS TO REMEMBER

- Cyber law is the law governing electronic communication devices, all aspects of transactions, and activities on and involving the Internet, World Wide Web, and cyberspace.
- As the users of cyberspace grow increasingly diverse, and the range of online interaction expands, there has also been steady increase in the cybercrimes being perpetrated—breach of online contracts, operation of online torts, crimes, etc. Hence arose the need to adopt a strict law by the cyber space authority to regulate criminal activities relating to cyberspace and to provide better administration of justice to the victims of cybercrimes.
- Lack of awareness among practitioners, constraints with jurisdiction, conflicts during collecting evidence, and lack of provisions for handling certain crimes are some of the legal issues associated with cyber laws.
- The objective of cyber security is to establish rules and measures to counter attacks and crimes carried out on the Internet.
- The risk reduction strategies of cyber law include promoting cyber security research and development, gaining intelligence about threats, improving firewalls, developing and using protocols and algorithms, introducing authentication aspects, focusing on cloud and mobile security, and cyber forensics.
- Existing initiatives for cyber security include the National Informatics Centre (NIC), Indian Computer Emergency Response Team (ICERT), Indo–US Cyber Security Forum, and alliances of cyber security with other nations such as US, China, and North Korea.

KEY TERMS

Access It means gaining entry into, instructing, or communicating with the logical, arithmetical, or memory function resources of a computer, computer system, or computer network.

Addressee It means a person who is intended by the originator to receive the electronic record but does not include any intermediary.

Affixing electronic signature It means adoption of any methodology or procedure by a person for the purpose of authenticating an electronic record by means of an electronic signature.

Asymmetric crypto system It means a system of a

secured key pair consisting of a private key for creating a digital signature and a public key to verify the digital signature.

Certifying authority It means a person who has been granted a licence to issue an electronic signature certificate under Section 24.

Communication device It means cell phones, personal digital assistance (sic), or combination of both or any other device used to communicate, send, or transmit any text, video, audio, or image.

Computer network It means the interconnection of one or more computers or computer systems or communication devices through (i) the use of satellite, microwave, terrestrial line, wire, wireless, or other communication media; and (ii) terminals or a complex consisting of two or more interconnected computers or communication devices, whether or not the interconnection is continuously maintained.

Computer resource It refers to computer, communication device, computer system, computer network, data, computer database, or software.

Cyber cafe This means any facility from where access to the Internet is offered by any person in the ordinary course of business to the members of the public.

Cyber law Cyber law is a branch of law used to describe the legal issues related to the use of communication technology or inter-networked information technology, particularly 'cyberspace', that is, the Internet.

Cyber security It refers to protecting information, equipment, devices, computer, computer resource, communication device and information stored therein from unauthorized access, use, disclosure, disruption, modification, or destruction.

Digital signature It refers to authentication of any electronic record by a subscriber by means of an electronic method or procedure in accordance with the provisions of Section 3.

Electronic form This refers to any information generated, sent, received, or stored in media, magnetic, optical, computer memory, micro film, computer generated micro fiche, or a similar device.

Electronic record This refers to data, record, or data generated, image or sound stored, received or sent in an electronic form or micro film or computer generated micro fiche.

Electronic signature It refers to authentication of any electronic record by a subscriber by means of the electronic technique specified in the Second Schedule and includes digital signature.

Intermediary This refers to any person who on behalf of another person receives, stores, or transmits that record or provides any service with respect to that record and includes telecom service providers, network service providers, Internet service providers, web hosting service providers, search engines, online payment sites, online-auction sites, online marketplaces and cyber cafes.

Secure system It refers to computer hardware, software, and procedure that (a) are reasonably secure from unauthorized access and misuse; (b) provide a reasonable level of reliability and correct operation; (c) are reasonably suited to performing the intended functions; and (d) adhere to generally accepted security procedures.

MULTIPLE-CHOICE QUESTIONS

1. The objective of cyber security is to establish measures against _____
   (a) criminals
   (b) crimes
   (c) attacks
   (d) (b) and (c)

2. The risk associated with cyber law can be minimized by _____
   (a) introduction of authentication aspects
   (b) gathering information about criminals
   (c) focusing on cloud computing
   (d) all of these

3. How many phases are present to achieve national cyber security?
   (a) 4
   (b) 6
   (c) 5
   (d) 7

4. The aim of the National Cyber Security Policy is to promote _____ in cyber security.
   (a) information system
   (b) R&D
   (c) both (a) and (b)
   (d) neither (a) nor (b)

5. _____ means the key of a key pair used to verify a digital signature and listed in the digital signature certificate.
   (a) Public key
   (b) Private key
   (c) Key pair
   (d) Hash key

6. The virtual world encompassing the Internet is called _____.
   (a) virtual space
   (b) computer space
   (c) cyberspace
   (d) system space
7. Cyber law governs all aspects of transactions and activities involving the _____
   (a) Internet
   (b) World Wide Web
   (c) cyberspace
   (d) all of these
8. Originally, the Internet was developed for _____
   (a) resource sharing
   (b) information sharing
   (c) resource and information sharing
   (d) none of these
9. The techno-savvy environment has made the Internet highly transaction with _____.
   (a) e-business and e-commerce
   (b) e-commerce and e-governance
   (c) e-governance and e-procurement
   (d) e-business, e-commerce, e-governance, and e-procurement
10. The conventional methods of transactions in business has been replaced with _____
    (a) Class A digital signatures
    (b) e-Mudhra
    (c) Internet banking
    (d) digital signatures and e-contracts

REVIEW QUESTIONS

1. Define cyber law.
2. Define cyber security.
3. What are the needs for cyber law?
4. Mention the advantages of cyber security.
5. Discuss the legal issues associated with cyber laws.
6. Explain the terms and terminologies associated with cyber laws.
7. What are the strategies involved in cyber security?
8. Explain the risk reduction strategies of cyber laws.
9. Explain briefly the initiatives to promote cyber security.

APPLICATION EXERCISES

1. List any five cybercrimes you have learnt in earlier chapters and justify the need for cyber laws to handle them.
2. For the cybercrimes listed in Q. 12.1, mention some of the legal issues, if applicable.
3. For the set of cybercrimes listed in Q. 12.1, mention some of the security measures that can be adopted to minimize the risks with cyber laws.

BIBLIOGRAPHY

1. Rajkumar Adukia (12 March 2014), *Overview of Cyber Laws in India*, available at: www.lawyersclubindia.com/ articles/ Overview-of-Cyber-Laws-in-India--6039.asp (Accessed 23 November 2017)
2. Data Security Council of India, A NASSCOM Initiative (2011), *Cybercrime Investigation Manual*, available at: https://uppolice.gov.in/.../28_5_2014_17_4_36_Cyber_Crime_Investigation_Manual (Accessed 23 November 2017)
3. Rohas Nagpal, *IPR & Cyberspace – Indian Perspective, Courseware of Asian School of Cyber Laws*, available at: http:// osou.ac.in/eresources/ introduction-to-indian-cyber-law.pdf (Accessed 23 November 2017)
4. Upcounsel (2011), *Cyber Law: Everything you Need to Know*, available at: https://www.upcounsel.com/cyber-law (Accessed 23 November 2017)

Answers to Multiple-choice Questions

| 1. (d) | 2. (a) | 3. (c) | 4. (b) | 5. (a) |
| 6. (c) | 7. (d) | 8. (c) | 9. (d) | 10. (d) |

Cyber Laws in India and Case Studies

13

Learning Objectives

This chapter provides an overview of the cyber laws in India and some case laws. The objective of this chapter is to provide a deeper insight into the laws that are part of the Information Technology Act 2000 (ITA 2000) and Indian Penal Code (IPC) to handle cybercrimes. The chapter presents the IT Act 2000 and the salient features of the Information Technology Amendment Act 2008 (ITAA 2008). It presents the laws categorized as crime against individual, property, and nation with case laws as applicable. Besides this, the cyber laws associated with ensuring cyber security are presented. The other laws related to cyber security and cybercrime investigation are discussed. The amendments carried out to the Indian Evidence Act 1872 (IEA) and the Banker's Book Evidence Act 1891 are furnished in this chapter. The reader will be familiar with the following after studying the chapter:

- The relationship among cybercrime, cyber security, and cyber laws
- IT Act 2000
- IT Amendment Act 2008
- Cyber laws for cybercrime against individual, property, and nation
- Cyber laws for cyber security and other applicable laws
- Amendments made to IEA and the Banker's Book Evidence Act 1891

13.1 CYBER LAWS, CYBERCRIME, AND CYBER SECURITY

The domains associated with cyber laws are cybercrime and cyber security. Cyber security protects individuals, organizations, and their businesses from cybercrime. Every country in the world looks for ways to promote cyber security and to prevent cybercrime. In this regard, the Government of India enacted the Information Technology Act in 2000 with the goal to ensure safe browsing on the Internet and other related activities.

13.2 CYBER LAWS IN INDIA

Cyber laws provide legal protection to anyone using the Internet for personal, business, or other activities. Cyber laws are otherwise called laws of the Internet. The Indian laws and Acts that address the various aspects of cybercrimes are as follows:
1. Information Technology Act 2000
2. Information Technology Amendment Act 2008
3. Indian Penal Code 1860
4. The Indian Evidence Act 1872
5. The Indian Telegraph Act 1885
6. Bankers' Book of Evidence Act 1891

The following rules and regulations are also covered under cyber laws:
1. Information Technology (Certifying Authorities) Rules 2000
2. Information Technology (Security Procedure) Rules 2004

3. Information Technology (Certifying Authority) Regulations 2001

However, the primary source of cyber law in India is the Information Technology Act 2000 (IT Act).

13.3 INFORMATION TECHNOLOGY ACT 2000

The Information Technology Act 2000 (IT Act) came into force on 17 October 2000. The main purpose of the Act is to provide legal recognition to electronic commerce and to facilitate filing of electronic records with the government. The Information Technology Act 2000 regulates the use of computers, computer systems, and computer networks, as well as data and information in the electronic format. This legislation has dealt with various aspects pertaining to electronic authentication, digital (electronic) signatures, cybercrimes, and liability of network service providers.

The Preamble to the Act states that it aims at providing legal recognition for transactions carried out by means of electronic data interchange and other means of electronic communication, commonly referred to as *electronic commerce*, which involves the use of alternatives to paper-based methods of communication and storage of information and aims at facilitating electronic filing of documents with government agencies. The IT Act of 2000 was developed to promote the IT industry, regulate ecommerce, facilitate e-governance, prevent cybercrimes, and to foster security practices. The IT Act 2000 consists of 90 Sections spread over 13 Chapters and has two Schedules.

ITA 2000 was amended in December 2008 as the IT (Amendment) Act 2008 (ITAA 2008), by the Information Technology Amendment Bill which omitted some of the Sections and Schedules and was passed in the Lok Sabha on 22 December 2008, and in the Rajya Sabha on 23 December 2008. It received the assent of the President on 5 February 2009 and was notified with effect from 27 October 2009. ITAA 2008 has created a strong data protection regime by mandating reasonable security practices to protect sensitive personal information and several provisions for handling cybercrimes such as identity theft and cyber terrorism. The Indian Penal Code and the Indian Evidence Act were also amended to include cybercrimes and digital evidences covered by ITA 2000. Table 13.1 provides a glimpse of the IT Act 2000.

Table 13.1 Glimpse of ITA 2000

	IT Act 2000
Section 43	Penalty and compensation for damage to computer, computer system, etc.
Section 66	Computer-related offences
Section 66C	Punishment for identity theft
Section 66D	Punishment for cheating by personation by using computer resource
Section 66E	Punishment for violation of privacy
Section 66F	Cyber terrorism relevant case
Section 67	Punishment for publishing or transmitting obscene material in electronic form (Pornography)
Section 67B	Punishment for publishing or transmitting of material depicting children in sexually explicit act, etc., in electronic form

13.3.1 Scheme of IT Act 2000

The Scheme of IT Act 2000 is as follows:

Chapter I	–	Preliminary
Chapter II	–	Digital signature and electronic signature (Sections 3 and 3A)
Chapter III	–	Electronic governance (Sections 4–10A)
Chapter IV	–	Attribution, acknowledgement, and dispatch of electronic records (Sections 11–13)
Chapter V	–	Secure electronic records and secure electronic signatures (Sections 14–16)

Chapter VI	–	Regulation of certifying authorities (Sections 17–34)
Chapter VII	–	Electronic signature certificates (Sections 35–39)
Chapter VIII	–	Duties of subscribers (Sections 40–42)
Chapter IX	–	Penalties, compensation, and adjudication (Sections 43–47)
Chapter X	–	The cyber appellate tribunal (Sections 48–64)
Chapter XI	–	Offences (Sections 65–78)
Chapter XII	–	Intermediaries not to be liable in certain cases (Section 79)
Chapter XIIA	–	Examiner of electronic evidence (Section 79A)
Chapter XIII	–	Miscellaneous (Sections 80 to 90)

Since Chapters IX and XI–XIII alone are related to cybercrime and cyberspace, they are discussed in detail in the following sections.

13.3.2 Salient Features of the Information Technology (Amendment) Act 2008

Some of the salient features of the Information Technology (Amendment) Act 2008 are summarized here:

1. The Act applies to any offence or contravention committed outside India by any person, irrespective of his/her nationality, if the Act or conduct constituting the offence or contravention involves a computer, computer system, or computer network located in India (Section 75).
2. Certain documents and transactions such as negotiable instruments (excepting a cheque), power of attorney, trust(s), will, and contract for sale of immoveable property are excluded from the purview of this Act.
3. Statutory requirements have been prescribed for retention of electronic records in a format that captures the information accurately and which facilitates tracking back. Intermediaries are liable for penal provisions for non-compliance under 67C of the ITAA 2008.
4. Controller of certifying authorities is responsible for the issuing of licences to certifying authorities who in turn are licensed to issue digital signatures (Section 18).
5. Dishonest and fraudulent contraventions of Acts defined under Section 43 of the ITAA2008 are offences under Section 66 of the ITAA 2008. If the Acts are simply contraventions, then they will be dealt with by the adjudicating officers designated by the government under Section 46 of the IT Act. An adjudicating officer can adjudicate and award a compensation of up to ₹5 crores.
6. Officers of the rank of police inspectors and above are empowered to investigate offences under ITAA 2008.
7. As per the Information Technology (Procedure and Safeguards for Interception, Monitoring and Decryption of Information) Rules 2009, the Secretary in the Ministry of Home Affairs in the Government of India and the Secretary of the Home Department in the respective state/union territory governments are authorized to order the interception, monitoring, or decryption of information from any computer resource(s).
8. As per the Information Technology (Procedure and Safeguards for Blocking for Access of Information by Public) Rules 2009, the central government can designate an officer of the central government (not below the rank of a joint secretary) to issue directions for blocking public access of any information in computer resources (Section 69A of the ITAA 2008).
9. Computer offences as per the ITAA 2008 include the following:
   (a) Computer-related offences (include source code tampering, unauthorized access, disruption, damage, etc., of computer resources) defined under Section 65, 66, and 66 B to D.
   (b) Obscenity and related offences as defined in Sections 66E, 67, 67A, and 67B
   (c) Threat to unity and integrity of India (cyber terrorism), Section 66F
   (d) Power to issue directions by competent authorities to block access, monitor traffic, etc., under Sec 67C, 69, 69A, 70, and 70B.

(e) CERT-In designated as the national nodal agency for critical information infrastructure protection—all the offences with up to three years punishment has been made bailable and, as such only sections 66F, 67A, 67B, 69, 69A, and 70 of the ITAA are non-bailable.

13.4 CYBERCRIMES AND CYBER LAWS

Cybercrimes are categorized into three as discussed in Chapter 2 and are as follows:
1. *Crime against individual* occurs online and affects people. It includes cyber harassment and stalking, distribution of child pornography, various types of spoofing, credit card fraud, human trafficking, identity theft, etc.
2. *Crime against property* occurs online, against property such as a computer or a server. It includes DDoS attacks, hacking, virus transmission, copyright infringement, and IPR violations, among others.
3. *Crime against nation* is an attack on the nation's sovereignty and an act of war. It includes hacking, accessing confidential information, cyber warfare, cyber terrorism, pirated software, etc.

Laws have been enacted to handle all classes of crimes. Most of the cybercrimes are addressed by the IT Act 2000 and the Indian Penal Code (IPC). Cybercrimes dealt with the IT Act include the following:
1. Sec. 65, Tampering with computer source documents
2. Sec. 66, Hacking computer systems and data alteration
3. Sec. 67, Publishing obscene information
4. Sec. 70, Unauthorized access of protected systems
5. Sec. 72, Breach of confidentiality and privacy
6. Sec. 73, Publishing false digital signature certificates

and that under the IPC include the following:
1. Sending threatening messages by email, Indian Penal Code (IPC) Sec. 503
2. Sending defamatory messages by email, IPC Sec. 499
3. Forgery of electronic records, IPC Sec. 463
4. Bogus websites and cyber fraud, IPC Sec. 420
5. Email spoofing, IPC Sec. 463
6. Web-jacking, IPC Sec. 383
7. Email abuse, IPC Sec. 500

Besides this, there are Special Acts to deal with cybercrimes, which are as follows:
1. Online sale of arms under the Arms Act 1959
2. Online sale of drugs under the Narcotic Drugs and Psychotropic Substances Act 1985

The following sections present the laws in detail with a sample case. Table 13.2 highlights some of the offences and the penalty under the IT Act 2000.

Table 13.2 Offences punishable under IT Act 2000

Section under IT Act 2000	Offence	Penalty
Sec. 43	Damage to computer, computer system, etc.	Compensation not exceeding one crore rupees to the person so affected
Sec. 43A	Body corporate failure to protect data	Compensation not exceeding five crore rupees to the person so affected
Sec. 44(a)	Failure to furnish document, return, or report to the controller or the certifying authority	Penalty not exceeding one lakh and fifty thousand rupees for each such failure

(*Contd*)

Table 13.2 (Contd)

Section under IT Act 2000	Offence	Penalty
Sec. 44(b)	Failure to file any return or furnish any information, books or other documents within the time specified	Penalty not exceeding five thousand rupees for every day during which such failure continues
Sec. 44(c)	Failure to maintain books of account or records	Penalty not exceeding ten thousand rupees for every day during which the failure continues
Sec. 45	Where no penalty has been separately provided	Compensation not exceeding twenty five thousand rupees to the person affected by such contravention or a penalty not exceeding twenty five thousand rupees
Sec. 65	Tampering with computer source documents	Imprisonment up to three years, fine which may extend up to two lakh rupees, or both
Sec. 66	Sending offensive messages through communication service, etc.	Imprisonment for a term which may extend to three years and with fine
Sec. 66B	Retains any stolen computer resource or communication device	Imprisonment for a term which may extend to three years, fine which may extend to rupees one lakh, or both
Sec. 66C	Fraudulent use of electronic signature	Imprisonment for a term which may extend to three years and shall also be liable to fine which may extend to rupees one lakh
Sec. 66D	Cheats by personating by using computer resource	Imprisonment for a term which may extend to three years and shall also be liable to fine which may extend to one lakh rupees
Sec. 66E	Publishing obscene images	Imprisonment which may extend to three years, fine not exceeding two lakh rupees, or with both
Sec. 66F	Cyber terrorism	Imprisonment which may extend to imprisonment for life
Sec. 67	Publishes or transmits unwanted material	Imprisonment for a term which may extend to three years and with fine which may extend to five lakh rupees and in the event of a second or subsequent conviction with imprisonment of either description for a term which may extend to five years and also with fine which may extend to ten lakh rupees
Sec. 67A	Publishes or transmits sexually explicit material	Imprisonment for a term which may extend to five years and with fine which may extend to ten lakh rupees and in the event of the second or subsequent conviction with imprisonment of either description for a term which may extend to seven years and also with fine which may extend to ten lakh rupees
Sec. 67B	Abusing children online	Imprisonment for a term which may extend to five years and with a fine which may extend to ten lakh rupees and in the event of second or subsequent conviction with imprisonment of either description for a term which may extend to seven years and also with fine which may extend to ten lakh rupees
Sec. 67C	Preservation of information by intermediary	Imprisonment for a term which may extend to three years and shall also be liable to fine

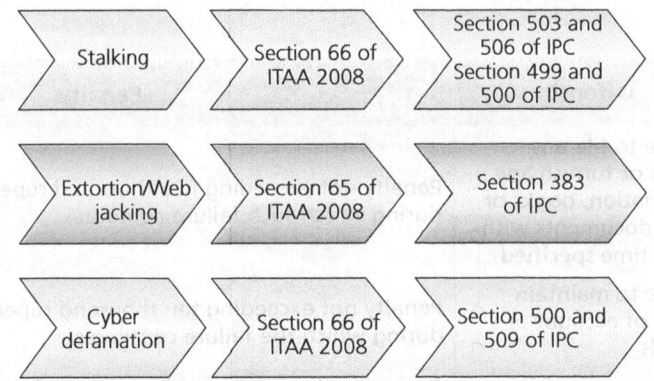

Fig. 13.1 Cybercrimes against individual and applicable cyber laws

13.5 CRIME AGAINST INDIVIDUAL

Some of the cybercrimes against individual are shown in Fig. 13.1, with applicable laws under ITAA 2008 and IPC.

13.5.1 Cyber Defamation

Cyber defamation (Case 1) refers to publishing offensive content against an individual with the intention to spoil the reputation and modesty of a woman with the use of electronic communication devices (ECDs) and the Internet. It is punishable under Sections 500 and 509 of the IPC (Boxes 13.1–13.3).

Box 13.1 Section 509 IPC

Word, gesture, or act intended to insult the modesty of a woman

Whoever intending to insult the modesty of any woman, utters any word, makes any sound or gesture, or exhibits any object, intending that such word or sound shall be heard, of that such gesture or object shall be seen, by such woman, or intrudes upon the privacy of such woman, shall be punished with simple imprisonment for a term which may extend to one year, or with fine, or with both.

Box 13.2 Section 500 IPC

Punishment for defamation

Whoever defames another shall be punished with simple imprisonment for a term which may extend to two years, or with fine, or with both.

Case 1: Cyber Defamation

Abhishek, a teenaged student, was arrested by the Thane police in India following a girl's complaint about tarnishing her image on social networking site Orkut.

Abhishek had allegedly created a fake account in the name of the girl with her mobile number posted on the profile. The profile had been set in such a way that it drew lewd comments from many who visited her profile.

The Thane Cyber Cell tracked down Abhishek from the false email ID that he had created to open the account.

Box 13.3 Section 66 IT Act

Section 66 of the Information Technology Act was struck down by the Supreme Court of India as unconstitutional as it hit the root of liberty and freedom of expression guaranteed under Article 19(1)(a) of the Constitution of India, the two cardinal pillars of democracy. This section had been widely misused by the police to arrest innocent persons for posting critical comments about social and political leaders on social networking sites.

13.5.2 Cyber Stalking

Cyber stalking (Case 2) involves repeatedly following a person, harassing over phone calls and with written messages, and threatening either the individual or his/her family using the Internet, email, and other ECDs. It is punishable under Sections 503 and 506 of the IPC if the messages are of threatening nature. If defamatory, it is punishable under Sections 499 and 500 of IPC (Box 13.4–13.6).

Box 13.4 Section 499 IPC

Defamation

Whoever by words either spoken or intended to be read, or by signs or by visible representations, makes or publishes any imputation concerning any person intending to harm, or knowing or having reason to believe that such imputation will harm the reputation of such person, is said, except in the cases hereinafter expected, to defame that person.

Box 13.5 Section 503 IPC

Criminal intimidation

Whoever threatens another with any injury to his person, reputation or property, or to the person or reputation of anyone in whom that person is interested with, intent to cause alarm to that person, or to cause that person to do any act which he is not legally bound to do, or to omit to do any act which that person is legally entitled to do, as the means of avoiding the execution of such threat, commits criminal intimidation.

Explanation-A threat to injure the reputation of any deceased person in whom the person threatened is interested, is within this section.

Box 13.6 Section 506 IPC

Punishment for criminal intimidation

Whoever commits the offence of criminal intimidation shall be punished with imprisonment of either description for a term which may extend to two years, or with fine, or with both If threat be to cause death or grievous hurt, etc.—And if the threat be to cause death or grievous hurt, or to cause the destruction of any property by fire, or to cause an offence punishable with death or [imprisonment for life], or with imprisonment for a term which may extend to seven years, or to impute, unchastity to a woman, shall be punished with imprisonment of either description for a term which may extend to seven years, or with fine, or with both.

> **Box 13.7 Section 383 IPC**
>
> **Extortion**
>
> Whoever intentionally puts any person in fear of any injury to that person, or to any other, and thereby dishonestly induces the person so put in fear to deliver to any property or valuable security, or anything signed or sealed which may be converted into a valuable security, commits 'extortion'.

> **Case 2: Cyber Stalking**
>
> In the first successful prosecution under the California (USA) cyber stalking law, prosecutors obtained a guilty plea from a 50-year-old former security guard who used the Internet to solicit rape of a woman who rejected his romantic advances.
>
> He terrorized the 28-year-old victim by impersonating her in various Internet chat rooms and online bulletin boards, where he posted, along with her telephone number and address, messages that she fantasized about being raped. On at least six occasions, sometimes in the middle of the night, men knocked at the woman's door saying that they wanted to rape her.

13.5.3 Web Jacking

This sort of crime involves the perpetrators forcefully taking control of a website by cracking the administrative privileges so that the actual owner of the website has no more control over it (Case 3). It is punishable under Section 65 of the ITAA 2008 and Section 383 of the IPC (Boxes 13.7 and 13.8).

> **Box 13.8 Section 66E ITAA 2008**
>
> **Punishment for violation of privacy**
>
> Whoever intentionally or knowingly captures, publishes, or transmits the image of a private area of any person without his or her consent, under circumstances violating the privacy of that person, shall be punished with imprisonment which may extend to three years or with fine not exceeding two lakh rupees, or with both.
>
> Explanation — For the purposes of this Section —
>
> (a) "Transmit" means to electronically send a visual image with the intent that it be viewed by a person or persons.
> (b) "Capture," with respect to an image, means to videotape, photograph, film, or record by any means.
> (c) "Private area" means the naked or undergarment clad genitals, pubic area, buttocks, or female breast.
> (d) "Publishes" means reproduction in the printed or electronic form and making it available for public.
> (e) "Under circumstances violating privacy" means circumstances in which a person can have a reasonable expectation that
>   (i) He or she disrobes in privacy, without being concerned that an image of his private area was being captured or
>   (ii) Any part of his or her private area would not be visible to the public, regardless of whether that person is in public or private place.

Case 3: Web Jacking

In an incident reported in the USA, the owner of a hobby website for children received an email informing her that a group of hackers had gained control over her website. They demanded a ransom of 1 million dollars from her. The owner, a school teacher, did not take the threat seriously. She felt that it was just a scare tactic and ignored the email.

The hackers web jacked her website and subsequently altered a portion of the website that was titled 'How to have fun with goldfish'. In all the places, 'goldfish' was replaced with the word 'piranhas'. Piranhas are tiny but extremely dangerous flesh-eating fish. Many children who visited the popular website, unfortunately followed the instructions and tried to play with piranhas bought from pet shops. They were seriously injured.

13.5.4 Violation of Privacy

Privacy violation is punishable under Section 66E of ITAA 2008.

13.6 CRIME AGAINST PROPERTY

Some of the cybercrimes against property are shown in Fig. 13.2 along with applicable laws.

13.6.1 Theft of Data, Viral Attack, Hacking, Denial of Service Attack, and Cyber Bullying

Section 43 of ITAA 2008 (Box 13.9) addresses the civil offence of theft of data and to enforce civil liability. Criminality in the offence of data theft is dealt with in Sections 65 and 66 of ITAA 2008. Writing a virus program or spreading a virus or a worm mail, a bot, a Trojan, or any other malware in a computer network or causing a denial of service attack (DoS) in a server are all punishable under Section 43 and attract civil liability by way of compensation (Cases 4 and 5).

Case 4: Viral Attack

Berhampur City Hospital in Odisha's Ganjam district fell prey to WannaCry virus and ransomware attack in May 2017. The systems in the hospital were managed by the National Informatics Centre (NIC). The viral attack had encrypted the data of e-Aushadhi portal and Hospital Information Management System (HIMS). The alleged hacker had demanded the payment of a ransom of $300 to restore the old files of the system and a warning that the data in the system would be destroyed if there had been any delay in payment.

Case 5: Distributed Denial-of-Service Attack

Internet service providers (ISPs) in Mumbai were targeted by a distributed denial of service (DDoS) attack. It is the world's largest attack against ISPs and also the largest cyber-attack against India. As a result, legitimate internet access by all the customers of the ISP was prevented and the users experienced slowing down of Internet speed. The perpetrators of the attack were suspected to have originated from Eastern Europe and China.

Investigation of DDoS attacks necessitates access to several numbers of routers to locate the perpetrators. Besides this, the perpetrator might have initiated the attack from botnet that is not necessarily within India. Indian laws namely ITA 2000 and IPC 1860 are sufficient to deal with this attack if it had been initiated from a botnet or a person within India. However laws enabling investigation overseas and extradition of the criminal from abroad, which are usually in the form of treaties between countries are required otherwise.

Thus, the corporate responsibility for data protection is greatly emphasized by Section 43A, whereby corporates are under an obligation to ensure adoption of reasonable security practices. The central government vide its notification dated 11 April 2011 has categorized password, details of bank accounts or card details, medical records, etc., as sensitive personal data and has created awareness about data privacy and the need for data protection and has insisted on the role of the top management and the Information Security Department in organisations in ensuring data protection, especially while handling the customers' and other third party data.

The following are punishable under Section 43 and Section 66 of ITAA 2008 (Boxes 13.9–13.11):
1. Data theft that is carried out with a criminal intention
2. Hacking, virus, worms, and logic bombs
3. Cyber bullying which is considered to be the unauthorized use or access to facilities and is treated like hacking. Cyber bullying can be defined as any communication posted or sent online by a minor, by instant messenger, email, social networking site, website, diary site, online profile, interactive game, handheld device, cell phone, or other interactive device that is intended to frighten, embarrass, harass, or otherwise target another minor.

> **Box 13.9 Section 43 ITAA 2008**
>
> **Penalty and compensation for damage to computer, computer system, etc.**
>
> If any person without the permission of the owner or any other person who is in charge of a computer, computer system, or computer network
>
> (a) Accesses or secures access to such computer, computer system, or computer network or computer resource
> (b) Downloads copies or extracts any data, computer database, or information from such computer, computer system, or computer network, including information or data held or stored in any removable storage medium
> (c) Introduces or causes to be introduced any computer contaminant or computer virus into any computer, computer system, or computer network
> (d) Damages or causes to be damaged any computer, computer system or computer network, data, computer database, or any other programs residing in such computer, computer system, or computer network
> (e) Disrupts or causes disruption of any computer, computer system, or computer network
> (f) Denies or causes the denial of access to any person authorized to access any computer, computer system, or computer network by any means
> (g) Provides any assistance to any person to facilitate access to a computer, computer system, or computer network in contravention of the provisions of this act, rules, or regulations made thereunder
> (h) Charges the services availed by a person to the account of another person by tampering with or manipulating any computer, computer system, or computer network
> (i) Destroys, deletes, or alters any information residing in a computer resource or diminishes its value or utility or affects it injuriously by any means
> (j) Steals, conceals, destroys or alters or causes any person to steal, conceal, destroy or alter any computer source code used for a computer resource with an intention to cause damage, shall be liable to pay damages by way of compensation to the person so affected.

> **Box 13.10 Section 43A ITA 2000**
>
> **Compensation for failure to protect data**
>
> Where a body corporate, possessing, dealing, or handling any sensitive personal data or information in a computer resource which it owns, controls, or operates, is negligent in implementing and maintaining reasonable security practices and procedures, and thereby causes wrongful loss or wrongful gain to any person, such body corporate shall be liable to pay damages by way of compensation, to the person so affected.

Box 13.11 Section 66 ITAA 2008

Computer-related offences

If any person, dishonestly, or fraudulently, does any act referred to in Section 43, shall be punishable with imprisonment for a term which may extend to three years or with fine which may extend to five lakh rupees, or with both. Explanation: For the purpose of this section: (a) The word "dishonestly" shall have the meaning assigned to it in Section 24 of the Indian Penal Code; (b) The word "fraudulently" shall have the meaning assigned to it in Section 25 of the Indian Penal Code.

Box 13.12 Section 65 ITA

Tampering with computer source documents

Whoever knowingly or intentionally conceals, destroys, or alters or intentionally or knowingly causes another to conceal, destroy, or alter any computer source code used for a computer, computer program, computer system, or computer network, when the computer source code is required to be kept or maintained by law for the time being in force, shall be punishable with imprisonment up to three years or with fine which may extend up to two lakh rupees, or with both.

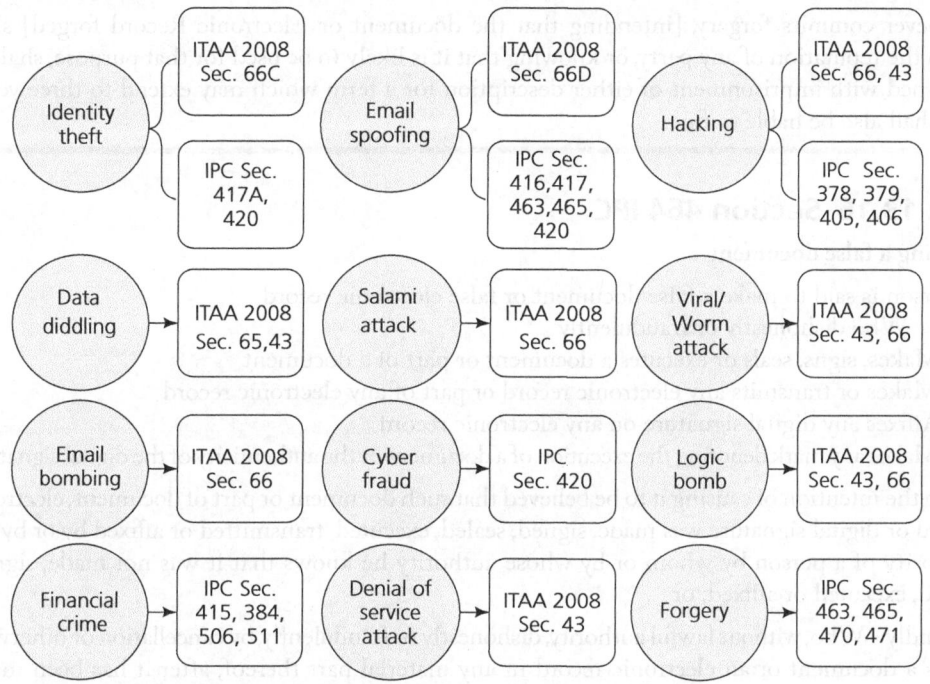

Fig. 13.2 Cybercrimes (shown in round bubbles) against property and applicable cyber laws

13.6.2 Forgery

Fabrication of an electronic record or committing forgery (Case 6) attracts punishment under Section 65 of ITAA 2008 (Box 13.13). Forgery may include counterfeit currency notes, postage and revenue stamps, stamp papers, mark sheets and other certificates that are prepared by offenders employing sophisticated computers, printers, and scanners. Further, forgery is punishable under IPC Sections 463, 464, 468, and 469 (Box 13.14–13.17).

Box 13.13 Section 463 IPC

Forgery

Whoever makes any false documents or electronic record part of a document or electronic record with, intent to cause damage or injury, to the public or to any person, or to support any claim or title, or to cause any person to part with property, or to enter into any express or implied contract, or with intent to commit fraud or that fraud may be committed, commits forgery.

Box 13.14 Section 468 IPC

Forgery for purpose of cheating

Whoever commits forgery, intending that the [document or Electronic Record forged] shall be used for the purpose of cheating, shall be punished with imprisonment of either description for a term which may extend to seven years, and shall also be liable to fine.

Box 13.15 Section 469 IPC

Forgery for purpose of harming reputation

Whoever commits forgery, [intending that the document or Electronic Record forged] shall harm the reputation of any party, or knowing that it is likely to be used for that purpose, shall be punished with imprisonment of either description for a term which may extend to three years, and shall also be liable to fine.

Box 13.16 Section 464 IPC

Making a false document

A person is said to make a false document or false electronic record
First - Who dishonestly or fraudulently
(a) Makes, signs, seals or executes a document or part of a document
(b) Makes or transmits any electronic record or part of any electronic record
(c) Affixes any digital signature on any electronic record
(d) Makes any mark denoting the execution of a document or the authenticity of the digital signature,

With the intention of causing it to be believed that such document or part of document, electronic record or digital signature was made, signed, sealed, executed, transmitted or affixed by or by the authority of a person by whom or by whose authority he knows that it was not made, signed, sealed, executed or affixed; or

Secondly - Who, without lawful authority, dishonestly or fraudulently, by cancellation or otherwise, alters a document or an electronic record in any material part thereof, after it has been made, executed or affixed with digital signature either by himself or by any other person, whether such person be living or dead at the time of such alteration; or

Thirdly- Who dishonestly or fraudulently causes any person to sign, seal, execute or alter a document or an electronic record or to affix his digital signature on any electronic record knowing that such person by reason of unsoundness of mind or intoxication cannot, or that by reason of deception practised upon him, he does not know the contents of the document or electronic record or the nature of the alterations.

> **Case 6: Forgery**
>
> In October 1995, the Economic Offences Wing of Crime Branch, Mumbai (India), seized over 22,000 counterfeit share certificates of eight reputed companies worth ₹34.47 crores. These were allegedly prepared using desktop publishing systems.

13.6.3 Data Diddling

Data diddling refers to attempts made towards illegal and unauthorized alteration of data. It is punishable under Sections 43 and 65 of ITAA 2008 (Case 7).

> **Case 7: Data Diddling**
>
> In a renowned company, the accounts executive was alleged of changing the employees' time sheet information before that data was entered into the payroll application. This resulted in paying more than what the employee had actually worked for and a huge revenue loss for the company.
>
> In another incident, one of the employees of a company allegedly altered the presentation of his colleague with junk data with the intention to harm him as he was jealous about him. This led to situation where the colleague was expelled from the company.

13.6.4 Email Bombing

Email bombing refers to sending an enormous amount of email messages targeting a server with the intention to crash it (Case 8). It is a variant of DoS attack where a legitimate user is denied the server because of flooding of information to the server by an attacker.

> **Case 8: Email Bombing**
>
> An ex-employee of a company wanted to take revenge against the company and the former employer. He decided to launch an email bombing attack, which is a kind of denial of service attack using an application. Accordingly, he sent millions of email message to the ex-employer's email account, which led to the crash of company's email server.

13.6.5 Possession of Stolen Electronic Communication Devices

Data or an ECD if in the possession of a person other than the owner is punishable under Section 66B of ITAA 2008 and Section 411 of IPC.

> **Box 13.17 Section 66B ITAA 2008**
>
> Punishment for dishonestly receiving stolen computer resource or communication device
>
> Whoever dishonestly receives or retains any stolen computer resource or communication device knowing or having reason to believe the same to be stolen computer resource or communication device, shall be punished with imprisonment of either description for a term which may extend to three years or with fine which may extend to rupees one lakh, or with both.

> **Box 13.18 Section 66C ITAA 2008**
>
> Punishment for identity theft
>
> Whoever fraudulently or dishonestly make use of the electronic signature, password, or any other unique identification feature of any other person, shall be punished with imprisonment of either description for a term which may extend to three years and shall also be liable to fine which may extend to rupees one lakh.

> **Box 13.19 Section 417 A IPC**
>
> **Punishment for cheating using digital signature of another person**
>
> Whoever cheats by using the digital signature, password or any other unique identification feature of any other person, shall be punished with imprisonment of either description for a term which may extend to three years and shall also be liable to fine.

> **Box 13.20 Section 419 A IPC**
>
> **Punishment for cheating by impersonation using communication network or computer resource**
>
> Whoever, by means of any communication network or computer resource, cheats by personation shall be punished with imprisonment of either description for a term which may extend to five years and shall also liable to fine.

13.6.6 Identity Theft and Password Theft

Identity theft may be any one of the four types: financial identity theft, criminal identity theft, identity cloning, and business/commercial identity theft. The crimes associated with identity theft include illegal immigration, terrorism, espionage, blackmail, attack of payment systems like credit/debit card processing, medical insurance, etc. (Case 9)

Amendments to IPC to introduce two new sections 417A and 419A to cover identity theft were recommended by the expert committee on the Amendments to IT Act 2000. Hence any act punishable under Section 66C of ITAA 2008 (Box 13.18) is punishable under Sections 417A and 419A of IPC (Boxes 13.19 and 13.20).

13.6.7 Financial Crime

Section 66D under ITAA 2008 is used to handle cybercrimes where money is the motive. Thus financial crimes such as cyber cheating, credit/debit card frauds, money laundering, hacking, accounting scams involving bank servers, and computer manipulation which are on the rise due to the Internet and mobile banking and online trading are punishable as per this section. Salami attack is a kind of financial crime and is punishable under Section 66 of ITAA 2008 (Box 13.21).

Any dishonest concealment of facts is a deception by IPC Section 415 (Box 13.22) and so financial crimes such as cyber cheating, credit card frauds, money laundering, hacking, accounting scams, and computer manipulation are punishable under this section (Case 10).

> **Case 9: Identity Theft**
>
> The biggest case of identity theft ever recorded took place in August 2009. Eleven people, including a US secret service informant, were charged in connection with the hacking of nine major retailers and the theft and sale of more than 41 million credit and debit card numbers. This data breach is believed to be the largest hacking and identity theft case ever prosecuted by the US Department of Justice, which announced that the suspects were charged with conspiracy, computer intrusion, fraud, and identity theft. Three of those charged were US citizens, whereas the others were from places such as Estonia, Ukraine, Belarus, and China.

> **Box 13.21 Section 66D ITAA 2008**
>
> **Punishment for cheating by personation by using computer resource**
>
> Whoever by means of any communication device or computer resource cheats by personation, shall be punished with imprisonment of either description for a term which may extend to three years and shall also be liable to fine which may extend to one lakh rupees.

Box 13.22 Section 415 IPC
Cheating

Whoever by deceiving any person, fraudulently or dishonestly induces the person so deceived to deliver any property to any person, or to consent that any person shall retain any property, or intentionally induces the person so deceived to do or omit to do anything which he would not do or omit if he were not so deceived, and which act or omission causes or is likely to cause damage or harm to that person in body, mind, reputation or property, is said to "cheat".

Box 13.23 Section 416, 417, and 419 IPC
Cheating by personation

A person is said to "cheat by personation" if he cheats by pretending to be some other person, or by knowingly substituting one person for another, or representing that he or any other person is a person other than he or such other person really is.

Section 417: Punishment for cheating

Whoever cheats shall be punished with imprisonment of either description for a term which may extend to one year, or with fine or with both.

Section 419: Punishment for cheating by personation

Whoever cheats by personation shall be punished with imprisonment of either description for a term which may extend to three years, or with fine, or with both.

Case 10: Financial Crime

The Hyderabad police in India arrested an unemployed computer operator and his friend, a steward in a prominent five-star hotel, for stealing and misusing credit card numbers belonging to hotel customers. The steward noted down the various details of the credit cards, which were handed by clients of the hotel for paying their bills. Then, he passed all the details to his computer operator friend who used the details to make online purchases on various websites.

Case 11: Email Spoofing

A branch of the erstwhile Global Trust Bank in India experienced a major issue. Numerous customers decided to withdraw all their money and close their accounts.
An investigation revealed that spoofed emails had been sent to many of the bank's customers stating that the bank's financial position was unsound and therefore could close operations at any time.

Box 13.24 Section 420 IPC
Cheating and dishonestly inducing delivery of property

Whoever cheats and thereby dishonestly induces the person deceived any property to any person, or to make, alter or destroy the whole or any part of a valuable security, or anything which is signed or sealed, and which is capable of being converted into a valuable security, shall be punished with imprisonment of either description for a term which may extend to seven years, and shall also be liable to fine.

> **Case 12: Email Fraud**
>
> In 2005, an Indian businessman received an email from the Vice President of a major African bank offering him a lucrative contract in return for a kickback of ₹1 Million.
> The businessman had many telephonic conversations with the sender of the email. He also verified the email address of the Vice President from the website of the bank and subsequently transferred the money to the bank account mentioned in the email. It later turned out that the email was a spoofed one and was actually sent by an Indian based in Nigeria.

13.6.8 Email Spoofing

Email spoofing (Case 11) is punishable under Section 66D of ITAA 2008 as an offence. Since spoofing involves the use of identity of another source to disguise as well as forgery, it is punishable under IPC Sections 416, 417, 463, 465 and 419 (Box 13.23).

13.6.9 Email Fraud

This class of crime covers all fraudulent activities in which email is involved. The Nigerian 419 scam is the best example (Case 12). Email fraud is punishable under IPC Section 420 (Box 13.24).

13.6.10 Copyright Infringement Crimes

Crimes under this category include software piracy, copyright infringement, trademark violations, and theft of computer source code (Case 13). Such crimes are punishable under Sections 63 and 63A of ITAA 2008 and under IPC Section 51 (Boxes 13.25–13.27).

> ### Box 13.25 Section 51 IPC
> **When copyright infringed**
>
> *Copyright in a work shall be deemed to be infringed*
>
> (a) when any person, without a licence granted by the owner of the copyright or the Registrar of Copyrights under this Act or in contravention of the conditions of a licence so granted or of any condition imposed by a competent authority under this Act
>
>   (i) does anything, the exclusive right to do which is by this Act conferred upon the owner of the copyright; or
>
>   (ii) permits for profit any place to be used for the communication of the work to the public where such communication constitutes an infringement of the copyright in the work, unless he was not aware and had no reasonable ground for believing that such communication to the public would be an infringement of copyright; or
>
> (b) when any person
>
>   (i) makes for sale or hire, or sells or lets for hire, or by way of trade displays or offers for sale or hire, or
>
>   (ii) distributes either for the purpose of trade or to such an extent as to affect prejudicially the owner of the copyright, or
>
>   (iii) by way of trade exhibits in public, or
>
>   (iv) imports into India, any infringing copies of the work:
>
> PROVIDED that nothing in sub-clause (iv) shall apply to the import of one copy of any work for the private and domestic use of the importer.
>
> Explanation: For the purposes of this section, the reproduction of a literary, dramatic, musical or artistic work in the form of a cinematograph film shall be deemed to be an "infringed copy".

Box 13.26 Section 63 ITA 2000

Offence of infringement of copyright or other rights conferred by this Act

Any person who knowingly infringes or abets the infringement of

(a) the copyright in a work, or
(b) any other right conferred by this Act, except the right conferred by Section 53A,

shall be punishable with imprisonment for a term which shall not be less than six months but which may extend to three years and with fine which shall not be less than fifty thousand rupees but which may extend to two lakh rupees:

PROVIDED that where the infringement has not been made for gain in the course of trade or business the court may, for adequate and special reasons to be mentioned in the judgment, impose a sentence of imprisonment for a term of less than six months or a fine of less than fifty thousand rupees.

Explanation-Construction of a building or other structure which infringes or which, if completed, would infringe the copyright in some other work shall not be an offence under this section.

Box 13.27 Section 63A ITA 2000

Enhanced penalty on second and subsequent convictions

Whoever having already been convicted of an offence under section 63 is again convicted of any such offence shall be punishable for the second and for every subsequent offence, with imprisonment for a term which shall not be less than one year but which may extend to three years and with fine which shall not be less than one lakh rupees but which may extend to two lakh rupees:

PROVIDED that where the infringement has not been made for gain in the course of trade or business the court may, for adequate and special reasons to be mentioned in the judgement, impose a sentence of imprisonment for a term of less than one year or a fine of less than one lakh rupees:

PROVIDED FURTHER that for the purpose of this Section, no cognisance shall be taken of any conviction made before the commencement of the Copyright (Amendment) Act, 1984 (65 of 1984).

13.6.11 Sale of Illegal Articles on the Internet

This refers to the sale of narcotic drugs which may pose a potential threat to the health and safety of people and patients, weapons, wildlife, and lottery, etc., through the Internet (Case 14). Such sale is facilitated by online payment gateways. Such an act is punishable under The Narcotic Drugs and Psychotropic Substances Act 1985 (NDPS Act). However, if such websites claiming to make sale are operated from outside India, the same can be blocked by the Indian authorities under ITAA 2008. Further, such sort of money making is not permissible as per the Indian foreign exchange laws through legal channels.

Case 13: Copyright Infringement

A software professional from Bangalore (India) was booked for stealing the source code of a product being developed by his employers. He started his own company and allegedly used the source code to launch a new software product.

Case 14: Online Sale of Illegal Articles

In March 2007, the Pune rural police cracked down on an illegal rave party and arrested hundreds of illegal drug users. The social networking site Orkut.com is believed to have been one of the modes of communication for gathering people for the said illegal drug party.

13.7 CRIME AGAINST NATION

Some of the cybercrimes against nation are shown in Fig. 13.3 along with applicable laws.

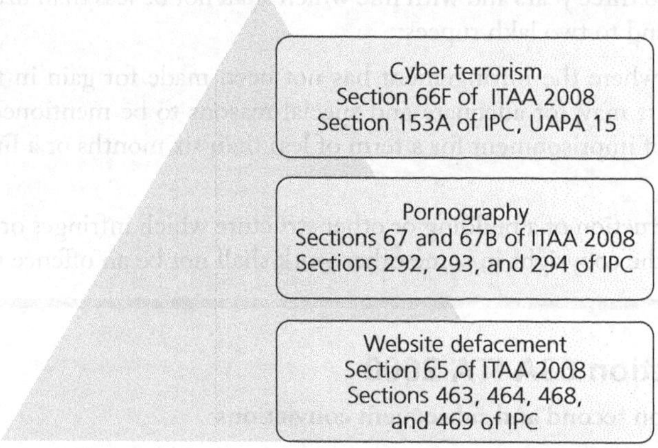

```
          Cyber terrorism
          Section 66F of ITAA 2008
          Section 153A of IPC, UAPA 15

          Pornography
          Sections 67 and 67B of ITAA 2008
          Sections 292, 293, and 294 of IPC

          Website defacement
          Section 65 of ITAA 2008
          Sections 463, 464, 468,
          and 469 of IPC
```

Fig. 13.3 Cybercrimes against nation and applicable cyber laws

13.7.1 Cyber Terrorism

It refers to crimes committed in cyberspace, especially disruptive activities which target social, religious, and political beliefs, thereby spoiling the peace and well-being of the society, a community, or a nation (Case 15). Cyber terrorism is punishable under Section 66F of ITAA 2008 (Box 13.28).

Section 66 focuses on mens rea whereby the destruction, deletion, alteration of data to diminish the utility and value of data so as to cause loss to others is considered an offence. All the Acts under Section 66 are cognizable and non-bailable. Offences under Section 43 which pose civil liability are handled by this section, if done with a criminal intention.

Case 15: Cyber Terrorism

In 2001, in the backdrop of the downturn in US–China relationships, Chinese hackers released the Code Red virus into the wild. This virus infected millions of computers around the world and then used those computers to launch DoS attacks on US websites, prominently the website of the White House.

Case 16: Web Defacement

Mahesh Mhatre and Anand Khare were arrested in 2002 for allegedly defacing the website of the Mumbai Cyber Crime Cell. They had allegedly used password cracking software to crack the FTP password for the police website. They then replaced the home page of the website with pornographic content. The duo was also charged with credit card fraud for using 225 credit card numbers belonging to American citizens.

13.7.2 Website Defacement

This class of crime refers to the replacement of the home page of a website with another page which is defamatory in nature. Government and religious sites witness web defacement by hackers. They may even be defaced for political reasons (Case 16).

> **Box 13.28 Section 66F ITAA 2008**
>
> **Punishment for cyber terrorism**
>
> (1) Whoever:
>
> (A) With an intent to threaten the unity, integrity, security, or sovereignty of India or to strike terror in the people or any section of the people by
>   i. Denying or cause the denial of access to any person authorized to access computer resource
>   ii. Attempting to penetrate or access a computer resource without authorization or exceeding authorized access
>   iii. Introducing or causing to introduce any Computer Contaminant and by means of such conduct causes or is likely to cause death or injuries to persons or damage to or destruction of property or disrupts or knowing that it is likely to cause damage or disruption of supplies or services essential to the life of the community or adversely affect the critical information infrastructure specified under section 70
>
> (B) Knowingly or intentionally penetrates or accesses a computer resource without authorization or exceeding authorized access, and by means of such conduct obtains access to information, data, or computer database that is restricted for reasons of the security of the State or foreign relations; or any restricted information, data, or computer database, with reasons to believe that such information, data, or computer database so obtained may be used to cause or likely to cause injury to the interests of the sovereignty and integrity of India, the security of the State, friendly relations with foreign states, public order, decency or morality, or in relation to contempt of court, defamation or incitement to an offence, or to the advantage of any foreign nation, group of individuals, or otherwise, commits the offence of cyber terrorism.
>
> (2) Whoever commits or conspires to commit cyber terrorism shall be punishable with imprisonment which may extend to imprisonment for life.

> **Box 13.29 Section 67 ITAA 2008**
>
> **Punishment for publishing or transmitting obscene material in electronic form**
>
> Whoever publishes or transmits or causes to be published in the electronic form, any material which is lascivious or appeals to the prurient interest or if its effect is such as to tend to deprave and corrupt persons who are likely, having regard to all relevant circumstances, to read, see, or hear the matter contained or embodied in it, shall be punished on first conviction with imprisonment of either description for a term which may extend to three years and with fine which may extend to five lakh rupees and in the event of a second or subsequent conviction with imprisonment of either description for a term which may extend to five years and also with fine which may extend to 10 lakh rupees.

Web defacement is punishable, just as forgery, and is dealt with in Sections 463, 464, 468, and 469 of the IPC and Section 65 of ITAA 2008 (Box 13.29).

13.7.3 Pornography

This sort of crime makes use of ECDs to develop, publish, print, and distribute obscene content via the Internet. This crime is punishable under Section 67 of ITAA 2008 (Boxes 13.30–13.32) and Sections 292, 293, and 294 of IPC (Boxes 13.33 and 13.34).

Box 13.30 Section 67A ITAA 2008

Punishment for publishing or transmitting of material containing sexually explicit act, etc., in electronic form

Whoever publishes or transmits or causes to be published or transmitted in the electronic form any material which contains sexually explicit act or conduct shall be punished on first conviction with imprisonment of either description for a term which may extend to five years and with fine which may extend to 10 lakh rupees and in the event of second or subsequent conviction with imprisonment of either description for a term which may extend to seven years and also with fine which may extend to 10 lakh rupees.

Box 13.31 Section 67B ITAA 2008

Punishment for publishing or transmitting of material depicting children in sexually explicit act, etc., in electronic form

Whoever:

(a) Publishes or transmits or causes to be published or transmitted material in any electronic form which depicts children engaged in sexually explicit act or conduct

(b) Creates text or digital images, collects, seeks, browses, downloads, advertises, promotes, exchanges, or distributes material in any electronic form depicting children in obscene or indecent or sexually explicit manner

(c) Cultivates, entices, or induces children to online relationship with one or more children for and on sexually explicit act or in a manner that may offend a reasonable adult on the computer resource

(d) Facilitates abusing children online

(e) Records in any electronic form own abuse or that of others pertaining to sexually explicit act with children, shall be punished on first conviction with imprisonment of either description for a term which may extend to five years and with a fine which may extend to 10 lakh rupees and in the event of second or subsequent conviction with imprisonment of either description for a term which may extend to seven years and also with fine which may extend to 10 lakh rupees.

Exceptions: Provided that provisions of Sec 67, 67A and this Section does not extend to any book, pamphlet, paper or writing, drawing, painting representation or figure in electronic form – if it is for public good, in the interest of science, literature, art or learning or other objects of general concern; or, if it is kept for bonafide heritage or religious purposes.

Explanation: For the purposes of this section, "children" means a person who has not completed the age of 18 years.

Box 13.32 Section 67C ITAA 2008

Preservation and retention of information by intermediaries

(1) Intermediary shall preserve and retain such information as may be specified for such duration and in such manner and format as the Central Government may prescribe.

(2) Any intermediary who intentionally or knowingly contravenes the provisions of Sub-section (1) shall be punished with an imprisonment for a term which may extend to three years and shall also be liable to fine.

Screening videographs and photographs of illegal activities through the Internet come under this category; making pornographic videos or MMS clippings, or distributing such clippings through mobile or other forms of communication through the Internet will be handled under Section 67 of ITAA 2008 (Case 17).

> **Box 13.33 Section 292 IPC**
>
> Sale, etc., of obscene books, etc.
>
> For the purposes of sub-section (2), a book, pamphlet, paper, writing, drawing, painting representation, figure or any other object, shall be deemed to be obscene if it is lascivious or appeals to the prurient interest or if its effect, or (where it comprises two or more distinct items) the effect of any one of its items, is, if taken as a whole, such as to tend to deprave and corrupt persons who are likely, having regard to all relevant circumstances, to read, see or hear the matter contained or embodied in it.
>
> 3(2) Whoever
>
> (a) Sells, lets to hire, distributes, publicly exhibits or in any manner puts into circulation, or for purposes of sale, hire, distribution, public exhibition or circulation, makes, produces or has in his possession any obscene book, pamphlet, paper, drawing, painting, representation or figure or any other obscene object whatsoever, or
>
> (b) Imports, exports or conveys any obscene object for any of the purposes aforesaid, or knowing or having reason to believe that such object will be sold, let to hire, distributed or publicly exhibited or in any manner put into circulation, or
>
> (c) Takes part in or receives profits from any business in the course of which he knows or has reason to believe that any such obscene objects are, for any of the purposes aforesaid, made, produced, purchased, kept, imported, exported, conveyed, publicly exhibited or in any manner put into circulation, or
>
> (d) Advertises or makes known by any means whatsoever that any person is engaged or is ready to engage in any act which is an offence under this section, or that any such obscene object can be procured from or through any person, or
>
> (e) Offers or attempts to do any act which is an offence under this section, shall be punished
>
> 4 [on first conviction with imprisonment of either description for a term which may extend to two years, and with fine which may extend to two thousand rupees, and, in the event of a second or subsequent conviction, with imprisonment of either description for a term which may extend to five years, and also with fine which may extend to five thousand rupees].
>
> 5 Exception-This section does not extend to
>
> (a) Any book, pamphlet, paper, writing, drawing, painting, representation or figure
> (i) The publication of which is proved to be justified as being for the public good on the ground that such book, pamphlet, paper, writing, drawing, painting, representation or figure is in the interest of science, literature, art of learning or other objects of general concern, or
> (ii) Which is kept or used bona fide for religious purposes
>
> (b) Any representation sculptured, engraved, painted or otherwise represented on or in-
> (i) Any ancient monument within the meaning or the Ancient Monuments and Archaeological Sites and Remains Act, 1958 (24 of 1958), or
> (ii) Any temple, or on any car used for the conveyance of idols, or kept or used for any religious purpose.
>
> **IPC Section 292A**
>
> **Printing, etc., of grossly indecent or scurrilous matter or matter intended for blackmail**
> Whoever

(Contd)

Box 13.33 (Contd)

(a) Prints or causes to be printed in any newspaper, periodical or circular, or exhibits or causes to be exhibited, to public view or distributes or causes to be distributed or in any manner puts into circulation any picture or any printed or written document which is grossly indecent, or in scurrilous or intended for blackmail, or

(b) Sells or lets for hire, or for purposes of sale or hire makes, produces or has in his possession, any picture or any printed or written document which is grossly indecent or is scurrilous or intended for blackmail; or

(c) Conveys any picture or any printed or written document which is grossly indecent or is scurrilous or intended for blackmail knowing or having reason to believe that such picture or document will be printed, sold, let for hire distributed or publicly exhibited or in any manner put into circulation; or

(d) Takes part in, or receives profits from, any business in the course of which he knows or has reason to believe that any such newspaper, periodical, circular, picture or other printed or written document is printed, exhibited, distributed, circulated, sold, let for hire, made, produced, kept, conveyed or purchased; or

(e) Advertises or makes known by any means whatsoever that any person is engaged or is ready to engage in any Act which is an offence under this section, or that any such newspaper, periodical, circular, picture or other printed or written document which is grossly indecent or is scurrilous or intended for blackmail, can be procured from or through any person; or

(f) Offers or attempts to do any act which is an offence under this section shall be punished with imprisonment of either description for a term which may extend to two years, or with fine, or with both.

Provided that for a second or any subsequent offence under this section, he shall be punished with imprisonment of either description for a term which shall not be less than six months and not more than two years.

Explanation I - For the purposes of this section, the word scurrilous shall be deemed to include any matter which is likely to be injurious to morality or is calculated to injure any person:

Provided that it is not scurrilous to express in good faith anything whatever respecting the conduct of-

(i) A public servant in the discharge of his public functions or respecting his character, so far as his character appears in that conduct and no further; or

(ii) Any person touching any public question, and respecting his character, so far as his character appears in that conduct and no further.

Explanation II - In deciding whether any person has committed an offence under this section, the Court shall have regard inter alia, to the following considerations-

(a) The general character of the person charged, and where relevant the nature of his business

(b) The general character and dominant effect of the matter alleged to be grossly indecent or scurrilous or intended for blackmail

(c) Any evidence offered or called by or on behalf of the accused person as to his intention in committing any of the acts specified in this section.

> **Box 13.34 Sections 293 and 294 IPC**
>
> **Section 293: Sale, etc., of obscene objects to young persons**
>
> Whoever sells, lets to hire, distributes, exhibits or circulates to any person under the age of twenty years any such obscene object, as is referred to in IPC Section 292 or offers of attempts so to do, shall be punished (on first conviction with imprisonment or either description for a term which may extend to three years, and with fine which may extend to two thousand rupees, and, in the event of a second or subsequent conviction, with imprisonment of either description for a term which may extend to seven years, and also with fine which may extend to five thousand rupees)
>
> **Section 294: Obscene acts and songs**
>
> Whoever to the annoyance of others- (a) Does any obscene act in any public place, or (b) Sings, recites or utters any obscene song, ballad or words, in or near any public place, shall be punished with imprisonment of either description for a term which may extend to three months, or with fine, or with both.

> **Case 17: Cyber Pornography**
>
> A school student from Delhi (India), who was regularly teased for having a pock-marked face, used a free hosting provider to create http://www.amazinggents.8m.net. He regularly uploaded morphed photographs of teachers and girls from his school onto the website. He was arrested when the father of one of the victims reported the case to the police.

13.8 CYBER LAWS FOR CYBER SECURITY

Cyber laws have been enacted to define the rank of the investigating officers under Section 80 of ITAA2008. Further, the powers to ensure cyber security way of interception or monitoring or decryption, blocking from public access and monitoring and collecting the traffic data for cyber security are defined in Sections 69, 69A, and 69B respectively (Box 13.36). Besides this, establishment of Computer Emergency Response Team of India (CERT-IN) as the national nodal agency for maintaining cyber security is defined under Section 70 (Box 13.37). Further, laws have been enacted to punish in case of breach in security. Table 13.3 presents some of the offences against cyber security and the penalty under the IT Act 2000. The associated laws are presented in this section.

Table 13.3 Summary of laws that focus on cyber security

Section under IT Act 2000	Offence	Penalty
Sec. 70	Unauthorized access to protected systems	Imprisonment for a term which may extend to ten years and shall also be liable to fine
Sec. 71	Misrepresentation to the controller or the certifying authority for obtaining licence or electronic signature certificate	Imprisonment for a term which may extend to two years, or with fine which may extend to one lakh rupees, or with both
Sec. 72	Breach of confidentiality and privacy	Imprisonment for a term which may extend to two years, or with fine which may extend to one lakh rupees, or with both
Sec. 72A	Disclosure of information in breach of contract	Imprisonment for a term which may extend to three years, or with a fine which may extend to five lakh rupees, or with both

(Contd)

Table 13.3 (Contd)

Section under IT Act 2000	Offence	Penalty
Sec. 73 & 74	Publishing false digital signature certificates	Imprisonment for a term which may extend to two years, or with fine which may extend to one lakh rupees, or with both

Under this ITAA 2008 Section, the nominated government official will be able to listen in to all phone calls, read the SMSes and emails, and monitor the websites that one visited, subject to adherence to the prescribed procedures and without a warrant from a magistrate's order.

Box 13.35 Section 68 ITA 2000

Power of controller to give directions

(1) The Controller may, by order, direct a Certifying Authority or any employee of such authority to take such measures or cease carrying on such activities as specified in the order if those are necessary to ensure compliance with the provisions of this act, rules, or any regulations made thereunder.

(2) Any person who intentionally or knowingly fails to comply with any order under Sub-section (1) shall be guilty of an offence and shall be liable on conviction to imprisonment for a term not exceeding two years or to a fine not exceeding one lakh rupees or to both.

Box 13.36 Section 69 ITA 2000

Powers to issue directions for interception or monitoring or decryption of any information through any computer resource

(1) Where the Central Government or a State Government or any of its officer specially authorized by the Central Government or the State Government, as the case may be, in this behalf may, if is satisfied that it is necessary or expedient to do in the interest of the sovereignty or integrity of India, defence of India, security of the State, friendly relations with foreign states, or public order or for preventing incitement to the commission of any cognizable offence relating to above or for investigation of any offence, it may, subject to the provisions of Sub-section (2), for reasons to be recorded in writing, by order, direct any agency of the appropriate government to intercept, monitor, or decrypt or cause to be intercepted or monitored or decrypted any information transmitted, received, or stored through any computer resource.

(2) The procedure and safeguards subject to which such interception or monitoring or decryption may be carried out, shall be such as may be prescribed.

(3) The subscriber or intermediary or any person in charge of the computer resource shall, when called upon by any agency which has been directed under Sub-section (1), extend all facilities and technical assistance to:

   (a) Provide access to or secure access to the computer resource containing such information; generating, transmitting, receiving, or storing such information

(Contd)

Box 13.36 (Contd)

(b) Intercept or monitor or decrypt the information, as the case may be
(c) Provide information contained stored in computer resource

The subscriber or intermediary or any person who fails to assist the agency referred to in Sub-section (3) shall be punished with an imprisonment for a term which may extend to seven years and shall also be liable to fine.

Section 69A ITA 2000

Power to issue directions for blocking of public access to any information through any computer resource

(1) Where the Central Government or any of its officer specially authorized by it in this behalf is satisfied that it is necessary or expedient so to do in the interest of sovereignty and integrity of India, defence of India, security of the State, friendly relations with foreign states, or public order or for preventing incitement to the commission of any cognizable offence relating to above, it may subject to the provisions of Sub-sections (2) for reasons to be recorded in writing, by order direct any agency of the Government or intermediary to block access by the public, or cause to be blocked for access by public any information generated, transmitted, received, stored, or hosted in any computer resource.
(2) The procedure and safeguards subject to which such blocking for access by the public may be carried out shall be such as may be prescribed.
(3) The intermediary who fails to comply with the direction issued under Sub-section (1) shall be punished with an imprisonment for a term which may extend to seven years and also be liable to fine.

Section 69B ITA 2000

Power to authorize to monitor and collect traffic data or information through any computer resource for cyber security

(1) The Central Government may, to enhance Cyber Security and for identification, analysis and prevention of any intrusion or spread of computer contaminant in the country, by notification in the official Gazette, authorize any agency of the Government to monitor and collect traffic data or information generated, transmitted, received, or stored in any computer resource.
(2) The intermediary or any person in-charge of the computer resource shall when called upon by the agency which has been authorized under Sub-section (1), provide technical assistance and extend all facilities to such agency to enable online access or to secure and provide online access to the computer resource generating, transmitting, receiving, or storing such traffic data or information.
(3) The procedure and safeguards for monitoring and collecting traffic data or information, shall be such as may be prescribed.
(4) Any intermediary who intentionally or knowingly contravenes the provisions of Sub-section (2) shall be punished with an imprisonment for a term which may extend to three years and shall also be liable to fine.

Explanation: For the purposes of this section,
(I) "Computer Contaminant" shall have the meaning assigned to it in Section 43
(II) "Traffic data" means any data identifying or purporting to identify any person, computer system, or computer network or location to or from which the communication is or may be transmitted and includes communications' origin, destination, route, time, date, size, duration, or type of underlying service or any other information.

Box 13.37 Section 70 ITA 2000

Protected system

(1) The appropriate government may, by notification in the Official Gazette, declare any computer resource which directly or indirectly affects the facility of Critical Information Infrastructure, to be a protected system.

Explanation: For the purposes of this section, "Critical Information Infrastructure" means the computer resource, the incapacitation or destruction of which, shall have debilitating impact on national security, economy, public health, or safety.

(2) The appropriate government may, by order in writing, authorize the persons who are authorized to access protected systems notified under Sub-section (1).

(3) Any person who secures access or attempts to secure access to a protected system in contravention of the provisions of this section shall be punished with imprisonment of either description for a term which may extend to 10 years and shall also be liable to fine.

(4) The Central Government shall prescribe the information security practices and procedures for such protected system.

Section 70A: National Nodal Agency

(1) The Central Government may, by notification published in the official Gazette, designate any organization of the Government as the national nodal agency in respect of Critical Information Infrastructure Protection.

(2) The national nodal agency designated under Sub-section (1) shall be responsible for all measures, including Research and Development relating to protection of Critical Information Infrastructure.

(3) The manner of performing functions and duties of the agency referred to in Sub-section (1) shall be such as may be prescribed.

Section 70B: Indian Computer Emergency Response team to serve as national agency for incident response

The Indian Computer Emergency Response team shall serve as the national agency for performing the following functions in the area of Cyber Security:

— Collection, analysis, and dissemination of information on cyber incidents

— Forecast and alerts of cyber security incidents

— Emergency measures for handling cyber security incidents

— Coordination of cyber incidents response activities

— Issue guidelines, advisories, vulnerability notes, and white papers relating to information security practices, procedures, prevention, response, and reporting of cyber incidents

— Such other functions relating to cyber security as may be prescribed

13.9 OTHER CYBER LAWS ASSOCIATED WITH CYBERCRIME AND CYBERSPACE

Other laws which are important in the context of cybercrime and cyber space are summarized in this section (Boxes 13.38–13.40).

Box 13.38 Sections 71 and 72 ITA 2000

Section 71: Penalty for misrepresentation

Whoever makes any misrepresentation to, or suppresses any material fact from, the Controller or the Certifying Authority for obtaining any license or Electronic Signature Certificate, as the case may be, shall be punished with imprisonment for a term which may extend to two years or with fine which may extend to one lakh rupees, or with both

Section 72 : Breach of confidentiality and privacy

Save as otherwise provided in this act or any other law for the time being in force, any person who, in pursuant of any of the powers conferred under this act, rules, or regulations made thereunder, has secured access to any electronic record, book, register, correspondence, information, document, or other material without the consent of the person concerned discloses such electronic record, book, register, correspondence, information, document, or other material to any other person shall be punished with imprisonment for a term which may extend to two years or with fine which may extend to one lakh rupees, or with both.

Box 13.39 Section 72A–75 ITA 2000

Section 72A: Punishment for disclosure of information in breach of lawful contract

Save as otherwise provided in this act or any other law for the time being in force, any person, including an intermediary who, while providing services under the terms of lawful contract, has secured access to any material containing personal information about another person, with the intent to cause or knowing that he is likely to cause wrongful loss or wrongful gain discloses, without the consent of the person concerned, or in breach of a lawful contract, such material to any other person shall be punished with imprisonment for a term which may extend to three years or with a fine which may extend to five lakh rupees, or with both.

Section 73: Penalty for publishing electronic signature certificate false in certain particulars

(1) No person shall publish an Electronic Signature Certificate or otherwise make it available to any other person with the knowledge that:
   a. The Certifying Authority listed in the certificate has not issued it,
   b. The subscriber listed in the certificate has not accepted it,
   c. The certificate has been revoked or suspended, unless such publication is for the purpose of verifying a digital signature created prior to such suspension or revocation.

(2) Any person who contravenes the provisions of Sub-section (1) shall be punished with imprisonment for a term which may extend to two years or with fine which may extend to one lakh rupees, or with both.

Section 74: Publication for fraudulent purpose

Whoever knowingly creates, publishes, or otherwise makes available an Electronic Signature Certificate for any fraudulent or unlawful purpose shall be punished with imprisonment for a term which may extend to two years or with fine which may extend to one lakh rupees, or with both

(Contd)

Box 13.39 (Contd)

Section 75: Act to apply for offence or contraventions committed outside India

(1) Subject to the provisions of Sub-section (2), the provisions of this act shall apply also to any offence or contravention committed outside India by any person irrespective of his/her nationality. (2) For the purposes of Sub-section (1), this act shall apply to an offence or contravention committed outside India by any person if the act or conduct constituting the offence or contravention involves a computer, computer system, or computer network located in India.

Box 13.40 Section 76–80 and 84B ITA 2000

Section 76: Confiscation

Any computer, computer system, floppies, compact disks, tape drives, or any other accessories related thereto, in respect of which any provision of this act, rules, orders, or regulations made there under has been or is being contravened, shall be liable to confiscation: Provided that where it is established to the satisfaction of the court adjudicating the confiscation that the person in whose possession, power, or control of any such computer, computer system, floppies, compact disks, tape drives, or any other accessories relating thereto is found is not responsible for the contravention of the provisions of this act, rules, orders, or regulations made there under, the court may, instead of making an order for confiscation of such computer, computer system, floppies, compact disks, tape drives, or any other accessories related thereto, make such other order authorized by this act against the person contravening of the provisions of this act, rules, orders, or regulations made there under as it may think fit

Section 77: Compensation, penalties, or confiscation not to interfere with other punishment

No compensation awarded, penalty imposed, or confiscation made under this act shall prevent the award of compensation or imposition of any other penalty or punishment under any other law for the time being in force.

Section 77A: Compounding of offences

A Court of competent jurisdiction may compound offences other than offences for which the punishment for life or imprisonment for a term exceeding three years has been provided under this act.

Provided that the court shall not compound such offence where the accused is by reason of his previous conviction, liable to either enhanced punishment or to a punishment of a different kind. Provided further that the court shall not compound any offence where such offence affects the socioeconomic conditions of the country or has been committed against a child below the age of 18 years or a woman.

The person accused of an offence under this act may file an application for compounding in the court in which offence is pending for trial and the provisions of Section 265B and 265C of Code of Criminal Procedures, 1973 shall apply.

Section 77B: Offences with three years imprisonment to be cognizable

Notwithstanding anything contained in Criminal Procedures Code, 1973, the offence punishable with imprisonment of three years and above shall be cognizable and the offence punishable with imprisonment of three years shall be bailable.

(Contd)

Box 13.40 (Contd)

> **Section 78: Power to investigate offences**
>
> Notwithstanding anything contained in the Code of Criminal Procedure, 1973, a police officer not below the rank of Inspector shall investigate any offence under this Act.
>
> **Section 79A: Central government to notify the examiner of electronic evidence**
>
> The Central Government may, for the purposes of providing expert opinion on electronic form evidence before any court or other authority specify, by notification in the official Gazette, any department, body, or agency of the Central Government or a State Government as an Examiner of Electronic Evidence.
>
> **Section 80: Power of police officer and other officers to enter, search, etc.**
>
> (1) Notwithstanding anything contained in the Code of Criminal Procedure, 1973, any police officer, not below the rank of an Inspector or any other officer of the Central Government or a State Government authorized by the Central Government in this behalf may enter any public place and search and arrest without warrant any person found therein who is reasonably suspected of having committed or of committing or of being about to commit any offence under this Act.
>
> **Section 84B: Punishment for abetment of offences**
>
> Whoever abets any offence shall, if the act abetted is committed in consequence of the abetment, and no express provision is made by this act for the punishment for such abetment, be punished with the punishment provided for the offence under this Act.

13.10 SUMMARY OF CYBER LAWS IN INDIA

Table 13.4 presents a list of laws under the IPC, the IT Act 2000, and the ITAA 2008 for dealing with cybercrimes in India.

Table 13.4 Summary of Indian cyber laws

S. no.	Indian cyber laws	Sections
Cybercrimes under ITA Act 2008		**Section under ITA Act 2008**
1	Tampering documents stored in computers	Sec. 65
2	Hacking computer systems, alteration of data	Sec. 66
3	Dishonestly receiving stolen electronic communication devices	Sec. 66B
4	Identity theft	Sec. 66C
5	Impersonation	Sec. 66D
6	Violation of privacy	Sec. 66E
7	Cyber terrorism	Sec. 66F
8	Publishing obscene content in electronic form	Sec. 67
9	Publishing of material containing sexually explicit act in electronic form	Sec. 67A

(Contd)

Table 13.4 (Contd)

S. no.	Indian cyber laws	Sections
10	Publishing of material involving children in sexually explicit act in electronic form	Sec. 67B
11	Preservation and retention of information by intermediaries	Sec. 67C
12	Unauthorized access to protected system	Sec. 70
13	Breach of confidentiality and privacy	Sec. 72
14	Publishing false digital signature certificates	Sec. 73
15	Publication for fraudulent purpose	Sec. 74
16	Offences by companies	Sec. 85
IT Act and cybercrime preventive measures		**Section under ITA Act 2008**
1	Powers to issue directions for interception or monitoring or decryption of any information through any computer resource	Sec. 69
2	Power to issue directions for blocking for public access of any information through any computer resource	Sec. 69A
3	Power to authorize to monitor and collect traffic data or information through any computer resource for cyber security	Sec. 69B
4	Exemption from liability of intermediary in certain cases	Sec. 79
IT Act and punishments		**Section under ITA Act 2008**
1	Penalty and compensation for damage to computer, computer system, etc.	Sec. 43
2	Compensation for failure to protect data	Sec. 43A
3	Penalty for misrepresentation	Sec. 78
4	Act to apply for offence or contraventions committed outside India	Sec. 75
5	Compensation, penalties, or confiscation not to interfere with other punishment	Sec. 77
6	Compounding of offences	Sec. 77A
7	Offences with three years imprisonment to be cognizable	Sec. 77B
8	Exemption from liability of intermediary in certain cases	Sec. 79
9	Punishment for abetment of offences	Sec. 84B
10	Punishment for attempt to commit offences	Sec. 84C
IPC and cybercrimes		**Section under IPC**
1	Sending threatening messages by email	Sec. 503 IPC
2	Sending defamatory messages by email	Sec. 499 IPC
3	Bogus websites, cyber frauds	Sec. 420 IPC
4	Email spoofing	Sec. 463 IPC
5	Making a false document	Sec. 464 IPC
6	Forgery for purpose of cheating	Sec. 468 IPC
7	Forgery for purpose of harming reputation	Sec. 469 IPC

(Contd)

Table 13.4 (Contd)

S. no.	Indian cyber laws	Sections
8	Web jacking	Sec. 383 IPC
9	Email abuse	Sec. 500 IPC
10	Punishment for criminal intimidation	Sec. 506 IPC
11	Criminal intimidation by an anonymous communication	Sec. 507 IPC
12	Copyright infringement	Sec. 51, Sec. 63 IPC
13	Knowing use of infringing copy of computer programme to be an offence	Sec. 63B IPC
14	Obscenity	Sec. 292 IPC
15	Obscene acts and songs	Sec. 294 IPC
16	Sale, etc., of obscene objects to young person	Sec. 293 IPC
17	Printing etc., of grossly indecent or scurrilous matter or matter intended for blackmail	Sec. 292A IPC
18	Theft of computer hardware	Sec. 378 IPC
19	Punishment for theft	Sec. 379 IPC
20	Identity theft	Sec. 417A IPC and Sec. 419A IPC
Other cybercrimes and specific acts		Act
1	Online sale of drugs	Narcotic Drugs and Psychotropic Substances (NDPS) Act, 1985
2	Online Sale of Arms	Arms Act, 1959

13.11 AMENDMENTS TO THE INDIAN EVIDENCE ACT 1872 IN VIEW OF INFORMATION TECHNOLOGY ACT 2000

The Information Technology Act 2000 brought some changes to the Indian Evidence Act which plays a pivotal role in the administration of justice. Cybercrimes can be tracked from electronic records and in order to recognize them, changes have been made in the Evidence Act, which are as follows:

1. The definition of "Evidence" is changed as "all documents including electronic records produced for the inspection of the Court" from "all documents produced for the inspection of the Court" so as to include electronic records as evidence.
2. So as to recognize admission in electronic form, in Section 17 the definition is changed as oral or documentary or contained in electronic form.
3. Section 22A has been inserted to deal when all the oral admission as to contents of electronic records are relevant (Box 13.41).
4. Section 39 is substituted as new Section 39 to provide a rule relating to statements to be given in evidence to explain the nature of statements part of electronic records.
5. Section 47A deals with the opinion as to electronic signature where relevant (Box 13.42).
6. While Section 65 provides the detailed method of proving the contents of electronic records, Special provisions as to evidence relating to electronic record has been introduced with Section 65A (Box 13.43).
7. Special provisions relating to proof as to verification of electronic signature is introduced with 67A (Box 13.44).
8. Section 73A(a) is inserted to provide the manner in which digital signature of a person in enquiry must be proved (Box 13.45).

9. Section 81 A is inserted as a direction to Courts to presume the genuineness of electronic record (Box 13.46).
10. Section 85A is inserted to provide presumption as to draw an inference as to the conclusion of the agreement in electronic form with electronic signature of parties. Section 85B provides presumption as to electronic records and electronic signature as not having been altered. Section 85C provides electronic signature certificates by which inference can be drawn as to the correctness of information listed in the electronic signature certificate.
11. Section 88A is inserted so as to provide presumption as to electronic messages that the court may presume that an electronic message forwarded by the originator through electronic mail server to the addressee to whom the message as fed into his computer for transmission.
12. Section 88 A is inserted so as to provide presumption as to the genuineness of electronic record of five years which is produced from proper custody.

Box 13.41 Section 22A IEA 1872

When oral admissions as to contents of electronic records are relevant

Oral admissions as to the contents of electronic records are not relevant, unless the genuineness of the electronic record produced is in question.

Box 13.42 Section 47A IEA 1872

Opinion as to [electronic signature] when relevant

When the Court has to form an opinion as to the [electronic signature] of any person, the opinion of the Certifying Authority which has issued the [Electronic Signature Certificate] is a relevant fact.

Box 13.43 Section 65A IEA 1872

Special provisions as to evidence relating to electronic record

The contents of electronic records may be proved in accordance with the provisions of Section 65B.

Box 13.44 Section 67A IEA 1872

Proof as to [electronic signature]

Except in the case of a secure [electronic signature], if the [electronic signature] of any subscriber is alleged to have been affixed to an electronic record that fact that such [electronic signature] is the [electronic signature] of the subscriber must be proved.

Box 13.45 Section 73A(a) IEA 1872

Proof as to verification of digital signature

that person or the Controller or the Certifying Authority to produce the Digital Signature Certificate

> **Box 13.46 Section 81A IEA 1872**
>
> **Presumption as to gazettes in electronic forms**
>
> The Court shall presume the genuineness of every electronic record purporting to be the Official Gazette or purporting to be electronic record purporting to be the Official Gazette or purporting to be electronic record directed by any law to be kept by any person, if such electronic record is kept substantially in the form required by law and is produced from proper custody.

13. Section 131 is substituted by a new section which runs as under Production of documents or electronic records which another person having possession could refuse to produce. No one shall be compelled to produce documents in his possession or electronic records under his control, which any other person would be entitled to refuse to produce if they were in his possession or control unless such last mentioned person consents to their production.

It is certified that the above information is a true extract in printed form of the relevant data created in the usual and ordinary course of business and stored on the hard disk of the computer system installed at branch of the bank.

It is further certified that the access to the computer system and the data stored thereon in controlled by pre-defined user permissions exercised through unique ID and associated password;

That physical access to the computer/server room is prevented by locking the server room and the branch after office hours. Detection of any unauthorized changes in the data after day-end and before day-again activity is through procedures which are built into the application program. Unauthorized changes in the data during regular working hours are prevented/detected through verification of output with authorized input;

That in case of system failure, the data is retrieved from the backup kept on tape/floppy/cartridge/hard disk, which is under the control of system administrator/designated employee of the branch;

That backup is verified by the system during the process of transfer of data to backup media;

That backup devices and media are kept under lock and key which are in the custody of a designated staff member; and that physical and logical access control are in place as against tampering of the system.

It is further certified that to the best of our knowledge and, the computer system that generated and stored this information operated at the time of such generation/storage of the data and printout represent correctly the relevant data.

System Administrator   Branch Manager
Seals of the signatures may also be affixed.
(Name of responsible official)
Designation

Fig. 13.4 Certificate issued in accordance with Amendments to the Banker's Book Evidence Act 1891

13.12 AMENDMENTS TO THE BANKER'S BOOK EVIDENCE ACT 1891 IN VIEW OF INFORMATION TECHNOLOGY ACT 2000

Prior to the amendment of the Banker's Book Evidence Act 1891, Section 2 of this act provided that a copy of a bank statement would be admissible in the court only when it is certified to the effect. These certificates are to be issued by a person occupying a responsible position in relation to the operation of the relevant system or the management of the relevant activities, whichever is appropriate. It became difficult to issue such a certificate as banks started maintaining their record on computer.

Hence, the Banker's Books Evident Act 1891 was amended vide Third Schedule of the Information Technology Act 2000. After this amendment, printouts of the data stored in a floppy, disk, tape, or any other electromagnetic media have also been made admissible provided they are certified as per Section 2A of this Act and is shown in Fig. 13.4.

13.13 INDIAN LAWS RELATED TO INTELLECTUAL PROPERTY

Drastic changes in the international trade practices made it mandatory to have legislations in India to govern them. Though computer databases, software, and programs are protected under copyright laws in India, some of the laws related to Intellectual Property that were compiled under the agreement on Trade Related Intellectual Property Rights (TRIPS) are as follows:

- Patents Act 1970 that was amended to Patent Amendment Act 2005. The Act allows the filing of all product patents with a regulatory authority. It contains provisions to grant exclusive marketing rights (EMRs) for five years or till the patent is granted or rejected, whichever is earlier. It also provides for publication of all patent applications within 18 months of filing or priority date, whichever is earlier.
- Trade Marks Act1999 governs India's trade and commerce in the global arena.
- The Designs Act 2000 governs the artistic design of a product.
- The Geographical Indication of Goods (Registration and Protection) Act 1999 avoids misuse of Indian geographical indications.
- Copyright Act 1957 protects the right of text, audio, and video content owners.
- The Semiconductor Integrated Circuits Layout Design Act 2000
- The Protection of Plant and Varieties and Farmers Rights Act 2001
- The Biological Diversity Act 2002

13.14 INDIAN CASE LAWS

Some of the case laws and the court rulings in Indian courts associated with cybercrime and digital evidence are summarized in this section:

State of Maharashtra vs Dr Praful B Desai AIR 2003 SC 2053 The Supreme Court observed that video conferencing is an advancement of science and technology which permits seeing, hearing, and talking with someone who is not physically present, with the same facility and ease as if he was physically present. The legal requirement for the presence of the witness does not mean the actual physical presence. The court allowed the examination of a witness through video conferencing and concluded that there is no reason why the examination of a witness by video conferencing should not be an essential part of electronic evidence.

Bodala Murali Krishna vs Smt. Bodala Prathima 2007 (2) ALD 72 The court held that, "...the amendments carried to the Evidence Act by introduction of Sections 65-A and 65-B are in relation to the electronic record. Sections 67-A and 73-A were introduced as regards proof and verification of digital signatures. As regards presumption to be drawn about such records, Sections 85-A, 85-B, 85-C, 88-A and 90-A were added. These provisions are referred only to demonstrate that the emphasis, at present, is to recognize the electronic records and digital signatures, as admissible pieces of evidence."

Dharambir vs Central Bureau of Investigation 148 (2008) DLT 289 The court arrived at the conclusion that when Section 65-B talks of an electronic record produced by a computer referred to as the computer output, it would also include a hard disc in which information was stored or was earlier stored or continues to be stored. It distinguished as there being two levels of an electronic record. One is the hard disc which once used itself becomes an electronic record in relation to the information regarding the changes the hard disc has been subject to and which information is retrievable from the hard disc by using a software program. The other level of electronic record is the active accessible information recorded in the hard disc in the form of a text file, or sound file or a video file, etc. Such information that is accessible can be converted or copied as such to another magnetic or electronic device such as a CD or pen drive. Even a blank hard disc which contains no information but was once used for recording information can also be copied by producing a cloned hard disk or a mirror image.

Jagjit Singh vs State of Haryana (2006) 11 SCC 1 The speaker of the Legislative Assembly of the State of Haryana disqualified a member for defection. When hearing the matter, the Supreme Court considered the digital evidence in the form of interview transcripts from the Zee News television channel, the Aaj Tak television channel, and the Haryana News of Punjab Today television channel. The court determined that the electronic evidence placed on record was admissible and upheld the reliance placed by the speaker on the recorded interview when reaching the conclusion that the voices recorded on the CD were those of the persons taking action. The Supreme Court found no infirmity in the speaker's reliance on the digital evidence and the conclusions reached by him. The comments in this case indicate a trend emerging in Indian courts: judges are beginning to recognize and appreciate the importance of digital evidence in legal proceedings.

Twentieth Century Fox Film Corporation vs NRI Film Production Associates (P) Ltd AIR 2003 KANT 148 In this case, certain conditions have been laid down for video-recording of evidence:

1. Before a witness is examined in terms of the audio–video link, the witness is to file an affidavit or an undertaking duly verified before a notary or a judge that the person who is shown as the witness is the same person as who is going to depose on the screen. A copy is to be made available to the other side (Identification affidavit).
2. The person who examines the witness on the screen is also supposed to file an affidavit/undertaking before examining the witness with a copy to the other side with regard to identification.
3. The witness has to be examined during the working hours of Indian Courts. Oath is to be administered through the media.
4. The witness should not plead any inconvenience on account of time difference between India and USA.
5. Before examination of the witness, a set of plaint, written statement, and other documents must be sent to the witness so that the witness is acquainted with the documents and an acknowledgement is to be filed before the Court in this regard.
6. The learned judge is to record such remarks as is material regarding the demur of the witness while on the screen.
7. The learned judge must note the objections raised during recording of witness and decide the same at the time of arguments.
8. After recording the evidence, the same is to be sent to the witness and his signature is to be obtained in the presence of a notary public and thereafter it forms part of the record of the suit proceedings.
9. The visual is to be recorded and the record would be at both ends. The witness also is to be alone at the time of visual conference and notary is to certify to this effect.
10. The learned judge may also impose such other conditions as are necessary in a given set of facts.
11. The expenses and the arrangements are to be borne by the applicant who wants this facility.

Amitabh Bagchi vs Ena Bagchi AIR 2005 Cal 11 (Sections 65-A and 65-B of the Evidence Act 1872 were analysed.)

The court held that the physical presence of a person in court may not be required for the purpose of adducing evidence and the same can be done through a medium like video conferencing. Sections 65-A and 65-B provide provisions for evidences relating to electronic records and admissibility of electronic records, and that definition of electronic records includes video conferencing.

State (NCT of Delhi) vs Navjot Sandhu AIR 2005 SC 3820 There was an appeal against the conviction following the attack on Parliament on 13 December 2001. This case dealt with the proof and admissibility of mobile telephone call records. While considering the appeal against the accused for attacking the Parliament, a submission was made on behalf of the accused that no reliance could be placed on the mobile telephone call records, because the prosecution had failed to produce the relevant certificate under Section 65-B(4) of the Evidence Act. The Supreme Court concluded that a cross-examination of the competent witness acquainted with the functioning of the computer during the relevant time and the manner in which the printouts of the call records were taken were sufficient to prove the call records.

Anvar P.V. vs P.K. Basheer and Others, in Civil Appeal No. 4226 OF 2012 In a judgment of The Hon'ble Supreme Court delivered in Anvar P.V. vs P.K. Basheer and Others, in Civil Appeal No. 4226 OF 2012 decided on 18September 2014, it was held that the computer output is not admissible without Compliance of 65B. It overruled the judgment laid down in the State (NCT of Delhi) vs Navjot Sandhu alias Afzal Guru [(2005) SCC 600 by the two judge Benchof the Supreme Court. The court specifically observed that the Judgment of Navjot Sandhu [supra], to the extent, the statement of the law on adm issibility of electronic evidence pertaining to electronic record of this court, does not lay down correct position and is required to be overruled. This judgment has put to rest the controversies arising from the various conflicting judgments and thereby provided a guideline regarding the practices being followed in the various High Courts and the Trial Court as to the admissibility of the Electronic Evidences. The 14 legal interpretations by the court of the following Sections 22A, 45A, 59, 65A, and 65B of the Evidence Act have confirmed that the stored data in CD/DVD/pen drive is not admissible without a certificate u/s 65 B(4) of the Evidence Act and further clarified that in the absence of such a certificate, the oral evidence to prove existence of such electronic evidence and the expert view under Section 45A Evidence Act cannot be availed to prove authenticity thereof. It has been specified in the judgment that the genuineness, veracity, or reliability of the evidence is looked into by the court subsequently only after the relevance and admissibility are fulfilled. The need to ensure the source and authenticity, pertaining to electronic records is because it is more vulnerable to tampering, alteration, transposition, excision, etc. Without such safeguards, the whole trial based on proof of electronic records can lead to mockery of justice. The original recording in the digital voice recorders/mobile phones need to be preserved as they may get destroyed. In such a case, the issuance of certificate under section 65B(4) of the Evidence Act cannot be given. Therefore such CD/DVD is inadmissible and cannot be exhibited as evidence. The oral testimony or expert opinion is also barred and the recording/data in the CD/DVDs do not serve any purpose for the conviction.

Abdul Rahaman Kunji vs The State of West Bengal MANU/WB/0828/2014 The Hon'ble High Court of Calcutta while deciding the admissibility of the email held that an email downloaded and printed from the email account of the person can be proved by virtue of Section 65B r/w Section 88A of the Evidence Act. The testimony of the witness to carry out such procedure to download and print the same is sufficient to prove the electronic communication.

In Jagdeo Singh vs The State and Ors. MANU/DE/0376/2015 The Hon'ble High Court of Delhi, while dealing with the admissibility of intercepted telephone calls in a CD and CDR which were wit hout a certificate u/s 65B Evidence Act, observed that the secondary electronic evidence without certificate u/s 65B Evidence Act is inadmissible and cannot be looked into by the court for any purpose whatsoever. The person who wants to rely on emails must fulfil the conditions contained under subclause 2 of Section 65-B. This means that a person filing the printout of an email in court can rely upon it as original without the need to actually file the original soft copy of it. In thecase of Ark Shipping Co. Ltd. vs GRT Ship Management Pvt. Ltd (2008 (1) ARBLR 317 Bom. The Hon'ble Court extracted an affidavit under Sec. 65B by considering the fact and the circumstances of that case. Butin VodafoneEssar Ltd vs Raju Sud the court dispensed with the requirement under Sec. 65-B.

Mrs Havovi Kersi Sethna vs Mr Kersi Gustad Sethna AIR 2011 Bom. 283 In a case on the Law of Evidence for Tape Recorded Conversation and its Admissibility thereof in evidence, it is settled law that tape recorded conversation is admissible in evidence. The appreciation of evidence would require consideration of three requirements: identification, relevancy and accuracy. It is left to the Defendant to pass those tests. If the tests are not passed, the tape recorded conversation would be of no use in effect ultimately. Notice of motion disposed of.

Law of Evidence - Sealing of recorded conversation - Recording of Evidence Tape Recorded Conversation - Held, the requirement of sealing the recorded conversation would not be applicable in this case. This is a civil trial. There is no question of sealing of a conversation recorded by a party to the civil case himself. The sealing requirement is only in criminal trials. Notice of motion disposed of.

Tape recorded conversation is admissible in evidence provided the identification, relevancy, and accuracy of the same is proved. Requirement of sealing of recorded conversation used as evidence is not applicable to a civil trial but is applicable only to a criminal trial.

POINTS TO REMEMBER

- Cyber laws provide legal protection to anyone using the Internet for either business or other activities.
- Cyber laws are otherwise called laws of the Internet.
- Information Technology Act 2000 (IT Act) came into force on 17October 2000.
- Information Technology Act 2000 regulates the use of computers, computer systems, and computer networks and also data and information in the electronic format.
- Information Technology Amendment Act 2008 has created a strong data protection regime by mandating reasonable security practices to protect sensitive personal information and several provisions for handling cybercrimes such as identity theft and cyber terrorism.
- Computer-related offences (include source code tampering, unauthorized access, disruption, damage, etc., of computer resources) are defined under Section 65, 66, and 66 A–D.
- Obscenity and related offences are defined in Sections 66E, 67, 67A, and 67B.
- Threat to unity and integrity of India (cyber terrorism) are defined in Section 66F.
- Power to issue directions by competent authorities to block access, monitor traffic, etc., are given in Sections 67C, 69, 69A, 70, and 70B.
- Sections 66F, 67A, 67B, 69, 69A, and 70 of the ITAA 2008 are non-bailable.
- Cybercrimes dealt with in the IT Act include the following: Sec. 65, Tampering with computer source documents; Sec. 66, Hacking Computer Systems and Data Alteration; Sec. 67, Publishing obscene information; Sec. 70, Unauthorized access of protected systems, Sec. 72, Breach of confidentiality and privacy, and Sec. 73, Publishing false digital signature certificates.
- Cybercrimes dealt with the IPC include the following: Sending threatening messages by email, IPC; Sec. 503, Sending defamatory messages by email, IPC Sec. 499; Forgery of electronic records, IPC Sec. 463; Bogus websites and cyber fraud, IPC Sec. 420; Email Spoofing, IPC Sec. 463; Web jacking, IPC Sec. 383; and Email Abuse, IPC Sec. 500.
- The special Acts to deal with cybercrimes are as follows: Online sale of arms under the Arms Act 1959 and Online sale of drugs under the Narcotic Drugs and Psychotropic Substances Act 1985.

KEY TERMS

Cyber defamation This refers to publishing offensive content against an individual with the intention to spoil the reputation and modesty of a woman using ECDs and the Internet. It is punishable under Sections 500 and 509 of IPC.

Cyber stalking This means repeatedly following a person, harassing over phone calls and with written messages and threatening either the individual or his/her family using the Internet, email, and other ECDs.

Cyber terrorism This refers to crimes committed in cyberspace, especially disruptive activities that target the social, religious, and political beliefs thereby spoiling the peace and well-being of society, a community or a nation.

Data diddling This refers to attempts made towards illegal and unauthorized alteration of data.

Email bombing This refers to sending enormous amount of email messages targeting a server with the intention to crash it.

Pornography This means making use of ECDs to develop, publish, print, and distribute the obscene content through the Internet.

Sale of illegal articles on the Internet This refers to the sale of narcotic drugs which may pose a potential threat to the health and safety of people and patients, weapons, wildlife and lottery, etc., through the Internet.

Web jacking This refers to the perpetrators taking control of a website forcefully by cracking the administrative privileges so that the actual owner of the website has no more control over it.

Website defacement This refers to the replacement of a home page of a website with another page that is defamatory in nature.

MULTIPLE-CHOICE QUESTIONS

1. The rules and regulations of cyber laws are covered under _____.
   (a) Information Technology (Certifying Authorities) Rules 2000
   (b) Information Technology (Security Procedure) Rules 2004
   (c) Information Technology (Certifying Authority) Regulations 2001
   (d) All of these

2. Punishment for violation of privacy comes under _____.
   (a) Section 66A   (c) Section 66D
   (b) Section 66C   (d) Section 66E

3. _____ refers to sending enormous amount of email messages targeting a server with the intention to crash it.
   (a) Email bombing
   (b) Email chatting
   (c) Virus
   (d) Worm

4. IPC Section 468 refers to _____.
   (a) forgery for the purpose of harming reputation
   (b) forgery for the purpose of cheating
   (c) neither (a) nor (b)
   (d) both (a) and (b)

5. Web jacking and cyber defamation are the cybercrimes against _____.
   (a) property
   (b) individual
   (c) nation
   (d) (a), (b), and (c)

6. The IT Act of 2000 was developed _____.
   (a) to promote the IT industry
   (b) to regulate e-commerce
   (c) to facilitate e-governance
   (d) (a), (b), and (c)

7. The IT Act 2000 consists of _____ sections.
   (a) 90   (c) 70
   (b) 80   (d) 60

8. Cyber stalking is punishable under Sections 503 and 506 of IPC if the messages are _____ in nature.
   (a) threatening   (c) both (a) and (b)
   (b) defamatory   (d) none of these

9. _____ is used to handle cybercrimes where money is the motive.
   (a) Section 66C under ITA 2008
   (b) Section 66D of ITA 2008
   (c) Section 66C under ITA 2007
   (d) Section 66D under ITA 2007

10. Email fraud is punishable under IPC Section _____.
    (a) 220   (c) 120
    (b) 320   (d) 420

11. _____ is/are punishable under Section 66F of ITAA 2008.
    (a) Website defacement
    (b) Cyber terrorism
    (c) Both (a) and (b)
    (d) None of these

12. Under the _____ act, the nominated government official will be able to listen in to all phone calls, read SMSes and emails, and monitor the websites that one has visited, subject to adherence to the prescribed procedure and without a warrant from a Magistrate's order.
    (a) ITA 2008
    (b) ITA 2007

(c) ITAA 2008
(d) ITAA 2007
13. _____ is inserted to direction to Courts to presume the genuineness of electronic record.
   (a) Section 80 A
   (b) Section 81 A
   (c) Section 80B
   (d) Section 81B
14. _____ deals with publishing false digital signature certificates.
   (a) Sec. 71
   (b) Sec. 72
   (c) Sec. 73
   (d) Sec. 74
15. Cyber defamation is punishable under _____.
   (a) Section 500 of IPC
   (b) Section 502 of IPC
   (c) Section 504 of IPC
   (d) Section 500 and 509 of IPC

REVIEW QUESTIONS

1. Mention the three categories of cybercrime.
2. Define cyber defamation.
3. What does ITA 2000 Section 71 say about penalty for misrepresentation?
4. Mention the offence and laws that focus on cyber security.
5. Explain the crime against property.
6. Discuss the crimes against nation in details
7. Briefly explain the Indian case laws.
8. Explain the amendments to the Indian Evidence Act 1872 in view of the Information Technology Act 2000 in detail.
9. Briefly explain the amendments to the Banker's Book Evidence Act 1891 in view of the Information Technology Act 2000.
10. Give a short note on crime against individual.

APPLICATION EXERCISES

1. Hemanika, while surfing on the Internet, noticed her profile on a website beautifulwoman.com. She was shocked to see her photograph. Her image had been tarnished. It was mentioned that she was a liar and a woman of bad conduct. Hemanika immediately approached the online community on the website and requested them to remove it. They removed the profile immediately. However, the profile was visible again. Hemanika is now thinking of filing a defamation suit against the online community. Examine the liability of online community with reference to the jurisprudence of defamation cases.
2. You are asked to submit an assignment on some topic for which you have to surf various pages on the Internet. You are downloading the articles and using them without seeking the permission of the authors claiming that it's just for personal use. Do you think that what you are doing is right? Are you liable to copyright infringement? Discuss and justify your views with the various cyber laws learnt from this chapter.
3. X comes to college to meet his friend. X notices that the computer lab is open and the computer systems are ON with Internet connectivity as well. He wishes to send a mail from the computer along with an attachment to that email from his pen drive. You are the administrator of the computer lab and you are asked to take legal action against X. What will be your course of action for this unauthorized access? Examine the liability of X.

BIBLIOGRAPHY

1. Overview of Cyber Laws in India, available at: www.lawyersclubindia.com/articles/Overview-of-Cyber-Laws-in-India--6039.asp (Accessed 30 November 2017)
2. Cybercrime Investigation Manual (2011), Data Security Council of India, A NASSCOM Initia-

tive, available at: https://uppolice.gov.in/.../28_5_2014_17_4_36_Cyber_Crime_Investigation_Manual.... (Accessed 30 November 2017)
3. Chapter 19 Cyber Laws in India (2002), available at: www.iibf.org.in/documents/Cyber-Laws-chapter-in-Legal-Aspects-Book.pdf (Accessed 30 November 2017)
4. Prashant Mali, Text Book of Cyber Crime and Penalties [As per ITAA 2008 and IPC] Draft Version, available at: https://archive.org/details/ATextBookOfCyberCrimeAndPenalties (Accessed 30 November 2017)
5. Yash Parikh (2012), "About Information Technology and Related to IT" available at: http://yash0924.blogspot.in/ (Accessed 30 November 2017)
6. Cyber Law: Everything you Need to Know (2011), available at: https://www.upcounsel.com/cyber-law (Accessed 30 November 2017)
7. Centre for Research on Cyber Crime and Cyber Law (2015), available at: https://www.upcounsel.com/cyber-law (Accessed 10 December 2017)
8. Baisakhi Jena (2017), Indian Evidence Act and Cyber Crimes, available at: http://www.4n6hub.com/indian-evidence-act-and-cyber-crimes/ (Accessed 10 December 2017)
9. Dharmendra Rautray (2003), *India: Intellectual Property Laws of India - An Overview*, available at: http://www.mondaq.com/india/x/23429/Trademark/ Intellectual+Property+Laws+of+India+An+Overview (Accessed 10 December 2017)
10. PTI (May 2017), *Ransomware Attack: Odisha's Govt Hospital Falls Prey to WannaCry Virus*, available at: https://www.livemint.com/Technology/X76bZbPH4nN4w7MaXN6tZL Ransomware-attack-Odishas-govt-hospital-falls-prey-to-Wann.html (Accessed on 4 April 2018)
11. Asheeta Regidi (2016), *Internet Service Providers in Mumbai Targeted in DDoS Attack*, available at: https://www.firstpost.com/tech/news-analysis/internet-service-providers-in-mumbai-targeted-in-ddos-attack-3685981.html (Accessed on 4 April 2018)

Answers to Multiple-choice Questions

1. (d)	2. (d)	3. (a)	4. (b)	5. (b)
6. (d)	7. (a)	8. (a)	9. (b)	10. (d)
11. (b)	12. (c)	13. (b)	14. (c)	15. (d)

INTERNATIONAL CYBER LAWS AND CASE STUDIES   14

Learning Objectives

This chapter provides an overview of international cyber laws. The objective of this chapter is to provide an insight into the cyber laws of representative countries from different areas, such as the United States of America, the United Kingdom, the Netherlands, Malaysia, and Australia. This chapter presents the cyber laws in force in the said countries and explains how cyber security is achieved by them. A comparison of the cybercrime legislations in representative countries for certain specific cybercrimes is presented. Case studies, as applicable, are highlighted wherever necessary. The reader will be familiar with the following after studying the chapter:

- Cyber laws of the representative countries
- Comparison of cybercrime legislations
- Case studies

14.1 INTRODUCTION

This chapter focuses on the laws that regulate crimes committed in the virtual world or cyber space at the international level and in jurisdictions outside India. It details the summaries of national laws governing information technology, e-commerce, online privacy, and computer-related crimes of five countries, namely the United States, the United Kingdom, the Netherlands, Malaysia, and Australia. It also covers the legal issues that arise in connection with the use of information technology across national boundaries. It provides a brief introduction to this wide-ranging subject matter which has interrelated topics, namely Internet governance, e-commerce, data protection and privacy, cybercrime, and cyber warfare and terrorism.

14.2 CYBERCRIME LEGISLATION IN THE NETHERLANDS

The main Act to deal with computer crimes in the Netherlands is the Computer Crime (CC) Act of 1993. In July 1999, a follow-up bill was introduced in the Parliament, the Computer Crime II, which was intended to refine and update several provisions of the Computer Crime Act. The following subsections present the specific crime legislations:

14.2.1 Specific Cybercrime Legislations

The following are some of the legislations available for specific cybercrimes:

Hacking

Hacking is penalized in Art. 138a of the Dutch Criminal Code (DCC). It is only punishable if someone infringes a (minimal) security measure. The maximum penalty is six months imprisonment or a fine of 4,500 Euros for 'simple' hacking and four years imprisonment or 11,250 Euros if the hacker copies data, if he/she hacks via public telecommunications and uses the processing capability, or hacks a third computer. However a minimum level of security protection is required. The CCII Bill penalizes trespassing in a computer as such, and mentions the breach of a security measure as an example of such trespassing.

Table 14.1 Offences and punishment associated with illegal interception

Article	Offence	Punishment
139a DCC	Interception of direct communications or non-telecom data (for closed areas and restricted to intercepting conversations)	Maximum penalty is a prison sentence of six months or a fine of 11,250 Euros
139b DCC	Interception of direct communications or non-telecom data (for other areas and restricted to intercepting conversations)	Maximum penalty is a prison sentence of three months or a fine of 4,500 Euros
139c DCC	Interception of telecommunications by technical means (interception of all other forms of communications including data)	Maximum penalty is a year's imprisonment or a fine of 11,250 Euros
139d DCC	Prohibition to place eavesdropping devices	Maximum penalty is a year's imprisonment or a fine of 11,250 Euros
139e DCC	Prohibition to pass on eavesdropping equipment or intercepted data	Maximum penalty is a year's imprisonment or a fine of 11,250 Euros
Art. 441 DCC	Prohibition to advertise for interception devices	Maximum penalty is a year's imprisonment or a fine of 11,250 Euros

Illegal Interception

Illegal interception is penalized by Art. 139 DCC and is presented in Table 14.1.

Art. 139a is concerned with the interception by technical means of voice communications or data communications. There are exceptions for, among others, participants to the communications, employers, and security services.

Data Manipulation and Viruses

Intentional manipulation of computer-related data is penalized in Art. 350a DCC. This includes deleting, changing, and adding data. The maximum penalty is two years imprisonment or a fine of 11,250 Euros. If the manipulation is committed after entering the computer through a public telecommunications network and if it results in serious damage, the maximum penalty rises to four years or 45,000 Euros.

Non-intentional (negligent) manipulation of computer-related data is penalized by Art. 350b DCC if serious damage is caused with a maximum penalty of one month's imprisonment or a fine of 2,250 Euros.

Intentionally making *computer viruses* available or disseminating them is penalized by Art. 350a DCC with a maximum penalty of four years imprisonment or a fine of 45,000 Euros. The unintentional (negligent) making available or dissemination of computer viruses is penalized by Art. 350b DCC with a maximum penalty of one month's imprisonment or a fine of 2,250 Euros.

System Interference, e-bombs, and DoS Attacks

System interference is also called *computer sabotage*. It is penalized under various provisions depending on the character of the system and of the interference. If the computer and networks are for the common good, intentional interference is punishable if the system is impeded or if the interference causes general danger to goods, services, or people according to Art. 161 sexies DCC. The maximum penalty varies from six months to 21 months of imprisonment for impeding a system and up to fifteen years if the interference causes someone's death. Negligent system interference is criminalized by Art. 161septies DCC and carries a maximum imprisonment of three months to one year.

Art. 351 DCC prohibits *intentional computer sabotage* (destruction or damage) of computers and telecom systems for the common good, regardless of the effect, with a maximum punishment of imprisonment for three years or a fine of 11,250 Euros. Negligent sabotage of such computers is penalized by Art. 351b DCC with lesser penalties.

Another form of system interference, *e-bombs*, is penalized in the Computer Crime II Bill enacted. Art. 138b DCC originally prohibits sending via the public telecoms network of data that are intended to block the recipient's access to the telecommunications network or service. This, however, only covers email bombing of specific targets but not denial-of-service (DoS) attacks. In the amended CCII Bill, Art. 138b penalizes the 'intentional and unlawful hindering of the access to or functioning of an automated work by offering or sending data to it' thus covering DoS attacks. The new provision will have a maximum penalty of imprisonment for one year or a fine of 11,250 Euros.

Spam

Spamming is not criminalized in the Criminal Code. In private law, a specific provision on spam was inserted in the Civil Code. Art. 7:46h of the Civil Code broadly provides that in the context of distance selling, sending a commercial message is only allowed with prior consent of the consumer (opt-in) or, for existing customers, a continuing possibility of opting-out. Moreover, each message has to include the true identity of the sender and a valid address for opting out.

Spamming is regulated in Article 11.7 Telecommunications Act and contains administrative provisions that mirror the civil rules of Art. 7:46h of the Civil Code. A single spam-related penalization has been enacted as a Corollary: Art. 1 Sub 2 of the Economic Offences Act which penalizes the anonymous sending of commercial or charitable email as an infringement with a maximum of six months' imprisonment or a fine of 11,250 Euros.

Misuse of Devices

Art. 6c of Cybercrime Convention (CCC) has provisions to criminalize the specific types of misuse of devices, which are as follows:

Misuse of devices or goods   Art. 234 DCC penalizes misuse of devices (goods or data) that the perpetrator knows to be designated for committing aggravated forgery.

Card forgery   Art. 232 DCC penalizes the provision, possession, receiving, obtaining, transport, sale, or transport of a forged payment card with a maximum of four years imprisonment.

Devices for telecom fraud   Art. 326c DCC penalizes the public offering, possession with the goal of distribution or import, and making or having available for profit of devices or data that are ostensibly designated for committing telecommunications fraud or keeping for profit with a maximum of one year imprisonment. If this happens on a professional basis, the maximum penalty increases to three years imprisonment.

Devices for oral or wire interception   Art. 441a DCC penalizes the advertisement of interception devices with a maximum penalty of two months' imprisonment.

Devices for software-protection circumvention   Art. 32a Copyright Act penalizes the public offering, possession with the goal of distribution, import, transport, export, and keeping for profit with a maximum penalty of six months imprisonment. This holds true only if the devices are exclusively designed to circumvent software-protection measures.

The CCII Bill penalizes the misuse of devices as follows:

1. Art. 139d para. 2 and 3 DCC penalizes misuse of devices or access codes with intent to commit hacking, e-bombing or DoS attacks, or illegal interception with up to six months imprisonment. The punishment is raised to a maximum of four years if the intent is to commit aggravated hacking (as in Art. 138a).
2. Art. 161sexies Section 2 DCC penalizes misuse of devices or access codes with intent to commit computer sabotage with up to one year imprisonment or a fine of 45,000 Euros.

In these provisions, following the CCC, 'misuse of devices' covers the manufacture, sale, obtaining, importation, distribution, or otherwise making available or having in one's possession devices that are primarily made suitable or designed to commit a certain crime.

14.2.2 Traditional Laws to Prosecute Cybercrimes

The following are some of the traditional laws that are applicable to prosecute cybercrimes:

Forgery

Computer-related forgery falls within the scope of the traditional provision on forgery Art. 225 DCC. The maximum penalty is six years imprisonment or a fine of 45,000 Euros. As provided in Art. 232 DCC, the intentional forgery of a payment card or value card carries a maximum penalty of six years imprisonment or a fine of 45,000 Euros. The use of such a card is punishable with the same penalty.

The CC II Bill extends this provision to cover all kinds of chip cards that are available to the general public and that are designed for payments or for other automated service provisioning.

Fraud

Computer-related fraud is dealt under the traditional provision on fraud, Art. 326 DCC. The maximum penalty is three years imprisonment or a fine of 45,000 Euros.

Other fraud-related offences that also cover computer-related crimes are *extortion* (Art. 317 DCC) and *blackmail* (Art. 318 DCC). The provision on extortion was changed in 2004 to include obtaining of pin codes and other data under threat of violence.

Telecom fraud has been specifically penalized in Art. 326c DCC. The use of a public telecoms service through technical means or false signals with the intention of not fully paying for it is punishable with up to three years imprisonment or a fine of 45,000 Euros.

Content-related Offences and Child Pornography

Content-related offences are punishable regardless of the medium in which the content has been published. These offences include discrimination (Art. 137c-g DCC), defamation of royalty (Art. 111–113 DCC), defamation of friendly heads of state (Art. 118–119 DCC) and defamation, libel, and slander (Art. 261–271 DCC).

Child pornography is penalized in Art. 240b DCC with a maximum penalty of four years imprisonment or a fine of 45,000 Euros. This includes the manufacture, distribution, publicly offering, and possession of pictures that show a minor in a sexual act. Doing this on a professional or habitual basis raises the maximum penalty to six years imprisonment.

Copyright Infringement

Article 31 of the Copyright Act criminalizes intentional infringement of someone else's copyright, punishable with a maximum imprisonment of six months. Intentionally offering for dissemination, stocking for multiplication or dissemination, importing or exporting, or keeping for pursuit of gain of an object containing a copyright infringement is punishable with maximally one year imprisonment (Article 31a Copyright Act), which rises to four years imprisonment if done as a profession or business (Article 31b). Articles 34 through 35d contain further offences, the most important of which is the intentional altering of copyrighted works in a way that is potentially harmful to their maker (Article 34).

Liability of Internet Service Providers

The *liability of Internet Service Providers* (ISPs) for illegal or unlawful content has been regulated as a consequence of the Electronic Commerce Directive. The major portion concerns civil liability as regulated in Art. 6:196c of the Civil Code. 'Mere conduit' providers are not liable; caching providers are not liable if they do not change information and if they operate according to generally recognized procedures; and providers of information services are not liable if they have no knowledge of unlawful content and if they remove or make inaccessible the information as soon as they do gain knowledge.

Art. 54a DCC determines that intermediaries who offer a telecommunications service consisting of transport or storage of data shall not be prosecuted as such if they do all that can reasonably be asked of them to ensure that the data is made inaccessible in response to an order from the public prosecutor. This exempts the ISP from liability. The prosecutor requires a warrant from the investigating judge for such an order so that there is an independent check by the courts on whether the information at issue really is illegal or unlawful.

14.2.3 Powers for Search and Seizure

Traditional search provisions cover computer searches (Articles 96b, 96c, 97, and 110 Dutch Code of Civil Procedural (DCCP). The general seizure provisions (Art. 95, 96, 96a, and 104 DCCP) can be used to seize data storage devices. Data as such cannot be seized since they are not considered 'goods' but they may be copied by law enforcement officers during a search, comparable to the copying of, for example, fingerprint marks.

The Data Production Orders Act introduced in Art. 125i DCCP provides the power to search in order to secure data (replacing the old Art. 125i DCCP, supra, Section 3.1.1). Since in certain cases there is a need to 'seize' rather than merely copy data (e.g., child porn or a virus program), the CCII Bill introduced powers to 'make data inaccessible' with Art. 125o DCCP. This can be done with data that are the object or the means of a crime by first copying and then deleting the data on the original device or by encrypting them. The final deletion of the data—or the restoration—must be ordered by a judge in court by Art. 354 DCCP.

In 1993, the Netherlands enacted a power to search a network by the Computer Crime Act. Art. 125j DCCP allows the person who conducts a search to also search computer networks from computers located at the search premises. The network search, however, may only be conducted to the degree that the network is lawfully accessible to the people who regularly stay in those premises. However, the network search cannot go beyond the Dutch borders.

Two further ancillary powers were introduced by the Computer Crime Act to the search and seizure procedures. These enable the investigating officer to order the undoing of a security measure (Art. 125k DCCP) and to order the decryption or handing over of a decryption key, of encrypted data (Art. 125k DCCP). The orders may not be given to suspects (Art. 125m-old DCCP).

As general safeguards in the procedures for investigating computers and data, obligations exist to delete retrieved data as soon as they are no longer relevant to the investigation except if they have to be used for a different case or registered in a serious crime register (Art. 125n DCCP) and to inform the administrator of an automated work from which data has been copied (Art. 125m-old DCCP). This notification requirement is broadened by the Act on Data Production Orders to cover notification of suspects, the controller of the data, and the right-holders of the place searched except in cases in which notification is not reasonably possible (Art. 125m-new DCCP).

14.2.4 Other ICT-related Investigation Powers

Direct eavesdropping is regulated by Art. 126l DCCP. It allows the public prosecutor to order an investigating officer to record confidential communications with a technical device in cases for which pre-trial detention is allowed and that seriously infringe the rule of law. The prosecutor needs authorization from the investigating judge for this.

A similar power exists for investigating plotted *organized crime* (Art. 126s DCCP). The power includes entering any premises to place an eavesdropping device. If the premise is a dwelling this can only be done in cases of crimes with a maximum punishment of eight years imprisonment or more and the judge has to authorize it explicitly.

14.3 CYBER LAWS IN MALAYSIA

The Malaysian government has passed certain cyber laws to reduce and control Internet abuses. The cyber laws that have been enacted and enforced since 1997 are as follows:

Communications and Multimedia Act 1998

This Act forms the basis for all other cyber laws in Malaysia which is governed by a specialized body in Information and Communication Technology (ICT) called Communication and Multimedia Commission. It ensures secure information, reliable network, affordable service, and high level of user confidence in the ICT industry all over Malaysia. It explains the roles and responsibilities of ISPs. It also states that there will be no filtering in accessing the Internet in Malaysia. Sending threatening messages may amount to a criminal offence under the Communications and Multimedia Act 1998 depending on the content or Section 503 of the Penal Code (criminal intimidation).

Table 14.2 Offences and punishments dealt with under Computer Crimes Act (CCA) 1997

Offences	Penalty
Unauthorized access to computer material (Section 3 of CCA 1997)	Fine not exceeding RM 50,000 or imprisonment not exceeding five years or both (RM is the Malaysian Ringgit)
Unauthorized access with intent to commit or facilitate commission of further offence (Section 4 of CCA 1997)	Fine not exceeding RM 150,000 or imprisonment not exceeding 10 years or both.
Unauthorized modification of the contents of any computer (Section 5 of CCA 1997)	(a) Fine not exceeding RM 100,000 or imprisonment not exceeding seven years or both (b) Fine not exceeding RM 150,000 or imprisonment not exceeding 10 years or both if the Act done is with intention of causing injury as defined in the Penal Code.
Wrongful communication (Section 6 of CCA 1997)	Fine not exceeding RM 25,000 or imprisonment not exceeding three years or both.
Abetments and attempts (Section 7 of CCA 1997)	Fine to be as for the principal offence but imprisonment not to exceed one half of the maximum term for the principal offence.

Computer Crimes Act 1997

This Act was enforced on 1 June 2000 and gives protection against the misuses of computers and computer criminal activities (Box 14.1). This Act deals with offences relating to the misuse of computers which are as follows:
1. Unauthorized access to computer material
2. Unauthorized access with intent to commit other offences
3. Unauthorized modification of computer contents.

The Act criminalizes certain offences and provides punishment as listed in Table 14.2.

Copyright Act (Amendment) 1997

Copyright protection in Malaysia is governed by the Copyright Act 1987. This Act was amended in 1997. It protects right-protected works from unauthorized copying and/or modification. The Act was enforced from 1 April 1999. It outlines the nature of works eligible for copyright, the scope of protection, and the manner in which the protection is accorded (Box 14.2). A unique feature of the Act is the inclusion of provisions for enforcing the Act. A special team of officers is appointed to enforce the Act.

> ### Box 14.1 Prosecution under Computer Crime Act 1997
>
> On 9 February 2012, a system analyst with the Lembaga Tabung Haji's Information Technology Department was charged under Section 5(1) of the Computer Crimes Act 1997 for three counts of tampering with the contents of the pilgrim's database between June and July 2010. In May 2014, he was found guilty of the charge and was sentenced to three years jail and was fined a total of RM 60,000 along with a jail term of eight months.

> **Box 14.2 Copyright Infringement**
>
> *Honda EX-5 Dream Copyright Infringement*
>
> The Japanese Automotive Motor Company, Honda, approached the Malaysian High Court for an interim injunction against the infringement of copyright in design of their motorcycle HONDA EX-5 DREAM. Honda claimed that the Malaysian automobile company infringes the copyrights of the two-dimensional drawings and its three-dimensional form in their production of COMEL MANJA JMP-100 (GS-5) motorcycle originally embodied in the motorcycle branded under the model of HONDA EX-5 DREAM of Honda.

Since 1 October 1998, this Act has been enforced to help prevent online transaction frauds. It provides both licensing and the regulation of certification authorities (CAs). Signor identity certification and the digital signature are issued by the CA.

Electronic Commerce Act 2006

This Act gave legal recognition to electronic messages in commercial transactions. It also provided how legal requirements could be fulfilled by using electronic messages. This Act also allowed the use of electronic means and other related matters to facilitate commercial transactions. It has been effective since 19 October 2006.

Electronic Government Activities Act 2007

The Malaysian government has enforced an Act to facilitate electronic delivery on government services to the public. It came into force on 1 January 2008.

Payment Systems Act 2003

This Act was enacted by the government on 1 November 2003. It covers both operators and issuers of designated payment instruments (DPIs). It also contains provisions to allow Bank Negara Malaysia (BNM) to perform its roles.

Penal Code, Including Chapter on Terrorism and Cyber Terrorism   This is of utmost importance in Malaysia because not all cybercrimes can be enforced using all cyber laws. Therefore, the penal code is used as a backup to charge criminals involved in cybercrimes. Examples include online fraud, online gambling, and online pornography.

14.4 CYBERCRIME LAWS IN THE UK

Cyber law in the United Kingdom (CLUK) of 226 pages, published in 2010 by Wolters Kluwer Law & Business Press, is composed of two parts: general introduction and body. The general introduction has five parts: the general background of the country, telecommunication infrastructure, the information and communications technologies market, e-commerce—facts and figures, and e-government initiatives.

The CLUK body consists of seven parts covering the following: ICT marketing management, protection of intellectual property in the ICT sector, ICT contacts, electronic transactions, non-contractual liability, privacy protection, and computer-related crimes. The body mainly introduces the legislations of telecom regulations as to the deregulation of the telecom industry, the opening and competition of telecommunications, and British Telecom (BT) privatization and commercialization.

Cybercrimes may be either cyber-enabled crimes or cyber-centric crimes. Cyber-centric crimes include unauthorized access to computer systems (e.g., hacking and the interception of communications) and new crimes brought about through the existence of computers. On the other hand, cyber-enabled crimes are crimes that have always existed but benefit from the existence of computers, for example, fraud (Box 14.3).

> **Box 14.3 Hacking**
>
> On 12 August 2015, at approximately 3:30 a.m., the three most popular Facebook pages of the University of Michigan—Michigan Football, Michigan Basketball, and Michigan Athletics—were found to have malicious postings. The Department of Information Technology Services (ITS) was first alerted of the inappropriate content by the user community. By 5:00 a.m., the ITS in turn had notified the University Director of Social Media, the Office of Public Affairs and Internal Communication, and the Department of Public Safety and Security and alerted the University of Michigan social media director to implement password changes on all official accounts and immediately access all third party applications that had been granted access to publish on the pages' behalf. The University Director of Social Media issued an acknowledgment of the hacking in collaboration with the Public Affairs and Michigan Athletics. At 8:38 a.m., Facebook took control and unpublished the three compromised pages. Within 15 minutes, the university page administrators were given their credentials and the process of cleaning the accounts began. By 10:00 a.m., the pages had gone live once again.

The various Acts to handle cybercrimes in the UK are as follows:

Computer Misuse Act 1990

The Act was introduced to criminalize unauthorized access to computer systems and to deal with hackers as the existing legislation was inadequate. It handles three offences as shown in Table 14.3. However, this Act has been amended by the Police and Justice Act 2006 and by the Serious Crime Act 2015 which introduced 3ZA and 3A.

Table 14.3 Offences under the Computer Misuse Act 1990

Section	Offences	Punishment
1	Unauthorized access to computer material	Up to 6 months of potential prison sentence and a fine of 5000 pounds
2	Unauthorized access with intent to commit or facilitate commission of further offences	Up to 5 years of potential prison sentence and an unlimited fine
3	Unauthorized modification of computer material	Up to 10 years of potential prison sentence and an unlimited fine
Amendments by the Police and Justice Act 2006 and the Serious Crime Act 2015		
Section	Offence	Punishment
3ZA	Unauthorized acts causing or creating risk of serious damage.	Maximum sentence of life imprisonment
3A	Making, supplying, or obtaining articles for use in offence under Section 1, 3, or 3ZA.	Potential prison sentence of two years imprisonment

Serious Crime Act 2015

Serious Crime Act 2015 amends the Computer Crime Act 1990 under Sections 42 and 43 to add the following offences (Box 14.4):
1. Section 42: Offences under 3A of Computer Misuse Act 1990 no longer require intent.
2. Section 43: Offence is said to be committed even if the accused is outside the United Kingdom at the time of the offence and as long as the Act is illegal in that country too and the offender is a United Kingdom national.

Box 14.4 Viral Attack

The first prosecution of an individual for distributing a computer virus came in 1995. Christopher Pile, aka 'the Black Baron' pleaded guilty to eleven charges under Sections 2 and 3 of the Computer Misuse Act and received an 18-month prison sentence. Pile created the viruses Pathogen and Queeg. Both pieces of malware were implemented in his simulated metamorphic encryption generator (SMEG) polymorphic engine, making them hard to detect, and both were designed to trash substantial portions of a victim's hard drive. He planted the viruses on bulletin boards disguised as games and, in one case, as an antivirus program. It was estimated that the viruses caused damage amounting to £1 million.

Police and Justice Act 2006

The Police and Justice Act amends the Computer Misuse Act 1990 to include 'unauthorised acts with intent to impair operation of computer' which effectively adds DoS attacks as an offence. It also states that the offence is said to have been committed even if DoS is only temporary and that it need not be against a specific computer, program, or data. Table 14.4 highlights the amendments made to Sections 1 and 3 of the Computer Misuse Act 1990.

Table 14.4 Amendments to the Computer Misuse Act 1990 by the Police Justice Act 2006

Section	Offences	Punishment
1	Unauthorized access to computer material	Up to two years of potential prison sentence and a fine of 5000 pounds
3	Unauthorized acts with intent to impair or with recklessness as to impairing, operation of computer, etc.	Up to 10 years of potential prison sentence and an unlimited fine

Terrorism Act 2000

The Terrorism Act 2000 came into force on 20 July 2000 and has eight Parts as shown in Tables 14.5 and 14.6. The offences dealt with in this Act are summarized in Table 14.6.

Table 14.5 Parts in the Terrorism Act 2000

Part	Description	Sections
1	Introductory	1–2
2	Proscribed organizations	3–13
3	Terrorist property	14–31
4	Terrorist investigations	32–39
5	Counterterrorism powers	40–53
6	Miscellaneous	54–63
7	Northern Ireland	65–113
8	General	114–131

Table 14.6 Offences dealt with in the Terrorism Act 2000

Sections		Description
11	Membership	Membership of proscribed organizations
13	Uniform	Commitment of an offence in a public place while wearing an item of clothing or carrying or displaying an article belonging to a proscribed organization
15	Fundraising	Fundraising and receipt of money and property for the purposes of terrorism
16	Use and possession	Possession of money and/or property with the intention of using it for the purpose of terrorism
17	Funding arrangements	Entering into or becoming concerned in an arrangement as a result of which money or other property is made available or to be made available to the other knowing or having reasonable cause to suspect it will or may be used for the purpose of terrorism
18	Money laundering	If a person enters into or becomes concerned with 'an arrangement which facilitates retention or control by or on behalf of another person of terrorist property by concealment, by removal from the jurisdiction, by transfer to nominees or in any other way'
44	Search of vehicles—authorizations	Failure to stop a vehicle when required to do so by a constable in the exercise of his powers
48	Authorizations	Parking of a vehicle in contravention of a prohibition or restriction
54	Weapons training	Providing instructions or training for the purposes of committing terrorist offences, or even by making an invitation to receive instruction or training one or more specific persons
56	Directing terrorist organization	Directing the activities of a terrorist organization
57	Possession of terrorist purposes	Possessing an article in circumstances which give rise to reasonable suspicion that its possession is for the purpose connected with the commission, preparation, or instigation of an act of terrorism
58	Collection of information	Collecting or making a record of information which is likely to be useful to a person committing or preparing an act of terrorism or possessing a document or record containing information of that kind where record refers to photographic or electronic record
59–61	Inciting terrorism overseas	Inciting terrorism overseas or within England and Wales, Northern Ireland, and Scotland.

Regulation of Investigatory Powers Act 2000

Regulation of Investigatory Powers Act 2000 (RIPA) covers the recording of transmissions by the police for the purposes of law enforcement and covers details such as interception with and without a warrant. No one can intercept a communication except either a law enforcement agent with an appropriate authorization (such as a warrant) or a party to the communication. A person may record a phone conversation to which they are a party legally but cannot share that recording with a third party.

Extradition Act 2003

Since cybercrimes have no legal borders, this Act is essential to deal with such crimes internationally. This Act emphasizes that an offender cannot escape justice by crossing borders. It attempts to extradite the requested person to the issuing state unless and until there is no breach of fundamental rights.

> **Box 14.5 DDoS Attack**
>
> *Everyone is a Potential Target: DDoS Attack on Boston Children's Hospital (Radware, 2015)*
>
> In 2014, Boston Children's Hospital became the first health care organization to be targeted by a hacktivist group. Because the hospital used the same ISP as seven other health care institutions in the area, the organized DDoS attacks had the potential to bring down multiple branches of Boston's Critical Health Care infrastructure. The DDoS attack launched against Boston Children's Hospital began with a threat (pre-strike boxing) and then involved three major strikes such as low rate attacks, attacks rump up, and attacks peak. As soon as the hospital became aware of the initial threat, it activated its multidisciplinary incident response team.

Part 1 of this Act implements the European Arrest Warrant (EAW) and the surrender procedures between member states. The UK has extradition relations with over 100 territories around the world. In urgent cases, the person a country requests to surrender for prosecution or for punishment can be arrested before the receipt of an EAW. An EAW must be received for a court hearing that must be held within 48 hours of the arrest.

The Extradition Act 2003, Part 2, defines a significant number of countries for which extradition relations exist although they are non-EU and therefore do not fall under the EAW scheme.

A person who is arrested for extradition will face two hearings: an 'initial hearing' that confirms the person's identity, informs them of the process, and fixes a date for the second hearing if the person does not voluntarily consent to extradition; the second hearing, the 'extradition hearing', is to allow a judge to ensure there are no bars to extradition as there are certain protections built into the Act that may prevent extradition.

Protection from Harassment Act 1997

This Act was introduced to deal with stalking. The Act gives both criminal and civil remedies.

Section 1 of the 1997 Protection from Harassment Act deals with the following criminal offences:

1. Pursuing a course of conduct amounting to harassment
2. A more serious offence where the conduct puts the victim in fear of violence

Under Section 2, such an Act could lead to a criminal penalty where the offence is subject to a maximum penalty of six months imprisonment, an unlimited fine, or both, and is arrestable.

Section 3 of the Act provides civil remedies.

A victim can pursue civil remedies when there is not enough evidence for a criminal prosecution, but perhaps enough to obtain a civil order. In criminal cases, the matter must be proved beyond reasonable doubt. The other sections of this Act are presented in Table 14.7.

Table 14.7 Sections in the Protection from Harassment Act 1997

	Sections
4	Putting people in fear of violence
4A	Stalking involving fear of violence, or serious alarm or distress
5	Restraining orders on conviction
5A	Restraining orders on acquittal

Communications Act 2003

The Communications Act came into force in 2003, superseding the Telecommunications Act 1984 and has provisions including accessing the Internet with no intention of paying becoming a criminal offence, political advertising on TV and radio being prohibited, and provisions being made for the visually impaired and hearing challenged television viewers.

> **Box 14.6 Section 127 of Communications Act 2003**
>
> A person is guilty of an offence if he or she:
>
> 1. sends by means of a public electronic communications network a message or other matter that is grossly offensive or of an indecent, obscene, or menacing character
> 2. or causes any such message or matter to be sent
>
> A person is guilty of an offence if, for the purpose of causing annoyance, inconvenience, or needless anxiety to another, he or she:
>
> 1. sends by means of a public electronic communications network, a message that he knows to be false;
> 2. causes such a message to be sent; or
> 3. persistently makes use of a public communications network
>
> A person guilty of an offence under this section shall be liable, on summary conviction, to imprisonment for a term not exceeding six months or to a fine not exceeding level five on the standard scale, or to both.

Section 127 covers the improper use of public electronic communications network and replaced Section 43 of the Telecommunications Act 1984 (Box 14.6).

1. The Malicious Communications Act 1988

Under the Malicious Communications Act 1988 it is an offence for any person to send another person a letter, electronic communication or article of any description which intends to cause distress or anxiety to the recipient. The offence carries a maximum sentence of 2 years' imprisonment.

2. Privacy and Electronic Regulations (EC Directive) 2003

It was introduced to address the problem of spam and is enforced by the Information Commissioner's Office, the UK's independent authority set up to promote access to official information and to protect personal information.

According to the regulations, companies must get an individual's permission before sending email or SMS messages (the law also applies to telephone calls and faxes) or should have sent a notification and got his/her consent.

However, there are significant limitations. In the first place, the regulations only apply to messages sent to individuals' email addresses and not business addresses. The penalties are also limited when compared to penalties for offences covered by the Computer Misuse Act. Breaches of the regulations must be reported to the Information Commissioner's Office which is responsible for deciding whether or not to take the offending organization to court. The offending organization may be fined up to £5,000 in a magistrates' court or up to an unlimited amount if the case is referred to trial by jury.

Legislation of ICT Intellectual Property

The British copyright system has been in constant development. On 1 January 1998, the new intellectual property law was established in the UK.

1956 Copyright Law enlarged the scope of protection of intellectual property rights. UK enacted the Patent Law in 1977 so as to make the European patent approved throughout the UK (Box 14.7). The Patent Act specifies that the patent right can be empowered to the product inventor or process innovator on the condition that the product is newly invented, innovative, and applicable to industrial development.

> **Box 14.7 Copyright Infringement**
>
> Author created illustrations and was the first owner of the illustrations. He also owned a publishing company called Publishing Ltd. A marketing company called Marketing Co. promoted Publishing Ltd under a development agreement. Publishing Ltd under an IP assignment transferred the IP ownership of illustrations to Author. Author's clothing business sold T-shirts featuring his illustrations, upon assuming ownership. A company called Infringing Co. exploited this without authority to do so. Infringing Co. compiled a catalogue of custom printed T-shirts and also sold t-shirts featuring Author's illustrations. Author claimed Infringing Co. infringed its copyright by copying, issuing copies of articles with Author's illustrations to the public, and communicating to the public. The court concluded that Publishing Ltd was the original copyright owner and had assigned its rights to Author who was entitled to the IP ownership, exploit the illustrations.

The 1988 Copyright has provided a legal foundation for copyright design and patent law including the written or other forms of recording work such as computer derivative works. The literature amendment includes databases and computer programs previously prepared by programming information.

Provisions for Identity Theft

The three distinct Acts of law that cover identity crime in the UK are as follows:
1. The Fraud Act 2006
2. The Identity Documents Act 2010
3. The Forgery and Counterfeiting Act 1981

The Fraud Act 2006 has two relevant pieces of legislation relating to identity crime. The first is dishonestly making a false representation to make a gain for oneself or another or to cause loss to another or to expose another to a risk of loss [Fraud Act 2006 Sections 1(2a), (3) and (4) and 2]. The second is Possession etc., of articles for use in frauds (Fraud Act 2006, Section 6).

The Identity Documents Act 2010 [formerly the Identity Cards Act 2006, Section 25(5) and (7)] include an offence for possessing or controlling a false or improperly obtained ID card or which relates to another, or apparatus etc., for making false ID cards.

The Forgery and Counterfeiting Act 1981 also has two pieces of legislation relating to identity crime. The first offence is 'using a false instrument etc. in respect of scheduled drug' (Forgery and Counterfeiting Act 1981, Sections 3 and 4). The second is 'using a false instrument or a copy of a false instrument' (Forgery and Counterfeiting Act 1981, Sections 3 and 4).

14.5 CYBERCRIME LAWS OF THE UNITED STATES

Cybercrime laws of the United States are broadly classified into two: substantive cybercrime laws and procedural cybercrime laws. Title 18 of the United States Code (U.S.C.) is the main criminal code of the federal government of the United States. It deals with federal crimes and criminal procedures. The symbol § stands for section/subsection.

Substantive Cybercrime Laws

Substantive cybercrime laws encompass laws to handle and prohibit cybercrimes such as online identity theft, hacking, intrusion into computer systems, child pornography, intellectual property, and online gambling, among others. They are summarized in Table 14.8.

Table 14.8 Substantive cybercrime laws of the United States

Title and Section/Sub-section	Description
18 U.S.C. § 1028	Fraud and related activity in connection with identification documents, authentication features, and information
18 U.S.C. § 1028A	Aggravated identity theft
18 U.S.C. § 1029	Fraud and related activity in connection with access devices
18 U.S.C. § 1030	Fraud and related activity in connection with computers
18 U.S.C. § 1037	Fraud and related activity in connection with electronic mail
18 U.S.C. § 1343	Fraud by wire, radio, or television
18 U.S.C. § 1362	Malicious mischief related to communications lines, stations, or systems
18 U.S.C. § 1462	Importation or transportation of obscene matters
18 U.S.C. § 1465	Transportation of obscene matters for sale or distribution
18 U.S.C. § 1466A	Obscene visual representation of sexual abuse of children
18 U.S.C. § 2251	Sexual exploitation of children
18 U.S.C. § 2252	Certain activities relating to material involving the sexual exploitation of minors
18 U.S.C. § 2252A	Certain activities relating to material constituting or containing child pornography
18 U.S.C. § 2252B	Misleading domain names on the Internet to deceive minors
18 U.S.C. § 2252C	Misleading words or digital images on the Internet
18 U.S.C. § 2425	Use of interstate facilities to transmit information about a minor
18 U.S.C. § 2319	Criminal infringement of a copyright
17 U.S.C. § 506	Criminal offences (related to copyright)
47 U.S.C. 605	Unauthorized publication or use of communications
The Unlawful Internet Gambling Enforcement Act of 2006	

Procedural Cybercrime Laws

This includes laws presented in Table 14.9 which govern the authority to preserve and obtain electronic data from third parties including ISPs, to intercept electronic communications, and to search and seize electronic evidence.

Table 14.9 Procedural cybercrime laws of the United States

Title and Section/Subsection	Description
18 U.S.C. § 2510-2522	Interception of wire, oral, or electronic communication
18 U.S.C. § 2701-2712	Preservation and disclosure of stored wire and electronic communication
18 U.S.C. § 3121-3127	Pen registers, and trap and trace devices

14.5.1 Computer Fraud and Abuse Act

The first federal computer crime statute was the Computer Fraud and Abuse Act of 1984 (CFAA). It is the federal anti-hacking statute that prohibits unauthorized access to computers and networks. Originally, CFAA had major limitations which are as follows:

1. It required proof that the person accessed the computer without authorization. Thus, merely viewing data stored on the computer was not illegal even if access was gained without authorization.
2. CFAA could not prosecute crimes that involved the use of computers as it focused only on the mode of entry into the computer.

In 1994, the Computer Fraud and Abuse Act was amended to deal with the problem of 'malicious code' such as viruses, worms, and other programs designed to alter, damage, or destroy data on a computer. Earlier CFAA only focused on access of the computer system and not on how that computer system was used. The amended CFAA could prosecute those who transmitted a program, information, code or command to a computer or computer system with the intent to cause damage to the computer or information in the computer or prevent the use of the system without the knowledge or the authorization of the owners of that computer. CFAA was amended till 2008, far beyond the original intent (Boxes 14.8–14.10).

CFAA is also known as Title 18 U.S.C Section 1030. Section 1030(a) includes seven types of computer activities which CFAA criminalizes and are as follows:

1. Unauthorized access of a computer to obtain national security information with an intent to harm the United States or for the benefit of a foreign nation
2. Unauthorized access of a computer to obtain protected financial or credit information
3. Unauthorized access of a computer used by the federal government
4. Unauthorized access to a protected computer with the intent to defraud
5. Intentional damage of a protected computer
6. Fraudulent trafficking of computer passwords and any other information that could be used to gain access to a protected computer
7. Threatening a protected computer with the intent of extorting money or something else of value

Box 14.8 Viral Attack

The Medical Practice was hit by CryptoLocker, a type of ransomware virus. CryptoLocker rendered the Medical Practice inoperable for several days and crippled its technology for more than a week. The attack made its way onto one of the Practice's computers via an email attachment which had the appearance of a vendor invoice. It searched for files to encrypt on a computer. The encrypted files would be present in the computer as well as in the network that was accessible via mapped network drives. The files which were present on any drive or network could be located and accessed by the ransomware.

The Practice immediately went to paper format for scheduling, clinic notes, and prescriptions. Their IT department downloaded all of its backup data and then uploaded it to its server to replace the corrupted data. The entire process took several days as the backup data was stored offsite, which required transportation of the data; the data needed to be cleaned with antivirus software; and then settings and policies needed to be recreated. The Practice provided an electronic incident report to the FBI and a hard copy to the local police department.

They hired an IT service provider knowledgeable in health care data security and upgraded its antivirus software. Software was added to capture and back up all emails. An email filter was put in place to catch, clean, and filter incoming emails. They also added a local data backup system, which would allow the data replacement process to be completed in a few hours rather than days. The Practice Manager filed a report with the Practice's insurance company hoping to recover some of the lost revenue but the policy did not cover malware attacks.

Box 14.9 Financial Crime

The defendant ran a business called 'Business A'. He was charged for his association with an online marketplace on Tor network that was popular for the sale of illegal narcotics, stolen credit cards, etc. The investigation by FBI began in 2013. The defendant stole bitcoins from users and vendors by trying to access the online marketplace on the Tor network. In order to steal the users' login credentials, the defendant adopted two ways: posting fake links that directed the users to a fake login page that was hosted on a laptop at the defendant's house and port forwarding the users, upon clicking the actual marketplace (especially where the users log in), through the defendant's computer server. Once the defendant had access to a user's account, he would use a program to notify him when a deposit was made into the user's bitcoin wallet. The defendant then used an online bitcoin tumbling service when transferring the bitcoins to hide his trial. The bitcoins would then be deposited into the defendant's bitcoin wallet.

Box 14.10 Hacking

Peace Dumps Yahoo User Data on the Dark Web

The well-known cybercriminal, Peace, listed 200 million records of Yahoo user credentials for sale on the dark web. The data included usernames, passwords that were hashed using the MD5 algorithm, and date of birth. The data was apparently collected illegitimately during a 2012 hack. Peace demanded 3 bitcoins or about $1,860 USD for the hack.

A more detailed listing of Section 1030(a) is as follows:

1. Section 1030(a)(1) makes it illegal to access a computer without authorization or in excess of one's authorization and obtain information about national defence, foreign relations, or restricted data (data concerning the design, manufacture, or utilization of atomic weapons and production of nuclear material) and not the unauthorized possession of it or its transmission. Section 1030(a)(1) requires proof that the individual knowingly accessed the computer without authority or in excess of authorization for the purpose of obtaining classified or protected information.
2. Section 1030(a)(2) makes it illegal to intentionally access a computer without authorization or in excess of authorization in order to obtain records of a financial institution or to obtain personal records of consumers from a consumer reporting agency. This section also makes it illegal to obtain information from any department or agency of the United States or any protected computer that is involved in interstate or foreign communication. It can potentially cover email servers, routers, and even personal computers if it can be convincingly proven that they are used in interstate communication. The primary purpose of Section 1030(a)(2) is to protect the confidentiality of computer data.
3. Section 1030(a)(3) covers the unauthorized access to any federal government computer making it illegal to access any government computer without authorization. While Section 1030(a)(2) covers the access of federal government computers with the intent of obtaining protected information, Section 1030(a)(3) covers all unauthorized access to federal computers regardless of any information having been obtained or not.
4. Section 1030(a)(4) covers computer fraud and any use of a computer to commit a fraud and exempts frauds where the value that was obtained was the use of the computer itself and where the value of such use is not more than $5000 in any one-year period. Originally, CFAA did not consider trespass a crime, but Congress recognized that usage of a computer has value of its own but nonetheless included the $5000 threshold to prevent turning every case of trespass into a fraud felony.

5. Section 1030(a)(5) is the most widely used section of CFAA and covers hacking and malicious code such as viruses and worms that do or attempt to cause damage to protected computers. In essence, it is a crime to cause damage or attempt to cause damage to a computer which would result in financial losses of more than $5000, loss or alteration of medical data, physical injury to a person, a threat to public health safety or affect administration of justice, national defence, or national security.
6. Section 1030(a)(6) covers trafficking of passwords or similar information which can be used to access computers. This section makes it a crime if such trafficking affects interstate or foreign commerce or the computer in question is in use by the US government.
7. Section 1030(a)(7) makes it illegal to use interstate or foreign communication to threaten a protected computer with the intent of extorting money or other things of value. It also covers the use of any interstate or international communication method when used in transmitting threats against computers, computer networks, and their data and programs. This includes mail, telephone, or any computer communication.

The provisions of the Computer Fraud and Abuse Act 18 U.S.C. § 1030 are presented in Table 14.10.

Table 14.10 Provisions of the Computer Fraud and Abuse Act 18 U.S.C. § 1030

Section	Offence	Punishment
1030 (a)(1)	Obtaining national security information	Maximum prison sentence of 10 years and 20 years in case of second conviction
1030 (a)(2)	Accessing a computer and obtaining information	Maximum prison sentence from one to five years and 10 years in case of second conviction
1030 (a)(3)	Trespassing on a government computer	Maximum prison sentence of one year and 10 years in case of second conviction
1030 (a)(4)	Accessing a computer to defraud and obtain value	Maximum prison sentence of five years and 10 years in case of second conviction
1030 (a)(5)(A)	Intentionally damaging by knowing transmission	Maximum prison sentence of one or 10 years and 20 years in case of second conviction
1030 (a)(5)(B)	Recklessly damaging by intentional access	Maximum prison sentence of one or five years and 20 years in case of second conviction
1030 (a)(5)(C)	Negligently causing damage and loss by intentional access	Maximum prison sentence of one year and 20 years in case of second conviction
1030 (a)(6)	Trafficking of passwords	Maximum prison sentence of one year and 20 years in case of second conviction
1030 (a)(7)	Extortion involving computers	Maximum prison sentence of five years and 10 years in case of second conviction
1030 (b)	Attempt and conspiracy to commit such an offence	Maximum prison sentence of 10 years for attempt but no penalty specified for conspiracy in Section (C)

14.5.2 Provisions for Handling Cyber Stalking

18 USC § 2261A was the stalking statute in federal law. It was amended to include the use of an 'interactive computer service' to 'engage in a course of conduct that causes substantial emotional distress to that person or places that person in reasonable fear of the death of, or serious bodily injury to' the victim or a member of the victim's family. Even then the statue became inapplicable to many cyber stalking cases as it requires that the defendant physically travel across state lines (Box 14.11).

The federal government also regulates the use of interstate communications to convey threats in 18 USC § 875. Section (c) of that statute applies to some forms of cyber stalking.

> **Box 14.11 18 USC § 875 (c) Use of Interstate Communications to Convey Threats**
>
> Whoever transmits in interstate or foreign commerce any communication containing any threat to kidnap any person or any threat to injure the person of another, shall be fined under this title or imprisoned not more than five years, or both.

While the statute 18 USC § 875 does not require physical travel over state lines and can apply to any form of communication, it requires that the communication be a threat to kidnap or injure the victim. Thus, 18 USC § 875 is not an ideal solution for the problem of cyber stalking in federal law because in cyber stalking the perpetrator intends to stalk, harass, or threaten the victim in a series of events that terrifies or creates a sense of fear to the victim but never conveys an explicit threat.

47 USC § 223 is another federal statute that terms as crime using a telecommunications device to knowingly send anyone 'any comment, request, suggestion, proposal, image, or other communication which is obscene or child pornography, with intent to annoy, abuse, threaten, or harass another person'. The statute further states that a telecommunications device does not include an 'interactive computer service'. Thus, the statute could protect victims whose stalkers have used texting or phone calls to contact them. It does not help victims whose stalkers use social media, forums, and email to harass them.

14.5.3 Provisions to Handle Cyber Terrorism

The Department of Homeland Security (DHS) is a Federal Department employed by the United States government and was established in 2003. DHS is responsible for both the regulation and oversight of matters considered to be acts of terrorism jeopardizing the well-being and safety of the general American populace. The primary concerns of the department are as follows:

- Protecting American citizens from cyber terrorism which includes virtual attacks from within the nation and internationally
- Protecting and preserving the liberties and freedom established within the United States Constitution while maintaining digital, virtual, and electronic safety and national security

In order to further deter potential threats of cyber terrorism, the DHS enacted the Patriot Act which permits them to collect information to prevent terrorist activities. To prevent prospective acts of cyber terrorism, the collection of information is authorized to take place on a virtual level, in the event that a legitimate threat has been detected.

14.5.4 Electronic Communications Privacy Act

The Electronic Communications Privacy Act (ECPA) is an amendment to the federal wiretap law and was enacted in 1986. The Act made it illegal to intercept stored or transmitted electronic communication without authorization. ECPA sets out the provisions for access, use, disclosure, interception, and privacy protections of electronic communications that affect interstate or foreign commerce (Boxes 14.12 and 14.13).

The Act prohibits illegal access and certain disclosures of communication contents. In addition, ECPA prevents government entities from requiring disclosure of electronic communications by a provider such as an ISP without first going through a proper legal procedure.

ECPA was amended in 1994 by the Communications Assistance for Law Enforcement Act (CALEA). CALEA requires that the ISPs have capabilities in their networks that allow the law enforcement team to carry out electronic surveillance of specific individuals when there is a need. However, CALEA did not remove the need for a warrant before such surveillance could be carried out.

> **Box 14.12 Privacy Violation**
>
> A victim received two calls from different people who said that they got an SMS from their provider asking them to call the mobile number (victim's mother's mobile number) so as to rectify faults with it. The victim had reported a fault on her mother's mobile number to the service provider. The victim lodged a complaint as she wanted an explanation and an assurance that the issue would not occur anymore. After investigation, it was discovered that one of the staff members had accidentally copied and pasted the customer's mobile number into the SMS sent to other customers. The victim asked for compensation for breach of privacy. The provider offered $500 and removed the details of the victim's mother from its systems to ensure that the problem would not happen again.

> **Box 14.13 Identity Theft**
>
> *Largest Identity Theft in US History*
>
> In a case involving a $13 million scam, Amar Singh and his wife Neha Punjabi Singh were found guilty of identity theft and enterprises corruption charges. The two were part of a group of 111 people who were arrested for taking part in an operation that netted the millions. According to court documents, the cyber criminals would receive information about unknown people from various foreign countries such as Russia and China as well as via state-wide suppliers who would use a skimming device to swipe consumer credit card information at retail or food establishments and illegal identification gathered from websites.

14.5.5 Cyber Security Enhancement Act

Cyber Security Enhancement Act (CSEA) granted sweeping powers to the law enforcement organizations and increased penalties that were set out in the Computer Fraud and Abuse Act.

Prior to CSEA, ISPs were forbidden from knowingly divulging the personal details of their customers. For example, to gain the contents of an email stored on the ISP's servers, the government needed a search warrant.

CSEA reduced the extent of privacy of stored data in the following ways:

1. Allowing ISP to voluntarily hand over personal information about its customers to a government agent, not just law enforcement officials, if the ISP has a reason to believe that the information concerns a serious crime. Thus, the law enforcement team is allowed to gain access to data without a warrant, which they would have previously required.
2. Allowing ISPs to let the law enforcement intercept electronic communications on its computers if they believe that it belongs to a trespasser who is not authorized by the ISP to be on their computer, thus completely bypassing any need for a warrant.

The Act also permits harsher sentences of up to 20 years for individuals who knowingly or recklessly commit a computer crime that results in death or serious body injury. In addition, CSEA increases penalties for first-time interceptors of cellular phone traffic, thus removing a safety measure enjoyed by radio enthusiasts.

14.5.6 Digital Millennium Copyright Act

The Digital Millennium Copyright Act (DMCA) was enacted in 1998 to update world copyright laws to deal with the new technology. The DMCA prohibits circumventing a technological measure designed to protect a copyright. By technological measure DMCA means an access control technology which can take many forms such as copy protection on CDs, requiring CD-keys or product codes to use installed software and so on. As such, anyone attempting to disable or bypass such a technological measure would be in violation of the law.

> **Box 14.14 Copyright Infringement**
>
> *Apple vs Microsoft*
>
> The battle between the tech giants started with who invented the graphical user interface (GUI). It seemed that although Microsoft helped to develop Macintosh, Jean Louis Gassee, who had taken over from Steve Jobs at the time, refused to allow Microsoft to use their software. Bill Gates pressed on nonetheless, and decided to add additional features to early prototypes of the Macintosh. When Gassee noted the software, he was enraged. However, he didn't want a lawsuit and ended up agreeing to license the Mac's visual displays. However, Windows 2.0 turned out to be almost identical and Gassee believed it to be a breach of contract because he allowed their software to be used for 1.0 and not for future versions. So without warning, Apple filed a lawsuit against Microsoft in 1988. Apple's case included 189 contested visual displays that violated its copyright. In 1989, the court ruled that 179 of the 189 disputed displays were covered by the existing licence. Furthermore, the other ten were not violations of Apple's copyright due to the merger doctrine where the idea-expression divide limits the scope of copyright protection by differentiating an idea from the manifestation of that idea. The lawsuit was decided in Microsoft's favour in 1993.

DMCA prohibits the following:

1. Manufacture or sale of devices or programs whose primary purpose is to circumvent access control technology
2. Removal or alteration of information identifying the author, copyright holder, performer, or director of a work and the terms and conditions for use of a work for the purpose of facilitating copyright infringement

The Act provides civil remedies as well as criminal penalties for violating the copyright protection. DMCA grants several exceptions from its prohibition on circumventing access control measures.

The Act provides an exemption for the law enforcement and government agencies. In addition, DMCA specifically permits the manufacture and sale of technology whose sole purpose is to help parents control what their children view on the Internet.

The Act eliminates the liability of ISPs if infringing material is transmitted through their network and computers so long as the ISP has no control over the content of the materials on its network and no copy is maintained on the service provider's system. In addition, ISPs cannot be held liable for infringement if copies of such material are automatically made for the purpose of temporary storage, that is, caching, if infringing materials or a link to it are stored and the ISP is unaware of such infringing activity. It does not receive a financial benefit that can be attributed to the infringing activity in cases where the ISP has the right and the ability to control such an activity and acts expeditiously to disable or remove any infringing material when notified in writing of such an activity.

14.5.7 Traditional Laws to Prosecute Cybercrime

Some of the traditional laws that could be used to prosecute cybercrime are as follows:

Economic Espionage Act

The Economic Espionage Act (EEA) was enacted in 1996 to put an end to trade secret misappropriation. EEA makes it a crime to knowingly commit an offence that benefits a foreign government or a foreign agent. The Act defines as crime either if trade secrets are stolen knowingly or if any attempt of stealing is done with the intention to benefit someone other than the owner of the trade secrets. EEA defines stealing of trade secrets as copying, duplicating, sketching, drawing, photographing, downloading, uploading, altering, destroying, photocopying, replicating, transmitting, delivering, sending, mailing, communicating, or conveying trade secrets without authorization. The Act, while not specifically targeted at computer crimes, nonetheless covers the use of computers (Box 14.14).

Other Criminal Copyright Infringement Statute
The National Stolen Property Act and Wire Fraud Statute fall under this category.

Fraudulent Online Identity Sanctions Act
The Fraudulent Online Identity Sanctions Act (FOISA) attempts to tackle the problem of criminals registering online domains under false identification. FOISA has a provision that increases the jail term if a person has used a domain to commit a crime involving copyright and trademark infringement and has obtained the domain by providing false contact information to a domain name registrar. The law would not make providing false contact information to domain name registrars a crime by itself; only if that domain is then used in committing a crime would FOISA be used against the criminal.

Computer Software Privacy and Control Act
The Computer Software Privacy and Control Act (CSPCA) is meant to deal with the problem of spyware and adware. The Act would prohibit transmission of software that collects and transmits personal information about the owner or operator of the computer and monitors and transmits web pages accessed by the owner or the operator, or modifies default computer settings like that of the home page, unless the owner or the operator gives their consent and the software has an uninstall option built into it.

State Laws
Apart from the federal government, many states have also passed computer crime laws. For example, Virginia passed the Virginia Internet Policy Act, composed of seven bills: Virginia Computer Crimes Act; Encryption Used in Criminal Activity; Encryption Technology; Virginia Computer Crimes Act Penalties; Freedom of Information; Privacy Protection; and Child Pornography and Indecent Liberties with Children in 1999 (Box 14.15).

14.5.8 Summary of US Federal Cyber Laws
Table 14.11 presents some of the frequently committed cybercrimes and the applicable US cyber laws.

Table 14.11 Summary of US federal laws

Cybercrime	Applicable Laws
Internet fraud	15 U.S.C. § 45, 52 (unfair or deceptive acts or practices; false advertisements) 15 U.S.C. § 1644 (credit card fraud) 18 U.S.C. § 1028, 1029, 1030 (fraud in connection with identification documents and information, access devices, and computers) 18 U.S.C. § 1341 et seq. (mail, wire, and bank fraud) 18 U.S.C. § 1345 (injunctions against fraud) 18 U.S.C. § 1956, 1957 (money laundering)
Online child pornography, child luring and related activities	18 U.S.C. § 2251 et seq. (sexual exploitation and other abuse of children) 18 U.S.C. § 2421 et seq. (transportation for illegal sexual activity)
Internet sale of prescription drugs and controlled substances	15 U.S.C. § 45 et seq. (unfair or deceptive acts or practices; false advertisements) 18 U.S.C. § 545 (smuggling goods into the United States) 18 U.S.C. § 1341 et seq. (mail, wire, and bank fraud; injunctions against fraud) 21 U.S.C. § 301 et seq. (Federal Food, Drug, and Cosmetic Act) 21 U.S.C. § 822, 829, 841, 863, 951-971 (Drug Abuse Prevention and Control)
Internet sale of firearms	18 U.S.C. § 921 et seq. (firearms)

(Contd)

Table 14.11 (Contd)

Cybercrime	Applicable Laws
Internet gambling	15 U.S.C. § 3001 et seq. (Interstate Horseracing Act) 18 U.S.C. § 1084 (transmission of wagering information) 18 U.S.C. § 1301 et seq. (lotteries) 18 U.S.C. § 1952 (interstate and foreign travel or transportation in aid of racketeering enterprises) 18 U.S.C. § 1953 (interstate transportation of wagering paraphernalia) 18 U.S.C. § 1955 (prohibition of illegal gambling businesses) 28 U.S.C. § 3701, 3702, 3703, 3704 (professional and amateur sports protection)
Internet sale of alcohol	18 U.S.C. § 1261 et seq. (liquor traffic) 27 U.S.C. § 122, 204 (shipments into states for possession or sale in violation of state law)
Online securities fraud	15 U.S.C. § 77e, 77j, 77q, 77x, 78i, 78j, 78l, 78o, 78ff (securities fraud)
Software piracy and intellectual property theft	17 U.S.C. § 506 (criminal copyright infringement) 17 U.S.C. § 1201 et seq. (copyright protection and management systems) 18 U.S.C. § 545 (smuggling goods into the United States) 18 U.S.C. § 1341, 1343 (frauds and swindles) 18 U.S.C. § 1831 et seq. (protection of trade secrets) 18 U.S.C. § 2318, 2319, 2320 (trafficking in counterfeit labels for phonorecords, copies of computer programs or computer program documentation, or packaging and copies of motion pictures or other audiovisual works)

Box 14.15 Pornography

Largest Child Pornography cases in the US

Two men pleaded guilty to felony charges in a federal court for their participation in an international criminal network known as Dreamboard which produced and disseminated depictions of graphic child sexual abuse via the Internet. Dreamboard was a member-only online bulletin board created and operated to promote pedophilia and encourage the sexual abuse of young children in an environment designed to avoid law enforcement detection. According to court documents, Dreamboard members traded graphic images and videos of adults molesting children. The prospective members had to create and share child pornography to gain entry into the group and to maintain membership once accepted.

14.6 AUSTRALIAN LAWS RELATED TO PRIVACY AND CYBER SECURITY DOMAINS

Australian laws specific to privacy and cyber security domains as well as other federal laws are presented in the following subsections.

14.6.1 Legal, Legislative, and Regulatory Environment

The cyber security-related legal, legislative, and regulatory obligations applicable to all private organizations and government agencies are as follows:

Crimes Act 1914

This Act codifies offences and functions alongside state legislation. The Crimes Amendment (Fraud, Identity, and Forgery Offences) Act 2009 No 99 amended the Crimes Act with respect to fraud, identity, forgery, and other related offences. Part 4AA deals with fraud, 4AB with identity offences, Part 4AC with money laundering, Part AD with criminal destruction and damage, Part 4AE with offences relating to transport services, Part 5 with forgery, and Part 5A with false and misleading information.

Criminal Code Act 1995

The cyber security-related offences are addressed in the Criminal Code Act 1995 (Criminal Code). This legislation was enacted because of some of the popular viruses of the late 1990s, such as 'Melissa' and 'ILOVEYOU'. It abolished all common law offences and superseded the Crimes Act 1914. It applies to serious online harassment and online bullying behaviour. Under the Criminal Code Act 1995 (Cth) it is an offence to use the Internet, social media, or a telephone to menace, harass, or cause offence. The maximum penalty for this offence is three years imprisonment or a fine of more than $30,000.

Table 14.12 Offences created in Cybercrime Act 2001

Offence	Punishment
Unauthorized access, modification, or impairment to commit a serious offence	Maximum penalty equal to the maximum penalty for the serious offence
Unauthorized modification of data to cause impairment (especially hacking)	Maximum penalty of 10 years imprisonment
Unauthorized impairment of electronic communications	Maximum penalty of 10 years imprisonment
Possession of data with intent to commit a computer offence	Maximum penalty of three years imprisonment
Supply of data with intent to commit a computer offence	Maximum penalty of three years imprisonment
Unauthorized access to restricted data	Maximum penalty of two years imprisonment
Unauthorized impairment of data held in a computer disk, credit card, and so forth	Maximum penalty of two years imprisonment

Cybercrime Act 2001

This Act offers more comprehensive regulation of computer- and Internet-related offences such as unlawful access and computer trespass, damaging data and impeding access to computers, theft of data, computer fraud, cyber stalking and harassment, and possession of child pornography. It creates a number of investigation powers and criminal offences designed to protect the security, reliability, and integrity of computer data and electronic communications. Further, it enhances the applicability of the existing search-and-seizure provisions relating to electronically stored data.

The Cybercrime Act adds a new Part 10.7 to the Criminal Code Act 1995 (Cth). This Act creates seven computer offences, as listed in Table 14.12 (see Box 14.16).

Spam Act 2003

This Act establishes a scheme for the regulation of commercial email and other types of electronic messages. It restricts unauthorized, unsolicited electronic messages with some exceptions. Rules for consent, identification of sender, and the unsubscribe features are explained in this Act.

Spam Act is enforced by Australian Communications and Media Authority (ACMA). This Act prohibits sending of unsolicited commercial electronic messages via email, short message service (SMS), multimedia message service (MMS), and instant messages with an Australian link. If the spam originates in Australia or a spam is destined to a recipient in Australia, it is said to have an Australian link. Voice calls and fax messages are not covered by this Act. Instead these two are managed by the 'do not call register' maintained by ACMA. Australian businesses, especially repeat corporate offenders that fail to comply with this Act, can be fined up to AUD $1.1 million per day.

> **Box 14.16 Cyber Stalking**
>
> A victim used one particular chat room on a regular basis and there was an offender whom she got to know quite well. The next time they were chatting he started asking her very personal and gross questions. She realized his behaviour was most improper, and so ended the conversation straightaway and blocked him. However, he started harassing her and sending her disgusting emails, attaching explicit photos of himself. When she told him to stop this, he sent even more abusive messages and graphic photos. The victim's parents contacted the ISP and changed her email address as well as blocked him. As the offender knew the chat rooms she used regularly, he posted messages stating false and abusive points about her. Suddenly, the victim started getting threatening text messages from an unknown mobile number. Finally the victim's parents reported the issue to the police. The police discovered that various pieces of information were available about the victim on the Internet; these would have helped the offender to track her down. The victim had used her name, age, and photo on the chat profile, and this could be viewed by anyone. Her netball team had a website that listed when and where her games were each week, and her mobile number had been posted online because of a charity event she was organizing.

Government bodies, registered charities, political parties, and educational institutions can send messages without consent. All others have to comply with three main rules which are as follows:

Consent Only send commercial electronic messages with the addressee's consent, either expressed or inferred

Identification Include clear and accurate information about the person or business that is responsible for sending the commercial electronic message

Unsubscribe Ensure that a functional unsubscribe facility is included in all commercial electronic messages and deal with unsubscribe request promptly

The full definition of a commercial message according to ACMA is one that

1. offers, advertises, or promotes the supply of goods, services, land, business, or investment opportunities;
2. advertises or promotes a supplier of goods, services, land, or a provider of business or investment opportunities; and
3. helps a person dishonestly obtain property, for commercial advantage or other gains, from another person.

There are two types of consent: express consent and inferred consent. Express consent means the message receiver has deliberately and intentionally opted-in to receive electronic messages from the sender. Examples of express consent include the following: voluntarily providing an email address, ticking a box next to a consent acceptance page on a website, and providing a mobile number to receive such messages. Inferred consent means that there is already an established relationship with the message sender. This type of consent is tricky and the message sending organization must carefully note this before sending commercial messages.

All commercial electronic messages must have clear and accurate identification information so that the receiver can contact the sender. Identifying information can be the 'from' field or subject line of an email, a website address, identification of an SMS or MMS message, and the body of the message text.

A sender's commercial messages must have an 'unsubscribe' facility with clear instructions on how to unsubscribe. Some unsubscribe facilities can be a text message reply with 'Stop' to unsubscribe, preference change option on a website, or an unsubscribe link at the bottom of the commercial electronic message (Box 14.17).

> **Box 14.17 Theft of Data**
>
> The complainant stated that she attended the respondent's counselling service about family issues. The complainant alleged that when leaving the counselling service, the counsellor disclosed information about her personal issues to a friend waiting for her in the waiting room. The respondent believed the complainant had consented to any disclosure that may have occurred. The complainant disputed that she had provided consent for this disclosure. The matter was resolved by payment of $5000 compensation by the respondent for hurt and humiliation.

14.6.2 Other Federal Legislative Acts

The following Acts are important from a cyber security perspective:

Australian Security Intelligence Organization Act 1979

The Australian Security Intelligence Organization (ASIO) Act establishes and prescribes the ASIO's functions and powers. It includes provisions for computer access warrants, security assessments, and listening and tracking devices.

Copyright Amendment (Digital Agenda) Act 2000 (CADA)

The Copyright Amendment (Digital Agenda) Act 2000 (CADA) contains provisions to criminalize devices for the purpose of the circumvention, or facilitating the circumvention, of any effective technological protection measures (now Part V Division 2A of the Copyright Act). Activities protected by the amendments include the importation, and manufacture of circumvention devices and circumvention services (s 116A Copyright Act), removal or alteration of electronic rights management information (s 116B) and, commercial dealings with works whose electronic rights management information is removed or altered (s 116C). Altering electronic rights management information may include commercially distributing, importing a copy of a work, communicating the copy to the public, or using a copy knowing the electronic rights management information had been so removed or altered without the permission of the copyright holder (s 116C, the latter dealing also applies to s 116B).

Electronic Transactions Act 1999

It provides a regulatory framework that recognizes the importance of the information economy and facilitates the use of electronic transactions.

National Security Information (Criminal and Civil Proceedings) Act 2004 (Cth)

This Act prevents the disclosure of information in federal criminal and civil proceedings where the disclosure is likely to prejudice national security except where preventing the disclosure would seriously interfere with the administration of justice.

14.7 COMPARISON OF CYBERCRIME LEGISLATIONS

Table 14.13 summarizes the applicable cybercrime legislations of the representative countries. It can be seen that all the representative countries have enacted specific laws and acts to address almost all the cybercrimes except for a few which can be penalized under some other sections. This shows that the countries have focussed on cyber security. It is necessary for the countries to keep pace with the cybercrime trends and make sure that the cyber security policies are sufficient. If not, amendments to cyber laws have to be made every now and then.

Table 14.13 Comparison of cybercrime legislations in representative countries

Cyber-crimes	Applicable cybercrime legislations in different countries					
	India	Netherlands	Malaysia	United Kingdom	United States	Australia
Cyber bullying	Sections 43 and 66 of Information Technology Amendment Act 2008	–	No specific law that criminalizes bullying	The Protection from Harassment Act penalizes with a maximum penalty of six months imprisonment, an unlimited fine, or both, and is arrestable.	No specific law	Criminal Code Act 1995
Forgery	Sections 463, 465, 470, and 471 of Indian Penal Code	Art. 225 DCC penalizes with a maximum penalty of six years imprisonment or a fine of 45,000 Euros.	Computer Crimes Act 1997	Computer Misuse Act 1990	Computer Fraud and Abuse Act 1984 or 18 U.S.C. § 1030	Crimes Amendment (fraud, identity and forgery offences) Act 2009
Data diddling	Sections 43 and 66 of Information Technology Amendment Act 2008	–	Computer Crimes Act 1997	Data Protection Act 1998	–	–
Email bombing	Section 66 of Information Technology Amendment Act 2008	Article 138b DCC imposes a maximum penalty of imprisonment for one year or a fine of 11,250 Euros.	Computer Crimes Act 1997	Computer Misuse Act 1990	–	–
Identity theft	Section 66C of Information Technology Amendment Act 2008; Sections 417A and 419A of Indian Penal Code	Collecting data is punishable as hacking (Art. 138a DCC) or illegal interception (Art. 139c DCC) and using the data is punishable as fraud (Article 326 DCC), theft (forgery or impersonation)	Computer Crime Act 1997; Communications and Multi-media Act 1998; Digital Signature Act 1997; Electronic Transactions Act 2006	Computer Misuse Act 1990 (or) The Fraud Act 2006; The Identity Documents Act 2010; and The Forgery and Counterfeiting Act 1981	18 U.S.C. § 1028; 18 U.S.C. § 1028A	Crimes Amendment (fraud, identity, and forgery offences) Act 2009

(Contd)

Table 14.13 (Contd)

Cyber-crimes	Applicable cybercrime legislations in different countries					
	India	Netherlands	Malaysia	United Kingdom	United States	Australia
Theft of data	Sections 65 and 66 of Information Technology Amendment Act 2008	Article 138 DCC for hacking and Article 311C for aggravated theft	Computer Crimes Act 1997	Data Protection Act 1998	Computer Fraud and Abuse Act 1984 (or) 18 U.S.C. § 1030	Cybercrime Act 2001
Viral attack	Sections 43 and 66 of Information Technology Amendment Act 2008	Art. 350a DCC with a maximum penalty of four years imprisonment or a fine of 45,000 Euros	Computer Crimes Act 1997	Computer Misuse Act 1990	Computer Fraud and Abuse Act 1984 (or) 18 U.S.C. § 1030	Cybercrime Act 2001
Hacking	Sections 43 and 66 of Information Technology Amendment Act 2008; Sections 378, 379, 405, and 406 of Indian Penal Code	Article 138a DCC with six months imprisonment or a fine of 4,500 Euros for 'simple' hacking and four years imprisonment or 11,250 Euros if the hacker copies data, or if he/she hacks via public telecommunications and uses processing capacity or hacks onwards to a third computer.	Computer Crimes Act 1997	The Terrorism Act 2000	Computer Fraud and Abuse Act 1984 (or) 18 U.S.C. § 1030	Cybercrime Act 2001
Denial-of-service attack	Section 43 of Information Technology Amendment Act 2008	CCII Bill amended Article 138b DCC imposes a maximum penalty of imprisonment for one year or a fine of 11,250 Euros.	–	The Terrorism Act 2000	Computer Fraud and Abuse Act 1984 (or) 18 U.S.C. § 1030	Criminal Code Act 1995
Financial crime	Sections 415, 384, 506, and 511 of Indian Penal Code	Article 326 DCC penalizes with a maximum penalty of four years imprisonment	Communications and Multimedia Act 1998	Computer Misuse Act 1990	18 U.S.C. § 1956, 18 U.S.C. § 1957	Crimes Amendment (fraud, identity, and forgery offences) Act 2009

(Contd)

Table 14.13 (Contd)

Cyber-crimes	Applicable cybercrime legislations in different countries					
	India	Netherlands	Malaysia	United Kingdom	United States	Australia
Copyright infringement	Sections 63 and 63A of Information Technology Amendment Act 2008; Section 51 of Indian Penal Code	Article 31 of the Copyright Act criminalizes infringement with a maximum imprisonment of six months to one year imprisonment which rises to four years if done as a profession or business	Copyright Act (Amendment) 1997	Copyright Law 1956	Digital Millennium Copyright Act; 17 U.S.C. § 506; 17 U.S.C. § 1201	Copyright Amendment (digital agenda) Act 2000 (CADA)
Sale of illegal article	The Narcotic Drugs and Psychotropic Substances Act 1985; Arms Act 1959	–	Communications and Multimedia Act 1998	–	18 U.S.C. § 545, 18 U.S.C. § 1341, 21 U.S.C. § 301, 21 U.S.C. § 822, 829, 841, 863, 921, 951-971	–
Cyber terrorism	Sections 66 F of Information Technology Amendment Act 2008; Section 153A, UAPA 15-122 of Indian Penal Code	–	Penal Code	The Terrorism Act 2000	Patriot Act	–
Pornography	Sections 67 and 67B of Information Technology Amendment Act 2008; Sections 292, 293, and 294 of Indian Penal Code	Article 240b DCC penalizes with a maximum penalty of four years imprisonment.	Penal Code	–	18 U.S.C. § 2251 and 18 U.S.C. § 2421	Cybercrime Act 2001
Cyber stalking	Section 66 of Information Technology Amendment Act 2008; Sections 499, 500, 503, and 506 of Indian Penal Code	Article 285b DCC penalizes with a maximum imprisonment of three years	No specific law that criminalizes stalking	The Protection from Harassment Act penalizes with a maximum penalty of six months imprisonment, an unlimited fine, or both, and is arrestable.	18 U.S.C. § 875 and 47 U.S.C. § 223	Cybercrime Act

(Contd)

Table 14.13 (Contd)

Cyber-crimes	Applicable cybercrime legislations in different countries					
	India	Netherlands	Malaysia	United Kingdom	United States	Australia
Web jacking	Section 65 of Information Technology Amendment Act 2008; Section 383 of Indian Penal Code	Article 138a DCC with six months imprisonment or a fine of 4,500 Euros for 'simple' hacking and four years imprisonment or 11,250 Euros if the hacker copies data, or if he/she hacks via public telecommunications and uses processing capacity or hacks onwards to a third computer.	Computer Crimes Act 1997	The Terrorism Act 2000	Computer Fraud and Abuse Act 1984 or 18 U.S.C. § 1030	Cybercrime Act 2001

POINTS TO REMEMBER

Cybercrime legislation in the Netherlands
- The main Act regarding computer crime is the Computer Crime Act of 1993.
- Hacking is penalized in Art. 138a DCC.
- Interception of direct communications or non-telecom data transfer is penalized by Art. 139a DCC (for closed areas) and by Art. 139b CCC (for other areas).
- Interception of telecommunications is penalized by Art. 139c DCC. It concerns the interception, by technical means, of public telecommunications.
- Intentional manipulation of computer-related data is penalized under Art. 350a DCC.
- Non-intentional (negligent) manipulation of computer-related data is penalized under Art. 350b DCC.
- The intentional making available or dissemination of computer viruses is penalized under Art. 350a DCC.
- Misuse of devices, as provided for in Art. 6 CCC, is currently penalized in only a few special cases: payment cards, devices for telecom fraud, devices for oral or wire interception, and devices for software-protection circumvention.
- Computer-related forgery falls within the scope of the traditional provision on forgery Art. 225 DCC.
- Telecom fraud has been specifically penalized under Art. 326c DCC.
- Child pornography is penalized under Art. 240b DCC, with a maximum penalty of four years imprisonment or a fine of 45,000 Euros.
- The liability of ISPs for illegal or unlawful content has been regulated as a consequence of the Electronic Commerce Directive.
- In Dutch law, there is power to order production of data regulated by Art. 125i Dutch CCP.
- Traditional search provisions cover computer searches (Articles 96b, 96c, 97, and 110 CCP).
- The Bill on Data Ordering Provisions 2005 was introduced in Art. 125i of the DCCP, the power to search in order to secure data.
- Interception of public telecommunications is

regulated by Art. 126m DCCP.
- The power to order production of subscriber data in general is regulated by Art. 126nc and 126uc DCCP, which cover identifying information.

Cyber laws in Malaysia
- Communications and Multimedia Act 1998 is the main basis for other cyber laws in Malaysia. This Act ensures that information is secure, the network is reliable, and the service is affordable all over the world.
- The Computer Crimes Act 1997 is to overcome misuse of computers as it is an offence in Malaysia.
- Copyright Act (Amendment) 1997 is the amendment of Copyright Act 1987.
- Since 1 October 1998, the Digital Signature Act 1997 has been enforced to help prevent on-line transaction frauds.
- The Electronic Commerce Act 2006 gives legal recognition of electronic messages in commercial transactions.
- The Payment Systems Act 2003 covers both operators payments system and issuers of designated payment instruments (DPIs).

UK Cybercrime laws
- The Computer Misuse Act 1990 is a response to the growing concern that existing legislation is inadequate for dealing with hackers.
- The Police and Justice Act 2006 amends the Computer Misuse Act to include unauthorized acts with intent to impair operation of computer which effectively adds denial-of-service attacks as an offence.
- The Terrorism Act 2000 is in regard to the disruption of computer systems.
- The Regulation of Investigatory Powers Act 2000 (RIPA) is a long and complex Act which covers many things including recording transmissions for legal purposes and covers the Extradition Act 2003, Part 1, and implements the European Arrest Warrant (EAW) which allows extradition to 28 territories designated as Category 1 territories.
- The Patent Act specifies that the patent right can be empowered to the product inventor or process innovator on the condition that the product is newly invented, innovative, and applicable to industrial development.

Cybercrime laws of the United States
- Substantive cybercrime laws include laws prohibiting online identity theft, hacking, intrusion into computer systems, child pornography, intellectual property, and online gambling.
- The first federal computer crime statute was the Computer Fraud and Abuse Act of 1984 ('CFAA').
- The National Information Infrastructure Act (NIIA) was passed in 1996 to expand the CFAA to encompass unauthorized access to a protected computer in excess of the parties' authorization.
- The Electronic Communications Privacy Act (ECPA) was passed in 1986 as an amendment to the federal wiretap law.
- The Cyber Security Enhancement Act (CSEA) was passed together with the Homeland Security Act in 2002.
- The basic purpose of the Digital Millennium Copyright Act (DMCA) is to amend the Title of the United States Code and to implement the World Intellectual Property Organization (WIPO) Copyright Treaty and Performances and Phonograms Treaty which were designed to update world copyright laws to deal with the new technology.
- The Economic Espionage Act (EEA) was passed in 1996 and was created in order to put a stop to trade secret misappropriation.
- The Fraudulent Online Identity Sanctions Act (FOISA) attempts to tackle the problem of criminals registering online domains under false identification.
- The Computer Software Privacy and Control Act (CSPCA) is meant to deal with the problem of spyware and adware.

Australian laws related to privacy and cyber security domains
- The Cybercrime Act offers more comprehensive regulation of computer- and Internet-related offences such as unlawful access and computer trespass, damaging data and impeding access to computers, theft of data, computer fraud, cyber stalking and harassment, and possession of child pornography.
- The Spam Act establishes a scheme for the regulation of commercial email and other types of electronic messages.
- The Cybercrime Act 2001 was enacted because of some of the popular viruses of the late 1990s such as 'Melissa' and 'ILOVEYOU'.
- The Spam Act 2003 is enforced by the Australian

Communications and Media Authority (ACMA). Australian businesses that fail to comply with this Act can be fined up to AUD $1.1 million per day for repeated corporate offenders.
- The Telecommunications (Interception and Access) Act 1979 applies to all ISPs and telecommunications network providers.
- The Crimes Act 1914 codifies offences against the Commonwealth.
- The Criminal Code Act 1995 abolishes all common law offences and is gradually superseding the Crimes Act 1914.
- The Electronic Transactions Act 1999 provides a regulatory framework that recognizes the importance of information economy and facilitates the use of electronic transactions.
- The Intelligence Services Act 2001 provides parliamentary and judicial support for the Australian Secret Intelligence Service (ASIS) and the Australian Signals Directorate (ASD).
- The National Security Information (Criminal and Civil Proceedings) Act 2004 (Cth) prevents the disclosure of information in federal criminal and civil proceedings where the disclosure is likely to prejudice national security except where preventing the disclosure would seriously interfere with the administration of justice.

KEY TERMS

Australian Security Intelligence Organization (ASIO) Act 1979 This Act establishes and prescribes ASIO's functions and powers. It includes provisions for computer access warrants, security assessments, and listening and tracking devices.

Communications and Multimedia Act 1998 This Act ensures that information is secure, the network is reliable, and the service is affordable all over the world. It also ensures high level of user confidence in the information and communication technology industry.

Computer Crimes Act 1997 This Act gives protection against the misuse of computers and computer criminal activities such as unauthorized use of programs, illegal transmission of data or messages over computers, and hacking and cracking of computer systems and networks.

Computer Misuse Act 1990 This Act brings in three offences: (a) unauthorized access to computer material, (b) unauthorized access with intent to commit or facilitate commission of further offences, and (c) unauthorized acts with intent to impair or with recklessness as to impairing operation of computers, etc.

Computer Software Privacy and Control Act (CSPCA) This Act prohibits transmission of software that collects and transmits personal information about the owner or operator of the computer, monitors and transmits web pages accessed by the owner or the operator, or modifies default computer settings such as home page, unless the owner or the operator gives his/her consent and the software has an uninstall option built into it.

Copyright Act (Amendment) 1997 This Act protects the copyright works from unauthorized copying and/or alteration.

Crimes Act 1914 This Act codifies offences against the Commonwealth. It functions alongside state legislation and is gradually superseding the Criminal Code Act 1995.

Criminal Code Act 1995 This Act contains federal offences and abolishes all common law offences and is gradually superseding the Crimes Act 1914.

Cybercrime Act This Act offers more comprehensive regulation of computer- and Internet-related offences such as unlawful access and computer trespass, damaging data and impeding access to computers, theft of data, computer fraud, cyber stalking and harassment, and possession of child pornography.

Cyber Security Enhancement Act This Act grants sweeping powers to the law enforcement organizations and increased penalties that were set out in the Computer Fraud and Abuse Act.

Digital Millennium Copyright Act This Act also prohibits the manufacture or sale of devices or programs whose primary purpose is to circumvent access control technology.

Digital Signature Act 1997 This Act has been enforced to help prevent online transaction frauds.

Economic Espionage Act (EEA) This Act defines stealing of trade secrets as copying, duplicating, sketching, drawing, photographing, downloading, uploading, altering, destroying, photocopying, replicating, transmitting, delivering, sending, mailing, communicating, or conveying trade secrets without authorization.

Electronic Commerce Act 2006 This Act provides how legal requirements can be fulfilled by using electronic messages. This Act also allows the use of electronic means and other related matters to facilitate commercial transactions.

Electronic Communications Privacy Act (ECPA) This Act makes it illegal to intercept stored or transmitted electronic communication without authorization.

Electronic Transactions Act 1999 This Act provides a regulatory framework that recognizes the importance of the information economy and facilitates the use of electronic transactions.

Extradition Act Part 1 of this Act implements the European Arrest Warrant (EAW) which allows extradition to 28 territories designated as Category 1 territories. Part 2 of this Act defines a significant number of countries for which extradition relations exist, although they are non-EU and therefore do not fall under the European Arrest Warrant scheme.

Fraudulent Online Identity Sanctions Act This Act includes a provision that would increase jail term for people who provide false contact information to a domain name registrar and then use that domain to commit copyright and trademark infringement crimes.

Intelligence Services Act 2001 This Act provides parliamentary and judicial support for the Australian Secret Intelligence Service (ASIS) and the Australian Signals Directorate (ASD). This Act also grants powers to Australian Security Intelligence Organisation (AISO).

Intentional manipulation of computer-related data This includes deleting, changing, and adding data.

Interception of direct communications or non-telecom data This concerns the interception by technical means of voice communications or of data communications.

Interception of telecommunications This concerns the interception by technical means of public telecommunications.

National Security Information (Criminal and Civil Proceedings) Act 2004 (Cth) This Act prevents the disclosure of information in federal criminal and civil proceedings where the disclosure is likely to prejudice national security except where preventing the disclosure would seriously interfere with the administration of justice.

Police and Justice Act 2006 This Act makes explicit that the offence does not have to be against a specific computer, program, or data. Additionally it states that an offence is caused even if denial-of-service is only temporary.

Procedural cybercrime law This Act includes laws that govern the authority to preserve and obtain electronic data from third parties, including Internet service providers; authority to intercept electronic communications; and authority to search and seize electronic evidence.

Regulation of Investigatory Powers Act 2000 (RIPA) This Act states that it is illegal to intentionally intercept any communication transmitted by means of a public telecommunications system without lawful authority. It is also illegal to intentionally intercept any communication transmitted by means of a private telecommunications system if they do not have the right to control the operation of the system or does not have the expressed or implied consent of such a person to make the interception.

Spam Act 2003 This Act prohibits the sending of unsolicited commercial electronic messages via email, short message service (SMS), multimedia message service (MMS), and instant messages with an Australian link.

Substantive cybercrime law This Act includes laws prohibiting online identity theft, hacking, intrusion into computer systems, child pornography, intellectual property, and online gambling.

MULTIPLE-CHOICE QUESTIONS

1. The main Act regarding computer crime is the _____.

   (a) Computer Anti-Crime Act of 1993
   (b) Computer Crime Act of 1993

(c) Computer Crime Act of 1992
(d) Computer Anti-Crime Act of 1992
2. _____ includes deleting, changing, and adding data which is penalized in Art. 350a CC.
 (a) Intentional manipulation of computer-related data
 (b) Negligent manipulation of computer-related data
 (c) Computer virus
 (d) System interference, ebombs, and DoS attacks
3. The maximum penalty for computer-related forgery is _____ in the Netherlands.
 (a) six years imprisonment and a fine of 45,000 Euros
 (b) six years imprisonment or a fine of 45,000 Euros
 (c) six years imprisonment
 (d) a fine of 45,000 Euros
4. There is a power to order production of data in _____ law.
 (a) Indian   (c) African
 (b) American   (d) Dutch
5. The Act on Data Production Orders 2005 allows the ordering of _____.
 (a) identifying data by any investigating officer in case of a crime
 (b) other data by the public prosecutor in cases for which pre-trial detention is allowed
 (c) sensitive data by the investigating judge in case of a pre-trial detention crime
 (d) All of these
6. According to Art. _____, if the data is encrypted, the people targeted by the production order excluding suspects can be ordered to decrypt them.
 (a) 125nh CCP
 (b) 126nh CCP
 (c) 125ni CCP
 (d) 126ni CCP
7. _____ allows the person who conducts a search to also search computer networks from computers located at the search premises.
 (a) 125h CCP
 (b) 125i CCP
 (c) 125j CCP
 (d) 125k CCP
8. _____ Act is the main pillar for other cyber laws in Malaysia.
 (a) Communication and Multimedia Act 1998
 (b) Computer Crime Act 1997
 (c) Copyright Act 1997
 (d) Digital Signature Act 1997
9. Computer Fraud and Abuse Act (CFAA) is also known as _____.
 (a) Title 17 U.S.C Section 1030
 (b) Title 18 U.S.C Section 1030
 (c) Title 27 U.S.C Section 1040
 (d) Title 28 U.S.C Section 1040
10. Cyber Security Enhancement Act (CSEA) was passed together with the _____ in 2002.
 (a) Hilland Security Act
 (b) Hilland Protection Act
 (c) Homeland Security Act
 (d) Homeland Protection Act
11. The Digital Millennium Copyright Act (DMCA) prohibits _____ designed to protect a copyright.
 (a) circumventing a software measure
 (b) circumventing a hardware measure
 (c) circumventing a technological measure
 (d) none of these
12. The two types of consent are _____.
 (a) impress consent and inferred consent
 (b) express consent and inferred consent
 (c) impress consent and infringement consent
 (d) express consent and infringement consent
13. The Telecommunications (Interception and Access) Act 1979 applies to all _____.
 (a) Internet service providers
 (b) telecommunication network providers
 (c) both (a) and (b)
 (d) none of these

REVIEW QUESTIONS

1. What is interception of telecommunications according to the cybercrime legislation in the Netherlands?
2. What is the Communication and Multimedia Act 1998 according to the cyber laws in Malaysia?
3. What is Regulation of Investigatory Powers Act (RIPA) 2000?
4. Mention the applicable laws for the Internet sale

of prescription drugs and controlled substances.
5. List out the main rules to be followed by private bodies and non-educational institutions in Australia according to the Spam Act 2003.
6. What is the full definition of a commercial message according to the Australian Communications and Media Authority (ACMA)?
7. Mention the seven types of computer criminal activities according to the Computer Fraud and Abuse Act (CFAA).
8. Explain the cyber laws in place in Malaysia.
9. Describe the Computer Fraud and Abuse Act of the United States.
10. Discuss the traditional laws to prosecute cybercrimes.
11. List out and explain the other Federal Legislative Acts of Australia.
12. Explain in detail the Australian laws related to privacy and cyber security.
13. Discuss briefly the specific cybercrime legislation in the Netherlands.

APPLICATION EXERCISES

1. Compare how the cyber laws in Malaysia and the Netherlands deal with hacking.
2. From among the cyber laws learnt for five different countries, which legislation is superior and sound in your opinion and is the best in handling cybercrimes and why?

BIBLIOGRAPHY

1. Bert-Jaap Koops (2005), *Cybercrime Legislation in the Netherlands*, available at: http://www.cyberlawdb.com/gcld/wp-content/uploads/2010/04/cybercrime.pdf (Accessed 13 December 2017)
2. HollyGraceful (2016), *UK Cyber Crime Law*, available at: https://www.gracefulsecurity.com/uk-cyber-crime-law/ (Accessed 13 December 2017)
3. David Emm (2009), *Cybercrime and the Law: A Review of UK Computer Crime Legislation*, available at: https://securelist.com/cybercrime-and-the-law-a-review-of-uk-computer-crime-legislation/36253/ (Accessed 13 December 2017)
4. Michael Rappa, *U.S. Federal Cybercrime Laws*, available at: http://digitalenterprise.org/governance/us_code.html (Accessed 13 December 2017)
5. Mohamed Afique (2013), *Cyber Law in Malaysia*, available at: http://malaysiancyberwarriors.blogspot.in/2013/03/introduction-of-cyber-law-acts-in.html (Accessed 14 December 2017)
6. Babu Veerappa Srinivas (2015), *A Concise Guide to Various Australian Laws Related to Privacy and Cybersecurity Domains*, available at: https://www.sans.org/reading-room/whitepapers/legal/concise-guide-australian-laws-related-privacy-cyber-security-domains-36072 (Accessed 14 December 2017)
7. Al Rees (2006), *Cybercrime Laws of the United States*, available at: https://www.oas.org/juridico/spanish/us_cyb_laws.pdf (Accessed 14 December 2017)
8. CCID, *Laws of Malaysia Act 563 Computer Crimes Act 1997*, available at: http://ccid.rmp.gov.my/Laws/Computer_Crime_Act_1997.pdf (Accessed 14 December 2017)
9. EJCL, *Cybercrime Legislation in the Netherlands*, available at: https://www.ejcl.org/143/art143-10.pdf (Accessed 02 February 2018)
10. Super User, *Defamation Act, 1957*, available at: http://uumpress.uum.edu.my/index.php/bookstore/act/93-defamation-act-1957 (Accessed 02 February 2018)
11. Lawyerment (2014), *What is Computer Crimes Act 1997?*, available at: https://www.lawyerment.com/library/kb/Intellectual_Property/1360.htm (Accessed 02 February 2018)
12. Foongchengleong, *Computer Crimes Act 1997*, available at: http://foongchengleong.com/tag/computer-crimes-act-1997/ (Accessed 02 February 2018)
13. Forrestwilliamssolicitors, *Malicious Communications Act – A Case Study*, available at: https://forrestwilliamssolicitors.com/news/malicious-communications-act/ (Accessed 02 February 2018)
14. Legislation, *Defamation Act 2013*, available at: http://www.legislation.gov.uk/ukpga/2013/26/

pdfs/ukpga_20130026_en.pdf (Accessed 02 February 2018)
15. The Cybersmile Foundation, *Legal Perspective*, available at: https://www.cybersmile.org/advice-help/category/cyberbullying-and-the-law (Accessed 03 February 2018)
16. Services, *Defamation Act 2013*, available at: https://services.parliament.uk/bills/2012-13/defamation.html (Accessed 03 February 2018)
17. Kellywarnerla, *UK Defamation: Legal Overview*, available at: http://kellywarnerlaw.com/uk-defamation-laws/ (Accessed 03 February 2918)
18. Victoria Jones (2015), *What is the Communications Act 2003?*, available at: https://www.walesonline.co.uk/news/local-news/what-is-thecommunications-act-2003-8488958 (Accessed 03 February 2018)
19. David Wall, S. (2013), *Future Identities: Changing identities in the UK – The Next 10 years*, available at: https://www.gov.uk/government/uploads/system/uploads/attachment_data/file/275784/13-521-identity-related-crime-uk.pdf (Accessed 03 February 2018)
20. Kellywarnerlaw, *U.S. Defamation Laws & Standards*, available at: http://kellywarnerlaw.com/us-defamation-laws/ (Accessed 03 February 2018)
21. HG.org, *Is there a Law against Cyberstalking or Cyberharassment?*, available at: https://www.hg.org/article.asp?id=31710 (Accessed 03 February 2018)
22. Cyberstalking, *Federal Criminal Statutes*, available at: http://cyberstalking.web.unc.edu/federal-criminal-statutes/ (Accessed 03 February 2018)
23. Users, *Laws that May Apply to DDoS Attacks*, available at: http://users.atw.hu/denialofservice/ch08lev1sec2.html (Accessed 04 February 2018)
24. Criminal, *Cyberbullying*, available at: http://criminal.findlaw.com/criminal-charges/cyber-bullying.html (Accessed 04 February 2018)
25. Identity-theft, *What you Need to Know about Electronic Forgery*, available at: https://identity-theft.laws.com/electronic-forgery (Accessed 04 February 2018)
26. Cyber Laws, *Understanding Cyber Terrorism*, available at: https://cyber.laws.com/cyber-terrorism (Accessed 04 February 2018)
27. Electronic Frontiers Australia (2006), *Defamation Laws & the Internet*, available at: https://www.efa.org.au/Issues/Censor/defamation.html (Accessed 04 February 2018)
28. ACORN, *Cyber-bullying*, available at: https://www.acorn.gov.au/learn-about-cybercrime/cyber-bullying (Accessed 04 January 2018)
29. Legislation, *Crimes Amendment (Fraud, Identity and Forgery Offences) Act 2009No 99*, available at: https://www.legislation.nsw.gov.au/acts/2009-99.pdf (Accessed 04 February 2018)
30. Legislation, *Copyright Amendment (Online Infringement) Act 2015*, available at: https://www.legislation.gov.au/Details/C2015A00080 (Accessed 04 February 2018)
31. The National Archives, *Terrorism Act 2000*, available at: https://www.legislation.gov.uk/ukpga/2000/11/contents (Accessed 20 February 2018)
32. Chatterjee (2000), *The Terrorism Act 2000: An Analysis*, Amicus Curiae, Issue 39, pp. 19–25, available at: http://sas-space.sas.ac.uk/3727/1/1302-1392-1-SM.pdf (Accessed 20 February 2018)
33. House of Lords (2015), *Extradition: UK Law and Practice*, available at: https://publications.parliament.uk/pa/ld201415/ldselect/ldextradition/126/126.pdf (Accessed 23 February 2018)
34. www.parliament.uk (2017), *The Protection from Harassment Act 1997*, available at: http://researchbriefings.parliament.uk/ResearchBriefing/Summary/SN06648 (Accessed 23 February 2018)
35. PLC IPIT & Communications (2013), *Defamation Act 2013: Summary of Main Provisions*, available at: https://uk.practicallaw.thomsonreuters.com/8-526-7636?transitionType=Default&contextData=(sc.Default)&firstPage=true&bhcp=1 (Accessed 23 February 2018)

Answers to Multiple-choice Questions

1. (b)	2. (a)	3. (b)	4. (d)	5. (d)
6. (b)	7. (c)	8. (a)	9. (b)	10. (c)
11. (c)	12. (b)	13. (c)		

Appendix

The following example explains the procedures involved in case analysis.
Disclaimer: The chosen case is an actual one. Some of the specific details have been intentionally omitted.

INVESTIGATION REPORT

Case Brief

The *Subject*, who was a reputed apparels merchant, had in place an e-commerce payment gateway of a reputed merchant service, facilitated through the Pay seal application, to allow customers make online purchases. This enabled the merchant to accept cards for payments. The cybercrime involved came to light when the merchant received chargeback (Chargeback is a form of customer protection provided by the banks issuing credit cards, so as to allow cardholders to file a complaint regarding fraudulent transactions when discrepancies are noticed in the accounts statement. If the transaction is known to be fraudulent on investigation, the bank would have to refund the original value to the cardholder. If the cardholder cannot prove the transaction to be legitimate, the entire amount of the transaction, along with an additional fee, will be debited from his/her account.) for many transactions worth ₹17,71,464 during a six-month period from October 2010. The bank had debited the merchant account for the chargeback received from various issuing banks.

The most common reason for chargeback involves fraudulent transactions, where the credit card is used without the authorization and consent of the cardholder. In such a case, the merchant is held solely responsible. Hence, a case was registered based on a complaint from the *Subject*.

With the help of a cyber-forensics team, all the suspected transactions and the associated customers (the addresses of the fraudsters) were traced. It was revealed that many international card transactions were carried out by a group of suspected users based in Coimbatore and Tirupur. Further, the shipping address provided by the fraudsters in this case matched the ID proof collected from the suspects while handling a similar case in 2008, against the merchant M/s. X in which the team had collected the ID proofs of a few suspected users during the time of delivery of the product. This led to the suspicion that the same people were involved in this fraud as well.

Computer type	: Generic laptop with serial #123456789
Operating system	: Microsoft Windows 7
Offence	: Credit card fraud
Case investigation officer	: Marks
Evidence number	: 123456
Chain of custody	: See attached form
Where examination took place	: Cyber Forensic Lab, Government of Tamil Nadu, Chennai
Tools used	: Encase v.6.6

Investigation

The investigating officer (IO) examined the complaint and collected relevant details and documents pertaining to the cybercrime.

Acquisition The IO collected the IP address details from the Internet service provider (ISP). The suspects were located from the physical address provided by the ISP. The logs from the payment gateway and the transaction log from the bank were obtained. Credit card information was obtained from the system. A generic laptop, with serial #123456789 was seized from the suspect. The hardware configuration of the system was documented and a duplicate of the hard drive (HD) was created in a manner that protected and preserved the evidence. The CMOS information, including time and date, was documented.

Examination The log from the ISP was examined to locate the date and time of commission of the crime. The log from the bank and the payment gateway were examined for transactions involving international credit card numbers of the suspect; these were preserved. The directory and file structures, including file dates and times, were recorded. A file header search was conducted to locate all the credit card details. The files were reviewed and those containing information related to the credit card services were preserved. Image files were reviewed and those containing the images of credit cards were preserved. The last accessed time and date of the files indicated when the files were last accessed.

Documentation and Reporting

The IO recorded the findings of the investigation in a report and examined various witnesses. The *Suspect* details are as follows:
1. The *Suspect*, a marine engineer, was the mastermind of the entire episode. He worked at a BPO in Bangalore and then at a call centre in Coimbatore. Through his contacts, he obtained the address of the websites selling the compromised card data of international banks. The mode of communication was 'chatting'. The applications used were Yahoo Messenger and ICQ Instant Messenger. This was confirmed from Windows artifacts.
2. The websites accessed by him to purchase the card data were http://ltdcc.com/ and https://www.v-market.org/login/. He had good knowledge of card payment operations, Viz. 3D, the secure operations of various banks, types and categories of cards, chargeback, insurance cover for disputes, currency conversion, validity and reliability of websites, and the card data sold through them.
3. Through this website, anyone could purchase card data using digital currency. Liberty Reserve was the digital currency used to purchase this data.
4. The *Suspect* created a Liberty Reserve account in his name on the website http://www.libertyreserve.com/.
5. He then converted Indian rupees to digital currency through the website http://cashforaction.com/ and the amount converted was directly transferred to his LR account on http://www.libertyreserve.com/.
6. After this currency conversion, the *Suspect* purchased the compromised card data from the websites http://ltdcc.com/ and https://www.v-market.org/login/.
7. He had checked the type of card (Business/Platinum/Silver) and the bank name through the websites www.binbase.com and www.binchecker.com, and purchased the card data. This helped him carry out non-3D secure transactions (using business cards) that had high credit limit.
8 .The acquired credit card data of international banks were used to purchase various products online, from different websites, by providing fake identity and address details.
9. A scanned copy of the front and reverse of the credit cards, and fake international identity proofs were purchased through the website www.scanlab.name.
10. The fraudster managed to convince the merchants by providing the bank name and a scanned copy of the card if they asked for it, based on alerts provided by the acquirer on suspicious transactions.
11. All the products were dispatched through XXXX Courier. The *Suspect* personally visited the courier office to collect the deliveries.
12. He sold the products brought through fraudulent transactions in the secondary market.

FINAL REPORT

The final report prepared by the IO, along with the forensic analysis report, other relevant documents, and the statements of the witnesses were filed before the appropriate court for prosecution against the *Suspect*.

CYBER FORENSIC ANALYSIS REPORT

One 20-GB HD was obtained from the IO through the appropriate court with a case brief and a requisition to determine the relevant evidence, along with another 20-GB new HD to take a forensic copy.

Memorandum for : xxxxxxxxxxxxxxxxx

Subject : Forensic media analysis report

Subject : Plastic card fraud case number 123456

Preliminary

The following were the steps carried out in the forensic lab before analysis:

1. The hash value was computed for the original 20-GB HD of the laptop seized from the *Suspect*. The hash value computed was xxxxxxxxxxxxxxxxxxxxxxxxx.
2. The hash value was computed for the fresh HD. The hash was NULL, thus implying that the HD contained no data.
3. Forensic copying was carried from the original HD to the new HD.
4. The hash value of the forensic copy, xxxxxxxxxxxxxxxxxxxxxxxxx, was computed to ensure that it matched the hash value computed for the original HD. This implied that an exact copy of the contents had been created (i.e., the data in the original HD).
5. Encase V6.6 was used to analyse the forensic copy.
6. After obtaining the relevant and necessary evidence (image and text files) from the forensic copy, the same steps were repeated on the original copy to acquire the evidence.
7. All the steps carried out were documented, along with the evidences gathered, then and there.

Summary of Findings

1. 15 files containing the images of credit cards were recovered.
2. 15 files containing text associated with credit card services were recovered.

Items Analysed

Tag number : 123456

Item description : One generic laptop with serial #123456789

Details of Findings

Findings in the generic 20-GB HD, model ABCDE, serial# 1234ABCD5678, recovered from tag number 12345, one generic laptop, serial #123456789 are as follows:

1. The examined HD was found to contain the Microsoft Windows 7 operating system.
2. The directory D:\AAAA\CREDITCARD\ was found to contain 15 files comprising images of credit cards. The file directory of the 15 files disclosed that the file creation dates and times were between October 2010 and March 2011, and the last accessed date was March 2011.
3. The directory D:\PERSONAL\ contained 15 files that included credit card services.
4. The directory D:\PERSONAL\CREDENTIALS contained passwords and usernames that were used to carry out fraudulent transactions as well as the details of registration to procure the credit card data and digital currency.

All the aforementioned files and their contents are attached as a hard copy annexure.

Thomas   Released by _____
Computer forensic examiner

INDEX

A

Access 394
Acquisition 187, 233
Acquisition of evidence 274, 280
   Android operating system 290
   Base station 289
   Controlled boots 281
   Digital cameras 293
   Email and Internet 286
   Financial/Banking institutions 293
   Forensic boot disk 281
   Internet service providers 293
   iOS 291
   iTunes backup 291
   Jail breaking 291
   Live systems 283
   Logical file collection 285
   Logical methods 291
   Mobile operating systems 290
   Mobile phone 288
   Network drives imaging 285
   Non-detachable hard disk drive 284
   Optical media 292
   PDA 288
   Physical methods 291
   Research in motion 290
   SD cards 290
   SIM card 290
   Social media 288
   Social networking sites 293
   Standalone hardware device 284
   Switched-off systems 281
   Symbian OS 290
   Tablets 291
   Time zone conversion 293
   USB drives 292
   Website domain hosting providers 293
   Windows phone 290
Actusreus 46
Admissibility of digital evidence 352
   Admissible 350
   Authentic 351
   Believable 351
   Complete 351
   Reliable 351

Admissibility of electronic records 354
   Section 22A 356
   Section 45A 356
   Section 65A 354
   Section 65B 354
Adware 96
Affixing electronic signature 394
Analysing hard drive with WinHex 312
   Acquiring forensic copy of drive 312
   Analysing hard disk 316
   Analysing slack space and free space 327
   Computing hash 314
   File carving 328
Analysis of evidence 301
Anonymizers 57
Application layer 26
   Remote procedure call 26
   Sockets 26
Application layer protocols 26
   Domain name system 27
   File transfer protocol 27
   Hyper text transfer protocol 27
   Post office protocol 27
   Simple mail transfer protocol 27
Asymmetric crypto system 395
Australian laws 460
   Australian Security Intelligence Organization Act 1979 463
   Copyright Amendment Act 2000 463
   Crimes Act 1914 460
   Criminal Code Act 1995 461
   Cybercrime Act 2001 461
   Electronic Transactions Act 1999 463
   National Security Information Act 2004 463
   Spam Act 2003 461

B

Backdoors 69
Blockchain 129
Bot 69
Botnet 69
Browser artifacts 219
   Apple Safari 220
   Firefox 219
   Google Chrome 220

Internet Explorer 219
Opera 220

C

Capturing of forensic copy with Forensic Toolkit Imager 302
Capturing hard drive 303
Capturing main memory 302
Categories of evidence with respect to law 357
   Direct evidence 358
   Illustrative evidence 358
   Indirect evidence 358
   Substantive evidence 358
Certifying authority 395
Chain of custody 277
Classification of cybercriminals 55
   Activists 55
   Coders 55
   Cyberpunks 55
   Getaways 55
   Internals 55
   Nation state actors 55
   Professionals 55
   Toolkit newbies 55
Communication device 395
Computer 395
Computer forensics investigations 152
   Computer emergency response team 153
   First responder 152
Computer network 395
Computer-related offences 51
   Computer-related forgery 52
   Data diddling 52
   Identity theft 52
   Impersonation 52
   Misuse of devices 52
   Pharming 52
   Remote commands 52
   Salami slicing attack 52
Computer resource 395
Computer security incident response team 177
   Forensic readiness 178
Computer system 395
Copyright and trademark-related offences 51
Copyright infringement crimes 414
Courtroom presentation system 363
Credit card fraud 48
Crime against individual 405
   Cyber stalking 405
   Section 66 IT Act 405
   Violation of privacy 406
   Web Jacking 406
   Cyber defamation 404
Crime against nation 416
   Pornography 417
   Website defacement 416
Crime against property 407
   Data diddling 411
   Email bombing 411
   Email fraud 414
   Email spoofing 413, 414
   Financial crime 412, 413
   Forgery 409
   Identity theft and password theft 412
   Sale of illegal articles on the Internet 415
   Theft of data 407
Cryptocurrency 122
   Bitcoin 124
   Dash 124
   Desktop wallet 123
   Ether 124
   Ethereum 125
   Golem 124
   Hardware wallet 123
   Lisk 125
   Lite coin 124
   Maid safe coin 124
   Mobile wallet 123
   Monero 124
   Online wallet 123
   Paper wallet 123
   Ripple 124
Cyber cafe 395
Cybercrime 44, 149
   Cyber forensics 149
   Cyber security 149
Cybercrime against individual 47, 78, 371
   Hacking using key logger 373
   Impersonation for purpose of cheating 375
   Phishing fraud 372
   Transmission of sexually explicit material through Internet 376
   Trolling on social media 377
   Credit card fraud 48
   Cyber extortion 81
   Cyber harassment 80
   Cyber stalking 79
   Delusional stalker 79
   Erotomanic stalker 79
   Extortion 48
   Facebook stalking 48
   Harassment 48
   Internet grooming 47, 79
   Internet troll 48
   Intimate stalker 79
   Online pedophilia 82
   Pedophilia 48
   Predatory stalker 80
   Pyramid scheme fraud 48

Index

Stalking 47
Trolling stalker 80
Vengeful stalker 79
Cybercrime against nation 52, 109, 386
  Content-related offences 53
  Pornography 53
  Racist and xenophobic material 53
  Blocking of websites 388
  Complex coordinated 109
  Content-related offences 111
  Cyber laundering 110
  Cyber terrorism 109
  Cyber warfare 110
  Email spam 54, 111
  Pornography 111
  Preparation of forged counterfeits using computers/printers/scanners 386
  Religious offences 53
  Simple unstructured 109
  Spread of false and defamatory information 53, 111
Cybercrime against property 48, 83, 379
  Data theft 383
  Data theft by ex-employee 385
  Hacking 384
  Online lottery scam 379
  Swindling of money by bank employee 381
  Theft in ATM 380
  Adware 103
  Armored virus 95
  Audio and video piracy 105
  Botnet 100
  Bot worm 91
  Browser hijacking software 96
  Call spoofing 88
  Child identity theft 107
  Client–server overuse 104
  Cluster viruses 94
  Companion virus 95
  Computer intrusion 90
  Computer-related forgery 106
  Computer-related offences 105
  Computer threats 97
  Copyright- and trademark-related offences 103
  Counterfeiting 104
  Cracking 48
  Cross-site scripting 49
  Data diddling 106
  Data interference 90
  Data piracy 104
  DDoS attack 97
  DoS attack 97
  Email bombing 99
  Email spamming 99
  Email spoofing 88
  Email worm 91
  Ethical worm 91
  Exploit kits 49
  Fast infectors 95
  File system virus 94
  Financial identity theft 107
  Hacking 48
  Hard-disk loading 104
  Identity theft 106
  Illegal access 48
  Illegal interception 87
  Impersonation 106
  Instant messaging worms 91
  IP spoofing 88
  List linking 99
  Logical threat 90
  Logic bomb 93
  Macro virus 95
  Malvertising 100
  Malware 96
  Mass mailing 99
  Medical identity theft 107
  Misuse of devices 107
  Online piracy 104
  Pharming 108
  Phreaking 87
  Polymorphic virus 95
  Porn-dialer 103
  Porn-downloader 103
  Porn-tool 103
  Pornware 103
  Potentially unwanted programs 101
  Ransomware 94
  Riskware 103
  Rogue security software 102
  Rootkit 96
  Salami slicing attack 108
  Service-specific phishing 86
  Skimming 89
  Slow infector 95
  SMS phishing 86
  SMS spoofing 88
  Softlifting 104
  Software piracy 103
  Sparse infector 95
  Spoofing 88
  Spyware 96, 102
  SQL injections 49
  Stealth virus 94
  System interference 97
  Theft of FTP passwords 49
  Trademark infringement 104
  Trademark-related offences 104
  Trojan-banker 92

Trojan-DDoS 92
Trojan-Downloader 92
Trojan-Dropper 92
Trojan-FakeAV 92
Trojan-GameThief 92
Trojan Horse 92
Trojan-IM 92
Trojan-Mailfinder 93
Trojan-Ransom 93
Trojan-SMS 93
Trojan-Spy 93
URL spoofing 88
Virus 94
Voice phishing 86
Web jacking 49
Worm 91
XSS attack 49
ZIP bombing 99
Cybercrime against property 83
   Clone phishing 85
   Cracking 83
   Data espionage 84
   Deceptive phishing 84
   Evil twin wi-fi attack 86
   Hacking 83
   Illegal access 83
   Illegal data acquisition 84
   Pharming 85
   Phishing 84
   Spear phishing 85
   Whaling attacks 85
Cybercrime associated with mobile ECD 54
   Bluetooth mobile hacking 54
   Crimes with calls 54
   Handset theft 54
   MMS crime 55
   SIM card cloning 55
   SMS-related crimes 54
Cybercrime law in the UK 445
   Communications Act 2003 449
   Computer Misuse Act 1990 446
   Extradition Act 2003 448
   Legislation of ICT Intellectual Property 450
   Police and Justice Act 2006 447
   Privacy and Electronic Regulations (EC Directive) 2003 450
   Protection from Harassment Act 1997 449
   Provisions for Identity Theft 451
   Regulation of Investigatory Powers Act 2000 448
   Serious Crime Act 2015 446
   Terrorism Act 2000 447
   The Malicious Communications Act 1988 450
Cybercrime Laws of the United States
   Computer Fraud and Abuse Act 452

Computer Software Privacy and Control Act 459
Cyber Security Enhancement Act 457
Digital Millennium Copyright Act 457
Economic Espionage Act 458
Electronic Communications Privacy Act 456
Fraudulent Online Identity Sanctions Act 459
Other Criminal Copyright Infringement Statute 459
Procedural Cybercrime Laws 452
Provisions for Handling Cyber Stalking 455
Provisions to Handle Cyber Terrorism 456
State Laws 459
Substantive Cybercrime Laws 451
Cybercrime legislation in Netherlands 439
   Card forgery 441
   Data manipulation and viruses 440
   Devices for telecom fraud 441
   Hacking 439
   Illegal interception 440
   Intentional manipulation of computer-related data 440
   Misuse of devices 441
   Non-intentional (negligent) manipulation of computer-related data 440
   Spam 441
   System interference 440
Cyber forensic examiners 361
Cyber forensic suite 235
Cyber laws for cyber security 421
Cyber laws in India 399
Cyber laws in Malaysia 443
   Communications and Multimedia Act 1998 443
   Computer Crimes Act 1997 444
   Copyright Act (Amendment) 1997 444
   Electronic Commerce Act 2006 445
   Electronic Government Activities Act 2007 445
   Payment Systems Act 2003 445
Cyber security 393, 395
Cyber terrorism 109
Cyber war 121

D

Dark web 139
Data 395
Database forensics 158
   Proactive approach 158
   Reactive approach 158
Data diddling 411
Data interference 50
   Logic bombs 50
   Ransomware 50
   Trojan horse 50
   Viruses 50
Data link layer 14
   Error control 14
   Flow control 15

Functions 14
DDoS attacks 58
Deep web 138
   Internet 138
   Surface Web 138
Deep web source repository 141
Digital evidence 184
   Non-volatile evidence 184
   Volatile evidence 184
Digital evidence as alibi 224
   Quick format 225
Digital evidence collection form 278
   Authentication of evidence 279
   Evidence safe 279
Digital evidence from standalone computers 187
Digital evidence on the Internet 223
   Evidence at application layer 224
   Evidence at network and transport layer 224
   Evidence from email 223
   Peer-to-peer networking 224
   Traceroute 224
Digital signature 69, 395
Disk forensics 155
Distributed denial-of-service attack 407
DoS attack, 58
Drive-by downloads 134
Drive imaging and validation tools 239

E

Electronic form 395
Electronic or digital evidence 269
   Primary evidence 269
   Secondary evidence 269
Electronic record 351, 395
Electronic signature 395
Email forensics 164
   Email protocols 165
Email tracking and tracing 336
   Email tracing 338
   Email tracing with online emailtracer 342
   Email tracking 336
   Email tracking with emailtracker pro 338
Evidence collection procedure 185
   Analysis 186
   Archival 187
   Chain of custody 186
   Identification 185
   Presentation 186
   Preservation 186
Evidence from mobile devices 222
Expert report 360
Extended file allocation table file system 198
   Allocation bitmap 199
   Allocation bitmap directory entry 200

Data region 199
Directory entries 199
Entry type 199
File directory entry 201
File system area 198
Root directory 199
Stream extension directory entry 201
TexFAT padding directory entry 200
Up-case table 199
Up-case table directory entry 200
Volume GUID directory entry 200
Volume label directory entry 199
Windows CE access control table directory entry 200
ext family of file systems 209
   Data unit 212
   ext3 209
   ext4 210
   Extended file system 209
   extents 210
   fdisk 212
   Filenames 212
   Group descriptor tables 212
   inodes 209
   Journal 212
   Journaling file systems 209
   Metadata (inode) 212
   mount 212
   Second extended file system 209
   stat 212
   Superblock 210
Extraction 234

F

FAT file system 195
   Boot record 195
   Cluster 195
   Data area 197
   Directory structure 196
   File allocation table 197
   File slack 198
   Formatting FAT file system 196
   RAM slack 198
   Residual slack 198
   Slack 198
File system 194
Firewall 36
   Application-level gateway 36
   Circuit-level gateway 36
   Network-level 36
   Stateful multilayer gateways 36
Forensic analysis 155
   Data recovery 155
   String and keyword searching 155
   System file analysis 155

Timeline analysis 155
Volatile evidence analysis 155
Forensic analysis tools for mobile devices 252
Forensic computing lab 173
Forensic examination process 154
   Acquisition 154
   Evaluation 154
   Extraction 154
   Identification 154
   Interpretation 154
   Presentation 154
   Preservation 154
Forensic hardware 252
   Forensic Computers Inc. 252
   Forensic systems 252
   Forensic Write Blockers 252
   FRED Forensic Network 252
Forensic hardware for mobile devices 255
   CellDEK 255
   Cellebrite 256
Forensic tool for integrity verification and hashing 240
   CRCMD5 241
   DiskSig 241
   HashCalc 241
   HashMyFiles 241
   MD5summer 241
Forensic tools 232
   Hardware forensic tools 233
   Software forensic tools 233
Forensic tools for analysing network 245
   Firewalk 245
   NetAnalysis 246
   Network Mapper 245
   OpenVPN 245
   Packet Tracer 245
   Snort 246
   Tripwire 245
   Wireshark 245
Forensic tools for analysis of registry 243
   RegRipper 243
   Regshot 243
Forensic tools for data recovery 241
   Byte Back 242
   IsoBuster 242
   Recuva 241
Forensic tools for email analysis 256
Forensic tools for encryption/decryption 243
   Encrypted Disk Detector 244
   VeraCrypt 243
Forensic tools for other media 251
   Digital Cameras 251
   Non-detachable hard disk drives 251
   USB drives 251
Forensic tools for password recovery 244

   ElcomSoft 244
   Ophcrack 244
   Passware Kit Forensic 244
Forensic tools for ram analysis 242
   DumpIt 242
   Live RAM Capturer 242
   Magnet RAM Capture 243
   Volatility 242
Forensic tools for UNIX system analysis 250
Forensic training and certifications 258
   AccessData Certified Examiner 258
   Advanced Information Security 259
   ASCLD/LAB 261
   Certified Computer Examiner 259
   Certified Computer Forensics Examiner 261
   Certified Electronic Evidence Collection Specialist 261
   Certified Forensic Computer Examiner 260
   Certified Hacking Forensic Investigator 259
   Certified Information Systems Auditor 260
   Certified ProDiscover Examiner 260
   Computer and Mobile Forensic Boot Camp 261
   EnCase Certified Examiner Programme 260
   GIAC Certified Forensic Analyst 260
   GIAC Certified Forensics Examiner 260
   Professional Certified Investigator 261
   SANS GIAC 261
Forensic utility for metadata processing 246
   Metadata Assistant 246
   PhotoMe 246
Fourth Amendment 279
Free and open-source forensic suite 235
   Autopsy 236
   Computer Aided Investigative Environment 237
   Digital Evidence and Forensic Toolkit 238
   Hiren's Boot CD 238
   Nirsoft 235
   OSForensics 235
   PALADIN 237
   Sleuth Kit 236
   Windows Sysinternals Live 236
Free and open-source forensic tools for mobile devices 252
   AFLogical 254
   BitPim 252
   MOBILedit Forensic Express 253
   Mobile Phone Examiner Plus 253
   SIMCon 253
Function 395

G
GPS forensics 163

H
Handling of digital evidence 294

Index

Forensic duplication 295
Hierarchical file system 213
Host-based IDS 37

I

Illegal data acquisition 49
   Data espionage 49
Illegal interception 50
   ATM hacking 50
   Skimming 50
   Spoofing 50
Incident 174
   Incident handling 175
   Incident reporting 176
   Incident response 176
   Incident response policy 177
Information 395
Information Technology Act 2000 400
   Scheme of IT Act 2000 400
Integrity preservation 233
Intermediary 396
Internet key exchange 30
Intrusion detection system 36
   Host-based IDS 37
   Hybrid IDS 38
   Network-based IDS 37
IPSec 29
   Communication modes 29
   Operations 29
IPSec communication 29
   IPSec authentication header 29
   IPSec encapsulating security payload 30
IPC Section 292 419
IPC Section 383 406
IPC Section 415 413
IPC Section 420 413
IPC Section 463 410
IPC Section 464 410
IPC Section 468 410
IPC Section 469 410
IPC Section 499 405
IPC Section 500 404
IPC Section 503 405
IPC Section 506 405
IPC Section 509 404
IPC Section 51 414
IPC Sections 293 421
ITA 2000 Section 43A 408
ITA 2000 Section 63 415
ITA 2000 Section 63A 415
ITA 2000 Section 68 422
ITA 2000 Section 69 422
ITA 2000 Section 69A 423
ITA 2000 Section 69B 423

ITA 2000 Section 70 424
ITA 2000 Section 70B 424
ITA 2000 Section 71 425
ITA 2000 Section 72 425
ITA 2000 Section 72A 425
ITA 2000 Section 73 425
ITA 2000 Section 76 426
ITA 2000 Section 78 427
ITA 2000 Section 79A 427
ITA 2000 Section 80 427
ITA 2000 Section70A 424
ITA Section 65 409
ITAA 2008 Section 66 409
ITAA 2008 Section 66B 411
ITAA 2008 Section 66C 411
ITAA 2008 Section 66D 412
ITAA 2008 Section 66F 417
ITAA 2008 Section 67 417
ITAA 2008 Section 67A 418
ITAA 2008 Section 67B 418
ITAA 2008 Section 67C 418
ITC 2000 Section 77B 426

K

Key pair 396

L

LAN technologies 8
   Ethernet 8
   Fast ethernet 8
   Giga ethernet 8
   Virtual LAN 8
   Wireless fidelity 9
Linux artifacts 221

M

Macintosh artifacts 220
Malvertising 134
Malware 57
Malware forensics 160
Memory forensics 172
   RAM analysis 172
   RAM artifacts 172
Mens rea 46
Mesh topology 11
   Full mesh 11
   Partial mesh 11
Miscellaneous tools 247
   AnaDisk diskette analysis tool 248
   CD/DVD inspector 247
   CopyQM Plus 248
   DiskScrub data overwrite utility 248
   DiskSearch pro 248
   DM 249

dtsearch 249
FileCNVT 249
Filter_I 247
GetFree 247
GetSlack 247
GetTime 248
Mandiant Redline 247
M-Sweep Data Scrubber 249
Net Threat Analyzer 249
NTI-DOC 248
Palm dd 249
Paraben Porn Stick 250
QuickView Plus FileViewer 250
Seized 248
SIFT 249
Snagit 250
TeleDisk 249
TextSearch Plus 248
WinHex 249
X-Ways forensics 249
Mobile forensics 160
Mules 55

N

Network forensics 156
Networking architecture 1
  Hybrid 2
  Peer-to-peer 1
  Server-based 1
Networking devices 6
  Bridge 6
  Gateway 7
  Hub 6
  Modem 7
  Repeater 6
  Routers 7
  Switch 7
Networking technologies 2
  Internetwork 4
  Local area network 3
  Metropolitan area network 3
  Wide area network 4
Networking topologies 9
  Bus 9
  Daisy chain 12
  Hybrid 12
  Mesh 11
  Point-to-Point 9
  Ring 11
  Star 11
  Tree 11
Network Layer 17
  Internetworking 19
  Network addressing 17
  Routing algorithms 19

Routing in network layer 17
Network layer protocols 19
  Address resolution protocol 20
  Internet control message protocol 20
  Internet protocol version 4 20
  Internet protocol version 6 21
  Reverse address resolution protocol 20
Network models 4
  Internet model 6
  OSI model 4
Network security at application layer 33
  DNSSec 35
  PGP 34
  S/MIME 34
Network security at transport layer 30
  HTTPS 32
  Layer 30
  SSH 33
  SSL 31
  TLS 31
New technology file system 202
  Orphaned files 209

O

Offences against individual 273
  Child abuse and pornography 273
  Domestic violence 274
  Email threats 274
  Harassment 274
  Homicide 274
  Network intrusion 274
  Stalking 274
Offences against nation 274
  Narcotics 274
Offences against property 274
  Computer fraud 274
  Financial fraud and counterfeiting 274
  Identity fraud 274
  Software piracy 274
  Telecommunication fraud 274
Onion router 141
Operating systems and their boot processes 188
  Extensible firmware interface 189
  Linux 189
  Macintosh 190
  Microsoft family 188
  MS - DOS 188
  Unified extensible firmware interface 189
  Windows 7 189
  Windows 8 189
  Windows 10 189
  Windows Vista 189
  Windows XP 189
Order of volatility of digital evidence 272

Originator 396
Other cyber laws associated with cybercrime and cyberspace 424
Other storage media 194
   Flash memory 194
   Floppy disks 194
   Optical storage media 194

P

Pay per install 134
Peer-to-peer 1
Phishing 57
Physical evidence 271
Physical layer 13
   Channel capacity 13
   Multiplexing 13
   Switching 14
   Transmission media 13
Physical threat 90
Piggybacking 71
Possession of stolen electronic communication devices 411
Presenting digital evidence 360
   Reporting 360
   Testimony 360
Pre-trial preparation 360
Private key 396
Proprietary forensic suites 238
   Computer online forensic evidence extractor 238
   Paraben tools 238
   Vogon 239
Proprietary forensic tools for mobile devices 254
   Oxygen forensic detective 254
   Paraben's device seizure 255
   PIN code 254
   PUK code 254
   XAMN 254
   XRY 255
Public key 396
Public prosecutor 362

R

RAM analysis with volatility 307
Ransomware 131
   AIDS 131
   Cryptodefense 131
   CryptoLocker 131
   CryptoWall 131
   CTB-Locker 132
   Encrypting ransomware 132
   External Rock 132
   Extortion 131
   LockerPin 132
   Locky 132
   Petya 132
   Reveton 131
   Samsam 132
   Scareware 132
   Screen lockers 132
   TeslaCrypt 132
   TROJ.RANSOM.A 131
   WannaCry 132
Ransomware-as-a-service 134
Reconstruction 235
Reporting 235
Role of forensic analyst in analysis 343
Rootkit 58
Rules of admissibility 357
   Authenticated 357
   Relevant 357

S

Salami attack 58
Search and seizure 275
Secondary evidence 270
Section 65B 355
Secure system 396
Security 150
   Application security 150
   Computing security 150
   Data security 150
   Financial security 150
   Human security 150
   Information security 150
   IT security 150
   Legal security 150
   National security 150
   Network security 150
   Physical security 150
   Public security 150
Seizure 187
Seizure memo 276
Silk road 142
Sniffer 59
Sources of evidence 187
Spit 72
Steganography 58
Storage medium 190
   Disk drive 190
   Extended partition 192
   Extended primary partition 193
   GPT partitions 193
   Hard disk drive 190
   Hidden partitions 193
   Hidden sector data 191
   Host protected areas 193
   Logical block addressing 191
   Master boot record 192

Master partition table 193
Partitioning 191
Primary partition 192
Sectors 190
Solid state drive 190
Subscriber 396
Switching 14
  Circuit switching 14
  Message switching 14
  Packet switching 14
System interference 51

T

TCP/IP protocol suite 12
Testimony 362
The Indian Evidence Act 1872 429
The Indian Evidence Act 1872 Section 22A
  Indian case laws 432
  Indian laws related to intellectual property 432
Tor browser 140
Traditional laws to prosecute cybercrimes
  Child pornography 442
  Copyright infringement 442
  Forgery 442
  Fraud 442
  Liability of internet service providers 442
Traditional laws to prosecute cybercrimes 442
Transmission control protocol 22
  Bandwidth management by TCP 23
  Characteristics 22
  Congestion control 24
  Connection management by TCP 23
  Crash recovery 25
  Error control and flow control in TCP 24
  TCP header 22
  Timer management 24
Transport layer 21
Types of cybercrime 47

U

User datagram protocol 25
  Characteristics 25
  UDP header 26

V

Validation and discrimination 234
  CRC32 234
  MD5 234
  SHA 234

Vandalism 72
Viral Attack 407

W

Warrantless searches 279
  Automobile 280
  Border searches 280
  Consent 280
  Exigent circumstances 280
  Open field 280
  Plain view 280
  Search incident to arrest 280
  Stop-and-frisk 280
  Subpeona 280
Whole disk encryption 221
Windows artifacts 217
  Access control lists 218
  Backup and restore 218
  Cloud storage facility 218
  Email 218
  Event logs 217
  Home groups 218
  Internet history artifacts 218
  Jumplists 217
  Link files 217
  RAM files 217
  Recycle bin 217
  Sticky notes 218
  Thumbnail cache 217
  Virtualization 217
  Volume snapshot service 218
  Wireless network history 219
Windows registry 215
Wireless forensics 157
Working with autopsy 328
  Adding data source 329
  Analysis basics 330
  Analysis of deleted files with autopsy 331
  Creating case 329
  Example use cases 330
  Ingest inbox 330
  Ingest modules 329
  Known bad hash files 331
  Media—images and videos 331
  Reporting 330
  Timeline 330
  Web artifacts 331

About the Authors

Dr Dejey is Assistant Professor at Department of Computer Science and Engineering, Anna University Regional Campus, Tirunelveli, which is one of the regional centres of Anna University, Chennai. She has completed her Bachelor's degree in 2003 (Computer Science and Engineering) and Master's degree (Computer Science and Engineering) as a first ranker and gold medallist in 2005. She was awarded Junior Research Fellowship in Engineering and Technology by the University Grants Commission, India in 2008 and was awarded PhD in 2011. She has more than 13 years of active teaching and research experience. Her research interests include image, signal, and video processing, data hiding, computer networks, and computer forensics. She has published around 40 research articles in reputed journals and conference proceedings and two book chapters. She has filed two Indian Patents, which are in the pipeline. She is a member of the Institution of Engineers (India), IEEE, and Indian Science Congress and a life member of the Indian Society for Technical Education.

Dr S. Murugan, IPS, a senior police officer, presently working as Inspector General of Police in Tamil Nadu has rich experience in handling cybercrime investigations and has supervised high-profile sensational cybercrime cases for the last 18 years. He had a five-year tenure in CBI, which gave him an excellent opportunity to handle many cybercrime investigations. In recognition of his contribution to the Police department, he was decorated with the Tamil Nadu Chief Minister's Medal for Outstanding Devotion to Duty in 2005 and President of India Police Medal in 2013. His academic qualifications range from Masters in Economics, Management and Computer Applications to a doctorate in Cybercrime, which focused on frauds in plastic money from the University of Madras. His thirst to learn and disseminate technical details of cyber forensics led him to pursue CFCE (Certified Forensics Computer Examiner) from the International Association of Computer Investigative Specialists (IACIS) USA, under the Government of India sponsorship in 2010. Since then, he has continuously qualified himself with a CFCE certification for 7 consecutive years, in order to keep pace with the international developments in the area of cyber forensics and is a member of the IACIS. He is a regular guest faculty for Tamil Nadu Police Academy, Tamil Nadu Judicial Academy, CBI Academy New Delhi, and National Judicial Academy Bhopal, besides delivering technical lectures on cyber forensics and cyber security in various universities across the country. He has published research articles in the field of plastic card frauds, cyber security issues, and dark net.

Related Titles

Digital Image Processing (2e)
S Sridhar
(9780199459353)

This second edition of *Digital Image Processing* is designed as a textbook for undergraduate engineering students of Computer Science, Information Technology, Electronics and Communication, and Electrical Engineering. The book provides a comprehensive coverage of the fundamental and advanced concepts of digital image processing.

New to this Edition
- A chapter on wavelet transforms and multiresolution analysis which focuses on the wavelet transform-based image processing as well as wavelet-based image compression
- Topics such as image security, visual effects, Radon transform, digital image forensics, and computer vision
- Pedagogical features such as crossword and word search problems

Soft Computing with MATLAB Programming
N.P. Padhy and Simon
(9780199455423)

Soft Computing with MATLAB Programming is a textbook designed for undergraduate students of computer science, information technology, electrical and electronics, and electronics and communication engineering as well as those pursuing an MCA degree. It aims to familiarize students with soft computing fundamentals and approaches to solving real-world problems.

Features
- swarm intelligence systems with their fundamental concepts of exploration and exploitation
- Artificial Bee Colony algorithm and Cuckoo ...ithm with suitable applications
- ...n 25 MATLAB programs with step-... ...nd 75 solved problems

Software Testing 2e
Chauhan
(9780199465873)

This second edition of *Software Testing* serves as a textbook for students of computer science, information technology, and computer applications. The book focuses on software testing, not only as a phase of software development life cycle (SDLC), but also as a complete process to fulfil the demands of quality software.

New to the Second Edition
A chapter on Agile Testing focusing on the agile testing methodology which has gained importance in recent years
Strengthened coverage of dynamic testing techniques, with the inclusion of robust worst-case testing method, orthogonal array testing strategy, predicate coverage, and path sensitization
Test case prioritization techniques based on data flow, module-coupling slice, and program structure analysis
Testing techniques such as reliability testing and system testing based on use-cases
Additional examples on black-box and white-box testing techniques

Professional Ethics 2e
R Subramanian
(9780199475070)

The second edition of *Professional Ethics* is a comprehensive textbook designed for budding engineers and managers. The book has been written in a simple manner with plenty of examples and case studies that will enable the readers to effectively resolve the ethical issues they will face in their professional lives.

New to this Edition
New sections on impact of social media, nature of values, personal ethics, value sciences, and so on
Appendix 1 which provides a set of practical exercises at the end of the book for instructors to conduct in the class
Multiple choice questions and more review exercise questions at the end of every chapter

Other Related Titles

...puter Networks
...day Cryptography
...ddiqui: *Natural Language* ...g *& Information Retrieval*

9780195671544 Padhy: *Artificial Intelligence & Intelligent Systems*